PACIFIC SYMPOSIUM ON
BIOCOMPUTING 2009

PACIFIC SYMPOSIUM ON BIOCOMPUTING 2009

Kohala Coast, Hawaii, USA
5–9 January 2009

Edited by

Russ B. Altman
Stanford University, USA

A. Keith Dunker
Indiana University, USA

Lawrence Hunter
University of Colorado Health Sciences Center, USA

Tiffany Murray
Stanford University, USA

Teri E. Klein
Stanford University, USA

World Scientific

NEW JERSEY • LONDON • SINGAPORE • BEIJING • SHANGHAI • HONG KONG • TAIPEI • CHENNAI

Published by

World Scientific Publishing Co. Pte. Ltd.

5 Toh Tuck Link, Singapore 596224

USA office: 27 Warren Street, Suite 401-402, Hackensack, NJ 07601

UK office: 57 Shelton Street, Covent Garden, London WC2H 9HE

British Library Cataloguing-in-Publication Data
A catalogue record for this book is available from the British Library.

BIOCOMPUTING 2009
Proceedings of the Pacific Symposium

ISBN-13 978-981-283-692-2
ISBN-10 981-283-692-6

Printed in Singapore by Mainland Press Pte Ltd

PACIFIC SYMPOSIUM ON BIOCOMPUTING 2009

This year, PSB remains at its most common venue, the Fairmont Orchid on the Big Island of Hawaii. We are now in our fourteenth year. A brief review of our field shows a clear trend: sequence analysis is back! The introduction of "next generation" sequencing technologies has taken many by surprise. These technologies are creating volumes of data that are amazing, and that stress out most existing software and hardware infrastructures. Reports of the demise of sequence analysis as an active research area are premature. Some say that bioinformatics is like waves on the beach—a new wave hits the beach regularly, but only some of the water actually sinks into the sand and stays a while—most of it goes back into the sea. So it is for bioinformatics, where waves of scientists arrive as exciting new data sets emerge (e.g. sequence data brought combinatorial algorithms and microarray data brought statisticians), some of these remain in the field, while others recede after the initial thrill wears out. Well, for those who do sequence analysis, the second wave has broke. More importantly, these data promise to enrich our ability to measure genotype, ask important questions about its relationship to phenotype, and bring the $1000 genome into view.

We would like to thank our keynote speakers. Dr. Drew Endy, Department of Bioengineering, Stanford University, will talk about "Post-Synthesis Genetics & Engineering Biological Simplicity." Our keynote in the area of Ethical, Legal and Social implications of technology will be Greg Hampikian, Professor of Biology and Criminal Justice Director of the Idaho Innocence Project, Boise State University, who will talk about "DNA Don't Lie: How Bioinformatics freed some of my best friends, and sent the guilty to prison."

PSB provides sessions focusing on hot new areas in biomedical computation. These sessions are usually conceived during earlier PSB meetings, as emerging fields are identified and targeted. The sessions often define a new scientific niche and bring together leaders in these exciting new areas. If you have an idea for a new session, contact the organizers at the meeting or by e-mail.

The diligent efforts of a dedicated group of researchers has led to an outstanding set of sessions, with associated introductory tutorials. These organizers provide the scientific core of PSB, and their sessions are as follows:

Gil Alterovitz, Judith A. Blake, Jonathan M. Dreyfuss, Drew Endy, and Marco F. Ramoni. *Biomolecular Networks: From Analysis To Synthesis*

Bernard Moret, Webb Miller, Pavel Pevzner, and David Sankoff. *Computational Challenges in Comparative Genomics: Gene Family Evolution, Ancestral Genome Reconstruction, and Applications to Human Genetics*

Tanya Berger-Wolf, Teresa Przytycka, and Mona Singh. *Dynamics of Biological Networks*

Roy Kerckhoffs, James Bassingthwaighte, Jeff Reinbolt, and Peter Arzberger. *Emerging Challenges in Multi-Scale Modeling in Biology*

Michael Ochs, John Quackenbush, Ramana Davuluri, and Habtom Ressom. *Knowledge-Driven Analysis and Data Integration for High-Throughput Biological Data*

Yves A. Lussier, Younghee Lee, Predrag Radivojac, Yanay Ofran, Atul Butte, and Maricel Kann. *Molecular Bioinformatics for Disease*

PSB 2008 will also feature two workshops: *Open Science: Developing tools and approaches to facilitate open exchange between scientists and data* by Cameron Neylon and Shirley Wu and *Post-transcriptional gene regulation: RNA-protein interactions, RNA processing, mRNA stability and localization* by Benjamin Blencowe, Steven Brenner, Timothy Hughes, and Quaid Morris.

Tiffany Murray provided outstanding support of the peer review process, assembly of the proceedings, and meeting logistics. We thank the National Institutes of Health, National Science Foundation, and the International Society for Computational Biology for travel grant support. We also acknowledge the many busy researchers who reviewed the submitted manuscripts on a very tight schedule. The partial list following this preface does not include many who wished to remain anonymous, and of course we apologize to any who may have been left out by mistake.

Aloha! Pacific Symposium on Biocomputing Co-Chairs,
September 30, 2008

Russ B. Altman
Departments of Bioengineering, Genetics & Medicine, Stanford University
A. Keith Dunker
Department of Biochemistry and Molecular Biology, Indiana University School of Medicine
Lawrence Hunter
Department of Pharmacology, University of Colorado Health Sciences Center
Teri E. Klein
Department of Genetics, Stanford University

Thanks to the reviewers...

Finally, we wish to thank the scores of reviewers. PSB requires that every paper in this volume be reviewed by at least three independent referees. Since there is a large volume of submitted papers, paper reviews require a great deal of work from many people. We are grateful to all of you listed below and to anyone whose name we may have accidentally omitted or who wished to remain anonymous.

Zaky Adam
Edo Airoldi
Mario Albrecht
Rommie Amaro
Peter Arzberger
Jean-Philippe Aumasson
Chris Bakal
Kim Baldridge
Vikas Bansal
Zviv Bar-Joseph
James Bassinghtwaighte
Mike Beer
Getachew Befekadu
Anne Bergeron
Tanya Berger-Wolf
Madhu Bhattacharjee
Ghislain Bidaut
Elhanan Borenstein
Guillaume Bourque
Yana Bromberg
Philipp Bucher
William Bush
Stuart Campbell
Vincent Carey
Rong Chen
James Costello
Lenore Cowen
Nadia El-Mabrouk
Eleazar Eskin
Debashis Ghosh
Piotr Gmytrasiewicz
Chittibabu Guda
Roderic Guigo
Katia Guimaraes
Habiba Habiba
Matt Hahn
Xiaoxu Han
Sridhar Hannenhalli
Midori Harris
Reece Hart
Andrew Hill
Zhangzhi Hu
Yang Huang
Kun Huang
Ilya Ioschikhes
Laxmanan Iyer
Maricel Kann
John Karro
Roy Kerckhoffs
Roy Kerckhoffs
Abbas Khalili
Ashfaq Khokhar
Seungchan Kim
Carl Kingsford
Judith Klein-Seetharaman
Jan Kors
Andrei Kossenkov
Tahsin Kurc
Alain Laederach
Jens Lagergren
Mayank Lahiri
Jim Leebens-Mack
Harold Lehmann
Yong Li
Jung-Chi Liao
David Liben-Nowell
David Liberles
Randy Linder
Gang Liu
Jian Ma
Thomas Mailund
Andrew McCulloch
Michael McGeachie
Nikhil Medhekar
Bernard Moret
Kevin Murphy
Taro Muso
Guy Nimrod
William Noble
Guillaume Obozinski
Michael Ochs
Yanay Ofran
Paivi Onkamo
Ron Pinter
Teresa Przytycka
Yanjun Qi
Amina Qutub
Ami Radunskaya
Archana Ramesh
Marco Ramoni
Jeff Reinbolt
Habtom Ressom
Marylyn Ritchie

Louis Rossi
Sushmita Roy
Karen Sachs
Frank Sachse
David Sankoff
Nigam Shah
Michael Showe
Mona Singh
John Spouge
Alexander Statnikov
Joshua Stuart
Thomas Stützle
Fengzhu Sun
Mahlet Tadesse
Haixu Tang
Jijun Tang
Eric Tannier
Glenn Tesler
John Tyson
Rency Varghese
Olga Vitek
Greg Vogt
Slobodan Vucetic
Yingwei Wang
Priyankara Wikramasinghe
Damian Wojtowicz
Ray Wollman
Stefan Wuchty
Xuhua Xia
Wenzhong Xiao
Andrew Xu
Jason Xuan
Xinan Yang
Yuzhen Ye
Yi-Kuo Yu
Elena Zaslavsky
Yuji Zhang
Liqing Zhang
Yu Zhang
Louxin Zhang
Liqing Zhang
Xiuwei Zhang
Jie Zheng
Elena Zotenko

CONTENTS

DYNAMICS OF BIOLOGICAL NETWORKS

KNOWLEDGE-DRIVEN ANALYSIS AND DATA INTEGRATION FOR HIGH-THROUGHPUT BIOLOGICAL DATA

BIOMOLECULAR NETWORKS: FROM ANALYSIS TO SYNTHESIS

GIL ALTEROVITZ

Division of Health Sciences and Technology, Harvard University/Massachusetts Institute of Technology, Cambridge, MA 02139, USA. Children's Hospital Informatics Program, Boston, MA 02115, USA. Department of Electrical Engineering and Computer Science, Cambridge, MA 02139, USA. Harvard Partners Center for Genetics and Genomics, Harvard Medical School, Boston, MA 02115, USA.

JUDITH A. BLAKE

The Jackson Laboratory, Bar Harbor, ME 04609, USA

JONATHAN M. DREYFUSS

Harvard Partners Center for Genetics and Genomics, Boston, MA 02115, USA.

DREW ENDY

Department of Bioengineering, Stanford University, CA 94305, USA

MARCO F. RAMONI

Children's Hospital Informatics Program, Boston, MA 02115, USA. Division of Health Sciences and Technology, Harvard University/Massachusetts Institute of Technology, Cambridge, MA 02139, USA. Harvard Partners Center for Genetics and Genomics, Harvard Medical School, Boston, MA 02115, USA.

1. Introduction

The study of biological networks is a very challenging research area in biocomputing. This field is of utmost importance, however. These networks underlie much of the emerging fields of synthetic and systems biology, which are revolutionizing molecular biology. Here, we use the term "biological networks" both to refer to biomolecular networks (e.g., gene regulatory networks) and to supporting knowledge representation networks (e.g., Gene Ontology).

There have been numerous recent advancements in synthetic biology. In academia, the Registry of Standard Biological Parts has been developed based on the concept of BioBricks, allowing scientists to browse through standardized and interchangeable biological parts for building synthetic biological systems. In the corporate world, the company Codon Devices is aiming to commercialize DNA synthesis on demand. Another milestone in the field is the first description of the

RNA ribo-regulator and other biological circuit components, such as the genetic toggle switch. Such progress has led to the development of engineered simple genetic circuits that mimic other common devices like oscillators and feedback loops[1]. Other recent achievements include the development of non-native behaviors like optimized drug synthesis and programmed spatial formation[2].

Although scientists have made significant progress in the development of synthetic and systems biology, the fields must still overcome several challenges related to biological networks[3]. To this end, this session offers novel methodologies in two general areas: namely, in inferring network structure from vast amounts of data and in utilizing these networks for pressing applications.

2. Session Papers

Eliciting the structure of a biological network is a fundamental issue. **Kuchaiev and Przulj** introduce a novel generative model of protein-protein interaction networks that accepts only the high-confidence portion of a network but is able to reproduce other low-confidence network features. **Tari *et al.*** propose a new paradigm to determine network structure, which allows biologists to construct networks tailored to their specific needs using Medline abstracts. **Costello *et al.*** present a novel data-driven methodology that synthesizes an ontology from gene-gene interactions and annotations regarding those genes.

Biological networks are dynamic entities which are highly dependent on cellular state. **Sen *et al.*** produce gene regulatory networks that are intrinsically state-dependent and offer context-specific annotation in addition to gene-gene interactions. **Roy *et al.*** develop a novel technique to infer functional networks from condition-specific responses and examine gene ontology enrichment of their inferred networks among different cell types.

We also seek to understand how biological networks change over time and across species. **Cheng and Riedel** offer a methodology, based on a stochastic simulation with time-varying inputs, which allows insight and validation into the time-dependent dynamics of a system. **Tian and Samatova** develop a fast algorithm based on connected components that identifies maximally conserved regions across species.

Identification of biomarkers is a substantial application of biological networks. Using a network paradigm, **Dudley and Butte** identify protein biomarkers through a novel methodology which integrates biofluids proteome data and inter-disease genomic relationships. **Sachs *et al.*** extend the Bayesian Network methodology, often used to identify biomarkers, by allowing cyclic pathways in the biological system to be modeled.

Acknowledgments

Thank you to all the authors who submitted their work to this session and to the reviewers who graciously contributed their time and expertise to the peer review process.

References

1. F. J. Isaacs, D. J. Dwyer, and J. J. Collins, *Nat. Biotechnol.* **24**, 5 (2006).
2. S. Basu, Y. Gerchman, C. H. Collins, F. H. Arnold, and R. Weiss, *Nature* **434**, 7037 (2005).
3. G. Alterovitz and M. F. Ramoni, *Systems Bioinformatics: An Engineering Case-Based Approach.* (Artech House Publishers, London, U.K.; Boston, 2007).

STOCHASTIC TRANSIENT ANALYSIS OF BIOCHEMICAL SYSTEMS AND ITS APPLICATION TO THE DESIGN OF BIOCHEMICAL LOGIC GATES

BIN CHENG AND MARC RIEDEL

Electrical and Computer Engineering
University of Minnesota
200 Union Street S.E.
Minneapolis, MN 55455
E-mail: bincheng@umn.edu, mriedel@umn.edu

In order to better characterize the behavior of biochemical systems, it is sometimes helpful and necessary to introduce *time-dependent input signals*. If the state of a biochemical system with such signals is assumed to evolve deterministically and continuously, then it can be readily analyzed by solving ordinary differential equations. However, if it assumed to evolve discretely and stochastically, then existing simulation methods cannot be applied. In this paper, we incorporate conditions for *transient analysis* into stochastic simulation and we develop the corresponding simulation algorithm. Applying our method to examples, we demonstrate that it can yield new insights into the dynamics of biochemical systems; specifically, it can be used to verify the design of biochemical logic gates.

1. Introduction

Certain biochemical systems appear to exploit randomness, choosing between different outcomes with a probability distribution – in effect, hedging their bets with a portfolio of responses. Examples include the *pap pili* epigenetic response of bacteria [1], the lentiviral positive-feedback loop in the HIV virus [2], and the lysis/lysogeny switch of the *lambda* bacteriophage [3].

Gillespie proposed the **stochastic simulation algorithm** (SSA) to characterize discrete, random biochemical reactions [4]. The SSA tracks integer quantities of the molecular species in a biochemical system, executing reactions at random based on propensity calculations. Repeated trials are performed to characterize the evolution of the system. Gillespie demonstrated that the SSA has a firmer physical basis than continuous, deterministic methods and that it provides more accurate simulation results [5]. The SSA works well for systems in which the quantities of the species are small. However, for larger systems, the computation time becomes prohibitive. Improved algorithms have been proposed [6]. Also, approximation methods have been applied [7], [8], [9], [10], [11].

In a typical application of the SSA, the system is assumed to be closed: after the initial state is fixed, the time evolution of the system only depends on the internal reactions. But what if we want to characterize the behavior of a system when there exist external mechanisms that modify the quantities of species? For instance, we may need to study what happens when an external source injects or drains certain species into or from the system. The external change might occur periodically or it might be dependent on the trajectory of the system, e.g., the quantities of species might be limited by threshold conditions. Existing methods, such as Gillespie's SSA, cannot handle such behaviors.

For electrical circuits, transient analysis consists of a time sweep using numerical methods to solve differential equations, with the operating points solved by setting all the time derivatives to zero. Similarly, in a biochemical system, various input signals can be defined and the system can be solved through ordinary differential equations (ODEs) or differential algebraic equations (DAEs). Indeed, several authors have suggested using a standard electrical simulation tool called SPICE to model biochemical reactions [12], [13], [14], [15], [16]. However, this presumes that the biochemical systems under investigation are continuous and deterministic.

In this paper, we propose an approach called **stochastic transient analysis** (STA). It incorporates time-dependent variations in the quantities of species into stochastic simulation. We consider pulse, piecewise-linear, and sinusoidal signals. The method can be readily generalized to include other free-form functions. The signals are either *forced* or *injected*. Further, threshold-triggered signals are incorporated.

We apply the STA method to analyze a simple model called *Lotka*. We also apply it to analyze the time-dependent behavior of the inverter model proposed in [17]. We propose designs for other types of biochemical logic gates (AND, OR, NAND and XOR) and we apply our STA method to verify the behavior of these.

2. Stochastic Transient Analysis

The power of transient analysis resides in the fact that the time behavior of the system can be simulated with different types of input signals to accommodate different analysis scenarios.

2.1. *Categorizing Input Signals for Stochastic Transient Analysis*

Although we do not discuss transient analysis of electric circuits here, we borrow from this field standard forms in which input signals can be categorized: pulse, piecewise linear (PWL) and sinusoidal.

(1) $PULSE(v_1, v_2, t_d, t_r, t_f, p_w, P)$, where v_1 is the initial value, v_2 is the pulsed value, t_d is the delay time, t_r is the rise time, t_f is the fall time, p_w is the pulse width and P is the period.
(2) $PWL(t_1, v_1, t_2, v_2, \ldots)$, where v_1 is the input value at time t_1, v_2 the input value at time t_2, etc.
(3) $SIN(v_0, v_a, F, t_d, \theta, \phi)$, where v_0 is the offset, v_a is the amplitude, F is the frequency, t_d is the delay, θ is the damping factor and ϕ is the phase.

In addition to the *form* of the input signals, we distinguish between two ways of *imposing* them onto the system: they are either *forced* or *injected*. The former means that the quantity of the species is set by some external mechanism rather than by the internal reactions. The latter means that an external mechanism adds to or subtracts from the quantity of the species that are present.

Also, we allow for *threshold-triggered* input signals: these are signals that start or stop based on threshold quantities of species. This is useful in simulating boundary conditions, say between qualitatively different phases in the evolution of a biochemical system. For example, suppose that after reaching a threshold quantity of 1000 for a species, the system is deemed to go through a transition: above this threshold, an external mechanism begins injecting new species.

The format of an input signal for our STA method is summarized as follows. Here <> means that the contents are mandatory while [] means that they are optional. (Some details are omitted; these will be implementation-dependent.)

$$
\begin{aligned}
&< SPECIES > < INITIAL > \\
&[< FUNCTION > < FORCED \mid INJECTED >]\ [CONDITION]
\end{aligned} \tag{1}
$$

where $SPECIES$ is the species name; $INITIAL$ is the initial quantity; $FUNCTION$ is the definition of an input signal function; $FORCED$ or $INJECTED$ is the way that the input is imposed; and $CONDITION$ is the definition of a threshold condition on the quantity of a species or a set of species.

2.2. *Stochastic Transient Analysis Algorithm*

Our transient analysis method could be incorporated into any of the proposed stochastic simulation algorithms, for instance that described in [6]. For simplicity, we describe its implementation in terms of Gillespie's direct SSA, proposed in [4].

Suppose we have a biochemical system consisting of n different species interacting through m different reactions. The species are denoted by $x_i (i = 1, \ldots, n)$ and their corresponding quantities are $X_i (i = 1, \ldots, n)$. The reactions are denoted by $R_i (i = 1, \ldots, m)$ and the corresponding rates are $c_i (i = 1, \ldots, m)$. The system state at time point t is denoted by $S_t = (X_1(t), \ldots, X_n(t))$. At the start of the simulation, the system state is $S_0 = (X_1(0), \ldots, X_n(0))$. At each step of the simulation, the next reaction to fire as well as the time point at which this reaction fires are decided. According to Gillespie, these are computed as follows:

$$\sum_{\nu=1}^{\mu-1} a_\nu < r_1 a_0 \le \sum_{\nu=1}^{\mu} a_\nu, \tag{2}$$

$$\tau = (1/a_0) ln(1/r_2), \tag{3}$$

$$a_0 \equiv \sum_{\nu=1}^{m} a_\nu, \tag{4}$$

$$a_\nu \equiv h_\nu c_\nu (\nu = 1, \ldots, m), \tag{5}$$

where μ is the next reaction that fires; τ is the time elapsing from the present until the next reaction fires; r_1 and r_2 are two random numbers generated by a unit-interval uniform random number generator; h_ν is the number of distinct molecular reactant combinations available for reaction R_ν in state $(X_1, \ldots, X_n)$; and c_ν is the *rate* for reaction R_ν.

In transient analysis, different types of input signals can be specified.

- For forced signals, the value of the species is sampled from the function that is defined, taken at the current time point; the next reaction has no effect on its value.
- For injected signals, the amount to be injected is calculated by subtracting the function value at the current time point t_c from the function value at the next time point t_n:

$$\text{injected number} = \lfloor f(t_n) \rfloor - \lfloor f(t_c)) \rfloor. \tag{6}$$

  This amount is added to the quantity of the corresponding species together with the change in quantity that the next reaction produces.

If a threshold condition exists, then a forced or injected signal is only applied if the threshold condition is met.

We note that with this increased flexibility in the simulation of input signals, the amount of data that must be recorded increases. If implemented as described above, the amount of data generated increases proportionally with the number of trials. To mitigate against this, we propose the following data structure; it requires constant memory, independent of the number of trials.

Suppose T is the simulation time, and N is the total number of simulation steps. We separate T into N time segments equally: $[(i-1)T/N, iT/N](i = 1, \ldots, N)$. A *accumulated* state $S^{(i)}(i = 1, \ldots, N)$ is attached to each corresponding time segment. For each time point in each trial, if the time point falls in i-th time segment, then the corresponding state is added to the i-th accumulated state $S^{(i)}$. At the end of simulation, each accumulated state is averaged by dividing it by the total number of trials that landed in the corresponding time segment. The final results can be plotted by using N general states at N time points. The data structure is shown in Fig. 1.

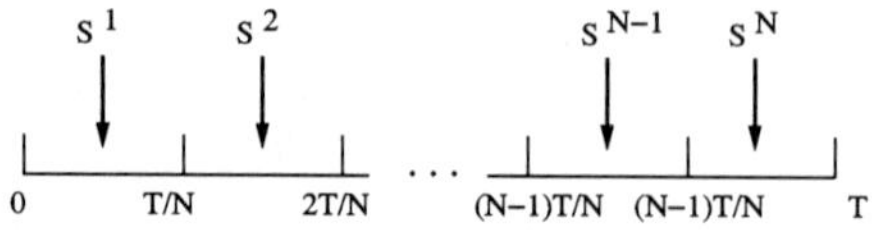

Figure 1. Time segmenting for multi-trial STA.

3. Stochastic Transient Analysis of the Lotka Model

In this section, we apply our method to a simple biochemical model called *Lotka*, described by the following set of coupled reactions:

$$x + y_1 \xrightarrow{c_1} x + 2y_1, \quad y_1 + y_2 \xrightarrow{c_2} 2y_2, \quad y_2 \xrightarrow{c_3} z \tag{7}$$

In [4], Gillespie analyzes the Lotka model. Some of the conclusions that he presents are qualitative deductions. For instance, he writes that "no matter what the state of the system is initially, it will eventually wind up in either the state $(y_1 = 0, y_2 = 0)$ or the state $(y_1 = \infty, y_2 = 0)$." No simulation results are provided to support this. Here we show how transient analysis can be used to elucidate such behaviors.

Note that we can characterize the situation where the model winds up in the state $(y_1 = 0, y_2 = 0)$ simply by simulating it for enough steps. However, we cannot do the same to characterize the situation where it winds up in the state $(y_1 = \infty, y_2 = 0)$. In this case, as y_1 approaches infinity, the time steps between reactions firing become infinitesimal small; eventually, an infinite number of reactions fire per time unit. Gillespie's SSA cannot handle such limiting conditions. However, with our transient analysis method, we can characterize such limits by defining threshold events. If we constrain the output of y_1 to be, say 20,000, as an approximation of infinity, we will get the waveform in Fig. 2(a).

Now suppose there is some external mechanism which forces y_1 to stay at a constant value of 1000. The waveform of Fig. 2(b) illustrates the simulation results for such forcing. Interestingly, we observe that y_2 still gets to zero, even though this takes much longer. We expect that the value of y_2 should be proportional to x, while the value of y_1 should not change with x. This can be verified through the simulation results depicted in the two waveforms of Fig. 2(d) and Fig. 2(e), where x is set to be a pulse function of time: $PULSE(10\ 30\ 50\ 0\ 0\ 50\ 100)$, as displayed in Fig. 2(c). Here threshold events are defined: once y_1 or y_2 reaches zero, more of the corresponding species is injected

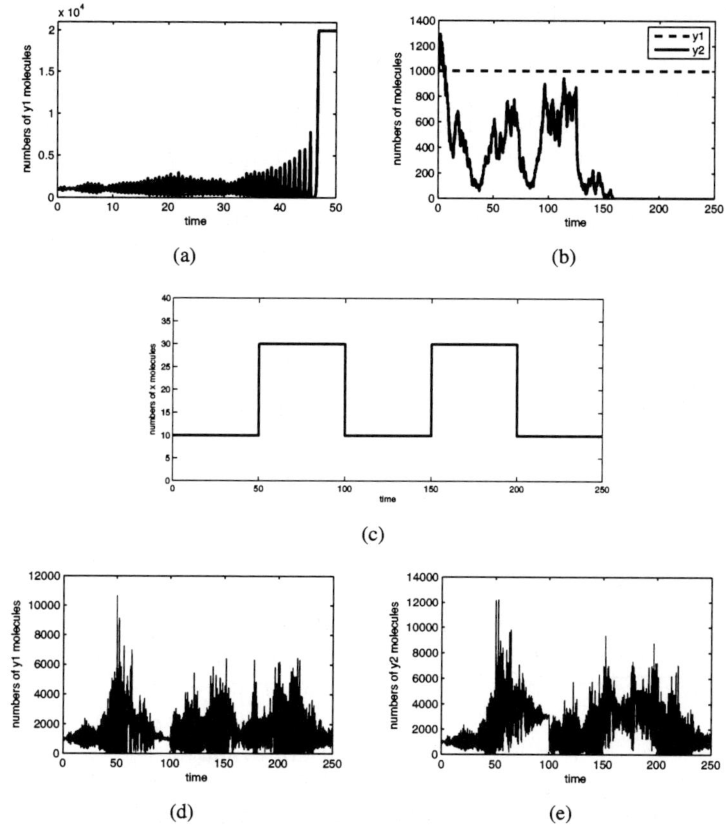

Figure 2. Stochastic transient analysis of the Lotka model, 3 trials. (a) System reaching state ($y_1 = \infty, y_2 = 0$); (b) y_1 is forced to a constant; (c) x is set to a pulse function of time; (d) y_1 does not change with x; (e) y_2 changes with x.

to keep the fluctuations going.

4. Stochastic Transient Analysis of Biochemical Logic Gates

In this section, we apply our STA method to the design of biochemical logic gates. First, we consider the inverter model proposed in [17]. Then, we propose a simple model for other logic gates: AND, NAND, OR and XOR. We simulate these using our STA method to characterize their behavior.

4.1. *Biochemical Inverter*

Some of the behaviors of gene regulation systems can be characterized as logical operations. For instance, RNA polymerase will stop transcribing a gene if there exists a repressor protein which binds to the operator of the gene's promoter. If one considers the concentration of the gene and the concentration of the repressor as two signals, then the relationship between them is like an inverter. In [17], the authors present the following set of coupled reactions modeling this behavior. They analyze their model, deterministically. Here we apply our stochastic transient analysis method.

$$
\begin{array}{lll}
a+a \xrightarrow{K_{dim(a)}} a2, & a2 \xrightarrow{K_{sngl(a)}} a+a, & z+z \xrightarrow{K_{dim(z)}} z2, \\
z2 \xrightarrow{K_{sngl(z)}} z+z, & gz+a2 \xrightarrow{K_{rprs(a2)}} gza2, & gza2 \xrightarrow{K_{dis(a2)}} gz+a2, \\
gza2+a2 \xrightarrow{K_{rprs(a4)}} gza4, & gza4 \xrightarrow{K_{dis(a4)}} gza2+a2, & a \xrightarrow{K_{dec(a)}} \phi, \\
a2 \xrightarrow{K_{dec(a2)}} \phi, & z \xrightarrow{K_{dec(z)}} \phi, & z2 \xrightarrow{K_{dec(z2)}} \phi, \\
gza2 \xrightarrow{K_{dec(ga2)}} gz, & gza4 \xrightarrow{K_{dec(ga4)}} gza2, & mrnaz \xrightarrow{K_{dec(mrna)}} \phi, \\
\end{array}
$$

$$
gz+rnap \xrightarrow{K_{xscribe}} gz+rnap+mrnaz, \quad mrna2+rnaa \xrightarrow{K_{xlate}} mrnaz+rrna+z \tag{8}
$$

In the model, the species a represents the input to the inverter, and the species z its output. In the simulation, we first provide a logical value '0' as the input, i.e., we set the quantity of the species a to zero. We expect to get logical '1' at the output; this corresponds to a quantity of about 12 of z.

To impose logic '1' at the input, the authors in [17] suggest that an externally-imposed drive is needed to increase the quantity of species a. Based on the strength of the external drive, a achieves an equilibrium at the bottom or at the top of its signal range. Accordingly, the transfer curve of the inverter can be drawn by changing the external drive from weak to strong, gradually. Here we consider a pulse input signal, shown in Fig. 3(a). We simulate the model using our STA method; the results are plotted in Fig. 3(b). These waveforms clearly depict the transient behavior of an inverter: when the input signal is low, the output signal is high and vice versa.

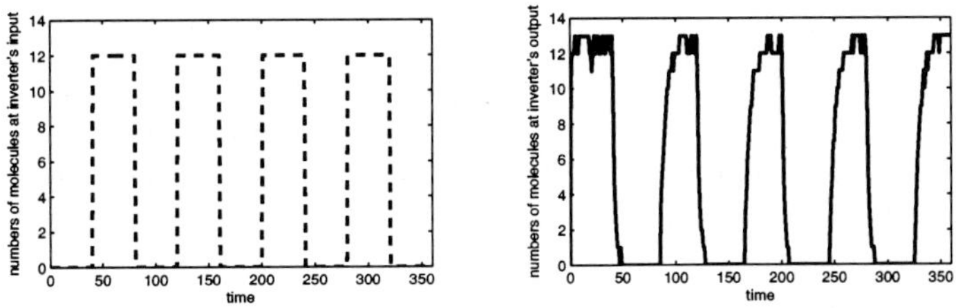

(a) (b)

Figure 3. Stochastic transient analysis of the inverter, 3 trials. (a) Input, (b) Output.

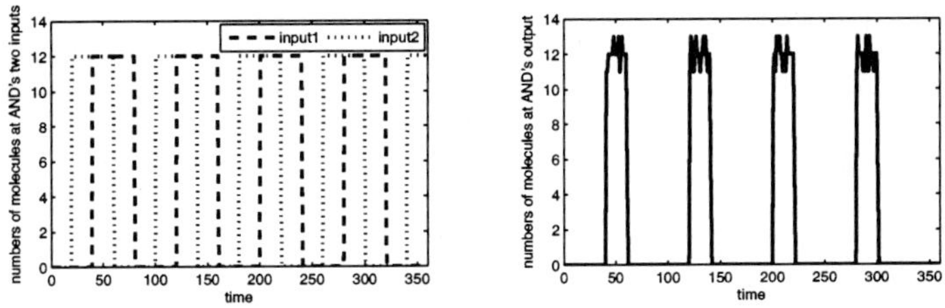

(a) (b)

Figure 4. Stochastic transient analysis of the AND gate, 3 trials. (a) Input; (b) Output.

4.2. *Logic Gates*

In [17], the approach taken to create a NAND gate is to "wire-OR" the outputs of multiple inverters by assigning them the same output gene. However no reaction model is given for the NAND gate. Here, we explicitly design logic gates with reaction models and then use STA to verify our models.

First we design the reaction model of an AND gate. This is given in (9). The reaction constants c_1 and c_2 can be adjusted to get the same quantity level for logic '1' as for the inverter ($c_2/c_1 \approx 12$). To verify the model, we apply as input signals two forced pulse signals with the same period but different phases, as shown in Fig. 4(a). The simulation result is plotted in Fig. 4(b). The waveform clearly depicts the transient behavior of an AND gate.

$$\begin{aligned} & x+y \xrightarrow{c_1} x+y+z \\ & z \xrightarrow{c_2} \emptyset \end{aligned} \quad . \tag{9}$$

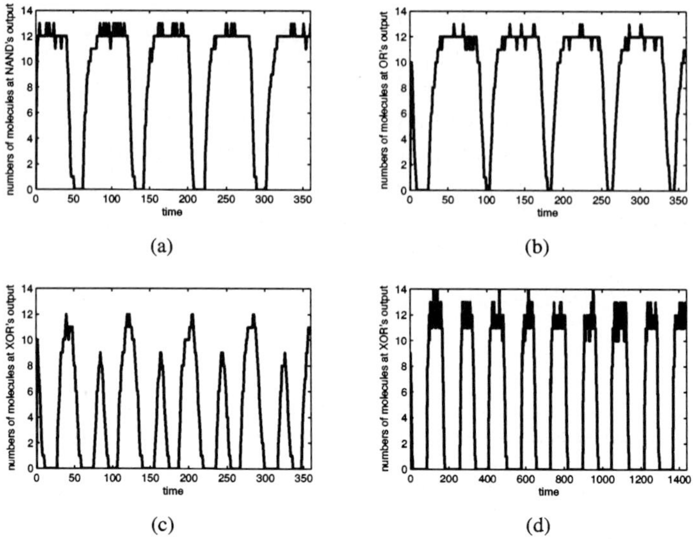

Figure 5. Stochastic transient analysis, 3 trials. (a) NAND, (b) OR, (c) XOR with high frequency, (d) XOR with low frequency.

Other logic gates are designed as follows.

- By simply connecting an inverter to the output of the AND gate, we obtain a NAND gate. Here the output of the AND gate acts as the external drive for the input of the inverter. Using the same input signals as for the AND gate simulation, we obtain the simulation results shown in Fig. 5(a).
- We can obtain an OR gate by hooking three inverters on an AND gate, one on each input and one on the output. The result is a total of 53 chemical reactions; these are not listed here. The simulation results are shown in Fig. 5(b), using the same input signals as for the AND gate.
- We can obtain an exclusive-OR (XOR) gate with two inverters, two AND gates, and one OR gate. The result is a total of 91 chemical reactions; these are not listed here. We show two different simulation results for the XOR gate: one in which the input signals are changing from low to high more rapidly, Fig. 5(c), and one where they are are changing less rapidly, Fig. 5(d). In Fig. 5(c), the output does not have enough time to reach the high level corresponding to logical '1'. In contrast, in Fig. 5(d), we obtain a clean XOR response.

5. Conclusions

The characterization of the XOR gate in the preceding section illustrates the sort of information that transient analysis provides: it allows us to characterize not only the input-output response of a system, but also its temporal dynamics. Our implementation of the method provides the flexibility to characterize such temporal dynamics for a variety of analysis scenarios. We are currently working on incorporating the method into simulation tools such as BioSPICE [18]. Also, we are applying it to problems in the computer-aided design of synthetic modules of biochemistry [19].

References

1. A. Hernday, B. Braaten, and D. Low. The Intricate Workings of a Bacterial Epigenetic Switch. *Advances in Experimental Medicine & Biology*, pages 547: 83–9, 2004.
2. L. Weinberger, J. Burnett, J. Toettcher, A. Arkin, and D. Schaffer. Stochastic Gene Expression in a Lentiviral Positive-Feedback Loop: HIV-1 Tat Fluctuations Drive Phenotypic Diversity. *Cell*, 122:169–182, 2005.
3. A. Arkin, J. Ross, and H. McAdams. Stochastic Kinetic Analysis of Developmental Pathway Bifurcation in Phage λ-Infected Escherichia Coli Cells. *Genetics*, pages 149:1633–1648, 1998.
4. D. Gillespie. Exact Stochastic Simulation of Coupled Chemical Reactions. *Journal of Physical Chemistry*, 81:No. 25, pp. 2340–2361, 1977.
5. D. Gillespie. Stochastic Simulation of Chemical Kinetics. *Annual Review of Physical Chemistry*, pages 58: 35–55, 2007.
6. M. Gibson and J. Bruck. Efficient Exact Stochastic Simulation of Chemical Systems with Many Species and Many Channels. *Journal of Physical Chemistry*, pages 105, 1876–1889, 2000.
7. D. Gillespie. Approximate Accelerated Stochastic Simulation of Chemically Reacting Systems. *J. Chem. Phys.*, pages 115(4), 1716–1733, 2001.
8. M. Rathinam, Y. Cao, L. Petzold, and D. Gillespie. Stiffness in Stochastic Chemically Reacting Systems: The Implicit Tau-leaping Method. *J. Chem. Phys.*, 2003.
9. Y. Cao, D. Gillespie, and L. Petzold. Avoiding Negative Populations in Explicit Tau Leaping. *J. Chem. Phys.*, pages 123, 054104–054112, 2005.
10. Y. Cao, D. Gillespie, and L. Petzold. Efficient Stepsize Selection for the Tau-leaping Method. *J. Chem. Phys.*, pages 124, 044109, 2006.
11. Y. Cao, D. Gillespie, and L. Petzold. The Adaptive Explicit-implicit Tau-leaping Method with Automatic Tau Selection. *J. Chem. Phys.*, pages 126(22), 224101–224109, 2007.
12. L. Nagel. SPICE2: A Computer Program to Simulate Semiconductor Circuits. *Memorandum No. UCB/ERL M520. Electronics Research Lab., College of Engi., UC, Berkeley*, 5 1985.
13. J. May and D. Mikulecky. Glucose Utilization in Rat Adipocytes. *Journal of Biological Chemistry*, pages 258, No. 8, pp. 4881–4888, 4 1983.
14. R. Seither, D. Hearne, D. Trent, D. Mikulecky, and D. Goldman. SPICE2 Network Simulation of Antifolate Effects on Purine and Pyrimidine Biosynthesis: Exploring the Role of Tetrahydrofolate Cofactor Depletion Versus Dihydrofolate Feedback Inhi-

bition. *Computera and Mathematics with Applications*, pages 20, No. 4–6, pp. 88–101, 1990.
15. R. Seither, D. Trent, D. Mikulecky, T. Rape, and D. Goldman. Effect of Direct Suppression of Thymidylate Synthase at the 5,l0-Methylenetetrahydrofolate Binding Site on the Interconversion of Tetrahydrofolate Cofactors to Dihydrofolate by Antifolates. *Journal Of Biological Chemistry*, pages 266, No. 8, pp. 18016–18023, 3 1991.
16. A. Brower, R. Ewing, R. Brower, and H. Abdel-Aty-Zohdy. SPICE for Modeling Biochemical Networks. *Circuits and Systems, 2005. 48th Midwest Symposium on*, 1:491–494, 8 2005.
17. R. Weiss, G. Homsy, and T. Knight Jr. Toward In-vivo Digital Circuits. *In Dimacs Workshop on Evolution as Computation Princeton NJ*, 1 1999.
18. S. Kumar and J. Feidler. BioSPICE: A Computational Infrastructure for Integrative Biology. *OMICS: A Journal of Integrative Biology*, pages 7(3): 225–225, 9 2003.
19. B. Fett, J. Bruck, and M. Riedel. Synthesizing Stochasticity in Biochemical Systems. *Proceedings of the 44th Annual Conference on Design Automation*, pages 640–645, 2007.

DATA-DRIVEN ONTOLOGIES

JAMES C COSTELLO [1,2*] DAN SCHRIDER [1] JEFF GEHLHAUSEN [1]
MEHMET DALKILIC[1,3]

[1] *School of Informatics,*
[2] *Department of Biology,*
[3] *Center for Genomics and Bioinformatics,*
Indiana University, Bloomington IN, 47405

Gene networks are important tools in studying gene-gene relationships and gene function. Understanding the relationships within these networks is an important challenge. Ontologies are a critical tool in helping deal with these data. The use of the Gene Ontology, for example, has become routine in methods for validation, discovery, *etc.* Here we present a novel algorithm that synthesizes an ontology by considering both extant annotation terms and also the connections between genes in gene networks. The process is efficient and produces easily inspectable ontologies. Because the relationships drawn between terms are heavily influenced by data, we call these "Data-Driven" Ontologies. We apply this algorithm to both discover new relationships between biological processes and as a tool to compare sets of genes across microrarray experiments. Supplemental data and source code are available at: `http://www.ddont.org`

1. Introduction

Researchers have unprecedented access to biological data organized in community supported and publicly available repositories; however, leveraging this enormous amount of data is neither routine nor simple.[1] Graph-theoretic and statistical approaches (often termed network approaches) are among the most popular means of studying complex biological relationships. Specific implementations of networks include, but are not limited to, coexpression networks,[2] protein-protein interaction networks,[3,4,5] and in the form of integrated functional linkage networks.[6,7,8] Extracting meaningful information from these networks has now become a challenge.

Ontologies are increasingly being employed to systematically organize data, thereby facilitating interpretation and exchange of information. The

*to whom correspondence should be sent. *email*: jccostel@indiana.edu

purpose of an ontology is, from observing the real world, to identify the pertinent existing instances and then to codify the relationships that exist between them.[9,10] In the context of biology, the Gene Ontology (GO)[11] is a well-known example; however, the GO is one of many potential and valid ontologies to describe any and all properties of genes. In short, there is no *canonical* ontological form. Indeed, multiple ontologies reveal relationships otherwise absent from a single ontology. Outside the GO, there are examples of ontologies that relate genes to diseases,[a] genes to organism anatomy,[b] and genes to gene expression data.[c]

In this study we present an algorithm that constructs an ontology by taking both gene network relationships and extant annotations on genes as inputs. We call this a "data-driven ontology" (hereafter abbreviated as DDOnt) to distinguish the process from building an ontology solely from human design. The DDOnts are "data-driven" in the sense that the relationships drawn between terms are influenced by the gene network connections. Our motivation is to use DDOnts to reveal potentially novel and interesting biological relationships that would not arise without the influence of gene networks. The complete deterministic, recursive algorithm is presented in Sec. 2.2.

We apply our method to three distinct, biologically motivated applications. First, given any gene network (*e.g.* relevance network,[2] integrated functional gene network,[7,6]) we synthesize a DDOnt to reveal novel relationships between biological processes. Second, given a set of genes of interest, we use gene network data to construct a DDOnt to find associated biological processes. Third, given a predefined set of genes, we explore how their associated biological processes are affected by varying experimental conditions as measured through microarray gene expression assays.

2. Methods

2.1. *Problem Statement*

Given a graph $G\langle V, E\rangle$ and a lexicon L, where lexical element $l_i \in L$ is mapped to 1 or more vertices of $G\langle V, E\rangle$, can we synthesize an ontology $\mathcal{O}$ where the relationships between lexical elements in L are determined through both annotation frequency and the edges in $G\langle V, E\rangle$?

$G\langle V, E\rangle$ is a set of genes $d_i \in V$ connected as edges $\{d_i, d_j\} \in E$ that

---

[a]http://diseaseontology.sourceforge.net/
[b]http://flybase.org/
[c]http://www.evocontology.org/

are determined from a dataset (*e.g.* coexpression correlation). Throughout all examples in this study we define the lexicon L as the set of terms in the GO category "biological process." When constructing $\mathcal{O}$, we follow two rules-of-thumb to draw connections between elements in L. 1) Lexical terms that are annotated to tightly connected genes in $G\langle V, E\rangle$ should be closely related in $\mathcal{O}$. 2) The prevalence of l_i is related to the distance of that term to a root, *i.e.* increasing prevalence means nearer to a root, decreasing prevalence means further from a root. We define a term's prevalence as a combination of two properties: (i) the count of genes annotated with l_i, and (ii) the connectivity of the genes annotated with l_i in $G\langle V, E\rangle$.

We have informally typed all directed edges that are drawn between lexical elements $\{l_i, l_j\}$ as: `is-related-to`. These edges are parent-child relationships where the parent is more prevalent (defined above) than the child.

2.2. *Ontology Construction Algorithm*

BUILDONTOLOGY is a deterministic, recursive algorithm that takes four inputs and produces an ontology $\mathcal{O}$ in the form of a tree.

The first input to BUILDONTOLOGY is $\mathbf{M}$. Given an annotated graph $G_{D,L}\langle V, E\rangle$ two matrices are constructed: one to reflect the relationships between nodes in the graph and the other to reflect the lexical annotations on the nodes. Matrix $\mathbf{M}_{D,D}$ is an adjacency matrix where $[a_{i,j}] = 1$ if $\{d_i, d_j\} \in E$; 0 otherwise. Matrix $\mathbf{M}_{D,L}$ is an incidence matrix where $[a_{i,j}] = 1$ if instance d_i is annotated with lexical term ℓ_j; 0 otherwise. We then set $\mathbf{M} \leftarrow \mathbf{M}_{D,D}\,\mathbf{M}_{D,L}$. We note that $\mathbf{M}$ captures the degree of connectivity of terms *via* the network. The second input to BUILDONTOLOGY is the set of candidate parent nodes P. This set can also be thought of as the complete set of leaf nodes in $\mathcal{O}$. Only leaf nodes are considered as potential parents to terms not added to $\mathcal{O}$, thus ensureing a tree-structured ontology is synthesized. The third input to BUILDONTOLOGY is $\mathcal{O}$ itself. On initialization $\mathcal{O}$ is empty and thus seeded with an artificial root node, termed "root," setting $P = \{\text{"root"}\}$. Upon completion of BUILDONTOLOGY, $\mathcal{O}$ will contain all the lexical terms from $G_{D,L}$. The fourth input is the *percent* value. As the algorithm progresses, the largest value within a matrix is found (*maxval* at line 6). Applying *percent* establishes the range $[(1 - \textit{percent}) \times \textit{maxval}, \textit{maxval}]$ and any columns in $\mathbf{M}$ that have a value falling within this range are selected (columns represent lexical terms).

BUILDONTOLOGY($\mathbf{M}$, P, $\mathcal{O}$, *percent*)

1 $p' \leftarrow \emptyset$, $C' \leftarrow \emptyset$, $maxval' \leftarrow 0$
2 **for** each $p \in P$
3 **do**
4 $\hat{\mathbf{M}} \leftarrow \mathbf{M}/(P/\{p\})$
/*$\mathbf{M}$ with all parent nodes in P (except for p) removed[d]*/
5 $\mathbf{M}' \leftarrow (\hat{\mathbf{M}}^\top \hat{\mathbf{M}})^2$
6 $maxval \leftarrow \text{Max}(\mathbf{M}'/\{p\})$
/*returns the maximum value in $\mathbf{M}'$ without column p[d]*/
7 $C \leftarrow \text{FindCloseColumns}(\mathbf{M}', maxval, percent)$
/*returns columns with a value $> (1 - percent) \times maxval$*/
8 **if** $(maxval > maxval')$
9 **then**
10 $maxval' \leftarrow maxval$, $C' \leftarrow C$, $p' \leftarrow p$
11 $\text{AddChildren}(\mathcal{O}, p', C')$
/*add the set of children C' to the parent p' in ontology $\mathcal{O}$*/
12 $P \leftarrow (P/\{p\}) \bigcup C'$
13 /*remove p, add children of p[d]*/
14 $\mathbf{M} \leftarrow \mathbf{M}/\{p'\}$
/*remove column p' from M[d]*/
15 **if** (all labels have been assigned)
16 **then return** $\mathcal{O}$
17 **else** BUILDONTOLOGY($\mathbf{M}, P, \mathcal{O}$)

The BUILDONTOLOGY algorithm proceeds by applying the rules-of-thumb mentioned in Section 2.1 until $\mathcal{O}$ is completely constructed. The algorithm proceeds by choosing the parent p' with the greatest value and the associated set of children C' (Lines 2 through 10). To do this, we first remove all potential parents in set P excluding p from $\mathbf{M}$, which produces $\hat{\mathbf{M}}$(Line 4). Next we calculate $\mathbf{M}'$ by multiplying the transpose of $\hat{\mathbf{M}}$ by $\hat{\mathbf{M}}$ to get a symmetric matrix, then square $\mathbf{M}'$ to amplify the disparities among relationships in $\mathbf{M}'$ (Line 5). From $\mathbf{M}'$, we find the maximum value present and return it as the variable *maxval* (Line 6). The set of children C of parent p are selected from $\mathbf{M}'$ if the column in $\mathbf{M}'$ has a value in the interval $[(1 - percent) \times maxval, maxval]$ (Line 7). This procedure is done

[d]We write set difference as A/B. To denote a similar matrix operation that removes a set of columns S from matrix $\mathbf{X}$, we write $\mathbf{X}/S$. For example, if $\mathbf{X} = [1\ 2\ 3\ 4]$, then $\mathbf{X}/\{2, 4\} = [1\ 3]$ (matrices are in bold text).

for each potential parent in P (Line 2) and the parent p with the greatest value is carried on through the algorithm (Lines 8 through 10). Upon completion of this loop, the selected parent p' and its set of children C' are added to $\mathcal{O}$ (Line 11). Once added to $\mathcal{O}$, p' is then removed from P and the children in C' are promoted to potential parents and appended to P (Line 12). p' is also removed from matrix $\mathbf{M}$ (Line 14). If all of the lexical terms have been assigned, then $\mathcal{O}$ is returned, otherwise we recursively apply this algorithm until all the terms are exhausted (Lines 7 through 17).

The complexity of this algorithm is $O(n^3)$ and performs well on the datasets tested in this study.

2.3. *Data*

2.3.1. *Microarray Coexpression Data*

The following 9 *S. cerevisiae* microarray datasets were downloaded from the Stanford Microarray Database (SMD): cell cycle (synchronized using CDC15 and alpha factor),[12] exposure to dithiothrietol (DTT),[13] diauxic shift,[14] exposure to gamma radiation,[15] sporulation,[16] heat stress,[13] starvation,[13] and exposure to H_2O_2.[13] All datasets were taken in their normalized form as log-transformed ratios and mapped to a unique yeast ORF ID. Spots not flagged as being problematic were averaged for both technical and biological replicates. Gene expression profiles with greater than 25% missing values for one gene across all conditions were removed, and any remaining expression profiles with missing values were imputed using KNNimpute.[17] For each gene expression profile per experiment, the difference between the maximum and minimum ratio value was calculated. Expression profiles that showed a difference less than 0.5 were removed.

The Pearson correlation coefficient was calculated using the expression profiles of each gene pair over in a given experiment. Significant correlations were determined through permutation testing, where the gene expression values within each condition are shuffled between genes. Correlation coefficients were calculated on the shuffled data to produce an empirical null distribution. The positively correlated gene pairs with a p-value < 0.01 were used and the other gene pairs were removed.

2.3.2. *Integrated Yeast Networks*

Two integrated yeast networks were used, namely YeastNet [6,18] and bioPIXIE (biological Process Inference from eXperimental Interaction

Evidence).[19] Both networks were built under a probabilistic framework using both yeast experimental data (*i.e.*, microarray, genetic interactions, protein interactions) and GO annotations. The value for a gene pair within the bioPIXIE network is a posterior probability from the Bayesian framework developed by Troyanskaya *et al.*,[20] while the values within YeastNet are weighted sums of log-likelihood scores as developed by Lee *et al.*.[6,18]

The bioPIXIE data[e] and YeastNet[f] data were downloaded June 2008. Any feature in the bioPIXIE or YeastNet integrated networks that did not match an ORF id in the *Saccharomyces* Genome Database gene registry were removed. Edges with values less than 0.3 in the bioPIXIE network and edges less than 1.0 in the YeastNet network were removed.

3. Results

3.1. *Ontology Properties*

We first demonstrate that the BUILDONTOLOGY algorithm performs well at recapitulating relationships between GO terms. As inputs, the algorithm takes both gene network data as $\mathbf{M}_{D,D}$ and gene annotation data as $\mathbf{M}_{D,L}$. To test the ability of the algorithm to reconstruct known relationships while ignoring any influence from a gene network, the initialization of $\mathbf{M}$ as input to BUILDONTOLOGY was changed from $\mathbf{M} \leftarrow \mathbf{M}_{D,D}\,\mathbf{M}_{D,L}$ to $\mathbf{M} \leftarrow \mathbf{M}_{D,L}$.

Using the Cytoscape[21] network visualization software and the MCODE graph clustering plug-in,[22] we extracted the top 20 subnetworks from both the bioPIXIE and YeastNet integrated gene networks. For each of these clusters, we built an ontology ignoring the gene network data. We consider a relationship between two terms in an ontology to be reasonably close if the shortest path length between these terms as defined in the GO is ≤ 3. This is based on the observation that when the shortest path between all pairwise GO biological process terms is calculated, more than 98% of the shortest paths are greater than a length of 3, with the average path length being ≈ 8.2. Roughly 46% of the GO terms across the tested ontologies were within a path length of 3 and the average path length between terms was 4.2 ($\sigma = 2.5$). In comparison, we then constructed DDOnts for the same 20 subnetworks in both the bioPIXIE and YeastNet networks while including the gene network relationships. For this set of ontologies, roughly 17% of the GO term connections were within a path length of 3 and the

[e] http://pixie.princeton.edu/pixie/
[f] http://www.yeastnet.org/

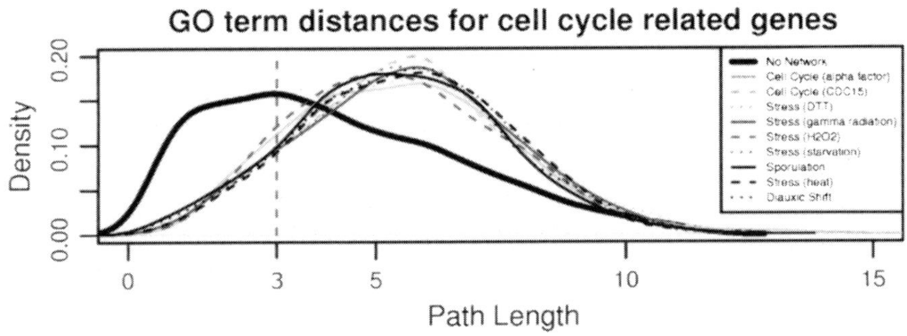

Figure 1. Distribution of the shortest path lengths between terms as measured in the Gene Ontology but assigned as a parent-child relationship in the Data-Driven Ontologies (DDOnt). All DDOnts were built from a common set of cell cycle related genes. The thick, black line represents an ontology built without gene network data, while the rest of the lines represent DDOnts constructed using microarray data. The change in distributions as compared to the ontology built without any network data was due solely to the gene-gene relationships in the microarray datasets. The path length cutoff of 3 is discussed in Section 3.1.

average path length was 5.4 ($\sigma = 2.1$). This comparison shows that the BuildOntology algorithm performed relatively well at recapitulating the term relationships defined in the GO, while including the network data places the terms further from their predefined positions in the GO revealing relationships directly influenced by the gene network data.

As a specific example of how the BuildOntology algorithm performs in relation to including or excluding network data, we first selected all genes annotated under the parent term *cell cycle* (GO:0007049) as listed in the SGD.[g] This set consisted of 386 genes annotated with 648 GO biological process terms. We then constructed an ontology from this set of terms and genes, while ignoring the network data. Next, we used 9 yeast microarray studies that cover a variety of experimental conditions to build gene networks (see Section 2 for datasets). Pearson correlation coefficients were calculated between gene expression profiles for each individual experiment and connections were drawn between a cell cycle gene pair if their coefficient was significant (see Section 2). This process results in 9 separate gene networks, which were used to build 9 DDOnts. The distribution of shortest path lengths for the 9 DDOnts and the ontology built without any gene network data are shown in Fig. 1. Including the gene network relationships

[g]checked on July, 2008

has a dramatic affect on the distribution of GO terms in the DDOnts.

3.2. *Data-Driven Ontologies for Discovery*

The bioPIXIE integrated gene network was used to demonstrate a specific application of a DDOnt to identify interesting relationships between biological processes. Using Cytoscape[21] with MCODE,[22] we extracted a series of subnetworks from bioPIXIE. An example subnetwork of 73 ORFs connected by 577 edges is shown in Fig. 2 **B**. Using this gene network and the 241 associated GO biological process annotations, we synthesized a DDOnt (Fig 2 **B**). The edge colors in the DDOnt correspond to the length of the shortest path between two terms within the GO. Edges colored grey are at a shortest path length of 1, 2, or 3, green 4, 5, or 6, red 7, 8, or 9, and blue at a length > 10. Areas of interest within the DDOnt consist of several terms with greater path lengths grouped together, which indicates potential novel relationships. The highlighted rosette in Fig. 2 **B** is an example. The parent node is *response to DNA damage stimulus* (GO:0006974). There are several examples of expected child terms, such as *double-strand break repair via break-induced replication* (GO:0000727), *DNA damage checkpoint* (GO:0000077), and *heteroduplex formation* (GO:0030491). Additionally, there are many terms that are connected to the parent node, but are located at a greater path length in the GO, such as *pre-replicative complex assembly* (GO:0006267), *regulation of chromatin assembly or disassembly* (GO:0001672), and *regulation of heterochromatin formation* (GO:0031445). Though these processes are located far apart in the GO, chromatin assembly and maintenance have long been linked to double-stranded breaks through recombination and cell cycle.[23] The reason these GO terms were put together in the arranged fashion can be seen from the network used in constructing the DDOnt. Nodes colored orange in Fig. 2 **A** are ORFs that are annotated with any of the GO terms contained in the rosette. These terms are highly connected and these relationships are taken into account in the constructing the DDOnt. If an ontology is built using the same algorithm, but excluding the gene network, this rosette is not formed and the GO terms are scattered throughout the ontology under parent nodes that are closer in path length.

We also applied the DDOnt to discover relationships from a set of predefined genes. We selected the 6 yeast ORFs annotated under the GO term *glucose transport* (GO:0015758). Next, we identified the immediate neighbor ORFs in the bioPIXIE network with an edge value > 0.5 and

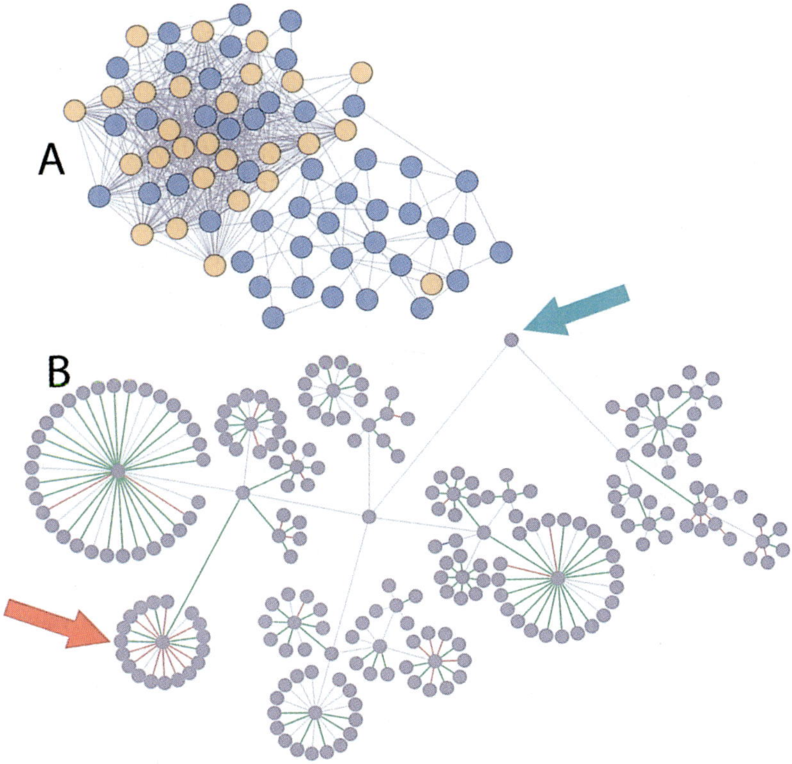

Figure 2. An example of a Data-Driven Ontology (DDOnt). The network of yeast genes (**A**) selected as a subnetwork of the bioPIXIE integrated gene network is used along with the Gene Ontology biological process annotations to synthesize the DDOnt (**B**) using the BuildOntology algorithm. The edge colors in **B** are discussed in Section 3.2. The root node in the ontology is designated by the teal arrow. The orange nodes in **A** are the genes that have been annotated with the rosette of terms in **B** designated by the red arrow. This rosette of terms are potentially novel relationships between biological processes where the parent node is *response to DNA damage stimulus* (GO:0006974).

constructed a network from this set of ORFs. This resulted in a gene network with 54 genes connected by 182 edges. This network and the associated GO terms were used to synthesize a DDOnt. Interestingly, the terms *proteolysis* (GO:0006508), *ubiquitin cycle* (GO:0006512), *phosphorylation* (GO:0016310), and *cell cycle phase* (GO:0022403) were placed under the parent term *monosaccharide transport* (GO:0015749). In the GO, these terms are located far apart, but their connections in the DDOnt reveal the links between cell cycle, proteolysis, phosphorylation, and glucose regulation. Of the set of six *glucose transport* (GO:0015758) annotated genes – RGT1, SKS1, MTH1, HXK2, HXK1, and GLK1 – four are kinases, thus supporting the relationship of phosphorylation. Of the immediate neighboring genes, GRR1, RSP5, and CDC34 are involved in ubiquitin-mediated protein degradation and RPN5 is a regulatory subunit of the 26S proteasome lid, thus supporting the proteolysis relationships. Lastly, several other immediate neighboring genes, CLN1 CLN2, CDC24, and GRR1 are all cell cycle related genes, thus supporting the cell cycle relationships. This is another example of how the construction of a DDOnt places GO terms together only because of the connections in a gene network. Although the relationships discussed are known, this example shows how potentially novel relationships can be discovered.

3.3. *Data-Driven Ontologies for Microarray Analysis*

Fluctuations in the behavior of sets of expressed genes under different experimental conditions is an underlying phenomenon studied through microarray experiments. Disentangling these effects is a major area of study. One property of DDOnts is to place terms closer to the root if they have high values. A term gets a high value from either being highly annotated or from the genes annotated with the term being highly connected. If a set of genes are fixed, thus fixing the GO term counts, and a DDOnt is built using two distinct gene networks, any differences between the two DDOnts is a result of the relationships in the gene networks exclusively. We can then find terms that are placed closer or farther from the root node to identify biological processes that are being affected through the relationships in gene networks.

We selected the 115 verified ORFs defined by the SGD to be involved in sporulation[g]. These 115 ORFs are annotated with 302 GO biological process terms. We first constructed a reference ontology using only the GO terms and ignoring any network data. Next, we constructed DDOnts for all 9 microarray datasets listed in Section 2. We then directly compared the 9 DDOnts to the reference ontology by measuring the distance of a GO term to the root. The GO terms where the distance from the root fluctuated most across the DDOnts were identified. Interestingly, mitosis, as represented by the terms *regulation of mitotic cell cycle* (GO:0007346), *chromosome segregation* (GO:0007059), and *mitosis* (GO:0007059), were

[g]checked on July, 2008

placed closer to the root node across all microarray DDOnts as compared to the reference ontology. An example of a general biological process that consistently moved farther from the root was conjugation; namely, *conjugation* (GO:0000746), *conjugation with cellular fusion* (GO:0000747), and *cellular morphogenesis during conjugation* (GO:0000767). In the reference ontology, these GO terms were all placed at a path length of 3 from the root, but when the microarray data were considered, these terms were consistently placed at a path length of 4 and 5 from the root. Both of these examples show that the relationships derived from gene expression profiles influence how a DDOnt is built, which reflects how biological processes are related in the experimental data.

4. Discussion

In this study we have demonstrated that ontologies (DDOnts) can be synthesized automatically from a gene network and a set of associated annotations. The presented algorithm is tractable and has been shown to effectively run on moderately sized data sets.

The scope of a constructed DDOnt is dependent on the input lexicon L and gene network. In this study we restricted L to be GO "biological process" terms, therefore the constructed DDOnts reflect biological processes. Similarly, the structure of a DDOnt is dependent on the input gene network. In the extreme cases where the network is either completely or sparsely connected, the gene network will have little effect on a DDOnt. The ideal input to the BUILDONTOLOGY algorithm is a well annotated gene network with connections between genes forming function specific clusters.

The results of constructing DDOnts from informative input data demonstrate that gene networks do have a major effect on DDOnt construction and edges between terms not closely related in the GO may represent interesting or even novel relationships. We also show that DDOnts constructed from different gene networks yield different perspectives. In summary, DDOnts appears to be useful tools in discovering and understanding biological relationships. Future work will include refining the algorithm, incorporating edge type information, and relaxing structural constraints (*i.e.* expanding directed acyclic graphs beyond the current tree structure).

All results and DDOnt source code can be found at: `http://www.ddont.org`

Acknowledgments

We thank the Center for Genomics and Bioinformatics for computer support, and H. Tang, P. Radivojac, B. Eads, and S. Beason for comments.

References

1. A. Lesk, Ed., *Database Annotation in Molecular Biology: Principles and Practice*, (John Wiley & Sons, LTD, 2005).
2. A. J. Butte, I. S. Kohane, *Pac. Symp. Biocomput.*, 418 (2000).
3. L. Giot, J. S. Bader, C. Brouwer, A. Chaudhuri, B. Kuang, Y. Li, Y. L. Hao, C. E. Ooi, B. Godwin, E. Vitols, et al., *Science* **302**, 1727 (2003).
4. T. Ito, T. Chiba, R. Ozawa, M. Yoshida, M. Hattori, *et al*, *PNAS USA* **98**, 2001 (4569-4574).
5. P. Uetz, L. Giot, G. Cagney, T. Mansfield, R. Judson, *et al.*, *Nature* **403**, 623 (2000).
6. I. Lee, S. V. Date, A. T. Adai, E. M. Marcotte, *Science* **306**, 1555 (2004).
7. C. L. Myers, D. Robson, A. Wible, M. A. Hibbs, C. Chiriac, C. L. Theesfeld, K. Dolinski, O. G. Troyanskaya, *Genome Biol.* **6**, R114 (2005).
8. I. Lee, B. Lehner, C. Crombie, W. Wong, A. Fraser, E. Marcotte, *Nat. Genet.* **40**, 181 (2008).
9. P. Simons, *Parts: A Study in Ontology (new edition)*, (Oxford University Press, USA, 2000).
10. S. Nirenburg, V. Raskin, *Ontological Semantics*, (MIT Press, 2004).
11. Gene Ontology Consortium, *Genome Res.* **11**, 1425 (2001).
12. P. T. Spellman, G. Sherlock, M. Q. Zhang, V. R. Iyer, K. Anders, *et al.*, *Mol. Biol. Cell* **9**, 3273 (1998).
13. A. P. Gasch, P. T. Spellman, C. M. Kao, O. Carmel-Harel, M. B. Eisen, *et al.*, *Mol. Biol. Cell* **11**, 4241 (2000).
14. M. J. Brauer, A. J. Saldanha, K. Dolinski, D. Botstein, *Mol. Biol. Cell* **16**, 2503 (2005).
15. A. P. Gasch, M. Huang, S. Metzner, D. Botstein, S. J. Elledge, P. O. Brown, *Mol. Biol. Cell* **12**, 2987 (2001).
16. S. Chu, J. DeRisi, M. Eisen, J. Mulholland, D. Botstein, *et al.*, *Science* **282**, 699 (1998).
17. O. Troyanskaya, M. Cantor, G. Sherlock, P. Brown, T. Hastie, R. Tibshirani, D. Botstein, R. Altman, *Bioinformatics* **17**, 520 (2001).
18. I. Lee, Z. Li, E. M. Marcotte, *PLoS ONE* **2**, e988 (2007).
19. C. Myers, D. Robson, A. Wible, M. Hibbs, C. Chiriac, C. Theesfeld, K. Dolinski, O. Troyanskaya, *Genome Biology* **6**, R114 (2005).
20. O. G. Troyanskaya, K. Dolinski, A. B. Owen, R. B. Altman, D. Botstein, *PNAS USA* **100**, 8348 (2003).
21. P. Shannon, A. Markiel, O. Ozier, N. Baliga, J. Wang, D. Ramage, N. Amin, B. Schwikowski, T. Ideker, *Genome Research* **13**, 2498 (2003).
22. G. Bader, C. Hogue, *BMC Bioinformatics* **4**, 2 (2003).
23. F. Paques, J. E. Haber, *Microbiol. Mol. Biol. Rev.* **63**, 349 (1999).

IDENTIFICATION OF DISCRIMINATING BIOMARKERS FOR HUMAN DISEASE USING INTEGRATIVE NETWORK BIOLOGY

JOEL T. DUDLEY AND ATUL J. BUTTE

Stanford Center for Biomedical Informatics Research, Departments of Medicine and Pediatrics, Stanford University School of Medicine, Stanford, CA 94305-5479, USA

There is a strong clinical imperative to identify discerning molecular biomarkers of disease to inform diagnosis, prognosis, and treatment. Ideally, such biomarkers would be drawn from peripheral sources non-invasively to reduce costs and lower potential for complication. Advances in high-throughput genomics and proteomics have vastly increased the space of prospective molecular biomarkers. Consequently, the elucidation of molecular biomarkers of clinical importance often entails a genome- or proteome-wide search for candidates. Here we present a novel framework for the identification of disease-specific protein biomarkers through the integration of biofluid proteomes and inter-disease genomic relationships using a network paradigm. We created a blood plasma biomarker network by linking expression-based genomic profiles from 136 diseases to 1,028 detectable blood plasma proteins. We also created a urine biomarker network by linking genomic profiles from 127 diseases to 577 proteins detectable in urine. Through analysis of these molecular biomarker networks, we find that the majority (> 80%) of putative protein biomarkers are linked to multiple disease conditions. Thus, prospective disease-specific protein biomarkers are found in only a small subset of the biofluids proteomes. These findings illustrate the importance of considering shared molecular pathology across diseases when evaluating biomarker specificity. The proposed framework is amenable to integration with complimentary network models of biology, which could further constrain the biomarker candidate space, and establish a role for the understanding of multi-scale, inter-disease genomic relationships in biomarker discovery.

1. Introduction

Perhaps one of the most compelling prospects of translational genomics is the potential for the discovery of novel molecular biomarkers of disease that offer early detection of pathogenesis, inform prognosis, guide therapy, and monitor disease progression. Despite expectations, the elucidation of accurate and discriminating disease biomarkers has proved challenging[1], and the widespread adoption of genomics-based biomarkers in the clinical management of disease remains to be realized[2]. There are many factors confounding the discovery and development of effective clinical biomarkers, including genetic variation between and among individuals and populations[3,4], deficiencies in biomolecule capture and quantification technologies[5], transient shifts in proteome composition due to acute-phase reactants and environmental stress[6-8], and logistical constraints related to associated costs and clinical acceptance[9,10]. Such confounding factors can contribute to appreciable clinical heterogeneity for a particular disease with regards to diagnosis, treatment, and outcome.

Despite the relatively limited impact of genomics on the development of clinical biomarkers to date, there has been notable success in applying genomics techniques to better clarify and characterize the clinical heterogeneity observed for many complex diseases. In particular, high-throughput gene-expression profiling using microarrays has proven successful as a means by which genome-scale events can be linked to clinical metrics. Ramaswamy *et al.* demonstrated that gene expression signatures could accurately differentiate adenocarcinoma subtypes[11]. Chen *et al.* used microarray profiling of lung cancer tissues to derive a prognostic five-gene expression signature associated with relapse and survival. Potti *et al.* derived a set of gene expression signatures that were successful in predicting response to chemotherapeutic agents[12]. Although significant, the impact of such findings remains far removed from the clinic, as they often require undesirably invasive procurement of patient tissues, improved handling of unstable molecules (e.g. RNA), and improved consistency of measurements. Such factors have consequently impeded the customary use of microarrays in most clinical settings.

The desire for minimally invasive biomarker strategies has put a focus on established clinical biofluids, such as blood and urine, as sources of putative molecular biomarkers. Both blood and urine are easily and inexpensively obtained from patients as a conventional facet of clinical care, therefore biomarker strategies leveraging these fluids are particularly amenable to current clinical protocol[13,14]. The advent of several blood plasma and urine proteome projects, with aims to identify the vast body of gene products comprising these biofluids, has generated new opportunities for genomics-based approaches to the elucidation of clinical molecular biomarkers[15,16]. Microarray analyses of blood and urine have identified expression signatures symptomatic of diseases such as rheumatoid arthritis[17], Alzheimer disease[18], Chronic Fatigue Syndrome[19], Huntington's disease[20], and glial brain tumors[21].

Disease conditions are most often signified by the dysregulation of complex biological pathways involving multiple, interacting gene products. Thus integrative approaches linking gene expression activity with proteomics and physiopathology are needed to identify highly discerning subsets of molecular biomarkers from the vast combinatorial space of candidates. One such approach is to frame the space of biomarker candidates within the context of inter-disease relationships. Traditional approaches to biomarker discovery are based on the implicit assumption that the heterogeneity of clinical disease classifications, which is often described using symptoms and anatomy, is reflected in the underlying molecular pathophysiology of the disease conditions. However, recent studies have shed light on widespread genomic and genetic correspondence between diseases previously thought to be dissimilar based on anatomy and manifest symptoms[22-24]. In fact, the similarity of responses across diseases and tissues raises concerns about the specificity of putative biomarkers derived under the consideration of only a single disease condition.

Here we propose an integrative, network-based model for protein biomarker prioritization that identifies protein biomarkers in blood and urine proteomes that exhibit high disease specificity using inter-disease relationships derived from gene expression profiles across hundreds of diseases and nearly ten thousand microarrays. We find that a majority of protein biomarkers detectable in blood and urine (>80%) exhibit non-discerning disease connectivity in the biomarker network, potentially impacting their clinical utility for a single disease. Our findings highlight the importance of integrating the context of broad inter-disease relationship profiling into future molecular biomarker discovery and prioritization efforts.

2. Methods

2.1. *Discovery and annotation of disease experiments*

Microarray experiments characterizing human disease conditions were automatically identified using a previously established method[25]. In brief, disease-associated microarrays were obtained from the NCBI Gene Expression Omnibus (GEO)[26] using text-mining techniques. We have previously shown that the experimental context for GEO Series (GSE), or collections of microarrays, can be obtained using MeSH terms from PUBMED records associated with GEO experiments. MeSH terms derived in this manner were evaluated for disease concepts using the Unified Medical Language System (UMLS)[27]. Each GSE determined to be relevant to a human disease was subject to automated annotation of the disease condition, the tissue or biological substance from which the samples were derived, and whether or not the experiment measured a normal control state complimentary to the annotated disease state by means of an additional text mining step[28]. We only retained microarray experiments in which the disease and normal conditions were measured in equivalent tissues in the same experiment. The disease and tissue annotations were manually reviewed in a post-processing step to ensure accuracy.

2.2. *Microarray data preparation and analysis*

For each microarray platform represented within the annotated disease experiments, we updated the mappings between the platform-specific probe identifiers and the Entrez GeneID identifiers in an automated manner using the Array Information Library Universal Navigator (AILUN) system[29]. For each disease experiment we derived a set of significantly differentially expressed genes using the Significance Analysis of Microarrays (SAMR) software[30]. The significant genes were determined by comparing a set of microarrays representing a normal biological state and a set representing the disease state. SAMR was configured to estimate the False Discovery Rate (FDR) using 1,000 rounds of randomized measurement permutations. Genes were considered to be significantly differentially expressed if the estimated fold-change was > 1.5, a customary fold-change threshold used with SAMR, and the estimated FDR was < 5%.

2.3. *Construction and analysis of the proteome biomarker networks*

A database of human blood plasma proteomes was constructed using data from the HUPO Plasma Proteome Project[15] (PPP) and a non-redundant list from the Plasma Proteome Institute[31]. Only the 3020 proteins from the high-confidence set of identified peptides in the HUPO PPP dataset were included in the analysis. Urine proteome data was obtained from the MAPU Proteome database[32] and the Urinary Exosome database[33]. The original data sets were parsed into a MySQL database and the International Protein Index (IPI) identifiers were mapped to Entrez GeneID identifiers using AILUN[29]. Disease-associated genes from microarray studies were associated with protein biomarkers using Entrez GeneID as the associative identifier. Networks were constructed such that diseases and genes (proteins) were nodes, and edges between gene and disease nodes were formed when a gene was found to be significantly differentially expressed in the disease state. The networks rendering and analysis was performed using the yED graph editor (http://www.yworks.com).

2.4. *Functional annotation enrichment analysis*

Functional annotation enrichment for disease-associated protein biomarkers was conducted using the DAVID system[34]. For each biomarker network, genes linked to at least one disease were considered to be the "gene list" and the entire list of gene identifiers associated with the respective proteomes were used as the background population. P-values were adjusted using Benjamini-Hochberg correction[35].

3. Results

Text-mining GEO for disease-associated experiments yielded 383 experiments, representing 238 unique diseases across 8,435 microarrays. In linking proteome biomarkers with disease, we find that 1,028 (38.5%) plasma and 577 (39.9%) urine proteins were found to be significantly differentially expressed in one or more of the 238 distinct disease states represented in the microarray data. Of those, 846 (82.2%) plasma and 490 (84.9%) urine proteins are significantly differentially expressed in more than one disease state. Thus, less than 20% of putative proteome disease markers exhibit specificity for a single disease.

Table 1. A subset of indiscriminate, highly-connected biomarker nodes and their disease targets.

Biomarker	Diseases
Plasma	
AZGP1	Cardiac hypertrophy, Spinal cord injury, Idiopathic cardiomyopathy, Idiopathic thrombocytopenic purpura, *E. coli* infection of the CNS, Hypercholesterolemia, Clear cell carcinoma of kidney, Hypertrophy, Glioblastoma, Adenoma of small intestine, Thrombocytopenia, Carcinoma in situ of small intestine, AML, Huntington's disease, Porcine nephropathy, Allergic asthma, Cirrhosis of liver, Adenovirus infection, Squamous cell carcinoma, Duchenne muscular dystrophy
CD46	Malignant neoplasm of prostate, Complex dental cavity, Fracture of bone, MODY, Dermatomyositis, Bacterial infection, Clear cell carcinoma of kidney, Spinal cord injury, Status epilepticus, Senescence, Fracture of femur, Barrett's ulcer of esophagus, Rheumatoid arthritis, Urothelial carcinoma, Astrocytoma, Glioblastoma, Congestive cardiomyopathy, Obesity, Lung transplant rejection
LAMA2	Breast cancer, Dermatomyositis, Malignant neoplasm of stomach, Acute lung injury, Malignant melanoma, Glioblastoma, Adenovirus infection, Duchenne muscular dystrophy, Acute promyelocytic leukemia, Senescence, Barrett's ulcer of esophagus, AML, Hypercholesterolemia, Hepatic lipidosis, Acute pancreatitis, Idiopathic thrombocytopenic purpura, Porcine nephropathy, Urothelial carcinoma, AIDS
Urine	
AKR1C1	Acute lung injury, Acute arthritis, Essential thrombocythemia, Ulcerative colitis, Lung transplant rejection, Malignant melanoma, Carcinoma in situ of small intestine, Dehydration, Adenoma of small intestine, Bacterial infection, Glioblastoma, Oligodendroglioma, Urothelial carcinoma, Progeria syndrome, Atrial fibrillation, Huntington's disease, SARS, Adenocarcinoma of lung
PRG4	Multiple benign melanocytic nevi, Urothelial carcinoma, Type 2 diabetes mellitus, Actinic keratosis, Adenocarcinoma of lung, Thrombocytopenia, Acute myeloid leukemia, Huntington's disease, Cardiomyopathy, Ventilator-associated lung injury, Macular degeneration, Congestive cardiomyopathy, Polycystic ovary syndrome, Dermatomyositis, Adenovirus infection, Acute pancreatitis
AQP2G4	Clear cell carcinoma of kidney, Dermatomyositis, Breast cancer, Duchenne muscular dystrophy, Hepatocellular carcinoma, Bacterial infection, Barrett's ulcer of esophagus, Helicobacter pylori GI infection, Macular degeneration, MODY, Urothelial carcinoma, AML, Crohn's disease, Ulcerative colitis, Epithelial proliferation

Table 2. Annotation enrichment for disease-associated biomarkers.

GO Term	P-value
Plasma	
(GO:0005975) carbohydrate metabolic process	3.1E-5
(GO:0019318) hexose metabolic process	1.1E-4
(GO:0006066) alcohol metabolic process	4.6E-4
(GO:0044459) plasma membrane part	5.3E-4
Urine	
(GO:0009308) amine metabolic process	7.7E-3
(GO:0044421) extracellular region part	1.4E-2
(GO:0050896) response to stimulus	1.8E-2

Among the biomarker proteins associated with disease conditions, we identified sets of enriched gene annotation terms, which were distinct for each biofluid (Table 2). Disease-associated plasma biomarker proteins were enriched for plasma membrane proteins, and proteins involved in sugar and carbohydrate metabolism. Disease-associated urine biomarker proteins were enriched for extracellular proteins, and proteins involved in amine metabolism and biotic stimulus response.

We found that a majority of diseases could not be linked to a disease-specific protein biomarker in either the blood plasma or urine proteomes. Among the distinct disease conditions represented in the microarray data, 136 (57.1%) were linked to plasma proteins, while 127 (53.4%) were mapped to urine proteins. Of these, 65.4% and 72.4% link exclusively to biomarkers shared by other diseases in plasma and urine respectively. A selection of disease conditions associated with multiple disease-specific biomarker proteins are listed in Table 3.

Table 3. A subset of diseases associated with multiple disease-specific protein biomarkers.

Disease	Disease-specific protein biomarkers
Plasma	
Idiopathic cardiomyopathy	*MACF1, SF3B2, RFX5, TLN1, FSHR, PCCA, PGK2, NEK1, RGS3, RGN, CYP3A43*
Thrombocytopenia	*CYLC2, PIGK, AASS, PANX2, DSPP, XPC, TBL1X, TCERG1*
Malignant melanoma	*PDE3A, CALR, PDCD6IP, CHAC, KIAA0586*
AIDS	*PAPPA, TRADD, KIAA0649, APRIN, MAP3K5*
Huntington's disease	*MAML1, PLGL, RNF10, KIAA0913, OAS1*
Urine	
Idiopathic cardiomyopathy	*DEFA3, ALDH1L1, CD177, TLN1, SLURP1, BPI, APOH, C8B*
Glioblastoma	*WISP2, PRDX3, TIMP2, ACO1*
Breast cancer	*ENPP4, PFKP, THBD, IGFALS*
Acute promyelocytic leukemia	*CSPG3, LGALS7, HSPA5*
Adenovirus infection	*VGF, AGA, UMOD*

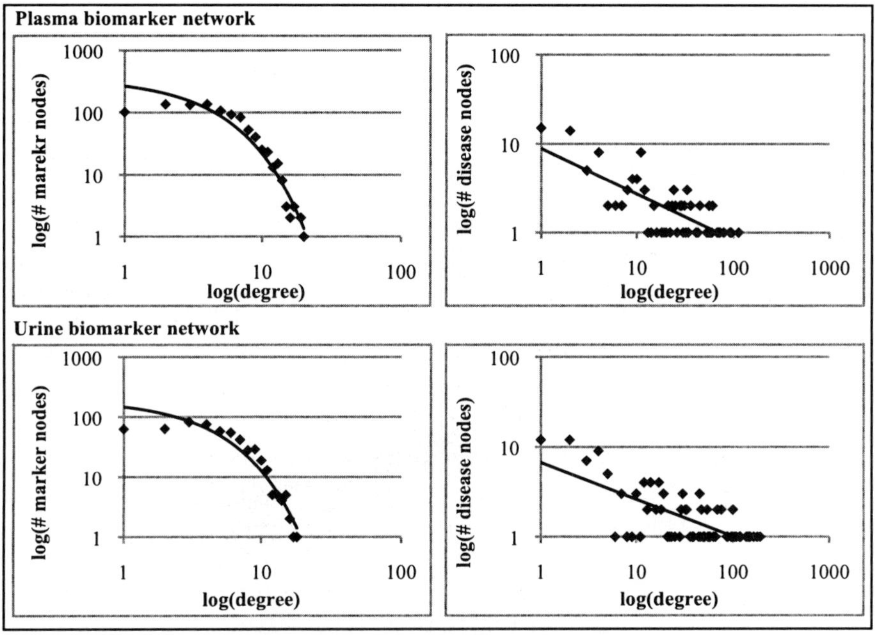

Figure 1. Independent log-log plots of node degree distributions for biomarker and disease nodes.

The mean disease linkage degree for a protein biomarker node was 5.09 in the plasma network and 5.06 in the urine network. The mean biomarker linkage degree for a disease node was 36.19 in the plasma network and 22.57 in the urine network. The distribution of disease connectivity across biomarker nodes was found to follow an exponential model in both the blood ($R^2 = 0.94$) and urine ($R^2 = 0.93$) networks, suggesting a scale-variance in attachment (Figure 1). The distribution of biomarker connectivity across disease nodes was found to follow a weak power-law model in both the blood plasma ($R^2 = 0.59$) and urine ($R^2 = 0.53$) biomarker networks, suggesting a scale-free property. This suggests that diseases with many biomarkers preferentially gain more biomarkers. The equivalent graph of the connectivity of the biomarkers matches an exponential curve. The shape of the curve actually splits into two parts. At low connectivity, biomarkers gain connections to diseases randomly as more diseases are added. At higher connectivity, biomarkers then gain connections to diseases preferentially if they are already connected. The two segments to the log-log plots of biomarker node degree distributions in Figure 1 thus suggest there are two separate populations of biomarkers.

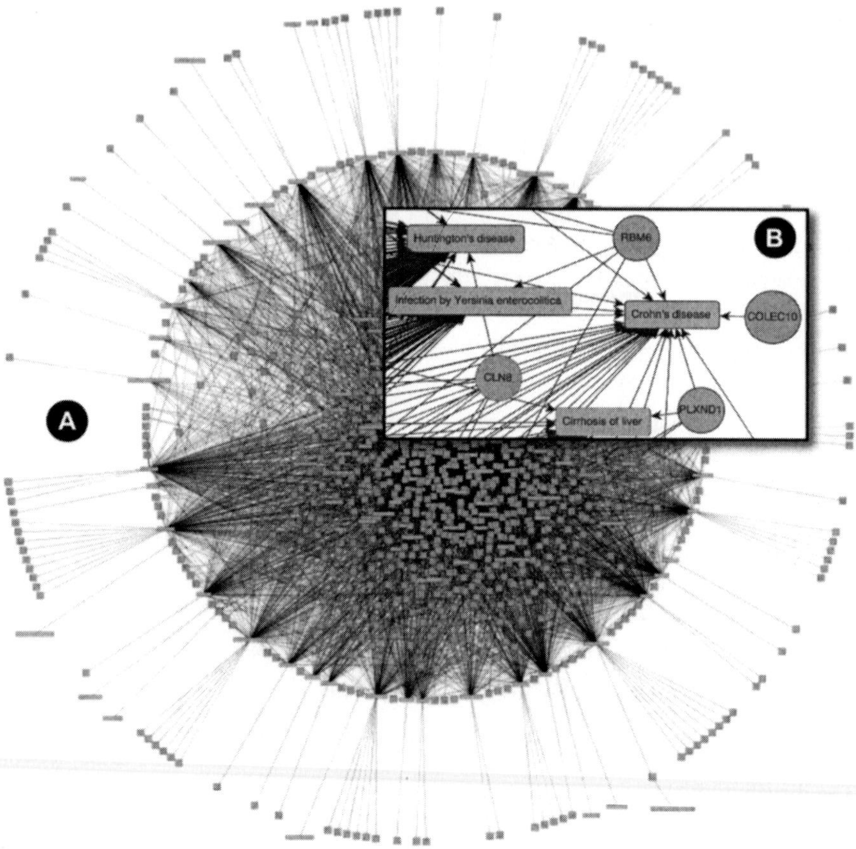

Figure 2. A rendering of the plasma biomarker network is shown (A). Disease-specific biomarkers (circles) are found extending from diseases (rectangles) at the periphery of the network. The inset image (B) illustrates a subset of the biomarker network showing *COLEC10*, a disease-specific biomarker for Crohn's disease, and non-specific biomarkers that connect Crohn's disease to other disease conditions.

4. Discussion

In this study we propose an integrative network model for biomarker prioritization using inter-disease relationships derived from microarray studies, and putative protein biomarkers from large-scale biofluids proteome studies. Unlike traditional biomarker prioritization approaches, our approach first considers all possible (i.e.

measureable) protein biomarkers in a biofluid proteome and places them within the context of inter-disease relationships across the broad spectrum of human disease to identify putative protein biomarkers that are likely to be highly discerning for a disease of interest.

Our approach finds validation in finding that a majority proportion (> 80%) of measurable proteins in both the blood plasma and urine proteomes are non-specific for any single disease condition. Interestingly, there are several diseases with more than one disease-specific biomarker, which could signify an opportunity to develop more robust, multi-marker assays for these diseases. Given the vast resources required for both identifying and biologically validating putative molecular biomarkers, these results suggest that it may be prudent to focus biomarker discovery efforts on the diseases enriched for disease-specific biomarker associations. Such enriched associations could indicate that a novel and discriminating pathway is involved in the pathogenesis of the disease, suggesting that validation efforts focused on such disease-specific associations could lead to the identification of highly discriminating upstream or downstream molecular biomarker candidates.

Although many of the discriminating disease-biomarker associations predicted by our approach remain to be biologically and clinically validated, there is, in several cases, a compelling degree of biological continuity between the predicted disease-specific biomarker and the understood molecular phenomena underlying the disease. One such example is our prediction that Collectin Sub-family Member 10 (*COLEC10*) as a putative disease-specific biomarker for Crohn's disease. Crohn's disease is a chronic, debilitating inflammatory bowel disorder that can affect any portion of the digestive tract[36]. Recent genome-wide association studies and other investigations into the pathogenesis of Crohn's disease have revealed a number of susceptibility genes[37,38] and the major role of the body's innate immune response against enteric microbiota[39,40]. Collectins have been implicated as significant regulators of the innate immune system, particularly with regards to host defense response to microorganisms[41]. Collectins are known to induce pro-inflammatory cytokines and participate in activation of the compliment system via the lectin pathway during the microorganism defense reponse[42]. Therefore *COLEC10* could serve as a novel biomarker that is sensitive to the episodic manifestations of Crohn's disease to inform ongoing disease management, whereas current biomarkers for the disease are primarily diagnostic[43].

Another interesting finding is the identification of GDP dissociation inhibitor 1 (*GDI1*) as a disease-specific biomarker for Hypercholesterolemia. *GDI1* is a known regulator of the GDP/GTP exchange reactions of *Rab* proteins and a participant in the vesicle mediated cellular transport[44]. *GDI* and *Rab* are also known to participate in the cellular transport of lipids, and *GDI/Rab* dysregulation has been observed in the presence of cholesterol accumulation[45].

We recognize several caveats in our approach. Foremost, our approach makes the naïve assumption that if a gene is significantly differentially expressed in a disease condition that this differential will be reflected in either blood plasma or urine regardless of the anatomical locus of the disease site. While quantifications of mRNA expression can be far removed from the modulation of protein fragments in biofluids, there is reason to believe that such an assumption can hold true in a sufficient number of cases. Interestingly, notable proportions of the proteins identified by blood plasma and urine proteome projects are annotated with Gene Ontology terms signifying intracellular localization, including: *intracellular part* (55.4%), *intracellular organelle part* (20.3%), *cytoskeleton* (9.6%), and *nuclear part* (6.4%). Such phenomenon may be accounted for by sufficient secretion of intracellular proteins inside small-membrane vesicles known as exosomes by various tissue types[46-50]. Furthermore, cells undergoing destruction as a consequence of pathogenesis are likely to emanate intracellular matter into biofluids. We also recognize that the specificity of a protein biomarker in our networks is subject to the availability of microarray data for any particular disease. The addition of novel disease conditions into future versions of the biomarker network could even further reduce the proportions of disease-specific protein biomarkers. This study is also limited by the quantity and quality of microarray datasets across diseases.

The framework proposed in this study is not intended to serve as an unequivocal means for biomarker elucidation. Rather we suggest that the integration of our approach with other forms of biomarker network biology is likely to lead to even more sophisticated approaches to informatics-based biomarker discovery. Alterovitz *et al.* proposed an information theoretic framework for biomarker discovery that identified high-quality peripheral biomarker candidates by identifying significant tissue-biofluid channels across a wide range of tissues and biofluids proteomes[51]. Our approach could be used in combination with their biofluids channel approach to find optimal intersections between disease-specificity space and biofluid-tissue interaction space to even further refine the scope of putative biomarker proteins for a particular disease condition.

5. Conclusion

The discovery of discerning molecular biomarkers for a disease condition of interest is encumbered by the vast combinatorial space of prospective candidate markers. Our work provides a novel framework for reducing the space of candidate markers by establishing an integrative, network-based model comprised of disease associated gene expression profiles and biofluids proteomes. While a more traditional biomarker discovery endeavor might start with the disease condition of interest to identify biomarker candidates in a "bottom-up" approach, we offer a "top-down" approach that begins with the broad space of human disease and full compliments of biofluid proteomes to quickly discern candidate protein biomarkers discriminately associated with the a disease condition. This work establishes the importance of genome-wide, inter-disease relationships in biomarkers discovery and paves the way for novel integrative methods that incorporate inter-disease network models to further refine biomarker discovery.

Acknowledgements

The work was supported by grants from the Lucile Packard Foundation for Children's Health, National Institute of General Medical Sciences (R01 GM079719), Howard Hughes Medical Institute, and the Pharmaceutical Research and Manufacturers of America Foundation. The authors would like to thank Alex Skrenchuk for computing support.

References

1. M. Baker, *Nature biotechnology* **23**, 297-304 (2005)
2. J. Bell, *Nature* **429**, 453-6 (2004)
3. P. D. Pharoah, *et al.*, *N Engl J Med* **358**, 2796-803 (2008)
4. J. P. Radich, *et al.*, *Genomics* (2004)
5. C. J. Johnson, *et al.*, *Proteomics* **8**, 715-30 (2008)
6. V. J. Nesatyy and M. J. Suter, *Mass spectrometry reviews* (2008)
7. D. L. Mendrick, *Toxicology* **245**, 175-81 (2008)
8. M. M. Nascimento, *et al.*, *Nephrol Dial Transplant* **19**, 2803-9 (2004)
9. S. A. Williams, *et al.*, *Nature reviews* **5**, 897-902 (2006)
10. R. Etzioni, *et al.*, *Nat Rev Cancer* **3**, 243-52 (2003)
11. S. Ramaswamy, *et al.*, *Proc Natl Acad Sci U S A* **98**, 15149-54 (2001)

12. A. Potti, *et al.*, *Nature medicine* **12**, 1294-300 (2006)
13. L. A. Liotta, *et al.*, *Nature* **425**, 905 (2003)
14. V. Thongboonkerd, *et al.*, *Contributions to nephrology* **141**, 292-307 (2004)
15. G. S. Omenn, *et al.*, *Proteomics* **5**, 3226-45 (2005)
16. J. Adachi, *et al.*, *Genome Biol* **7**, R80 (2006)
17. N. Olsen, *et al.*, *British Medical Journal* (2004)
18. O. C. Maes, *et al.*, *Neurobiol Aging* **28**, 1795-809 (2007)
19. N. Kaushik, *et al.*, *British Medical Journal* (2005)
20. F. Borovecki, *et al.*, *Proc Natl Acad Sci* (2005)
21. E. R. Smith, *et al.*, *Clin Cancer Res* **14**, 2378-86 (2008)
22. J. Lamb, *Nat Rev Cancer* **7**, 54-60 (2007)
23. K. I. Goh, *et al.*, *Proc Natl Acad Sci U S A* **104**, 8685-90 (2007)
24. E. Segal, *et al.*, *Nature genetics* **36**, 1090-8 (2004)
25. A. J. Butte and R. Chen, *AMIA Annu Symp Proc* 106-10 (2006)
26. T. Barrett, *et al.*, *Nucleic acids research* **33**, D562-6 (2005)
27. O. Bodenreider, *Nucleic acids research* **32**, D267-70 (2004)
28. J. Dudley and A. J. Butte, *Pac Symp Biocomput* 580-91 (2008)
29. R. Chen, *et al.*, *Nature methods* **4**, 879 (2007)
30. V. G. Tusher, *et al.*, *Proc Natl Acad Sci U S A* **98**, 5116-21 (2001)
31. N. L. Anderson, *et al.*, *Mol Cell Proteomics* **3**, 311-26 (2004)
32. Y. Zhang, *et al.*, *Nucleic acids research* **35**, D771-9 (2007)
33. T. Pisitkun, *et al.*, *Proc Natl Acad Sci USA* **101**, 13368-73 (2004)
34. G. Dennis, Jr., *et al.*, *Genome Biol* **4**, P3 (2003)
35. Y. Benjamini, *et al.*, *Behavioural brain research* **125**, 279-84 (2001)
36. D. C. Baumgart and S. R. Carding, *Lancet* **369**, 1627-40 (2007)
37. J. D. Rioux, *et al.*, *Nature genetics* **39**, 596-604 (2007)
38. J. C. Barrett, *et al.*, *Nature genetics* (2008)
39. C. A. Muller, *et al.*, *Cell Mol Life Sci* **62**, 1297-307 (2005)
40. M. T. Abreu and M. P. Sparrow, *Mt Sinai J Med, New York* **73**, 1067-73 (2006)
41. J. K. van de Wetering, *et al.*, *Eur. J. Biochem / FEBS* **271**, 1229-49 (2004)
42. H. P. Haagsman, *et al.*, *Neonatology* **93**, 288-94 (2008)
43. K. A. Papadakis, *et al.*, *Inflammatory bowel diseases* **13**, 524-30 (2007)
44. D. Bachner, *et al.*, *Human molecular genetics* **4**, 701-8 (1995)
45. C. Lebrand, *et al.*, *The EMBO journal* **21**, 1289-300 (2002)
46. M. P. Caby, *et al.*, *Int Immunol* **17**, 879-87 (2005)
47. H. G. Lamparski, *et al.*, *J Immunol Methods* **270**, 211-26 (2002)
48. X. Yu, *et al.*, *Cancer Res* **66**, 4795-801 (2006)
49. G. van Niel, *et al.*, *J Biochem* **140**, 13-21 (2006)
50. G. van Niel, *et al.*, *Gastroenterology* **121**, 337-49 (2001)
51. G. Alterovitz, *et al.*, *Pac Symp Biocomput* 231-42 (2008)

LEARNING THE STRUCTURE OF PROTEIN-PROTEIN INTERACTION NETWORKS*

OLEKSII KUCHAIEV

Department of computer science,
University of California, Irvine
CA, 92697-3425, USA
E-mail: oleksii.kuchaiev@uci.edu

NATAŠA PRŽULJ[†]

Department of computer science,
University of California, Irvine
CA, 92697-3425, USA
E-mail: natasha@ics.uci.edu

Modeling and analyzing protein-protein interaction (PPI) networks is an important problem in systems biology. Many random graph models were proposed to capture specific network properties or mimic the way real PPI networks might have evolved. In this paper we introduce a new generative model for PPI networks which is based on geometric random graphs and uses the whole connectivity information of the real PPI networks to learn their structure. Using only the high confidence part of yeast *S. cerevisiae* PPI network for training our new model, we successfully reproduce structural properties of other lower-confidence yeast, as well as of human PPI networks coming from different data sources. Thus, our new approach allows us to utilize high quality parts of currently available PPI data to create accurate models for PPI networks of different species.

1. Introduction

The majority of cellular functions are not carried out by single proteins, but by proteins acting together. Due to the recent advances in the experimental biological techniques such as yeast-2-hybrid[11], tandem affinity[12] purification and other high-throughput methods, there is a huge amount of protein-protein interaction (PPI) data publicly available. This amount

*This project was supported by the NSF CAREER grant IIS-0644424
[†]Corresponding author

of data requires new mathematical and computational approaches to be developed in order to model and analyze the complex networks that they form. An accurate model of PPI networks will allow better estimating all types of network statistics as well as generating synthetic networks of species for which protein-protein interactions have not been experimentally determined. This could help understand cellular processes and lead future biological experiments. Therefore, analyzing and modeling PPI networks has become a vibrant research area.

A PPI network is a graph with nodes (vertices) corresponding to proteins and edges (links) corresponding to interactions between the proteins. Many random graph models were proposed to model PPI networks. Some of the commonly used models are: Erdös-Rényi random graphs[1], Erdös-Rényi random graphs with the same degree distribution as in data, scale-free graphs[2], geometric random graphs[4,3] and stickiness-index-based models[5]. These models were designed to capture specific network properties such as the degree distribution, the clustering coefficient, the average diameter etc., or to model the way these networks might have evolved (e.g., preferential attachment[6], or gene duplication and mutation[7,15] models); usually, they do not exploit the entire connectivity information from the real PPI networks to learn the model's topological structure. We adopt an alternative approach and exploit the entire connectivity information of a real world PPI network to learn its structure. Thus, we do not force the model to reproduce some specific network properties, such as the degree distribution, the clustering coefficient, the diameter and such. Instead, since our new model learns its structure from the real data, most of these network properties are captured automatically.

2. Methods

Several studies have shown that *geometric random graphs* represent a good model for PPI networks[3,14,9]. A geometric random graph is a graph $G = (V, E)$, where V is a set of nodes distributed uniformly at random in some metric space and for any pair of nodes $x, y \in V$, there is an edge $(x, y) \in E$ if and only if the distance between nodes x and y, $d(x, y)$, in the metric space is less than or equal to some fixed parameter ϵ according to a chosen metric. Our new model is based on geometric random graphs. In our model, henceforth called *trained geometric model*[a], we do not distribute

[a]The Matlab code implementing the model is available upon request.

nodes in a metric space uniformly at random. Instead, we use available real world PPI data of high quality to learn the distribution of the points in a metric space. Having learned this distribution ($p_{learned}$), we generate model networks of arbitrary size by distributing points in the space according to the distribution $p_{learned}$ and connecting two nodes by an edge if they are close enough in space.

Hence, the crucial aspect of this model is the probabilistic distribution $p_{learned}$, which we use to generate model networks. To learn this distribution, we need to have the metric space and an example distribution of points in it that corresponds to real PPI data. Currently, it is hard even to hypothesize about the nature or dimensionality of space in which PPI networks reside. Thus, as a proof of concept, we choose 3-dimensional Euclidean unit cube as our metric space and the Euclidean metric. However, all of the techniques described below could be easily applied to any number of dimensions. Since real PPI networks contain only the connectivity information between the nodes, we need a technique that takes this information and embeds the network nodes into a metric space so that the topological structure of the network is preserved when viewed as a geometric graph constructed from the embedded nodes. That is, the spatial proximity of the nodes will correspond to the PPI network connectivity information as in a geometric random graph. For this purpose, we use the embedding algorithm introduced by Higham et al.[9].

Next, we model the distribution of the points in the space as a Mixture of Gaussians and learn the required parameters using Expectation-Maximization algorithm (EM)[10]. Mixture of Gaussians allows properties of complex distributions to be captured, and it allows easy sampling via the ancestral sampling technique[10]. The details are presented below.

2.1. *Embedding Algorithm*

The embedding algorithm introduced by D. Higham, M. Rašajski and N. Pržulj[9] is based on the ideas from Multi-Dimensional Scaling (MDS)[19]. It takes as an input a PPI network represented as an undirected unweighted simple graph $G = (V, E)$, and a dimensionality m of the Euclidean unit cube to be used for embedding. The algorithm outputs coordinates of each node in the space under the premise that the network connectivity information corresponds to Euclidean proximity, as in geometric random graphs. Conceptually, this algorithm consists of the following steps:

- $\forall$ pairs of nodes $i, j \in V$, calculate the shortest path distances between them, $d_G(i,j)$. Then choose a cutoff value K and define the matrix of pairwise distances using the formula: $d(i,j) = \sqrt{d_G(i,j)}$ iff $d_G(i,j) \leq K$ otherwise $d(i,j) = K_{max}$, for some K_{max}.

- Construct matrix A using double-centering process:

$$a_{ij} = -\frac{1}{2}(d_{ij}^2 - \frac{1}{n}\sum_{k=1}^{n} d_{ik}^2 - \frac{1}{n}\sum_{k=1}^{n} d_{kj}^2 + \frac{1}{n^2}\sum_{k=1}^{n}\sum_{l=1}^{n} d_{kl}^2)$$

- It can be shown that A has a Schur decomposition and therefore, the resulting embedding into the m-dimensional space is then specified by the eigenvectors of A, corresponding to the m largest eigenvalues. That is, for each protein i, it's coordinates in the m-dimensional Euclidean space are then equal to the i-th coordinates of the eigenvectors $v_1, ..., v_m$ of the matrix A. $(m \leq n)$

2.2. *Learning Network Structure*

To learn the distribution $p_{learned}$ we need to have a learning set $X_{learning}$ of 3-dimensional coordinates of n nodes in the Euclidean unit cube. These coordinates are the output of the embedding algorithm described in section 2.1 applied to the PPI network data that we wish to use for learning. To model the density $p_{learned}$ of the distribution of points in the space, we use a Mixture of Gaussians model[10]:

$$p_{learned}(x) = \sum_{k=1}^{d} \pi_k N(x, \mu_k, \sigma_k^2)$$

It is a linear combination of multi-dimensional (in our case 3 dimensional) Gaussian distributions. Coefficients $\pi_k, k = 1, ..., d$, represent a marginal distribution $p(z)$ ($\sum_{k=1}^{d} \pi_k = 1$ and $\forall k : \pi_k \in [0,1]$) over a latent d-dimensional binary variable z, which has "1-of-d" representation (a particular coordinate z_k is equal to 1 while the rest are equal to 0). This hidden variable indicates what mixture component explains a particular data point x; that is $p(x|z_k = 1) = N(x, \mu_k, \sigma_k^2)$. The parameters to be learned from the data are: π_k, μ_k and σ_k^2 for $k = 1, ..., d$. The number of mixtures d, is a parameter which we do not learn, but choose manually to control the complexity of the model. We use Expectation-Maximization (EM) algorithm to learn the parameters of the Gaussian Mixture model[10].

As in many machine learning problems, more complex models will always explain the learning data better then more simple ones, but too complex models might result in over-fitting and, as a result, will perform badly in the modeling stage. Thus, we need to choose d big enough to fit the learning data well and small enough to avoid over fitting and other possible issues, such as singularities in the covariance matrices σ_k^2.

For these purposes we use Bayesian Information Criterion (BIC). During our experiments, we found that for our particular case, standard BIC does not penalize the model for its complexity strongly enough; therefore, we increase the penalty for having more parameters two times and use the formula: $BIC = log(p(X|\theta)) - MlogN$. Here, X is our learning set, $log(p(X|\theta))$ is a log-likelihood function, θ is the parameters we learn (μ, σ^2 and π), M is number of free parameters in the model (size of θ), and N is a size of the learning set. By choosing different numbers of mixtures for our model (choosing different values of d) we vary the number of free parameters M of the model, which in our case equals $M = m^2+m+(d-1)*(m+m^2+1)$, where m is the dimensionality of the space we used for the embedding (m=3 in all our experiments). According to this criterion, we need to choose the model which has the highest value of BIC[10].

2.3. *Generating Model Networks*

After the model has learned the density function $p_{learned}(x)$, it could be used for generating model networks of arbitrary size. Suppose that we want to generate a network $Net = (V, E)$ with $|V| = n$ and $|E| = m$. We use an *ancestral sampling* technique to sample from distribution $p_{learned}(x) = p(x|z) * p(z)$, where marginal distribution $p(z)$ is given by the coefficients π_k, the conditional distribution of x given z is $p(x|z) = N(x, \mu_k, \sigma_k^2)$ and to sample n nodes from $p_{learned}(x)$ we repeat the following two consecutive steps n times: 1) sample the value of z from $(\pi_1, ..., \pi_k)$, 2) sample 3-dimensional coordinates of point x_i from $p(x|z_k = 1) = N(x, \mu_k, \sigma_k^2)$.

After this sampling procedure, we have n points distributed in 3-dimensional Euclidean unit cube according to the density $p_{learned}(x)$. Next, we adjust the value of parameter ϵ to obtain m pairs of nodes with distances between them not greater then ϵ. Then, we connect these m pairs of the nodes by edges and obtain a model network of the required size.

2.4. *Network Comparisons*

We compare the following global network properties of our model and the PPI data networks: the degree distribution, the average clustering coefficient and the diameter. However, these properties themselves do not tell us much about the structure of a network. For example, it is trivial to construct two networks with the same degree distribution but with vastly different local structure. The same is also true for the clustering coefficient and the average diameter. An additional issue with using these network properties for evaluating the performance of network models is the noise and incompleteness present in PPI networks. For example, if we take two networks with the same structure, by removing edges from them we will decrease the degrees of the nodes, the clustering coefficients and the average diameters of both networks and therefore, while some structure between the two networks might have remained similar, all three of the above properties might differ substantially.

To tackle these issues we use the *Graphlet Degree Distribution (GDD) Agreement* measure[14,13] introduced by Pržulj[14] to provide a detailed evaluation of structural similarities of large networks. We give a few definitions to explain this similarity measure. *A graphlet* is a small induced subgraph of a network[3]. There are 30 possible non-isomorphic graphlets on 2, 3, 4 and 5 nodes. To calculate GDD agreement between two networks, we need to calculate for each node in the network how many times it touches each of the 30 graphlets. From a topological point of view, it is relevant to distinguish between *automorphism orbits* of each graphlet. For example, in a 3-node path, the "end-nodes" are identical from the topological point of view (i.e., can be mapped to each other by an *automorphism*, an *isomorphism* of a graph with itself – see[14] for details), whereas the "middle node" is different; therefore, a 3-node path has two different automorphism orbits. There are 73 automorphism orbits for the 30 graphlets on 2 to 5 nodes. *A Graphlet Degree Distribution (GDD)* is a 73-component distribution of the automorphism orbits in a network. Its jth component, $d^j(k)$, is the sample distribution of the number of nodes in the network touching a particular graphlet (at automorphism orbit j) k times. Graphlet degree distribution has the degree distribution as its 1st component, which corresponds to the only automorphism orbit of a 2-node path (edge). The *GDD Agreement* is a similarity measure between graphlet degree distributions of two networks. It is a number between 0 and 1, meaning that two networks have similar GDDs if their GDD agreement is close to 1, and otherwise, their GDDs are different. It is important to note that graphlet degree distribution measures the local structure of a network, because it is based on small local neighborhoods of the nodes. It is especially a very strong measure of structural similarities of small-world networks (those with small diameters), since in these networks, 5-node graphlets reach all parts of the network. Since PPI networks have small diameters, GDD is good for measuring their structure.

We use GraphCrunch software package[13] to calculate GDD agreement and other network properties that evaluate the fit of model networks to the data.

3. Application

3.1. *Datasets and Learning*

We apply the above described approach to model PPI networks of yeast and human. These two particular species have been chosen because compared to other eukaryotic species, they have the most complete and accurate PPI network data available. For our learning set (henceforth $S_{learning}$), we use the largest connected component of the high confidence yeast network described in the study by Collins et al[16]. This part of the yeast *S. cerevisiae* PPI network is believed to be of quality comparable to the data produced by small-scale experiments. It has $1,004$ nodes and $8,323$ edges.

Table 1 presents real-world PPI networks that we modeled using our trained geometric model. All these networks contain only physical interactions between proteins. Using the notation from Table 1, note that $S_{learning} \subset$ YCH $\subset$ YCOL; that is, YCH is a high confidence part of the YCOL and $S_{learning}$ is the largest connected component of the YCH PPI network.

Table 1. Real-world PPI networks used to evaluate the performance of the trained geometric model.

Network	Nodes	Edges	Origin
YCH	1622	9074	high confidence yeast PPI network from Collins et al.[16]
YCOL	2390	16127	yeast PPI network from Collins et al.[16]
YBG	4716	32747	yeast PPI network from BIOGRID[17]
HBG	7930	23543	human PPI network from BIOGRID[17]
HRAD	9141	42456	human PPI network from Radivojac et al.[18]

We embed $S_{learning}$ into a 3-dimensional cube using the embedding algorithm described above. After that, we have 3-dimensional coordinates of the $1,004$ nodes of $S_{learning}$ PPI network; we present these coordinates in $X_{learning}$, a 3×1004 matrix, and use $X_{learning}$ to learn the distribution $p_{learned}(x)$. We got the highest value of the BIC for 7 mixture components in the model and therefore we put $d = 7$ in the distribution $p_{learned}(x)$.

3.2. *Analysis*

One of the most important network properties is the degree distribution. We compare the degree distributions of two networks using *Pearson correlation coefficient*. This coefficient ranges from −1 to 1: if it is close to 1 or −1, then there is a perfect linear correlation between two distributions (where nodes are ordered according to their degrees), and if it is close to 0, then there is no linear correlation. We measure this, not the difference such as sum of squared errors, because the latter is strongly affected by the amount of noise in PPI network data, while the correlation between distributions should be preserved even when some edges are missing (false negatives) or are mistakingly present (false positives). Figure 1 presents values of Pearson correlation coefficients of degree distributions between real PPI and model networks.

Table 2. Models used to model PPI networks.

Network	Model
ER	Erdös-Rényi random graph model[1]
ER_DD	ER model with the same degree distribution as in data
GEO	Geometric random graph model[3]
SF	Scale-free Barabasi-Albert preferential attachment model[2]
STICKY	Stickiness-index-based model[5]
TGEO	The new trained geometric model

Note that by construction, ER_DD model networks have exactly the same degree distribution as real PPI networks and therefore, their Pearson correlation coefficients with the PPI networks always equal to 1. It has been observed that PPI networks have power-law degree distributions[8], which gave rise to popularity of scale-free network models. However, as we can see in Figure 1, SF network model badly captures the degree distribution of the PPI network data compared to our new model, which has learned and captured the degree distribution from the part of real-world PPI network very well. As we can see in Figure 1, the new model significantly outperforms any other model in all cases. Note that we use only high quality part of yeast PPI network for learning, but this allowed us to generate model networks of high quality also for human PPI networks.

As illustrated in Figures 2 and 3, the trained geometric (TGEO) model is better fitting in all cases, except one, than the standard geometric random graph model (GEO). For BIOGRID[17] and Radivojac et al. data sets[18],

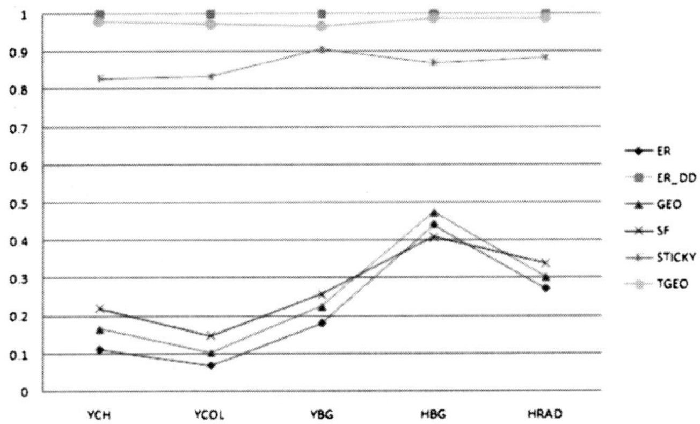

Figure 1. Pearson correlation coefficients of degree distributions between real PPI and model networks. Horizontal axis corresponds to the real PPI networks described in Table 1 and different labels correspond to different model networks described in Table 2.

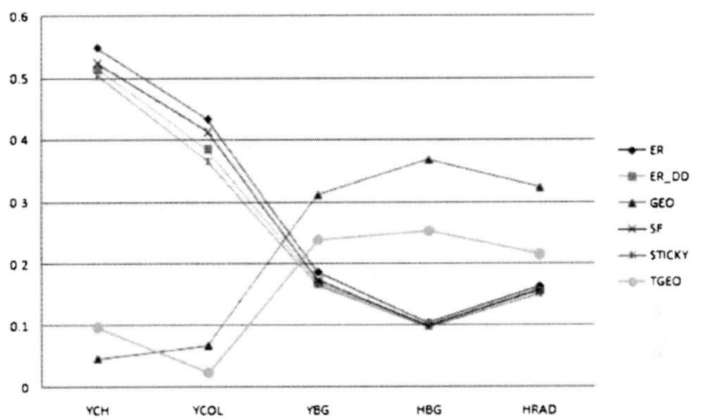

Figure 2. Differences in clustering coefficients of data and model networks. Horizontal axis corresponds to the real PPI networks described in Table 1 and different labels corresponds to different model networks described in Table 2.

GEO and TGEO models perform worse then other models. As described in section 2.4, this is likely caused by incompleteness and noise in the data. By incompleteness we do not mean the number of nodes (proteins) in the network, but presence or absence of real interactions among given nodes. Obviously, real yeast PPI network has some fixed diameter and clustering coefficient, which we can not measure now due to high noise levels in the data. Yeast networks coming from different sources (such as Collins et al.[16]

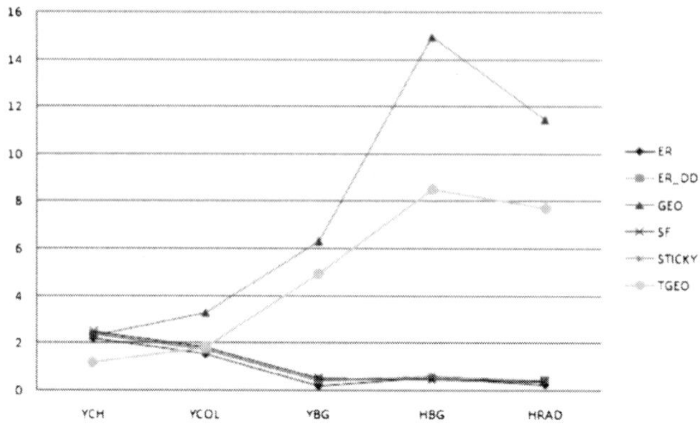

Figure 3. Differences in diameters of data and model networks. Horizontal axis corresponds to the real PPI networks described in Table 1 and different labels correspond to different model networks described in Table 2.

and BIOGRID[17]) have substantially different clustering coefficients and diameters (YCOL has the clustering coefficient of 0.44 and the diameter of 4.81, whereas YBG has the clustering coefficient of 0.19 and the diameter of 3.71). Thus, these two characteristics at the moment can not tell us much about the real structure of a PPI network.

Although global network properties are important, they might not capture the local structure of networks and are are also very sensitive to high levels of noise. Therefore, we are more interested in a more structurally constraining measure of network similarity, the GDD agreement[14]. Figure 4 presents GDD agreements between real PPI and model networks. It shows that our new trained geometric model (TGEO) fits all but one PPI network better than other models; the exception is yeast PPI network taken from BIOGRID[17], for which TGEO performs almost the same as STICKY model, but it captures the degree distribution much better then STICKY model (see Figure 1). Surprisingly, learning network structure from yeast PPI data gives better modeling insights into modeling human PPI networks than do other network models. This suggests that PPI networks of even such distant organisms as yeast and human exhibit similarities in their structures.

4. Discussion and Conclusions

Many network models have previously been introduced attempting to capture specific sets of PPI network properties, or to mimic the way in which

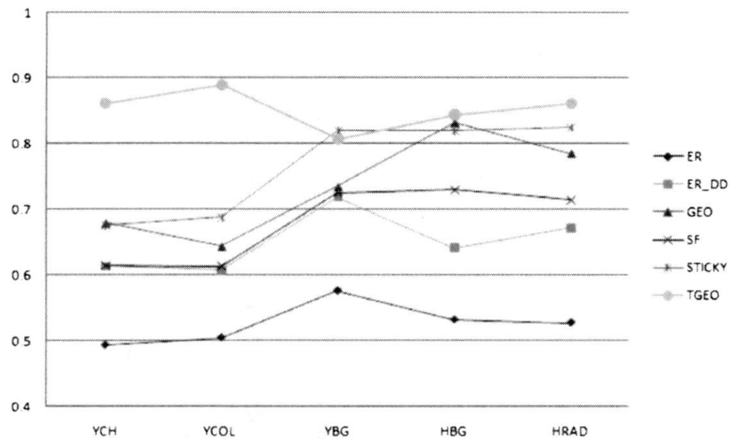

Figure 4. GDD-agreement between the data and model networks. Horizontal axis corresponds to the real PPI networks described in Table 1 and different labels correspond to different model networks described in Table 2.

these networks might have evolved. However, it is difficult to say to what extent network properties really describe network structure. Therefore, network models should use all of the available real network information, i.e., the entire network connectivity information, not only some of the network properties, to learn the structure of PPI networks. Using our new network embedding algorithm[9] and our previous observations that geometric random graphs provide a good model for PPI networks[3,14,9], we have reduced the problem of constructing a well-fitting model for PPI networks to a standard machine learning problem of density estimation. Our new *trained geometric model* uses parts of PPI networks of high quality to learn the network structure and uses this learned knowledge to generate model networks with arbitrary numbers of nodes and edges. Our experiments show that even if we use a learning PPI network from a simple eukaryotic organism (i.e., yeast) and then use our model trained in this way to model a PPI network of another higher eukaryotic organism (i.e., human), we get a substantial improvement in the fit of our new network model to PPI networks over all other currently commonly used random graph models. This suggest that PPI networks of even such distant species as yeast and human exhibit similar structural properties.

Currently, noise is one of the biggest challenges in analyzing and modeling PPI networks. We intentionally used only high confidence data from the study by Collins et al.[16] to learn real structural properties of PPI networks and thus to avoid learning the properties of the noise in PPI networks. Once the data of such quality becomes available for human and other species, it should be used to train species-specific models.

Our new network modeling approach can be applied not only to model PPI networks, but also to model other real-world networks. Since the distribution of points in a metric space is crucial for properties of geometric graphs, by learning this distribution from real-world networks we can generate model networks with different structural properties and in that way model networks arising in different applications.

References

1. Erdös,P. and Rényi,A., *Publ. Math.* **6**, 290-297, (1956).
2. Barabasi,A.L. and Albert,R., *Science* **286**, 509-512, (1999).
3. N. Pržulj, D. G. Corneil, and I. Jurisica, *Bioinformatics* **20**, 3508-3515, (2004).
4. M. Penrose. Geometric Random Graphs, *Oxford University Press*, (2003).
5. N. Pržulj and Des Higham, *Journal of the Royal Society Interface*, **3**, 10 (2006).
6. R. Albert and A. Barabási, *Reviews of Modern Physics*, **74**, 47-97 (2002).
7. A. Vaźquez, A. Flamminia, A Maritana, A. Vespignani, *Complexus*, **1**, 38-44 (2003).
8. Barabási AL, Oltvai ZN, *Nat Rev Genet.*, **2**, 5, (2004).
9. D. J. Higham, M. Rašajski, and N. Pržulj, *Bioinformatics*, **24**, 8 (2008).
10. C. Bishop, Pattern Recognition and Machine Learning, *Springler*, (2006).
11. T. Ito, K. Tashiro, S. Muta, R. Ozawa, T. Chiba, M. Nishizawa, K. Yamamoto, S. Kuhara, and Y. Sakaki, *Proc Natl Acad Sci U S A*, **97**, 3 (2000).
12. Gavin, A. C., Bosche, M., Krause, R., Grandi, P., Marzioch, M., Bauer, A., Schultz, J., Rick, J. M., Michon, A. M., Cruciat, C. M., Remor, M., Hofert, C., Schelder, M., Brajenovic, M., Ruffner, H., Merino, A., Klein, K., Hudak, M., Dickson, D., Rudi, T., Gnau, V., Bauch, A., Bastuck, S., Huhse, B., Leutwein, C., Heurtier, M. A., Copley, R. R., Edelmann, A., Querfurth, E., Rybin, V., Drewes, G., Raida, M., Bouwmeester, T., Bork, P., Seraphin, B., Kuster, B., Neubauer, G., and Superti-Furga, G. *Nature* **415** 6868 (2002).
13. T. Milenković, J. Lai, and N. Pržulj, *BMC Bioinformatics*, **9**, 70, (2008).
14. N. Pržulj, *Proceedings of the 2006 European Conference on Computational Biology, ECCB '06*, **23**, e177-e183, (2007).
15. N. Pržulj, *Proceedings of the 2007 IEEE Information Theory Workshop (ITW)* (2007).
16. Collins SR, Kemmeren P, Zhao XC, Greenblatt JF, Spencer F, Holstege FC, Weissman JS, Krogan NJ, *Mol Cell Proteomics.*, **3**, 6, (2007).
17. Stark, C. and Breitkreutz, B.J. and Reguly, T. and Boucher, L. and Breitkreutz, A. and Tyers, M., *Nucleic Acids Research*, **34**, D535-D539, (2006).
18. Radivojac, P., Peng, K., Clark, W.T., Peters, B.J., Mohan, A., Boyle, S.M. and Mooney, S.D. *Proteins*, **3**, 72, (2008).
19. Cox T. F. and Cox M. A. A. Multidimensional Scaling. *Chapman and Hall*, London. (1994).

INFERENCE OF FUNCTIONAL NETWORKS OF CONDITION-SPECIFIC RESPONSE - A CASE STUDY OF QUIESCENCE IN YEAST*

SUSHMITA ROY*, TERRAN LANE*, MARGARET WERNER-WASHBURNE‡ AND DIEGO MARTINEZ‡

**Department of Computer Science*
‡Department of Biology
1 University of New Mexico,
Albuquerque, NM 87131, USA

Analysis of condition-specific behavior under stressful environmental conditions can provide insight into mechanisms causing different healthy and diseased cellular states. Functional networks (edges representing statistical dependencies) inferred from condition-specific expression data can provide fine-grained, network level information about conserved and specific behavior across different conditions. In this paper, we examine novel microarray compendia measuring gene expression from two unique stationary phase yeast cell populations, quiescent and non-quiescent. We make the following contributions: (a) develop a new algorithm to infer functional networks modeled as undirected probabilistic graphical models, Markov random fields, (b) infer functional networks for quiescent, non-quiescent cells and exponential cells, and (c) compare the inferred networks to identify processes common and different across these cells. We found that both non-quiescent and exponential cells have more gene ontology enrichment than quiescent cells. The exponential cells share more processes with non-quiescent than with quiescent, highlighting the novel and relatively under-studied characteristics of quiescent cells. Analysis of inferred subgraphs identified processes enriched in both quiescent and non-quiescent cells as well processes specific to each cell type. Finally, SNF1, which is crucial for quiescence, occurs exclusively among quiescent network hubs, while non-quiescent network hubs are enriched in human disease causing homologs.

1. Introduction

Cellular adaptations essential for survival under changing enviromental conditions are driven by a complex, but coordinated set of interactions among genes, proteins and metabolites. Existing analyses of condition-specific

*This work is supported by NIMH (1R01MH076282-01, 1R01MH076282-03) NSF (IIS-070568, MCB0734918) and HHMI-NIH/NIBIB (56005678).

behavior typically identify genes differentially expressed across conditions. Fine-grained, interaction analysis among differentially expressed genes can provide deeper insight into human diseases such as cancers[1].

We define functional networks as networks with edges representing general, statistical dependencies among genes. We model functional networks using undirected, probabilistic graphical models, Markov random fields (MRFs). We present a new algorithm, Markov blanket search (MBS), for learning the MRF structure. MBS is based on Abbeel *et al.*'s theoretical work of *Markov blanket canonical parameterization* (MBCP)[2], which requires an exhaustive enumeration ($O(n^l)$) over variable subsets up to size l, where n is the number of variables. We establish an equivalence between the MB canonical parameters and *per-variable* canonical parameters[3], which requires enumeration over only singleton sets ($O(n)$), thus providing a tractable approach for learning genome-scale networks.

We apply our algorithm to two novel yeast (*S. cerevisiae*) microarray datasets measuring gene expression of quiescent and non-quiescent cells, isolated from glucose-starved stationary-phase cultures[4]. Quiescent cells play important roles in health and disease conditions of most living systems, but have been difficult to study due to their low metabolic activity[5]. The recent generation of microarray datasets for these cells[4], gives us the first opportunity to infer a functional network that provides a fine-grained characterization of yeast quiescence.

Algorithms for functional network inference can be broadly classified into: (a) pairwise models (capturing dependencies between only pairs of nodes), and (b) higher-order models (capturing dependencies among two or more nodes). Bayesian networks are higher-order, directed models[6,7,8], but the acyclic constraint of the graph structure cannot easily capture cyclic dependencies. Undirected graphical models can represent cyclic dependencies, but because network inference is much harder[2], higher-order dependencies are often approximated by lower-order (often pairwise) functions[9]. As biological networks are likely to have higher-order dependencies[10], higher-order models are more appropriate for modeling functional networks. Unlike Bayesian networks, our model captures cycles and, unlike pairwise models, we explicitly identify higher-order dependencies.

Bio-techniques for identifying condition-specific networks have been applied to transcriptional regulatory networks[11]. Computational identification of functional networks can provide a less expensive, complementary view of condition-specific networks. Some existing computational approaches integrate mRNA expression with known protein networks[1]. Other approaches identify cliques of co-regulated genes using mRNA expression. *De novo* functional network inference, can identify novel relationships absent from existing protein networks. Furthermore, identification of general statistical dependencies, including mRNA co-expression, can capture condition-specific responses involving the metabolic and proteomic levels.

We applied MBS to three microarray datasets measuring expression of quiescent, non-quiescent, and exponential yeast cells to genetic and chemical perturbations. We analyze the inferred networks to identify subgraphs that are common among, or that discriminate quiescent, non-quiescent cells, and exponential cells. In both quiescent and non-quiescent cells, fermentation and translation are down regulated, but not in exponential cells. The exponential and non-quiescent cells are more enriched in Gene Ontology[12] (GO) slim processes than quiescent cells, suggesting that the quiescent cells are under-studied novel cell types.

Hub analysis of the inferred networks identified SNF1, which is important for quiescence, to be present exclusively in quiescent cells. Hubs from non-quiescent cells are more enriched in yeast homologs of human disease genes. Overall, our results both agree well with existing knowledge and include novel findings that differentiate quiescent and non-quiescent cells.

2. Methods

2.1. *Markov random fields for biological networks*

A Markov random field (MRF) is an undirected, probabilistic graphical model that represents statistical dependencies among random variables (RVs), $\mathbf{X} = \{X_1, \cdots, X_n\}$. A MRF consists of a graph $\mathcal{G}$ and a set of potential functions $\psi = \{\psi_1, \cdots, \psi_m\}$, which together describe the structural and functional relationships among the RVs. Nodes correspond to continuous RVs encoding gene expression level, $X_i \in \mathcal{R}$.

Although MRFs capture both higher-order and cyclic dependencies, MRF structure learning is harder than in directed models[2]. Approaches to overcome this problem include dependency networks[13] and Markov blanket canonical parameterization (MBCP)[2]. MBCP requires the estimation of optimal Markov blankets (the set of immediate neighbors) for RV subsets, $\mathbf{Y} \subseteq \mathbf{X}$, $|\mathbf{Y}| \leq l$, where l is a pre-specified, maximum subset size.

We extend MBCP by establishing an equivalence between Markov blanket canonical parameters and *per-variable* canonical parameters[3]. As a consequence, we need to estimate Markov blankets of individual RVs instead of all subsets. Abbeel *et al*'s MBCP exhaustively enumerates over subsets up to size l taking $O(n^l)$ time. Instead, our approach enumerates over singleton sets resulting in a reduced complexity of $O(n)$, producing a more tractable structure learning approach.

Our algorithm also enforces structural consistency, which is not guaranteed in dependency networks and MBCP. In structurally consistent networks, for every pair $\{X_i, X_j\}$, X_i is in X_j's Markov blanket, if and only if, X_j is in X_i's Markov blanket. A structurally inconsistent network may not have an associated joint probability distribution making probabilistic inference difficult.

2.2. *Markov blanket search algorithm*

Our per-variable canonical parameterization enables structure search by identifying the best Markov blanket (MB) per RV[3]. We identify the best MB by minimizing conditional entropy[14,2], $H(X_i|\mathbf{M}_i)$ for each X_i given its MB, $\mathbf{M}_i$. We enforce structural consistency by computing the score gain on adding an edge, $\{X_i, X_j\}$. This combines the decrease in $H(X_i|\mathbf{M}_i)$ due to addition of X_j, with the conditional entropy change if X_j's MB was constrained to include X_i. Overall, MBS minimizes $\sum_{i=1}^{n} H(X_i|M_i)$ plus a regularization term, subject to the structure consistency constraint.

Let $\mathbf{M}_i^k$ denote the current best MB for X_i of size k, X_j denote a candidate for addition to $\mathbf{M}_i^k$, and $\mathbf{M}_j^k$ be the current best MB for X_j . Then the score gain is:

$$G_i = H(X_i|\mathbf{M}_i^k) - H(X_i|\mathbf{M}_i^k \cup \{X_j\}) + H(X_j|\mathbf{M}_j^k) - H(X_j|\mathbf{M}_j^k \cup \{X_i\}). \quad (1)$$

The MBS algorithm performs a greedy search to identify the best MB for each variable. Each search iteration uses a combination of *add* and *swap* operations. In the add stage of the $k+1^{th}$ iteration, we make one variable extensions to the current Markov blanket $\mathbf{M}_i^k$ of each X_i restricting it to at most $k+1$ RVs per MB. In the swap stage, we revisit all variables Z in the Markov blanket $\mathbf{M}_i^k$ of each X_i, and consider other RVs $Y \notin (\{X_i\} \cup \mathbf{M}_i^k)$, which if swapped in instead of Z, gives a score improvement. We assume all variables to have a Gaussian distribution. However, our general approach is applicable to other random variables, requiring only the estimation of conditional entropy.

2.3. *Data pre-processing*

We applied our algorithm to two yeast, *S. cerevisiae*, datasets from quiescent and non-quiescent cells[4], and one dataset from exponential cells[15]. We included only genes with $< 20\%$ missing data in all three datasets. As the quiescent and non-quiescent datasets had biological replicates, we filtered the genes further to discard not-reproducible genes. Our final datasets comprised $n = 2,818$ genes, with 170, 186, and, 300 measurements per gene in the quiescent, non-quiescent and exponential populations respectively.

2.4. *Analysis of inferred networks*

Generation of coarse modules: To obtain a high-level view of the inferred networks, we generated local subgraphs and clustered them into coarse modules. We excluded clusters of size < 5 and connected the remaining clusters into high-level graphs (Fig 1, Section 3.1).

We generated subgraphs by considering a node, and its neighbors reachable by r links. We refer to these subgraphs as 1-n subgraphs, denoting a $n$eighborhood reachable by traversing one link ($r = 1$). We computed a topological similarity for each pair of subgraphs, $\{S_i, S_j\}$, $t_{ij} = \frac{l_{ij}}{n_i+n_j}$, where l_{ij} is the sum of number of vertices common between S_i and S_j, and the number of edges across S_i and S_j. n_i and n_j are the vertex counts in subgraphs, S_i and S_j, respectively[16].

To obtain *coarse modular organization*, we first clustered the subgraphs using hierarchical clustering with average linkage[17]. We selected clusters to optimize between including majority of the genes, and to have clusters of size ≥ 5. This resulted in 230, 214 and 179 clusters, with $n = 2630, 2551$ and 2651 genes in quiescent, non-quiescent and exponential cells respectively. We then used topological similarity as edge weights for each pair of clusters.

To assess if the clusters were biologically meaningful, we computed an *annotation similarity*, f_{ij}, for each cluster pair, $\{C_i, C_j\}$. We obtained GO slim process enrichment vector, e_i per cluster. Each dimension $e_i(r)$ was the logarithm of p-value enrichment for each process term. The annotation similarity between C_i and C_j was the Pearson's correlation coefficient between e_i and e_j.

We developed a measure, *Annotation-Topological Similarity* (ATS), to assess if clusters that were topologically close were also similarly annotated. ATS is the Pearson's correlation coefficient between two vectors, v_A and v_T, each of length $\binom{|\mathbf{C}_x|}{2}$, where $\mathbf{C}_x$ is the set of clusters generated for population x. Each dimension $v_A(r)$ ($v_T(r)$) was the annotation (topological) similarity for r^{th} cluster pair, where $1 \leq r \leq \binom{|\mathbf{C}_x|}{2}$.

We define *relative enrichment* among two populations x and y that tests if x is equally, less or more annotated than y. Let p_y be the proportion of

y's clusters that are enriched ($< 10^{-2}$) in any slim term. Assuming r out of s clusters of x are enriched, we compute the probability of observing $\leq r$ out of s using the binomial with parameter p_y. The smaller the probability the more depleted is x as compared to y. Similarly, the probability of $\geq r$ out of s enriched clusters estimates how annotated x is as compared to y. Smaller the probability the more annotated x is w.r.t to y. We repeat this for testing y's relative enrichment w.r.t x. We do this analysis for all population pairs such as quiescent versus exponential, quiescent versus non-quiescent, etc.

Identification of up or down-regulated subgraphs: We analyzed the 1-n subgraphs for *significant up-regulation* or *down-regulation* at expression level using a similar approach to Chuang *et al.*[1] Each subgraph was assigned the average of the mean expression of the subgraph genes. For each dataset and subgraph size, we estimated a null distribution of subgraph expression by randomly sampling $s = 100,000$ subsets of all genes. A p-value < 0.05 was considered as significantly up or down-regulated.

Identification of conserved and specific subgraphs: To identify *conserved subgraphs* among two cell populations, A and B, we computed a match score for each 1-n subgraph, $S_i^A \in \mathbf{S}_A$ generated from A's network, using B's network structure. This score is the harmonic mean of recall, R_i^A, and precision, P_i^A, per subgraph S_i^A. For each $S_i^A \in \mathbf{S}_A$, $R_i^A = \frac{|E_i^A \cap E_i^B|}{|E_i^A|}$, where E_i^A is the edgeset of S_i^A in A's network, and E_i^B is the edge set among S_i^A's vertices in B's network. Similarly, $P_i^A = \frac{|E_i^A \cap E_i^B|}{|E_i^B|}$. The match of S_i^A in B's network is $F_i^A = \frac{2*P_i^A*R_i^A}{P_i^A+R_i^A}$. We assumed an $F_i^A > 0$ to indicate S_i^A is a conserved subgraph in B's network. The set of conserved subgraphs between A and B is $\mathbf{S}_A^M \cup \mathbf{S}_B^M$, where $\mathbf{S}_A^M \subseteq \mathbf{S}_A$ and $\mathbf{S}_A^M \subseteq \mathbf{S}_B$. Each $S_i^A \in \mathbf{S}_A^M$ has $F_i^A > 0$, when compared with B's network, and each $S_j^B \in \mathbf{S}_B^M$ has $F_j^B > 0$, when compared with A's network. A subgraph was considered *specific* to a particular population if it had a match score of zero for the remaining populations.

Gene ontology (GO) enrichment and false discovery rate: For each 1-n subgraph (or cluster), we used the hyper-geometric distribution to compute GO term enrichment. We sampled $s = 1000$ random subsets of size k from the $n = 2818$ genes, and computed their p-values for each term. The false discovery rate (FDR)[18] is associated with the number of terms enriched in a subgraph at a particular p-value. For example, if a subgraph of size k is enriched in u terms at $p < 10^{-4}$, FDR is $\frac{u'}{u}$, where u' is the average number of terms enriched in a random subgraph of size k at $p < 10^{-4}$. We used an FDR of ≤ 0.05 to identify significant enrichments.

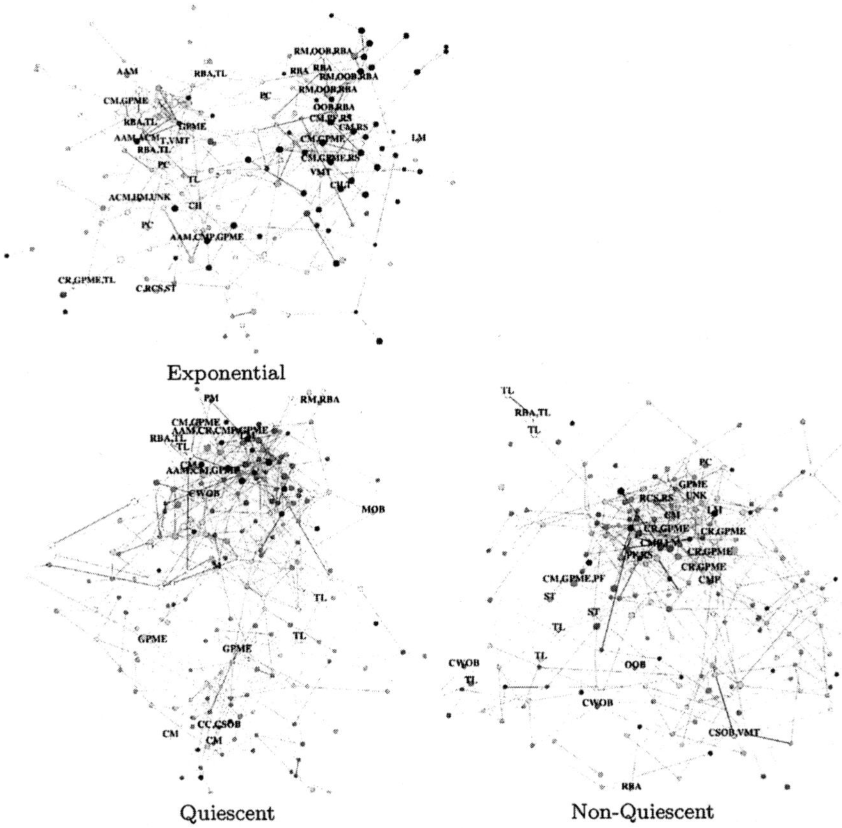

Figure 1. Coarse modular organization of networks inferred from three populations. Each node represents a cluster of genes. The node size is proportional to the gene count per cluster. The node shade indicates if cluster genes are more expressed (dark) or repressed (light). The edge thickness is proportional to similarity between two clusters. The nodes are labeled with processes enriched in the member genes. AAM: amino acid and deriv. ACM: aromatic comp CM: carb. meta CC: cell cycle CH: cellular homeostasis CR: cellular resp. CWOB: cell wall org. & biogen. CMP: cofactor metab. C: conjugation CSOB: cytoskeleton org. & biogen. GPME: gen. of pre. metab. &energy HM: heterocycle metab. LM: lipid metab. M: meiosis MOB: membr. org. & and biogen. OOB: organelle org & biogen. UNK &other PC: protein catab. PF: protein folding PM: protein modif RCS: resp. to chemical stimulus RS: response to stress RBA: ribosome biogen & assem RM: RNA metab ST: signal transduction TR: transcription TL: translation T: transport TP: transposition VMT: vesicle-mediated transport VM: vitamin metabolic process.

3. Results

3.1. *Modular organization in quiescent, non-quiescent and exponential cells.*

To obtain a high-level view of inferred networks, we clustered the subgraphs from inferred networks per population, followed by GO slim process enrichment of the clusters (Fig 1). We had high Annotation-Topological similarity (ATS) measure (see methods), for exponential (ATS=0.51) and non-quiescent cells (ATS=0.49), suggesting that similarly functioning genes are topologically close in the inferred networks. Quiescent cells had lower agreement (ATS=0.26), suggesting that gene expression is more informative for non-quiescent and exponential cells, than quiescent cells. Quiescent cells may employ additional mechanisms, including post-translational modification, to respond to stresses[19,4]. The presence of post-translational modification is further supported by up-regulation of a quiescent cluster enriched in protein modification (PM).

Relative enrichment analysis shows that both quiescent and non-quiescent cells have significantly fewer annotated clusters than exponential (Table 1). Quiescent is also less annotated than non-quiescent. This highlights the under-studied characteristics of quiescent cells, motivating further investigation of these cells.

Table 1. Relative enrichment of clusters. ↑ and ↓ denote enrichment and depletion p-value, respectively, of annotated clusters.

Population	Enriched/Total clusters	wrt EXP		wrt Q		wrt NQ	
		↑	↓	↑	↓	↑	↓
Exponential	32/179	–	–	3e-4	1.0	0.03	0.94
Quiescent	19/230	1	7e-5	–	–	1.0	0.02
Non-quiescent	27/214	0.04	0.96	0.01	0.99	–	–

3.2. *Fine grained analysis of the cell populations*

3.2.1. *Quiescent and non-quiescent cells show common global starvation response behavior.*

To identify similarly enriched processes, we obtained conserved subgraphs among two populations. For each subgraph, we compared GO process enrichment, and whether it agreed in expression – both up or both down-regulated – in the two populations being compared.

There were a large number of subgraphs common between quiescent and non-quiescent (Table 2). These subgraphs were enriched in glycolysis,

Table 2. Number of conserved subgraphs between pairs of populations. Enr implies enriched. Up and Down indicate significant up and down regulation, respectively.

Population pair	Subgraphs	Enr	Up	Down	Enr & Up	Enr & Down
Q-NQ	833	97	26	95	4	27
NQ-EXP	288	99	7	25	5	16
Q-EXP	311	98	0	27	11	24

fermentation, translation and fatty acid oxidation processes. However, only half agreed in expression. Several of these subgraphs had both up and down-regulated genes, resulting in only average expression of the entire subgraph. These heterogeneous dependencies indicate more complex relationships not likely to be captured by co-expression.

We also identified several subgraphs conserved between quiescent and non-quiescent populations, that were up-regulated, but did not have term enrichment. Genes from these subgraphs were associated with unknown biological process, further elucidating the importance of studying these cells.

Non-quiescent and exponential cells had several conserved subgraphs enriched in telomere maintenance, DNA packaging, chromatin assembly and mitotic recombination. These findings are consistent with previous knowledge of non-quiescent cells to have unstable genomes, and, therefore requiring these processes[5]. Comparison of quiescent and exponential cells did not identify any processes enriched in the up-regulated subgraphs. The processes enriched in down-regulated subgraphs included glycolysis, gluconeogenesis and ribosomal biogenesis.

Overall, the subgraph analysis suggests that quiescent and non-quiescent cells are more similar to each other, than each is to exponential cells. There are several subgraphs common to quiescent and non-quiescent cells, but not all agree in expression. The processes that are common between these cells suggest global environmental response as the cells transition from fermentable to non-fermentable carbon sources for energy.

Table 3. Subgraphs specific to individual populations. Same legend as Table 2.

Population	Subgraphs	Enr	Up	Down	Enr & Up	Enr & Down
Q	2295	70	174	160	7	17
NQ	2317	74	171	186	10	11
EXP	2570	232	327	206	54	44

3.2.2. *Differences in quiescent and non-quiescent cells suggest population-specific response*

We examined GO enrichment of subgraphs that occurred only in one population. Both quiescent and non-quiescent cells had fewer subgraphs with

Table 4. Processes exclusively up regulated in different populations

Population	Process	genes
Q	RAS signal transduction	IRA1, SPG3, YGR026W, BCY1, PFK2
	Sporulation	GPA2, GSC2, OSW2, CAF120, YOR277C
	de novo pyrimidine base biosynthetic process	ARF1,DIG1, URA1,URA3,YHR003C
NQ	hyperosmostic response	TRS120, MSB2, YHR100C, PBS2, RSF2
	regulation of DNA metabolic process	PIM1, DIG2, BCK2,SSL2, DPB11, PLB3, HST1, YNG1
EXP	ATP biosynthesis	COX20, QCR10,QCR8, ATP2, ATP7
	cell wall organization & biogenesis	AFR1, SKN1,GFA1, KTR2, DFG5
	amino acid biosynthesis	IDP1, ARO3, HOM2, YGL117W, YSC83, ARG4,SIP4, CPA2,ARG1, SER1, SSU1
	response to toxin	AAD10,AAD16, AAD4,BAP2,MID2, TAT1,TYR1

enriched processes than exponential (Table 3). The quiescent cells were exclusively enriched in sporulation and negative regulation of the RAS signal transduction pathway (Table 4). Down regulation of this pro-growth pathway indicates mechanisms to conserve energy expended in growth conditions. Furthermore, subgraph genes that are not annotated with signal transduction (SPG3, PFK2), are all important for stationary phase.

The non-quiescent cells exhibited processes involved in osmotic stress response and regulation of DNA recombination. This is consistent with these cells trying to cope with environmental changes and that they have unstable genomes. However, unlike quiescent cells, most of the processes up-regulated in non-quiescent cells, also occurred in exponential cells.

The exponential cells were enriched in response to chemical stresses, biosynthesis of amino acids and ATP biosynthesis. ATP biosynthesis was down-regulated in both quiescent and non-quiescent cells. The up-regulation of these energy producing pathways suggests that exponential cells expend a large amount of energy to make relevant mRNA in response to different stresses. In contrast, as quiescent cells are formed in response to a starvation condition, they are likely to sequester mRNA for rapid release in response to different stresses[19].

3.2.3. *Non-quiescent hubs are enriched in disease causing genes*

We analyzed the inferred networks to identify network hubs, nodes with degree ≥ 7. A significant overlap between quiescent and non-quiescent hubs (n=29) implied similarities among these cells due to global starvation response, consistent with Section 3.2.1.

Table 5. Hub nodes and their most enriched processes

Population	Hubs	Exclusive	Processes
Q	215	167	cell wall organization & biogenesis signal transduction, carbohydrate metabolism, organelle organization and biogenesis, generation of precursor metabolites
NQ	166	116	vesicle-mediated transport, response to stress, membrane organization & biogenesis
EXP	318	273	aminoacid & derivative process, cellular respiration ribosome biogenesis & assembly

The non-quiescent cells have been hypothesized as models for studying diseases in humans due to the instability of their genomes[4]. We asked if hubs from different cell populations were enriched in human disease causing gene homologs[20] (Table 6). Of the $n = 2818$ genes used to infer networks, there were $n = 225$ yeast genes, homologous to different human disease genes[a]. We found that hubs in non-quiescent cells are more likely to be enriched in disease homologs than either quiescent or exponential cells. This provides preliminary empirical evidence for the hypothesis that these cells can provide insight into human disease causing conditions.

We found network hubs from quiescent cells to be enriched (p-value <0.05) in signal transduction and cell wall biogenesis (Table 5). Among the quiescent hubs was SNF1, known to be crucial for the formation of quiescent cells. The non-quiescent hubs were enriched in stress response and vesicle mediated transport. Finally the exponential hubs were enriched in amino acid processes and cellular respiration. The enrichment of different processes further illustrates the underlying bio-chemical characteristics that discriminate these cells, and how they respond to different stresses.

Table 6. Enrichment of human disease gene homologs in hubs.

Population	Total Hubs	Homologous Disease Hubs	Pval
NQ	166	26	5e-4
Q	215	22	0.130
EXP	318	26	0.395

4. Conclusion

We have developed an undirected, probabilistic graph learning algorithm that can capture different types of dependencies that may exist in expression-based, functional networks. We use our algorithm to perform

[a]We downloaded human-yeast homologs from `http://www.biomart.org/index.html`

a network analysis of three yeast cell populations in starvation and exponential growth conditions. A high-level analysis of the modular structure of these networks suggests that quiescent cells are significantly under-annotated, highlighting the need to study these cells. Our analysis suggests that the non-quiescent cells share more characteristics with exponential cells as compared to quiescent. Analysis of individual subgraphs indicates that quiescent and non-quiescent cells exhibit similarities in their mechanisms to adapt to glucose starvation. However, there are processes specific to quiescent cells such as sporulation, which suggest alternative response mechanisms that might be active in these cells. Finally, we find that non-quiescent hubs are enriched in homologs of human disease genes. In summary, our network-based analysis has identified both previously known and novel biological processes that are important in these cells, giving a finer understanding of the mechanisms conserved and specific to these cells.

References

1. H.-Y. Chuang, E. Lee, Y.-T. Liu, D. Lee, T. Ideker, *Mol Syst Biol* **3** (2007).
2. P. Abbeel, D. Koller, A. Y. Ng, *JMLR* **7**, 1743 (2006).
3. S. Roy, T. Lane, M. Werner-Washburne, *Tech. Rep. TR-CS-2008-14*, University of New Mexico (2008).
4. A. D. Aragon, *et al.*, *Molecular Biology of the Cell* (2008).
5. C. Allen, *et al.*, *J Cell Biol* **174**, 89 (2006).
6. N. Friedman, *Science* **303**, 799 (2004).
7. J. Yu, A. A. Smith, P. P. Wang, A. J. Hartemink, *Bioinformatics* **20**, 3594+ (2004).
8. E. Segal, *et al.*, *Nat Genet* **34**, 166 (2003).
9. A. Margolin, *et al.*, *BMC Bioinformatics* **(Suppl 1): S7** (2005).
10. Y. Qi, H. Ge, *PLoS Computational Biology* **2**, e174+ (2006).
11. C. T. Harbison, *et al.*, *Nature* (2004).
12. M. Ashburner, *et al.*, *Nat Genet* **25**, 25 (2000).
13. D. Heckerman, D. M. Chickering, C. Meek, R. Rounthwaite, C. M. Kadie, *JMLR* **1**, 49 (2000).
14. T. M. Cover, J. A. Thomas, *Elements of information theory* (Wiley-Interscience, New York, NY, USA, 1991).
15. T. R. Hughes, *et al.*, *Cell* **102**, 109 (2000).
16. I. Lee, S. V. Date, A. T. Adai, E. M. Marcotte, *Science* **306**, 1555 (2004).
17. M. B. Eisen, P. T. Spellman, P. O. Brown, D. Botstein, *Proc Natl Acad Sci U S A* **95**, 14863 (1998).
18. E. I. Boyle, *et al.*, *Bioinformatics* **20**, 3710+ (2005).
19. A. D. Aragon, G. A. Quiñones, E. V. Thomas, S. Roy, M. Werner-Washburne, *Genome Biol* **7** (2006).
20. A. Hamosh, A. F. Scott, J. S. Amberger, C. A. Bocchini, V. A. McKusick, *Nucleic Acids Res* **33 Database Issue** (2005).

LEARNING CYCLIC SIGNALING PATHWAY STRUCTURES WHILE MINIMIZING DATA REQUIREMENTS

K. SACHS† AND S. ITANI†

karensachs@stanford.edu, ssolomon@mit.edu

J. FITZGERALD, L. WILLE AND B. SCHOEBERL

{jfitzgerald,lwille,bschoeberl} @merrimackpharma.com

M. A. DAHLEH, G. P. NOLAN

dahleh@mit.edu, gnolan@stanford.edu

† *These authors contributed equally to this work.*

Bayesian network structure learning is a useful tool for elucidation of regulatory structures of biomolecular pathways. The approach however is limited by its acyclicity constraint, a problematic one in the cycle-containing biological domain. Here, we introduce a novel method for modeling cyclic pathways in biology, by employing our newly introduced Generalized Bayesian Networks (GBNs). Our novel algorithm enables cyclic structure learning while employing biologically relevant data, as it extends our cycle-learning algorithm to permit learning with singly perturbed samples. We present theoretical arguments as well as structure learning results from realistic, simulated data of a biological system. We also present results from a real world dataset, involving signaling pathways in T-cells.

1. Introduction

Since the seminal work by Pe'er and Friedman, [8], Bayesian networks (BNs) have been used extensively in biology, to model regulatory pathways both in the genetic[13,8] and in the signaling pathway domain[10,14]. Bayesian network models encode probabilistic relationships among random variables in a domain, providing a framework for tasks such as structure learning. In a biological setting, the random variables represented are biologically important entities such as genes, small molecules and activated or phosphorylated proteins. The structure learning task consists of searching the space of possible structures to find the one that bests reflects probabilistic relationships in a biological dataset.

In spite of their usefulness, Bayesian network models are limited in their applicability in this domain because they are constrained to be acyclic, while positive and negative feedback loops abound in biological pathways. In particular, Bayesian network structure learning will *always* yield an inaccurate structure for any cycle containing pathway and, as a result, will fail in its predictive capacity (at minimum) for variables downstream of an incorrectly directed edge. When time course data are available, it is feasible to represent cycles by unrolling them in time, using a Dynamic Bayesian networks (DBNs), or Continuous Time Bayesian networks (CTBNs).[12,4] However, DBNs suffer from various computational challenges and necessitate timecourse data, which in some domains are not feasibly attainable in an applicable form (e.g.[10]). Therefore, it would be useful to find an approach for learning cyclic structures from static 'snapshot' data, collected at a single timepoint from a dynamic system.

We have recently developed a formalism for representing cyclic structures using Generalized Bayesian networks (GBNs), a form of Bayesian networks that we have generalized to encompass cycles.[1] This formalism enables structure learning in a cyclic domain, relying on perturbations which break the cyclic structure. Far from requiring an exhaustive set of perturbations, the algorithm is designed to minimize the number of interventions needed, requiring as few as merely one intervention per cycle for accurate structure learning.

Here, we present the first ever application of GBNs to biological signaling pathways. We apply the algorithm to realistic, biologically relevant data from a differential equation model of IGF signaling. Next, we substantially modify the structure learning algorithm to bring it incrementally closer to applicability in a biological domain, by minimizing the algorithm's data requirements. We then test this new algorithm on a reduced set of the synthetic data and compare its results. Finally, we perform structure learning on real data in which T-cell signaling molecules were measured using multidimensional flow cytometry from[10], and demonstrate that our new algorithm is able to elucidate cyclic structures in signaling pathways.

2. Background and Methods

We present background on BNs and GBNs, as well as the synthetic data used in this study.

2.1. Bayesian networks

Bayesian networks[15], represent probabilistic dependence relationships among multiple interacting components, illustrating the effects of pathway components upon each other in the form of an influence diagram- a graph (G), and a joint probability distribution. In the graph, the nodes represent variables (the biomolecules) and the (lack of) edges represent (conditional in)dependencies[15]. For each variable, a conditional probability distribution (CPD) quantitatively describes the form and magnitude of its dependence on its parent(s). Due to the factorization of the joint probability distribution, the graph must be *acyclic*, meaning that it must not be possible to follow a path from any node back to itself.[15]

2.2. Generalized Bayesian Networks

When building models of pathways, Bayesian network models have a number of strong advantages. They are flexible and interpretable, they can handle interactions of arbitrary complexity (given sufficient data), and they can smoothly incorporate both prior knowledge and interventional data in a principled way. However, they have one serious drawback for modeling biological systems: they are unable, as described above, to handle cycles in a static model. Because cycles abound in biological pathways, a static Bayesian network model usually cannot hope to capture all influence connections.

To address this problem and enable the use of Bayesian network models in a cyclic domain, we recently introduced Generalized Bayesian Networks (GBNs), a generalization of Bayesian networks to the cyclic domain.[1] In [1] we also present an algorithm in which the GBN formalism is used to recover causal structure given interventional (static) data, in acyclic *or* cyclic domains. The algorithm is briefly presented below.

GBN structure learning

Call the set of variables V and the subset of variables with interventions available I. Note that the inhibitors are *activity inhibitors*, typically small molecule inhibitors which perturb the activity- rather than the abundance- of a protein.

Algorithm: Learn Causal GBN structure

0: Start with a Causal GBN and an intervention set I.

1: [Probing experiments] Collect sets of i.i.d. samples under no-intervention and single-intervention data, i.e. when node i is intervened at, for each i in I.

2: Call subroutine 'detect descendants' to recover descendant information for all nodes in I. *Based upon response of variables to perturbations, further explained in subroutine below.*

3: Identify cycles. *Based upon perturbations which affect the abundance of the target variable. Because the inhibitors affect protein* **activity**, *an effect on abundance must be due to a loop from the target back to itself.*

4: Do Bayesian Network learning with the cycles broken (by interventions on nodes we call "**cycle breakers**"- a set of variables that include at least one representative from each cycle) and integrating the descendant information.

5: Determine the correct edges needed to close the cycles, by detecting the children of the cycle breakers.

6: Recover all missing edges in the DAG, and complete the Directed Cyclic Graph structure of the Causal GBN.

The subroutine used in step 2 is an important subroutine that uses the sensitivity of descendants to perturbations on their ancestors. This sensitivity can be described as the assumption that if a distribution of the parents of a variable change, then the distribution of the variable itself will change too. This idea was introduced in the context of GBNs[1]. To test whether a node was affected by a perturbation on another node (and thus deduce the anscestor/descendant relation), the following subroutine is used:

Subroutine: Detect descendants

0: Start with sets of n i.i.d. samples generated by a GBN, under no interventions as well as single-interventions at each i in I. Initialize a binary $|V| \times |I|$ descendant information matrix.

1: For each $j \in V$:

2:     Compute $\hat{\mathbb{P}}^n(X_j)$, the empirical marginal of X_j under no interventions.

3:     For each $i \in I$:

4:         Compute $\hat{\mathbb{P}}_i(X_j)$, the empirical marginal of X_j under the single-intervention i.

5:         Evaluate some distance between $\hat{\mathbb{P}}(X_j)$ and $\hat{\mathbb{P}}_i(X_j)$.

6:         If the distance exceeds a threshold, mark j as a descendant of i.

7:     Next i.

8: Next j.

9: Compute the transitive closure of the descendant information matrix, and return it.

The details of this algorithm are presented in[1] and it is proven that given enough data and interventions, this algorithm is guaranteed to recover the causal structure of the data. Note that step 4 removes all cycles to enable structure learning using standard BN. The algorithm developed in this work avoids this step, to avoid the requirement for conditions involving inhibitor combinations. As such, this modified algorithm is far better suited to the logistic reality of limited experimental data.

2.3. Synthetic Data and Model of IGF Signaling

To produce synthetic data, we use a mass action kinetic model describing the dynamics of the Insulin-like growth factor (IGF) signaling pathway.[5,6] IGF signaling is important in normal cell physiology, as well as pathological states such as cancer. A schematic representation is shown below. Mass action kinetic equations were used to create the model in MATLAB Sim-Biology v2.1.

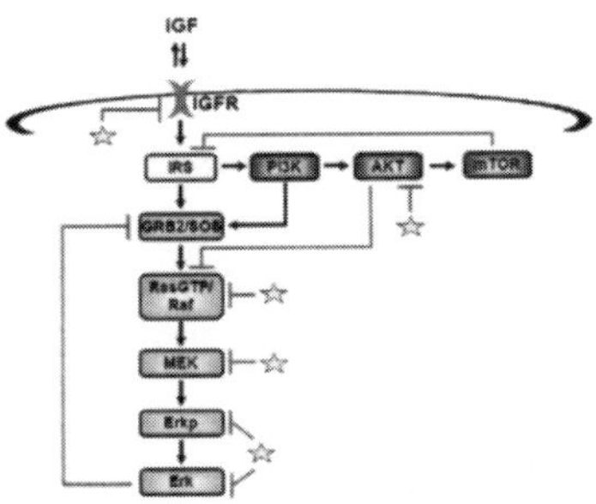

Figure 1. *True structure of the underlying dynamic system in IGF signaling.* Each node represents the active "on state" of the protein. Stars indicate phosphatases acting on the target molecules. Perturbations in the form of small molecule inhibitors are available for Mek, Akt, Pi3k, IGFR and mTor. The simulated data mimics these inhibitors by blocking enzyme activity.

There are three directed cycles in the model: IRS → PI3K → AKT → mTOR → IRS, GRB2/SOS → RasGTP → MEK → ERKp → ERK → GRB2/SOS, and GRB2/SOS → RasGTP → MEK → Erk → GRB2/SOS.

The stimulus employed is IGF, in addition, up to five perturbations are employed, at IGFR, MEK, PI3K, AKT, and mTOR, corresponding to actual existing small molecule inhibitors. All of the perturbations are activity inhibitions, that is, they inhibit the protein's activity, not permitting the targeted protein to phosphorylate other proteins. We generated measurements from four different time points, under 17 total conditions composed of IGF stimulus plus various combinations of inhibitors. For each condition, 1000 unique, randomly selected initial conditions (i.e. molecule concentrations) were employed- the equivalent of collecting 1000 unique cells in a flow cytometry experiment, or performing western blots on 1000 samples. Simulated "measurement noise" was also added.

The model was created by Jonathan Fitzgerald and colleagues at Merrimack Pharmaceuticals. It is a highly accurate imitation of the true biological system[19], and, accordingly, provides us with synthetic but realistic data, similar to the data one might acquire from a high throughput measurement technology (as in [10]). It is a flexible and realistic source of 'true to life' synthetic data, which, because it has a known ground truth model, provides an invaluable tool for assessing success of structure learning efforts.

3. Algorithm Description

In this section, we detail an algorithm for structure learning of cyclic networks with ***single*** perturbation data. This algorithm depends on first detecting the cycles in the network using perturbations, then using the data where each cycle is broken to recover its structure.

Algorithm: Learn Structure

0: Start with a set of variables V and a set of single-intervention variables $I \subset V$.
1: [Experiments] Collect sets of i.i.d. samples under no-intervention and single-intervention data, i.e. when node i is intervened at, for each i in I.
2: Call subroutine 'detect descendants' to recover descendant information for all nodes in I.
3: Identify one node per cycle from I, and the set of such nodes I_C. *Nodes in I_C are detected as self-descendants.*
4: Apply a BN structure learning algorithm to recover a DAG representation of the dependencies.
5: $\forall\, i \in I_C$
6: *Apply a BN algorithm on data with i inhibited.*
7: *Use conditional independency tests to prune out edges and determine path structure between nodes and their descendants.*
8: *Compare the different sub-graphs pertaining to i to recover the cycle structure.*
9: Next i.
10: Integrate all of the cycle structures with the result from step 4, and call the resulting structure G.

4. Algorithm Analysis

In this section we aim to study the algorithm we presented is Section 3, mainly focusing on the reasons it can be expected to have good performance as well as where it is expected to have limitations.

The main assumption needed for the algorithm to recover the true structure is that the BN structure learning algorithm recovers the correct structure of the non-cyclic part of the graph whenever it is applied. This means that the existence of a cycle shouldn't interfere with the conditional independencies between variables outside the cycle itself.

From the study in [18], it can be seen that the cycle does not usually affect the rest of the graph dramatically. This is mainly because the structure of the d-separation in a cycle (what dependencies/independencies conditioning on every set of variables induce) is the same as that of a loop with the same structure of the cycle except for one reversed edge. Thus the BN learning algorithm would tend to recover the whole structure with some reversed or missing edges from the cycle, and usually nodes that are outside the cycle will not be affected.

We therefore expect this algorithm to perform well, even compared to algorithms that use multiple-perturbation data. This is mainly because this algorithm reinforces and corrects the structure recovered by the BN algorithm. It does so using the descendant information from the perturbation analysis and the conditional independency analysis (step 7).

5. Results

In this section, we present structure learning results for BNs and GBNs, as well as the new GBN algorithm presented above. BN learning is implemented as in[10]. Data was discretized to 4 levels using interval discretization. We show the results obtained from the IGF-model based data with combinations of inhibitors, 17 conditions total, followed by conditions employing only single perturbations (6 conditions total), *up to* 1000 datapoints were used per condition. Lastly, we show results using a flow cytometry dataset of T-cell signaling molecules. The models presented are averaged over 20 individual results, edges with confidence > 0.8 are included. In the following graphs a dotted edge is an incorrect edge that was predicted, a black edge is a correct edge. For the GBN based algorithms, the degree of shifting for which a variable is considered to be a child of the perturbed variable is a model parameter. Here, we use a 20% shift as the cutoff, chosen based on an observed bimodality among the candidate children. Another good approach would be to determine the cutoff by randomizing the data and determining the magnitude of a null shift.

5.1. Multiple perturbations

The original GBN structure learning algorithm requires conditions in which multiple perturbations are applied simultaneously. Results from the BNs and the original GBN structure learning algorithm are shown below, both use 17 different conditions, each with different combinations of the 5 inhibitors. The BN results, shown on the left, find most of the model edges (missing the connection between mTor and IRS, as well as Ras/Raf and Mek, and shifting the $Erk \rightarrow Grb2Sos$ connection to Erkp), however, the model contains 10 additional, noncausal edges. The GBN results improve on the BN results substantially, missing only the $Erk \rightarrow Grb2Sos$ edge, which it too shifts to Erkp, and including only 2 extra noncausal edges.

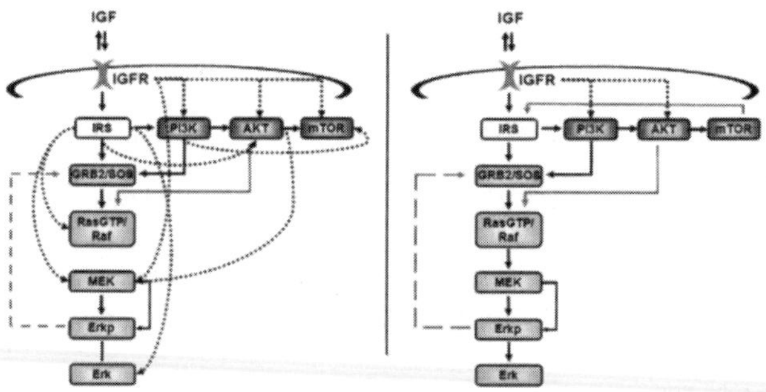

Figure 2. *BN and GBN results from synthetic data, using multiple perturbations per sample.* BN results are shown on the left, GBN results on the right. Broken arrows are noncausal. Missed edges are not indicated; the BN model misses 2 edges and shifts the $Erk \rightarrow Grb2Sos$ edge to Erkp, the GBN model misses no edges but shifts the$Erk \rightarrow Grb2Sos$ edge as well.

What causes the extraneous edges? We were originally surprised to see them, as the data comes from a clean, well defined system which, we had assumed, would be free of confounding elements that usually induce noncausal edges. Upon a closer look, we uncovered two likely culprits. One is the dynamics of the system, something we do not explicitly contend with in single timepoint, 'static' models, but which can have confounding effects induced by the *history* of a molecule being best represented by a different molecule, an effect called *entanglement* (see [7] for a discussion of this topic). A second is technically a causal effect, though it does not result from enzymatic alteration of one molecule by another. This is an effect that may be present when multiple molecules interact with the same intermediate molecule. Consider for instance the edge between IGFR and Pi3k, an extraneous, noncausal edge. Both IGFR and Pi3k interact with IRS, an interaction that takes IRS out of the pool of available, free IRS. Thus, if IGFR binds an IRS molecule, the IRS is no longer available to bind Pi3k (though the total amount of active IRS is unaffected, since IRS can be active while bound to its up or downstream partners). Though IGFR is not causally affected by IRS, it is nevertheless *competing* with Pi3k for IRS binding. IGFR binding affects the effective amount of available IRS that Pi3k "sees", inducing a dependence between IGFR and Pi3k that is independent of IRS abundance. We call this effect *occupancy*, because it is the result of limited available occupancy on the intermediate molecule- if the upstream molecule is bound, the downstream one cannot bind, and vice versa. The impact of *occupancy* in terms of whether an edge is likely to appear probably depends on the specifics of the interactions. for example, the duration of binding of each molecule, and the extent to which the intermediate molecule is present in excess or is in short supply.

5.2. Single perturbations

Above, we formulate a novel algorithm also based on GBNs, but able to perform structure learning from data with just one perturbation per sample. We present BN and GBN results below, both use data from 6 total conditions. The BN result includes many fewer edges than the one found with 17 conditions, possibly because the total number of datapoints is smaller (the BN scoring scheme penalizes complexity, so some edges appear only if sufficient data is available. Though we did not enforce that an equal number of datapoints be used regardless of the number of conditions, this would be a useful idea to try). The BN result contains 7 extra edges, fails to orient some connections, and misses three connections (missed connections not shown). The GBN model contains one extra edge, misses two connections and shifts the $Erk \rightarrow Grb2Sos$ edge to Erkp as before. We note that it is not clear to us why the edge between Erk and Grb2Sos is so commonly shifted to Erkp. Occupancy does not seem to be a factor, because binding times are short and Erk is not in short supply. It may be an effect of the dynamics of the system as discussed in [7]. Regarding data requirements, we use here all available single perturbation conditions, a total of 5 perturbational and 1 observational conditions. However, the learning results are robust down to just two conditions, albeit with a loss of some edges (results not shown). Note however that the descendent information from all 5 perturbations is employed. This points to the possibility that a low throughput approach (such as western blots) could be used to detect descendants, and a high throughput approach (such as flow cytometry) employed for just a small set of perturbation conditions. This approach may save resources and reduce expense while still yielding good results.

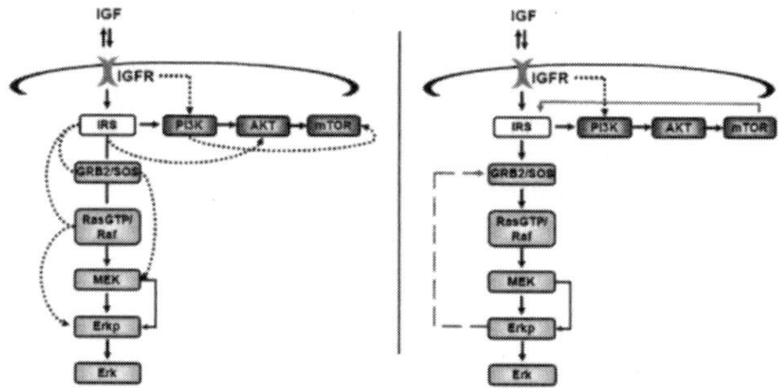

Figure 3. *BN and GBN results from synthetic data, using* one *perturbation per sample. BN results are shown on the left, GBN results on the right. Broken arrows are noncausal. Missed edges are not indicated; the BN model misses 4 edges and contains 7 extra edges, the GBN model misses 2 edges and shifts the* $Erk \rightarrow Grb2Sos$ *edge as before. It also contains one noncausal edge.*

5.3. Single perturbations with real data

To test our algorithm on real data, we employ a real-life dataset created using multidimensional flow cytometry, described in[10]. Measurements of T-cell signaling proteins are reported under observational as well as perturbational conditions with just one perturbation per condition. Perturbation conditions were available for Mek and Akt. The represented pathway is thought to contain (at least) two cycles: $Raf \rightarrow Mek \rightarrow p44/42(Erk) \rightarrow Akt \rightarrow Raf$, and $Raf \rightarrow Mek \rightarrow p44/42 \rightarrow Raf$ [10]. We focus on the proteins involved in these two cycles. BN and GBN results are shown below, with BN results on the left and GBN on the right. The BN result misses one edge and fails to orient two edges. The GBN result is nearly perfect, but it does fail to orient one edge.

6. Conclusions

In this paper, we demonstrate the first-ever application of cyclic-structure learning in signaling pathways using both synthetic and real life data, with

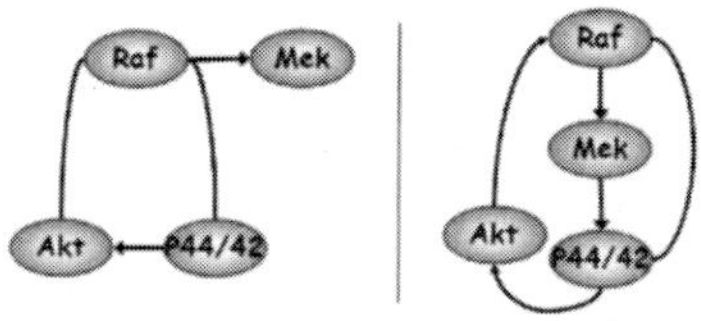

Figure 4. *BN and GBN results from real flow cytometry data, using* one *perturbation per sample. BN results are shown on the left, GBN results on the right. The BN model misses 1 edge and fails to orient 2 edges, the GBN model misses no edges but fails to orient one edge.*

score-based Bayesian networks. We present a novel structure learning algorithm grounded in the GBN formalism, and capable of handling single perturbations conditions, to reduce the algorithm's data demands. We test the original GBN algorithm on synthetic data from an accurate, differential equation model of IGF signaling. We then test the novel extended GBN algorithm formulated here, on both synthetic and real life data. In each case, our algorithm demonstrates clearly superior performance, in terms of elucidation of cyclic structures, correctly orienting model edges and even elimination of extraneous edges.

In our exploration using synthetic data, we discover unexpected edges and propose two main reason for their appearance- confounding effects of the dynamics of the system (discussed in a companion paper, [7]), and *occupancy* effects, based on multiple molecules binding to the same intermediate molecule, thus creating competition-like effects, even though not all of them may be causally affected by the intermediate molecule. This latter concept needs a more rigorous treatment, a topic that we will explore in future work. We also had available multiple timepoints from the synthetic data. For this study, a timepoint was selected arbitrarily, but the effect of timepoint selection will be discussed in [7], using the same synthetic dataset, along with the original GBN algorithm.

In the biological domain, we are often interested in a causal model, partly for the insight and understanding such a model conveys with respect to the modeled system, and partly for the possibility for system predictions which it enables. In disease states for instance, a characterization of the altered biological network can serve to guide therapeutic interventions. A truly causal model which includes correctly oriented edges is crucial- with it, a useful target can be identified and potentially detrimental effects can be avoided. Whereas previous attempts at modeling biological pathways with Bayesian networks have yielded useful results, the prevalence of cycles have confounded those efforts, compromising the causal nature of the learned models. With this work, by overcoming the acyclicity constraint, we have brought the structure learning capability incrementally closer to learning truly causal models.

7. Acknowledgements

This work was supported by a Leukemia and Lymphoma Society post doctoral fellowship to K.S., and NIH grants N01-HV-28183, U19 AI057229, 2 P01 AI36535, U19 AI062623, R01-AI065824, 1 P50 CA114747, 2P01 CA034233-22A1, and HHSN272200700038C, NCI grant U54 RFA-CA-05-024 and LLS grant 7017-6 to G.P.N.

References

1. S. Itani, M. Ohannessian, K. Sachs, G. P. Nolan and M. A. Dahleh, submitted *NIPS* (2008).
2. M. Calder, V. Vyshemirsky, D. Gilbert, R. Orton. (2006).
3. U. Nodelman, C. Shelton, D. Koller. *UAI* (2002).
4. U. Nodelman, C. Shelton, D. Koller. *UAI* (2003).
5. Carlson, C.J., Biochem Biophys Res Comm, 2004. 316(2): p. 533-9.
6. Moelling, K., et al., J Biol Chem, 2002. 277(34): p. 31099-106.
7. Itani, S., Sachs, K., Fitzgerald, J., Wille, L., Schoeberl, B., Nolan, G. and Dahleh, M., in preparation.
8. Friedman, N. and Linial, M. and Nachman, I. and Pe'er, D. (2000). *J Comput Biol*, 3-4, Volume 7.
9. Friedman, N. (2004). *Science*, 5659, Volume 303.
10. K. Sachs, O. Perez, D. Pe'er, D. A. Lauffenburger, and G. P. Nolan (2005). *Science.*
11. K. Sachs, D. Gifford, T. Jakkola, P. Sorger, and D. A. Lauffenburger(2002). *Science STKE.*
12. N. Friedman, K. Murphy, and S. Russell (1999). *Proceedings of the Fifteenth Annual Conference on Uncertainty in Artificial Intelligence*, pp. 139-147.
13. A. J. Hartemink, D. K. Gifford, T. S. Jaakkola, and R. A. Young (2001). *Pac Symp Biocomput.*
14. P. J. Woolf, W. Prudhomme, L. Daheron, G. Daley,and Q. and D. A. Lauffenburger (2004). *Bioinformatics.*
15. J. Pearl (1988). *Probabilistic Reasoning in Intelligent Systems.* Morgan Kauffman.
16. J. Pearl and T. S. Verma (1991). *Second KR*, pp. 441-452.
17. D. Heckerman, C. Meek and G. F. Cooper (1999). *Computation, Causation, and Discovery*, C. Glymour and G. F. Cooper, Eds., MIT Press, pp 141-166.
18. T. S. Richardson (1996). *UAI*, pp. 454–461.
19. Schoeberl B, Fitzgerald JB, Wille L, West K, Pace E, Harms B, Gibbons F, Donis E, Grantcharova V, Kumar A, Kudla A, Nielsen UB, Understanding IGF signaling dynamics through computational modeling, in preparation.

CONTEXT-SPECIFIC GENE REGULATIONS IN CANCER GENE EXPRESSION DATA[*†]

INA SEN[1,‡], MICHAEL P. VERDICCHIO[1,‡], SUNGWON JUNG[2], ROBERT TREVINO[1], MICHAEL BITTNER[2], SEUNGCHAN KIM[1,2]

[1]School of Computing and Informatics, Arizona State University
699 South Mill Avenue Suite 553; Tempe, AZ 85281, USA

[2]Computational Biology Division, Translational Genomics Research Institute
445 North Fifth Street; Phoenix, AZ 85004, USA

Learning or inferring networks of genomic regulation specific to a cellular state, such as a subtype of tumor, can yield insight above and beyond that resulting from network-learning techniques which do not acknowledge the adaptive nature of the cellular system. In this study we show that Cellular Context Mining, which is based on a mathematical model of contextual genomic regulation, produces gene regulatory networks (GRNs) from steady-state expression microarray data which are specific to the varying cellular contexts hidden in the data; we show that these GRNs not only model gene interactions, but that they are also readily annotated with context-specific genomic information. We propose that these context-specific GRNs provide advantages over other techniques, such as clustering and Bayesian networks, when applied to gene expression data of cancer patients.

1. Introduction

Under normal conditions, a cell maintains a specific state by tightly controlling various molecules using a variety of regulatory mechanisms. In the face of environmental changes, a cell adjusts its regulatory mechanisms accordingly. Mutation or other types of damage that alter these regulatory mechanisms may erode this control and cause the cell to transition into another state significantly different from the prior normal state [6]. If the normal state is taken to be "healthy" and the altered state is taken to be "tumor", for example, the regulatory functions must have been altered in significant ways to arrive at the "tumor" state. Since the way the system interprets and acts upon certain inputs is altered, we say that there is a change in *cellular context*.

---

* This work is supported by CA27502-23 (NIH), CA109552-01A1 (NIH), AI067773 (NIH), W81XWH-06-1-090 (DoD/CDMRP), CAA 0243-08 (SFAZ) and LM009706-01 (NIH).

† Code, scripts, supplementary information available online at http://sysbio.fulton.asu.edu/psb2009

‡ I. Sen and M. P. Verdicchio contributed equally to this study.

Although a tumor state of the cell is different from normal, the continuing proliferation and survival of cancer shows that such a state is indeed steady and maintained by complex regulatory behavior. If one can learn from the contextual information which regulating mechanisms differ from context to context, then one can potentially discover the mechanisms that initiate and maintain complex, hard-to-treat diseases, such as cancer.

High throughput data collection methods, including gene expression microarrays, provide vast amount of data to study various aspects of cellular processes. Many methods and techniques exist to discern and model the regulatory behavior of cells, and each certainly has distinct advantages and disadvantages. For instance, traditional clustering approaches like k-means or hierarchical clustering can help group samples or genes, revealing possible novel subtypes of diseases or subclasses of molecular functions. Bayesian networks have been employed as models of genomic regulation; however they inherently assume homogeneity of samples and thus cannot model different "cellular contexts", a serious limitation in non-homogeneous disease like cancer.

In this paper, we will first describe a mathematical model of contextual genomic regulation [3], a method based upon that model to identify *cellular contexts* [9], and then propose a novel method to construct context-specific gene regulatory networks. We apply the context mining method to gene expression data collected from a broad spectrum of cancer patients to reveal the modular and context-specific structure of gene regulatory networks hidden within the data. Finally, we conclude with the future direction of our work.

2. Methods

2.1. *Mathematical Model for Contextual Genomic Regulation*

It is important to select a mathematical model of a cell's regulatory activity that accounts for regulation which very actively adjusts to differing internal and external environmental factors. Rather than models which infer connections between single genes, or between genes and phenotypes, we wish to select a model which can find subsets of samples where it is possible to attribute the states of all the members of a set of controlled genes to a single gene, or to a small set of regulatory genes which have expression properties that could be the source of control.

Recently, Dougherty *et al.* [3] introduced a mathematical model to approximate contextual genomic regulation. Formally, the model assumes there are m sets, G_1, G_2, …, G_m, of *driver genes* and m corresponding sets, S_1, S_2, …,

S_m, of *driven genes*. For each set of driven genes S_j, there is a corresponding set G_j of driver genes regulating their behavior. G_1, G_2, …, G_m are not necessarily disjoint, and neither are S_1, S_2, …, S_m; thus some driver gene may regulate more than one driven set, and some driven gene may be regulated by more than one driver gene set.

Two parameters are essential to the definition of the Contextual Genomic Regulation (CGR) model. To define these parameters, consider a single set of driver genes G and its driven set of regulated genes S. For the set of drivers, still assuming a binary model (without loss of generality), there exists a state vector $\mathbf{Y} = (Y_1, Y_2, \ldots, Y_q)$ where Y_k ($1 \le k \le q$) gives the value of $g_k \in G$. Let regulation by the driver genes be such that for a state $\mathbf{y}$ of the driver gene state vector $\mathbf{Y}$ (for G), when $\mathbf{Y} = \mathbf{y}$, all genes in S take on the value 1 with high probability.

Without loss of generality, let $\mathbf{y}$ be the state in which all members of $\mathbf{Y}$ have the value 1, denoted by $\mathbf{1}$; we will consider two situations for G, namely the situation where $\mathbf{Y} = \mathbf{1}$ and the situation where $\mathbf{Y} \neq \mathbf{1}$. Similar to $\mathbf{Y}$ for G, let $\mathbf{X} = (X_1, X_2, \ldots, X_r)$ be the state vector for S where X_k ($1 \le k \le r$) gives the value of $s_k \in S$. In the first case, where $\mathbf{Y} = \mathbf{1}$, although the driver is *ON*, there may be other regulatory activities within the context affecting the driven genes. For any driven gene $s_k \in S$, the conditional probability of s_k being *ON* is stated

$$P(X_k = 1 \mid \mathbf{Y} = \mathbf{1}) = 1 - \delta_k \tag{1}$$

where δ_k depends on the extent that contextual effects diminish the influence of the driver on the driven gene s_k. Hence, we refer to δ as the *interference* parameter. Now if $\mathbf{Y} \neq \mathbf{1}$, then the probability that some driven state $X_k = 1$ depends on contextual effects alone and not the effects of drivers is given by

$$P(X_k = 1 \mid \mathbf{Y} \neq \mathbf{1}) = \eta_k \tag{2}$$

where η_k depends on the extent that contextual effects outside of the drivers activate the driven genes. Hence, we refer to η_k as the *crosstalk* parameter. Further considerations of the model, including prediction accuracy and error representation, are left to the original paper by Dougherty *et al* [3].

2.2. *Identification of Cellular Contexts*

Based on the above model, a cellular context is taken to be a set of genes, one or more of which function as drivers and the others as driven genes, which exhibit consistent transcriptional behavior across a subset of samples. Kim *et al.* [9] have proposed an algorithm to identify cellular contexts in gene expression data called *Cellular Context Mining* (CCM), where the genes in a context have significantly low interference and crosstalk values across the samples in the context. One major step in CCM, known as *in-silico* conditioning, is designed to be similar to a biologist manipulating the status of a gene or conditioning cells in an experiment with techniques including ectopic expression or gene silencing. With *in-silico* conditioning, the conditioning and the observations are not performed manually as the data is collected, but rather computationally after the data has been collected, hence the name. In this paper, we only consider a single gene driver at a time for conditioning, although the model allows for more. Each conditioning of a driver gene g_i on a value $Y_i = y_i$ yields a subset of samples M_i within which a set of driven genes $S_i = \{g_{i(1)}, \ldots g_{i(r)}\}$ appears to be tightly regulated, so a cellular context, therefore, is defined as $C_i = \{ G_i, \mathbf{Y}_i, S_i, M_i \}$[§]. Permutation tests were used to determine the most statistically significant contexts represented in the data. Note that each context defines regulatory relationships $G_i \rightarrow g \in S_i$, specific to M_i with G_i (drivers) conditioned on a value $Y_i = y_i$. These implicit relationships lead to the construction of context specific regulatory networks.

The advantage of the context mining method is that it is built upon a biologically-inspired mathematical model, which gives strong meaning to the direction of the edges, i.e. one driver gene controlling another. Also, CCM identifies each context with a corresponding driver gene and a set of samples, thereby ensuring the identification of a unique and statistically significant cellular context. Supplementary materials including an elucidating figure explaining CCM, a Java program implementing CCM, and other relevant materials are available at http://sysbio.fulton.asu.edu/psb2009. The CCM algorithm has been also parallelized to carry out analysis of large data sets [12].

2.3. *Context-Specific Gene Regulatory Networks*

This study asserts that the gene regulatory networks (GRNs) produced by cellular context mining exhibit biological advantages absent in related techniques. We first note that driver g_j in the context C_j might be driven by a g_i in another context C_i. The chaining of such regulatory relationships $g_i \rightarrow g_j$, in addition to implicit driver-driven relationships $g_i \rightarrow g \in S_i$, results in an interesting graphical structure, representing relationships between contexts. We call this a context-specific gene regulatory network (GRN) as each regulatory relationship $g_i \rightarrow g \in S_i$ is specific to corresponding subset of samples, M_i.

A context-specific GRN differs from other representations not in its graphical structure, but by the fact that contexts connected to one another in a

[§] The state vector of driven genes $\mathbf{X}_i$ is used for their identification, and not explicitly in C_i.

network differ in their sample composition. Formally, a context-specific GRN H is a pair $H = (V, E)$, where V is a set of gene-representing vertices and E is a set of edges oriented from genes designated as drivers to genes designated as driven; thus H is a directed graph structure, though not necessarily acyclic, since a driven gene in one context may be a driver in another.

Again, note that each edge e_{i*} is specific to only its corresponding subset of samples, M_i, where e_{i*} refers to $g_i \rightarrow g \in S_i$. This study shows that not only do context-specific GRNs report verifiable (and possibly novel) relationships between genes, but moreover the overall network structure groups itself into biologically meaningful and readily annotated *context clusters*. We applied this technique to the Target Now (TN) ** dataset, which includes gene expression profiles of 146 patients with refractory cancer. Amalgamation of identified cellular contexts yielded a context-based network structure. Biological annotation according to sample composition of context clusters and literature verification of gene-disease relevance was carried out. This is unique among network learning techniques, which we illustrate with a comparison to Bayesian network approaches.

2.4. *Bayesian Network Analysis*

The Bayesian network model is a popular tool for modeling GRNs. Here, we used a hybrid algorithm of *hierarchical clustering and order restriction* (H-CORE) [8] and *sparse candidate* (SC) [5] to learn genome-wide Bayesian network structures from a dataset. To compensate for the inadequate amount of observed data, we applied a k-fold bootstrapping in learning GRN with Bayesian network learning.

We built K subsets $\boldsymbol{D}_1, \boldsymbol{D}_2, \ldots, \boldsymbol{D}_K$ by randomly selecting $(k\text{-}1)/k$ proportion (for example, 90% if $k = 10$) of samples with replacement from $\boldsymbol{D}$. We applied the hybrid algorithm of H-CORE and SC to learn a Bayesian network structure H_i from each $\boldsymbol{D}_i$, and built K Bayesian network structures. The likelihood of an edge $g_i \rightarrow g_j$ to be in the final GRN H_{BN} was evaluated as follows:

$$L(g_i \rightarrow g_j) = \frac{\sum_{\forall H_l \text{ with } g_i \rightarrow g_j} \Pr(H_l \mid \boldsymbol{D}_l)}{\sum_{k=1}^{K} \Pr(H_k \mid \boldsymbol{D}_k)} \tag{3}$$

The final GRN H_{BN} only included the edges having likelihood L larger than 0.5.

** http://www.targetnow.com

3. Results

3.1. *Target Now Dataset*

In applying the method described above, we used the gene expression profile of the Target Now (TN) study. TN study aims to determine if patients with refractory cancer, who did not benefit from the standard types of treatment, could derive benefit from therapy with a drug not normally used for their particular form of cancer. The therapeutic to apply is one that has activity against a gene target that is found to be altered in that patient's cancer. The cancer patients contributing to the TN study all have late stage cancer. Late stage cancer is very frequently de-differentiated, having lost a great deal of the specialized functions present in the tissue from which it arose. Due to this biological simplification of the system, those genes whose abundance is found to be altered from the normal tissue of origin and whose change of abundance is found in other refractory cancers (of the same type or of different types) may be representatives of changes that are necessary to support a particular molecular subtype of cancer.

The TN dataset, which consists of 17,085 unique probes (Agilent-011521 Human 1A Microarray G4110A) from 146 patients with different types of refractory cancer, was used to learn context-specific GRNs. For each tumor type, its (normal) tissue of origin was used as a baseline and the ratio of the tumor to its tissue of origin was computed and the ratio value was ternarized using a statistical model [2]. The dataset was then pre-filtered based on transcription activity of each gene across the samples to be reduced to only 4,000 probes. The Target Now dataset sample distribution of the 146 samples between different cancer tumor types is listed in Table 1.

Table 1. TN Sample Distribution, number of samples associated with different cancer tumor types

Pancreas	20	Colon	7	Brain	4	Cervical	3	Esophagus	2
Ovarian	19	Kidney	6	Lung	4	Gallbladder	3	Skin	2
Melanoma	18	Salivary	6	Adipose	3	Rectal	3	T Cell	2
Breast	16	Adrenal	5	Bladder	3	Stomach	3	Thyroid	2

Single Sample: Appendix, Cartilage, Chondrosarcoma, Eccrine Adeno-carcinoma, Glioma, Gastric, Ileum, Lymphoma, Monocytes, Prostate, Uterus, Rhabdomyosarcoma, Synovial Cell Sarcoma, Skeletal Muscle, Testicular

3.2. *Context Clusters*

Running the context mining algorithm with a strict statistical significance threshold resulted in 205 contexts with p-value < 0.0005. Using these contexts, the method described to create context specific GRNs yielded a directed graph with 1,790 vertices (genes) and 9,566 edges (regulatory relationships), as shown in Figure 1 (in Sect. 3.3). This graph had an interesting property of being systematically fragmented into four separate *context clusters*, which were identified by locating the weakly connected components in the graph.

These context clusters provide a useful approach to interpreting the contexts found by CCM. The clusters typically display significant overlaps among their subsets of samples. This is due to complex inter-connections among drivers that result from particular common cellular processes being shared among them. Forming context clusters readily reveals these common cellular contexts on the basis of their more densely connected components.

When investigating four separate context clusters, we noticed the two largest context clusters consisted of densely connected parts loosely bound to one another. Seeking to further characterize the data on the basis of very dense connectivity, we investigated the connections within the two largest context clusters. In Figure 1, bottom right, we segregated the first large cluster into context clusters C and G. Context cluster G is easily separable as all its genes are neither under- nor over-expressed (unlike C), and only one edge exists between the context clusters C and G (C drives a gene also driven by G). These characteristics convinced us that C and G should be analyzed as separate context clusters. The weak connection may have been rooted in tissue of origin similarity, as between them they account for two-thirds of the pancreatic samples in the data with six members in each. Next, we segregated the top large cluster in Figure 1 into context clusters A, B and F. All driver-to-driver edges between A and B are oriented from A to B, implying a hierarchical regulatory relationship from A to B. Also, like C and G, their connection in the graph is explained by the fact that both A and B represent significant numbers of both breast and ovarian tumor types. Context clusters B and F share four edges, two involve genes driven by drivers in both B and F. The two remaining edges are both directed from F to drivers in B, indicating again a possible hierarchical regulatory relationship between the two.

On the basis of density of connection and directionality of control, we resolved the four original context clusters into seven biologically separable ones. Each of these seven context clusters are visible in Figure 1 and have the associated tumor types next to them. Enriched tumor types are highlighted in red and the numbers next to all tumor types correspond to the number of samples distinguished as significant by the scoring function (Equation 4, discussed in next section). See Table 1 for details of the dataset's sample composition.

3.3. *Enrichment Calculations*

3.3.1. *Sample Association to Context Cluster*

As a context cluster is a set of cellular contexts connected to one another through inter-context regulation, it would be informative to associate a set of samples to each context cluster based on its strength of association with the member contexts. Sample association to context clusters also allows annotation of the context cluster as a partial representative of cancer type. Since one context cluster is comprised of potentially many contexts, each representing a particular subset of biological samples, it is of interest which of those samples appears in more than one context in the context cluster. Samples were scored on the basis of occurrence within the context, over all the contexts found in the context cluster. A sample *s*, given a context cluster C consisting of *m* contexts $\{C_1, C_2, \ldots, C_m\}$, would have the scoring:

$$\mathrm{SAS}(s, C) = \sqrt[m]{\prod_{i=1}^{m} f_i(s)}, \quad \text{where } f_i(s) = \begin{cases} k_i/N, & s \in C_i \\ 1, & \text{otherwise} \end{cases} \tag{4}$$

where k_i is number of samples within context C_i and N is the total number of samples in the gene expression data. The sample which occurs in all contexts of the context cluster would then have the least score, and the sample which is not present in any of the contexts will have a score of 1. The samples with score less than 0.5 were associated to corresponding context cluster. Using only the selected samples, the distributions and tumor types were calculated across all context clusters. Figure 1 depicts the tumor types having non-zero sample counts corresponding to each context cluster.

3.3.2. *Tumor Type Enrichment*

After sample association to specific context cluster, each cluster was subjected to a statistical test for enrichment of specific types of tumors. The Yates corrected chi square test for significance was applied (some numbers were less than 5) to each tumor type-context cluster pair. Some of the significant results are summarized below in Table 2. Figure 1 highlights (in red) the tumor type considered enriched within the corresponding context cluster.

Intriguingly, context cluster A showed significant tumor enrichment of ovarian cancer, breast cancer and lung cancer. A literature survey shows breast cancer drugs are being used in the treatment of lung cancer [13], because of vital role of estrogen in lung development and subsequently cancer pathway.

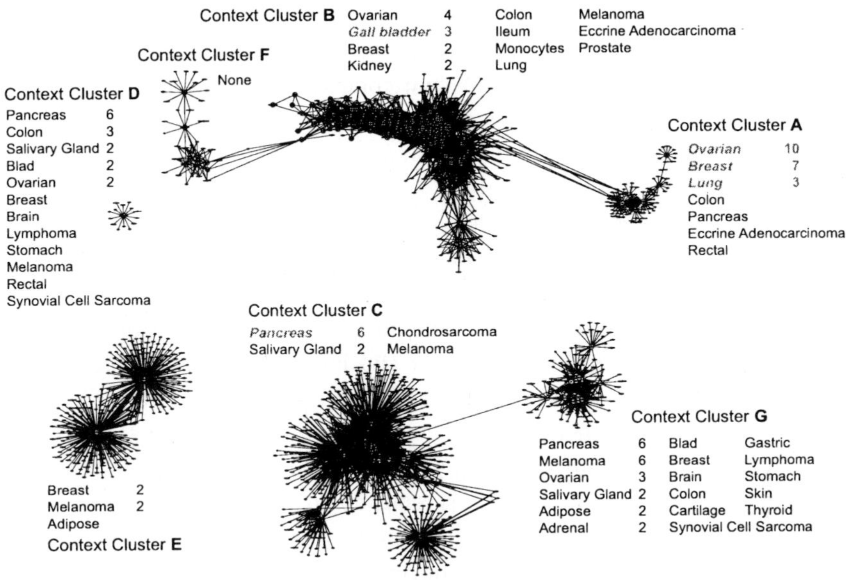

Figure 1. Context-clusters and context-specific GRNs – each context cluster is annotated with the corresponding set of samples and highlights significantly enriched tumor types in red (italicized). Table 1 shows tumor distribution in the dataset. Edges are oriented from driver genes (large vertices) to driven genes (small vertices). In color graph (available at the supplementary webpage), red, green and grey vertices represent up-, down- and no-change, respectively.

Table 2. Chi-square enrichment test p-values of tumor types in different context clusters

Context Cluster A		*Context Cluster B*		*Context Cluster C*	
Tumor Type	p-value	Tumor Type	p-value	Tumor Type	p-value
Ovarian	2.3e-05	Gallbladder	1.6e-04	Pancreas	8.2e-05
Breast	0.0057				
Lung	0.0120				

Conventional approaches such as clustering and Bayesian network learning provide some ability to observe sample enrichment, but they do so in ways that do not exploit the association of particular expression behaviors in subsets of the samples to the fullest extent. Since clustering and Bayesian network learning implicitly assume that the observed data is from a single distribution, their results are always diluted approximations relative to results that assume the observed data to have come from various different distributions and evaluate them in appropriate isolation.

We compared our method to some conventional clustering algorithms, i.e. hierarchical clustering and k-means clustering using similarity metrics of correlation and Euclidean distance, in Cluster version 3.0 [4], to group samples with similar gene expression profiles together. We were able to verify that in cases where a similar number of clusters (six or seven) were identified by conventional methods, the conventional clusters display significant overlap (ranging from 40% to 90% overlap) with context clusters in terms of samples (and thus tumor type enrichment). Conventional clustering algorithms do not however provide a quantitative evaluation with which to isolate vital gene markers or describe the genes' activity for the subtype of disease described by the sample subset. The context cluster approach has a distinct advantage of extracting relevant genes pertaining to the particular disease type.

Some examples of known gene interactions and relationships to diseases within context clusters were verified through a literature survey. Context cluster A involved breast cancer, ovarian cancer and lung cancer, and included genes such as TNFRSF1A, which is known to promote breast cancer [11]; CD74, usually expressed in ovarian and lung cancers, is being considered as a target for Multiple Myeloma treatment therapy [1]; HLA-DM, its expression when combined with that of HLA-DR, is considered to influence breast tumor progression and patient outcome [10]. Context cluster C, related to pancreatic cancer, contained GP2, a protein specifically expressed in pancreatic acinar cells and considered as a diagnostic marker in animals [7].

3.4. *Comparison of Context Mining and Bayesian Network Analysis*

Context clusters (A ~ G) were compared to H_{BN}, which is the result from the Bayesian network analysis. H_{BN} was composed of many subgraphs (H_{CC}s), which were connected components, and the top 32 large H_{CC}s were chosen as the target of comparison. The degree of overlap was evaluated for each pair (context cluster, H_{CC}) using the geometric mean of common gene ratios for the context cluster and the H_{CC}. A pair with a degree of overlap larger than 0.162 was determined to share a significant amount of genes after considering the empirical distribution of the degree of overlap. To figure out the difference between two results, enrichment analyses for GO terms were conducted for every context cluster and every H_{CC} using GoMiner [14].

Figure 2 compares context mining and Bayesian network analysis. The comparison revealed 10 H_{CC}s with a significant number of shared genes with at least one context cluster (Shared = $\{H_{CC,1}, H_{CC,2}, H_{CC,3}, H_{CC,7}, H_{CC,16}, H_{CC,17}, H_{CC,19}, H_{CC,26}, H_{CC,30}, H_{CC,32}\}$). When H_{CC}s in **Shared** were subject to GO term enrichment analysis, most of such H_{CC}s did not have any enriched GO term (8 of 10). On the contrary, other H_{CC}s with no significantly shared genes (**NonShared**) were often enriched with GO terms (19 of 22). The reason can be as follows: Bayesian network learning assumes that the observed data is from a

	HCC1	HCC2	HCC3	HCC4	HCC5	HCC6	HCC7	HCC8	HCC9	HCC10	HCC11
A	0.00	0.00	0.00	0.00	0.00	0.00	0.00	0.00	0.00	0.11	0.02
B	0.32	0.00	0.03	0.00	0.00	0.08	0.00	0.00	0.00	0.00	0.00
C	0.02	0.19	0.01	0.00	0.11	0.00	0.00	0.00	0.00	0.02	0.05
D	0.42	0.00	0.00	0.00	0.00	0.00	0.00	0.00	0.00	0.00	0.00
E	0.00	0.00	0.00	0.14	0.00	0.01	0.00	0.00	0.00	0.14	0.00
F	0.00	0.00	0.53	0.00	0.00	0.00	0.51	0.00	0.00	0.00	0.00
G	0.00	0.00	0.01	0.00	0.00	0.00	0.00	0.00	0.00	0.00	0.00

	HCC12	HCC13	HCC14	HCC15	HCC16	HCC17	HCC18	HCC19	HCC20	HCC21	HCC22
A	0.05	0.00	0.00	0.00	0.00	0.25	0.05	0.00	0.00	0.00	0.03
B	0.00	0.00	0.00	0.00	0.42	0.00	0.00	0.22	0.00	0.00	0.00
C	0.00	0.00	0.01	0.08	0.05	0.07	0.11	0.11	0.03	0.01	0.00
D	0.00	0.00	0.00	0.00	0.00	0.00	0.00	0.00	0.00	0.00	0.00
E	0.00	0.00	0.00	0.00	0.00	0.00	0.00	0.00	0.00	0.00	0.08
F	0.00	0.00	0.00	0.00	0.00	0.00	0.00	0.00	0.00	0.00	0.00
G	0.00	0.00	0.00	0.00	0.39	0.00	0.00	0.04	0.00	0.00	0.00

	HCC23	HCC24	HCC25	HCC26	HCC27	HCC28	HCC29	HCC30	HCC31	HCC32
A	0.00	0.00	0.00	0.11	0.00	0.00	0.00	0.00	0.03	0.00
B	0.00	0.00	0.00	0.00	0.00	0.00	0.00	0.14	0.00	0.17
C	0.01	0.00	0.00	0.01	0.00	0.08	0.07	0.00	0.00	0.00
D	0.00	0.00	0.00	0.00	0.00	0.00	0.00	0.00	0.00	0.00
E	0.00	0.00	0.00	0.17	0.00	0.00	0.00	0.00	0.02	0.00
F	0.00	0.00	0.00	0.00	0.00	0.00	0.00	0.00	0.00	0.00
G	0.00	0.00	0.00	0.00	0.00	0.00	0.00	0.31	0.00	0.21

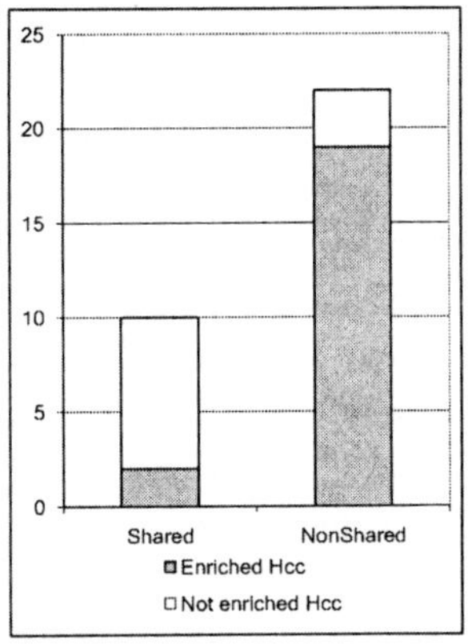

Figure 2. The left table shows the degree of overlap for each pair of a context cluster (A ~ G) and a subgraph ($H_{CC,1} \sim H_{CC,32}$) from Bayesian network analysis. A shaded cell in the table represents the pair with significantly shared genes. A shaded context cluster or a H_{CC} has at least one enriched GO term while non-shaded one has no enriched GO term. The right graph shows the portion of H_{CC}s with enriched GO terms for H_{CC}s with significantly shared genes (***Shared***) and H_{CC}s with no significantly shared genes (***NonShared***).

single distribution, and attempts to capture information consistent across all samples in the observed data. On the contrary, context mining captures information consistent in subset of samples. If an H_{CC} shares a significant amount of genes with a context cluster, it means that significant portion of information in the H_{CC} is consistent only in some subset of samples. This is against the viewpoint of Bayesian network learning, because the H_{CC} was built with a bias toward the process of capturing information consistent across all samples. For this reason, the shared genes in an H_{CC} may have inconsistent information with those unshared genes in the H_{CC}, eventually making it hard for the H_{CC} to have enriched GO terms. The conventional Bayesian network learning is therefore not an optimum choice for identifying context-specific information from some subset of samples.

This comparison and the tumor enrichment studies of context clusters in previous sections, shows that the context-specific GRN has a novel ability to represent context-specific information from subset of samples while conventional approaches assuming the entire sample set to be from single distribution have a harder time recognizing this kind of behavior.

4. Conclusions

This paper presents a novel approach to generate context clusters, i.e. disease-pertinent (cancer tumor type in case of TN dataset) gene regulatory networks, using cellular contexts through the CCM algorithm. This study asserts that these gene regulatory networks produced by CCM exhibit biological advantages absent in related techniques. The mapping of inter- and intra-context edges results in a graph of context clusters—the densely connected components—which is found to have interesting properties, such as specific cancer tumor type enrichment (used for context cluster annotation) and occurrence of genes relevant to the annotated disease. Comparison of this approach with conventional clustering algorithms demonstrated its advantage for relevant gene subset identification. When compared with Bayesian network analysis, we noted that context clusters can capture regulatory relationships specific to subset of samples while conventional Bayesian network learning rarely captures meaningful context-specific information.

We are currently working on multivariate *in-silico* conditioning in the context mining and incorporation of clinical annotation (if available) for learning and prediction purposes, in addition to the analysis of larger data sets such as GSK300 (https://cabig.nci.nih.gov/caArray_GSKdata/ – 300+ cancer cell line data) and expO data (http://expo.intgen.org/geo/home.do – 2000+ cancer patient samples).

References

1. Jack D. Burton *et al., Clinical Cancer Research* **10**, 6606-6611 (2004).
2. Yidong Chen et al., *Bioinformatics* **18**(9), 1207-1215 (2002).
3. E. R. Dougherty *et al., IEEE/ACM Transactions on Computational Biology and Bioinformatics*, doi:10.1109/TCBB.2007.70247 (2007).
4. M. B. Eisen *et al., PNAS* **v95** n25 (1998).
5. N. Friedman *et al., Proc. Fifth Conf. on Uncertainty in AI,* 206-215 (1999).
6. W. C. Hahn and R. A. Weinberg, *Nat Rev Cancer* **2** (331) (2002).
7. Y. Hao *et al., Archives Of Pathology And Laboratory Medicine* **128**;6, pages 668-74 (2004).
8. Sungwon Jung *et al., BioSystems* Vol. **90**, pp. 197-210 (2007).
9. S. Kim and I. Sen, *Comput Syst Bioinformatics Conf* **6**:169-79 (2007).
10. S. A. Oldford *et al., International Immunology* **18(11)** 1591-1602 (2006).
11. M. A. Rivas *et al., Exp. Cell Res.* PMID: 18061162 (2008).
12. W. Tembe *et al., IEEE Intl Workshop on Gen. Signal Processing and Statistics,* 1-4 (2008).
13. Olga K. Weinberg *et al., Cancer Res* **65**: 11287-11291 (2005).
14. Barry R. Zeeberg *et al., BMC Bioinformatics* **6**:168 (2005).

QUERYING PARSE TREE DATABASE OF MEDLINE TEXT TO SYNTHESIZE USER-SPECIFIC BIOMOLECULAR NETWORKS

LUIS TARI[1], JÖRG HAKENBERG[1], GRACIELA GONZALEZ[2], CHITTA BARAL[1]

[1]Department of Computer Science and Engineering,
Arizona State University, Tempe, AZ 85287, USA

[2]Department of Biomedical Informatics,
Arizona State University, Phoenix, AZ 85004, USA

{luis.tari, jhakenbe,graciela.gonzalez,chitta}@asu.edu

Curated biological knowledge of interactions and pathways is largely available from various databases, and network synthesis is a popular method to gain insight into the data. However, such data from curated databases presents a single view of the knowledge to the biologists, and it may not be suitable to researchers' specific needs. On the other hand, Medline abstracts are publicly accessible and encode the necessary information to synthesize different kinds of biological networks. In this paper, we propose a new paradigm in synthesizing biomolecular networks by allowing biologists to create their own networks through queries to a specialized database of Medline abstracts. With this approach, users can specify precisely what kind of information they want in the resulting networks. We demonstrate the feasibility of our approach in the synthesis of gene-drug, gene-disease and protein-protein interaction networks. We show that our approach is capable of synthesizing these networks with high precision and even finds relations that have yet to be curated in public databases. In addition, we demonstrate a scenario of recovering a drug-related pathway using our approach.

1. Introduction

Modeling large-scale biological knowledge in the form of networks is a common approach to advance our understanding of the mechanisms that govern the behavior of a cell. The steadily increasing amount of 'omics' data facilitates the synthesis of a wide variety of biological networks, from modeling physical protein-protein interactions [1], synthesizing networks of genes that share certain properties such as coexpression based on gene expression data [2], to proteins that share biological processes [3]. The synthesis of biological networks, such as gene-disease associations [4] and gene-drug interactions [5], provides insights to our understanding of the role of genetics in diseases. Descriptions of other kinds of biological networks can be found in [6].

As a way to share this wealth of biological knowledge, the data is made available in various databases, such as IntAct [7] and MIPS [8] for protein-protein interactions. Such interaction databases are typically curated manually by a team of scientists, aided by automated extractors or provided by external contributors in some cases. While interaction data from these databases are highly useful as a concise resource for biologists, the level of detail about the interactions is a priori defined by the databases. The interactions are often restricted to specific kinds of information so that information one might be interested, such as the structure or strength of the interactions, might not be encoded in the databases [9]. Biologists who use these interactions have to be aware of the limitations of the data, which can be unclear if the biologists are not familiar with the curation protocol for the particular database. In other words, biologists can only use the interaction data in a passive manner as they are not engaged in the curation process of the interactions. Biologists can perform filtering or visualization on the interactions provided by the databases as users, but not how the interactions are collected. Such passive use of interactions limits the applicability of the interaction data into research. On the other hand, Medline abstracts are publicly accessible and encode the necessary information to synthesize different kinds of biological networks. For instance, it was estimated that 270,000 Medline abstracts are classified as abstracts with mentions of human, mouse and yeast protein-protein interactions [10]. A more recent work found 150,000 protein-protein interactions in 1 million Medline abstracts [11].

Our goal is to provide a mechanism that allows users to synthesize biomolecular networks specific to their needs through queries against Medline abstracts. Unlike the traditional approaches in querying biomolecular networks that are synthesized from existing curated data, the networks generated from querying Medline abstracts can be more suitable to the users' needs. By using simple-to-use queries to our specialized database of Medline abstracts, these networks convey the information needed by the users, such as strength of the interactions, and such information might be missing in the networks that are synthesized from curated data. Users can specify precisely what kind of information they need in the networks through queries, such as preconditions of the interactions. In addition, users do not have to depend on the time-consuming curation process and synthesize biological networks from curated data that do not include the latest findings.

To implement this mechanism, Medline abstracts are parsed by a natural language parser to represent the syntactic structures of the sentences called parse trees. Entities such as genes, diseases and drugs are automatically identified with the use of entity recognizers as semantic information. The parse trees and the semantic information are then stored in a specialized relational database. Having the parse trees in the database enables the users to extract sentences using simplified linguistic queries. The resulting sentences allow the users to synthesize biomolecular networks specific to their needs. This new paradigm enables biologists to utilize information in Medline abstracts effectively and synthesize their own biomolecular networks.

Our proposed mechanism is different from automated extraction tools that perform biological relationship extraction. Typical relationship extractors such as iHop [12] rely on their own dictionaries of entities to identify their associations based on coccurrences. Other systems such as AliBaba [13] utilize linguistic patterns to extract relationships. From the user's point of view, such extraction systems are treated as black boxes and users cannot specify how the interactions should be obtained. Our method allows users to issue their own search criteria through queries to generate biomolecular networks of interest. A text analysis engine called TLM [14] is based on the idea of a retrieval system that retrieves sentences using textual patterns as queries. Unlike our approach, TLM does not utilize grammatical structures of the sentences to retrieve sentences. Interaction databases such as IntAct, MINT, MIPS for protein-protein interactions and PharmGKB [15] for gene-drug interactions rely on their curators to identify interactions, and users have no influence in the curation process. Proprietary pathway analysis tools such as Ingenuity IPA*, ActiveMotif† utilize their manually curated database to analyze experimental data, but the curation protocol is not accessible to the users. The use of query languages can be found in querying pathways, such as PQL [16], QPath [17] and PATIKA [18], and finding information that require multiple data sources, such as semCDI [19], GenoQuery [20], Cytoscape [21]. However, these query languages depend on curated data in order to return answers.

In this paper, we describe a new paradigm in how users can synthesize biomolecular networks. We illustrate how this approach can lead to the synthesis of gene-drug relationship networks, gene-disease association networks and protein-protein interaction networks.

2. Methods

Our proposed method is to place the user in control of synthesizing their own biomolecular networks through queries to our specialized database of Medline abstracts, and the results returned by the database are utilized to generate the resulting biomolecular networks. Suppose a user is interested in constructing a network of gene-drug relations, in which the drugs are metabolized by enzymes. The following query can be used:

```
<DRUG> _ metabolized by <GENE>
```

* Ingenuity Pathway Analysis Tool: http://www.ingenuity.com

† Active Motif: http://www.activemotif.com

The symbols `<DRUG>` and `<GENE>` infer that the sequences of words have to be a drug name and a gene/protein name in the matching sentences. The order of the tokens in the query matters, so that the above query specifies that the grammatical structures of the matching sentences include a syntactic dependency between the words "*metabolized*" and "*by*". Similarly, "*by*" has to be syntactically dependent on `<GENE>`. The operator _ is a wildcard operator that `<DRUG>` and "*metabolized*" may not have any syntactic dependency between them in the matching sentences. This query can retrieve support evidences such as "Diclofenac is widely used in the treatment of rheumatic diseases and is mainly *metabolized* in the liver *by* CYP2C9."(PMID: 8793607). The grammatical structure of the sentence reveals that there are syntactic dependencies between "*metabolized*" and "*by*", as well as "*by*" and "*CYP2C9*". By allowing users to perform their own queries, users can specify their own criteria in their target interactions. One way of specifying the strength of the interaction is to include the word "*extensively*" in the query as follows:

```
<DRUG> _ extensively metabolized by <GENE>
```

Here we are interested in drug-enzyme metabolic relations in which the strength of the interactions is described as "extensive". The support evidence "Tacrine is extensively *metabolized by* CYP1A2." (PMID:9209244) is an example retrieved by the query. There are cases when negative relations are reported in the literature. Our current system simply disregards sentences with words that indicate negation, such as "not", "no", so that sentences such as "Hesperetin was not metabolized by human CYP1A2" (PMID:10781868) are not retrieved as support evidences.

The essential component of our method is *parse trees* of Medline abstracts; parse trees are syntactic structures that represent the grammatical structures of sentences. Parse trees include *constituent trees* and *linkages*, in which constituent trees are hierarchical syntactic structures of sentences and linkages are composed of *links* that represent syntactic dependencies between pairs of words. These parse trees are generated automatically by the Link Grammar parser [22]. Such parse trees are ideal to be used for expressing linguistic patterns, which are commonly utilized in automated extraction systems. To store the parse trees, a database is needed to capture the hierarchical representation of abstracts, which include the sections of the abstracts such as title or body of the abstracts, parse trees and the semantic information of words. Semantic information includes the entity type of a sequence of words, such as whether it is a gene/protein name[‡], a drug name or a disease name. To cope with the high variation of gene names, an entity recognition system based on a statistical machine learning technique named BANNER [23] is utilized to identify gene names in text. Lists of drug and disease names from MeSH[§], DrugBank[**] and PharmGKB are employed to recognize drug and disease names. We called the database as the *parse tree database*, and the database is implemented using a relational SQL database. Since standard SQL queries are not ideal for expressing queries that involve linguistic patterns, we develop a query language called *parse tree query language* (*PTQL*) that are used to express linguistic patterns and query parse trees. The details of the PTQL query language and its implementation can be found in [24]. Similar to standard database query languages such as SQL, PTQL is designed to be used by developers and people who are familiar with linguistics. To facilitate the synthesis of biomolecular networks through querying of parse trees of sentences in Medline abstracts by biologists, we offer a simpler query language called *PTQL^{LITE}* that is not as expressive as PTQL but the syntax is close to keyword-based queries used in search engines. The sample queries shown in the beginning of this section are PTQL^{LITE} queries.

Figure 1 shows an overview of our approach in using PTQL^{LITE} queries to synthesize biomolecular networks. The *processor* utilizes the named entity recognizers and parses the MedLine abstracts, and stores the processed information in the parse tree database and the inverted index. The *middleware* handles the communication between the web interface and the parse tree database. The middleware takes PTQL^{LITE} queries as input and generates PTQL queries. The PTQL queries are translated into standard SQL queries before querying the parse tree database. Due to the complexity of the translated SQL queries, retrieving results from a large parse tree database can be slow. We increase the efficiency of our system by utilizing an off-the-shelve information retrieval (IR) system so that PTQL^{LITE} queries are first translated into IR queries to retrieve the matching sentences. The PTQL queries are applied to only the parse trees of the sentences retrieved by the IR system rather than the entire database of parse trees so that the process can be performed efficiently. We summarize the process of translating PTQL^{LITE} queries into SQL queries to retrieve answers by the middleware as follows:

---

‡ Gene, protein and enzyme names are indistinguishable by current automated entity recognizers, and sometimes even by human readers. From here on, we use "gene" to refer to genes/proteins/enzymes.

§ Medical Subject Headings (MeSH): http://www.nlm.nih.gov/mesh/

** DrugBank: http://www.drugbank.ca/

1. The *IR query generator* generates an IR query based on an $PTQL^{LITE}$ query provided by the user.
2. The IR query is used to retrieve relevant documents D and sentences S from the *inverted index*.
3. The *PTQL generator* translates the PTQL query into an SQL query and instantiate the query with document id $d \in D$ and sentence id $s \in S$.
4. The SQL query generated in Step 3 is applied to the *parse tree database*.
5. The *bionetwork generator* delivers the resulting network.

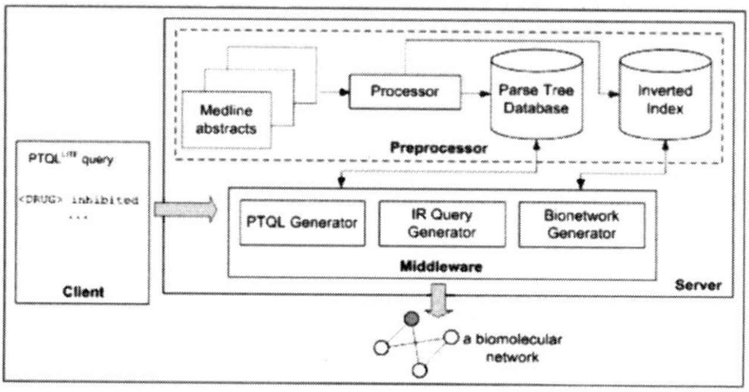

Figure 1 – A client-server system architecture for synthesizing bionetworks through querying parse trees of Medline abstracts. The middleware generates PTQL and IR queries based on the user's input, and retrieves information from the database and the index.

3. Synthesis of various biomolecular networks

In this section, we illustrate our approach with the synthesis of gene-drug relation, gene-disease association and protein-protein interaction networks. We evaluate our approach by using PharmGKB to verify the relations in the generated networks. As PharmGKB only covers part of the knowledge in the literature, we manually evaluated the correctness of the relations based on their corresponding supported evidences that were extracted by our approach.

3.1. *Gene-drug relationship networks*

Drug metabolism influences the effects of drug chemicals, and genetic variations can affect the effectiveness of drug metabolism. It is therefore essential to study the metabolic relations between enzymes and drugs. Here we illustrate the synthesis of a network of gene-drug relationships using our approach, specifically capturing the relations of drugs that are metabolized by enzymes. We use a collection of 13015 Medline abstracts from [25] that focus on topics about gene-drug relations to demonstrate the feasibility of our approach.

Sentences such as "*Triazolam is metabolized by CYP3A4*" (PMID:8612379) are typical examples of how gene-drug metabolic relations are described in biomedical articles. Biologists who are interested in such relations would use the following PTQL^{LITE} query to synthesize their networks:

`<DRUG> _ metabolized by <GENE>`

By default, the system filters out any sentences that infer negative relations. This query results a network of 141 genes and drugs with 138 relations generated from 178 supporting sentences, and each relation is supported by at least 1 sentence. The gene-drug network took about 10 seconds to be generated on a 2-GHz Intel DualCore CPU with 2 GB of RAM. To verify the correctness of the network, we used the gene-drug relations from PharmGKB [15], which is one of the largest curated databases of relations among genes, drugs and diseases that are publicly available. Among the 138 relations in the network, 43 of them can be found in PharmGKB. We further manually evaluated the correctness of the relations in the network based on their evidences that were extracted by our method. We observed that 122 out of 138 (i.e. precision of 88.41%) are indeed correct. We analyzed the incorrect relations and categorize the errors into two sources: (i) errors in extraction due to the sentence structure; (ii) errors due to recognition of entities. We list out some of these errors in Table 1. Example 1 in Table 1 is incorrect due to the fact that the clause describing the drug-enzyme metabolic relation for CYP2C9 is not in the same clause as the drugs lovastatin, simvastatin and atorvastatin. On the other hand, incorrect identification of drug names, such as recognizing "*important drugs*" and "*widely used drugs*" as drug names, leads to incorrect support evidences as shown in example 2 of Table 1. A careful, manual revision of the lexicon used for drugs would eradicate many of the type (ii) errors.

Table 1 – Support evidences that are extracted incorrectly by the query `<DRUG> _ metabolized by <GENE>`

	Gene/Drug	Incorrectly extracted evidence
1	CYP2C9/Lovastatin; CYP2C9/Simvastatin; CYP2C9/Atorvastatin	Lovastatin, simvastatin, and atorvastatin are substrates of CYP3A4, whereas *fluvastatin is* metabolized by CYP2C9. (PMID:11029845)
2	CYP2E1/important drugs	Among important drugs *metabolized by* CYP2E1 … (PMID:2134674)

We synthesize another network using the same PTQL^{LITE} query but specifying that the relations in the network have to be supported by at least 2 different publications. A smaller network of 33 vertices (10 genes and 23 drugs) with 27 edges is generated with this criterion, as shown in Figure 2. Such network allows the discovery of potential relations. For instance, the drugs omeprazole are metabolized by CYP3A4 and CYP2C19, and users might want to study a potential relation between CYP3A4 and CYP2C19. Table 2 lists some of the relations encoded by this network as well as the corresponding supported sentences. Among the 27 relations, only 2 of them are considered as an error in the extraction (i.e. precision of 92.59%). This experiment serves as a proof-of-concept that a biologist can easily synthesize a user-specific network with PTQL^{LITE} queries. This also shows that our approach can overcome the time-consuming process of expert curation, which generally results in a low coverage of the knowledge that has already been published in the literature.

Table 2 – A partial list of gene-drug relations generated by our approach using the pattern `<DRUG> _ metabolized by <GENE>`. Each gene-drug relation is supported by at least 2 different publications, and the relations are yet to appear in PharmGKB.

Gene/Drug	Support evidence
CYP3A4/Triazolam	Triazolam is *metabolized by* CYP3A4. (PMID:8612379)
CYP3A4/Terfenadine	… therapeutics (terfenadine and cyclosporine) known to be *metabolized by* CYP3A4 (PMID:10752642)
Alcohol dehydrogenase / Ethanol	Ethanol is *metabolized by* alcohol dehydrogenase in the human stomach. (PMID:9693201)
CYP1A2/Propafenone	Propafenone is mainly *metabolized by* CYP2D6 (PMID:10917404)
CYP2E1/Chlorzoxazone	Chlorzoxazone is mainly *metabolized* to 6-OHchlorzoxazone *by* CYP2E1. (PMID:7910460)

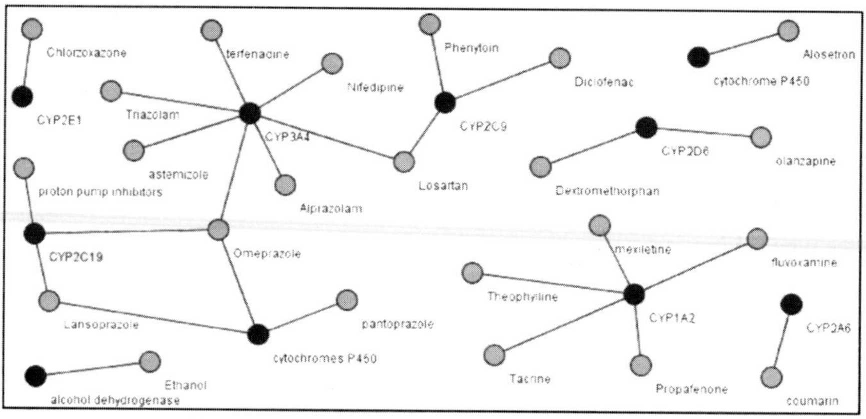

Figure 2 – A gene-drug network in which each edge represents a drug metabolized by an enzyme. Each edge is supported by at least 2 support evidences.

We illustrate our network synthesis approach with another example of drug-enzyme inhibition. The following PTQL^{LITE} query can be used:

```
<DRUG> _ inhibit <GENE> | <DRUG> _ inhibits <GENE> |
        inhibition of <GENE> by <DRUG> |
            <GENE> _ inhibited by <DRUG>
```

This query is composed of 4 subqueries, which are separated by the operator |, to capture the drug-enzyme inhibition relations. Our system essentially synthesizes the drug-enzyme inhibition network as the union of the relations resulted from the 4 subqueries. Using the criteria that each relation has to be supported by at least 2 publications, the resulting network is composed of 14 enzymes and 13 drugs with 19 relations. Among these 19 relations, 7 of the relations can be verified with the PharmGKB database. We further look into the other 12 relations, and realize that 9 relations are well supported by our extracted evidences, as shown in Table 3.

Table 3 – A list of correct drug-enzyme inhibitions and the corresponding support evidences. These correct relations are currently not contained in PharmGKB.

Gene/Drug	Support evidence
CYP3A4/Indinavir; CYP3A4/Nelfinavir; CYP3A4/Amprenavir	The HIV protease inhibitors amprenavir, indinavir, nelfinavir, ritonavir and saquinavir *inhibit* CYP3A4. (PMID:10926350)
CYP2D6/Terbinafine	Terbinafine *inhibits* CYP2D6. (PMID:11475469)
alcohol dehydrogenase/ 4-methylpyrazole	... 4-methylpyrazole to *inhibit* alcohol dehydrogenase. (PMID:2994256)
CYP2D6/Quinidine	*Inhibition of* CYP2D6 *by* quinidine ... (PMID:10510150)
catechol-O-methyltransferase/ tolcapone	*Inhibition of* catechol-O-methyltransferase *by* tolcapone has been shown .... (PMID:9343116)
Thiorphan/Bradykinin	... formation of the major metabolite bradykinin 1-7 was *inhibited by* thiorphan. (PMID:1629199)
benzoyl-Gly-His-Leu/ captopril	The metabolism of benzoyl-Gly-His-Leu was completely *inhibited by* captopril (PMID:7588745)

3.2. *Gene-disease relationship networks*

We illustrate how we can synthesize gene-disease association network using our approach. We use the following query to construct such gene-disease network:

```
<GENE> _ associated with <DISE> |
     <GENE> _ risk of <DISE>
```

Using the same 13015 Medline abstracts that were used in constructing gene-drug networks as described in the previous subsection, a network with 88 genes and diseases with 76 gene-disease relations is generated with the above query. Each of the relations in this network is supported by at least 1 publication. The evaluation of the network using PharmGKB shows that 7 of the relations can be confirmed as correct. Our manual evaluation shows that 54 of the 76 relations (i.e. precision of 71.05%) are correct by analyzing the extracted support evidences. We conclude that 11 of the incorrect relations are due to errors from the entity recognizers. For instance, *ACE inhibition* was incorrectly recognized as a gene name when in fact it is considered as a drug/treatment (even though *ACE* itself is a gene name). The rest of the incorrect relations are caused by incorrect extraction. We also synthesize another network using the criteria of at least 2 publications as support for the relations in the network. This results a small network with 11 genes and diseases with 6 relations. One of the reasons for such a small network is that our current system does not utilize normalization techniques to realize that terms such as "*vitamin D receptor*" and "*VDR*" refer to the same entity. Table 4 lists out some of these associations.

Table 4 – A list of gene-disease associations and the corresponding support evidences.

Gene/Disease	Support evidence
VDR/Osteoporosis	To determine whether a polymorphism of the VDR gene, already *associated with* osteoporosis .... (PMID:9259424)
UGT1A1/Gilbert's syndrome	The presence of an additional TA repeat in the TATA sequence of UGT1A1 has been *associated with* Gilbert's syndrome. (PMID:10340924)
VDR/ Hyperparathyroidism	Polymorphism of the VDR gene has recently been shown to be related to bone mineral density, and also *associated with* hyperparathyroidism .... (PMID:10508794)

3.3. *Protein-protein interaction networks*

We constructed a network of protein-protein interactions using the BioCreative 2 dataset [26]. The task in the BioCreative 2 IPS benchmark is to find protein-protein interactions for which a text provides evidence for a physical interaction between the proteins. A sample query is as follows:

`<GENE> _ binds with <GENE>`

We generated 11208 PTQL queries from the BioCreative 2 training dataset, and achieved a precision of 83.6% and recall of 58.6%.

4. Scenario

We illustrate a scenario on how users who are not familiar with linguistic structures can utilize our system to synthesize networks. Suppose the user is interested in synthesizing a pathway about the drug tamoxifen, the following query can be issued to first find all sentences that describe associations between genes and tamoxifen.

`<GENE> _ <DRUG="tamoxifen">|<DRUG="tamoxifen"> _ <GENE>`

The retrieved sentences contain cooccurrences of genes and tamoxifen. The user can examine some of these sentences and refine the relations with this query:

```
<DRUG="tamoxifen"> _ involvement of <GENE> |
Binding _ <DRUG="tamoxifen"> _ <GENE>
```

A pathway that involves the genes CYP3A4, CYP2B6 and CYP2D6 with tamoxifen can then be synthesized. This pathway is supported by the evidences as shown in Table 5, and can be verified with PharmGKB. This example illustrates the feasibility of pathway synthesis using our approach.

5. Conclusion

We demonstrate that our approach is capable of synthesizing biological networks with high precision without the use of curated data. This new paradigm of network synthesis tailors the specific needs of the users. Future work includes normalization of entities to handle name variations, and parse all Medline abstracts so that networks can be synthesized with respect to the latest findings. Inclusion of query templates that are typically used in describing gene-drug, gene-disease and protein-protein relations will be provided through the interface, so that synthesizing biological networks with our approach can even be simpler. A prototype based on 13015 Medline abstracts from [25], mainly focusing on gene-drug relations, is available at http://cbioc2.eas.asu.edu/netsynthesis.

Table 5 – A list of support evidences for the pathway that involves the drug tamoxifen

PMID	Support evidence
7748182	*Binding of* tamoxifen correlated with CYP3A4 and CYP2B6 content.
9037249	The proportion of activity inhibited by quinidine correlated positively with total microsomal tamoxifen 4-hydroxylation activity, indicating a major *involvement of* CYP2D6 in this reaction.

Acknowledgement

The authors would like to acknowledge the support of NSF through grant number 0412000 and a grant from SFAz. We also appreciate the helpful comments provided by the reviewers of this paper.

References

1. P. Uetz, L. Giot, G. Cagney, et al. A comprehensive analysis of protein-protein interactions in S. cerevisiae. *Nature,* **403**, 6770, 623-627 (2000).
2. V. van Noort, B. Snel and M. A. Huynen. The yeast coexpression network has a small-world, scale-free architecture and can be explained by a simple model. *EMBO Rep.,* **5**, 3, 280-284 (2004).
3. L. Tari, C. Baral and P. Dasgupta. Understanding the global properties of functionally related gene networks using GO. *PSB,* 209-220 (2005).
4. J. Lim, T. Hao, et al. A PPI network for human inherited ataxias and disorders of Purkinje cell degeneration. *Cell,* **125**, 4, 801-814 (2006).
5. G. Giaever, P. Flaherty, et al. Chemogenomic profiling: identifying the functional interactions of small molecules in yeast. *PNAS,* **101**, 3 (2004).
6. X. Zhu, M. Gerstein and M. Snyder. Getting connected: analysis and principles of biological networks. *Genes Dev.,* **21**, 9, 1010-1024 (2007).
7. H. Hermjakob, L. Montecchi-Palazzi, C. Lewington, et al. IntAct: an open source molecular interaction database. *NAR,* **32**, suppl 1, D452-455 (2004).
8. H. W. Mewes, D. Frishman, U. Guldener, et al. MIPS: a database for genomes and protein sequences. *Nucleic Acids Res,* **30**, 31-34 (2002).

9. M. J. Betts and R. B. Russell. The hard cell: from proteomics to a whole cell model. *FEBS Lett.,* **581**, 15, 2870-2876 (2007).
10. I. Donaldson, J. D. Martin, B. d. Bruijn, et al. PreBIND and Textomy - mining the biomedical literature for protein-protein interactions using a support vector machine. *BMC Bioinformatics,* **4**, 11 (2003).
11. K. Fundel, R. Kuffner and R. Zimmer. RelEx--Relation extraction using dependency parse trees. *Bioinformatics,* **23**, 3, 365-371 (2007).
12. R. Hoffmann and A. Valencia. A gene network for navigating the literature. *Nat. Genet.,* **36**, 7, 664 (2004).
13. C. Plake, T. Schiemann, M. Pankalla, et al. AliBaba: PubMed as a graph. *Bioinformatics,* **22**, 19, 2444-2445 (2006).
14. J. D. Martin. Fast and Furious Text Mining. *IEEE Data Eng. Bull.,* **28**, 4, 11-20 (2005).
15. M. Hewett, D. E. Oliver, D. L. Rubin, et al. PharmGKB: the Pharmacogenetics Knowledge Base. *NAR,* **30**, 1, 163-165 (2002).
16. U. Leser. A query language for biological networks. *Bioinformatics,* **21**, Supplement 2, ii33-ii39 (2005).
17. T. Shlomi, D. Segal, E. Ruppin, et al. QPath: a method for querying pathways in a PPI network. *BMC Bioinformatics,* **7**, 199-207 (2006).
18. E. Demir, O. Babur, U. Dogrusoz, et al. PATIKA: an integrated visual environment for collaborative construction and analysis of cellular pathways. *Bioinformatics,* **18**, 996-1003 (2002).
19. E. P. Shironoshita, Y. R. Jean-Mary, R. M. Bradley, et al. semCDI: A Query Formulation for Semantic Data Integration in caBIG. *J. Am. Med. Inform. Assoc.,* **15**, 4, 559-568 (2008).
20. F. Lemoine, B. Labedan and C. Froidevaux. GenoQuery: a new querying module for functional annotation in a genomic warehouse. *Bioinformatics,* **24**, 13, i322-9 (2008).
21. P. Shannon, A. Markiel, O. Ozier, et al. Cytoscape: a software environment for integrated models of biomolecular interaction networks. *Genome Res.,* **13**, 11, 2498-2504 (2003).
22. D. Grinberg, J. Lafferty and D. Sleator. A Robust Parsing Algorithm For Link Grammars. **CMU-CS-TR-95-125** (1995).
23. R. Leaman and G. Gonzalez. BANNER: An executable survery of advances in biomedical named entity recognition. *Pacific Symposium of Biocomputing (PSB),* 652-663 (2008).
24. P. H. Tu, C. Baral, et al. Generalized text extraction from molecular biology text using parse tree database querying. **TR-08-004** (2008).
25. J. T. Chang and R. B. Altman. Extracting and characterizing gene-drug relationships from the literature. *Pharmacogenetics,* **14**, 9, 577-586 (2004).
26. M. Krallinger, F. Leitner and A. Valencia. Assessment of the Second BioCreative PPI task: Automatic Extraction of Protein-Protein Interactions. *Proceedings of the Second BioCreative Challenge Workshop,* 41-54 (2007).

PAIRWISE ALIGNMENT OF INTERACTION NETWORKS BY FAST IDENTIFICATION OF MAXIMAL CONSERVED PATTERNS*

WENHONG TIAN[1,2†], NAGIZA F. SAMATOVA[1,2]

[1]Department of Computer and Mathematics Division , Oak Ridge National Laboratory, Oak Ridge, TN 37831 and [2]Computer Science Department, North Carolina State University, Raleigh, NC 27696, USA.

A number of tools for the alignment of protein-protein interaction (PPI) networks have laid the foundation for PPI network analysis. They typically find conserved interaction patterns by various local or global search algorithms, and then validate the results using genome annotation. The improvement of the speed, scalability and accuracy of network alignment is still the target of ongoing research. In view of this, we introduce a connected-components based algorithm, called HopeMap for pairwise network alignment with the focus on fast identification of maximal conserved patterns across species. Observing that the number of true homologs across species is relatively small compared to the total number of proteins in all species, we start with highly homologous groups across species, find maximal conserved interaction patterns globally with a generic scoring system, and validate the results across multiple known functional annotations. The results are evaluated in terms of statistical enrichment of gene ontology (GO) terms and KEGG ortholog groups (KO) within conserved interaction patters. HopeMap is fast, with linear computational cost, accurate in terms of KO groups and GO terms specificity and sensitivity, and extensible to multiple network alignment.

1. Introduction

Protein-protein interactions (PPI) are of central importance for virtually every process in a living cell. Information about these interactions can improve our understanding of diseases and provide the basis for new therapeutic approaches. A fundamental goal in systems biology is to understand how proteins in the cell interact with each other. However, finding all protein interactions is both costly and labor-intensive. High-throughput experimental techniques (e.g., yeast two-hybrid, co-immunoprecipitation) are generating PPI networks data, which is being collected and stored in public databases, such as DIP [1] and SNDB [2].

* This research was supported by the "Exploratory Data Intensive Computing for Complex Biological Systems" project from U.S. Department of Energy (Office of Advanced Scientific Computing Research, Office of Science). The work of NFS was also sponsored by the Laboratory Directed Research and Development Program of Oak Ridge National Laboratory, which is managed by UTBattelle for the LLC U.S. D.O.E. under contract no. DEAC05-00OR22725.

†Corresponding author: wenhong_tian@hotmail.com

Using PPI network data, integrated probability models are being developed to predict protein-protein interactions [3, 4]. Researchers [3-10] are trying to identify conserved patterns, such as ortholog groups and functionally similar pathways or protein complexes across species. The problem of identifying conserved network regions across species, known as the network alignment problem, is NP-hard [3-10]. This challenge attracts many researchers to find efficient heuristic solutions for the problem.

Most of previous network alignment tools focus on finding conserved functional patterns across species [3, 5, 6, 9] or on maximizing the overall number of matches between PPI networks globally [4]. The existing tools first find the conserved interaction regions by various search algorithms and then validate the results' accuracy against gene ontology (GO) [11] or KEGG ortholog (KO) groups [12] annotation.

In this paper, we explore a complementary approach to comparative analysis of PPI networks. Our focus is on fast identification of maximal conserved patterns across species with high accuracy validated against genome annotation. We propose an iterative connected-components-based algorithm with linear cost. This alignment algorithm combines information on HOmologs (genomic similarity), Physical interactions conservation and Equivalent functions (HOPE); we call our algorithm HopeMap.

1.1. Major Contributions

Our approach and focus are different from existing tools. The existing tools first find the conserved interaction regions by all kinds of search algorithms and then validate the results by gene annotations (e.g., GO, KO). Observing that the number of true homologs across species is relatively small compared to the total number of proteins in all species, we start with highly homologous clusters across species, find conserved interaction regions iteratively with a generic scoring system, and validate the results across multiple known functional annotations. Using known homolog clustering results, such as KO groups reduces the computational cost of finding sequence similar proteins by all kinds of searches in the existing tools. Applying connected-components based algorithm to find conserved regions across species assures a fast (if not the fastest) approach to identify maximal conserved patterns across species simultaneously; it is also a parameter-free clustering algorithm, unlike most of the existing tools that require setting up different sets of parameters for different alignments [8]. An iterative process can be applied to refine the identified regions. The generic scoring function is an open system; it currently combines evolutionary evidence such as genomic, interaction and functional similarities; it can incorporate, if available and necessary, more features in the future.

1.2. Related Work

The network alignment has been formulated formally [3-10]. While there are some variations among the tools, the key idea is similar: to combine gene sequence information and PPI network information to find conserved interaction regions across species. Each PPI network may be represented as an undirected graph $G = (V, E)$, where V is the set of nodes and E is the set of edges (G may be a weighted graph, i.e., a weight measure $w(e)$ may be associated with each edge e in E). A biological interpretation of network alignment is to find functional homologues across different organisms.

The general problem, known as the problem of finding common subgraphs across multiple networks, is NP-hard; heuristic methods combining sequence, interaction and functional similarities have been developed to tackle it. *A network alignment graph* can be built across two or more species based on protein sequence similarity, interaction conservation, and functional coherence. The nodes in the alignment graph represent sets of proteins, ideally one from each species, and edges for the conserved PPIs across the compared species. The heart of network alignment algorithms is to find conserved interaction regions across homologs among different species.

The history and some future directions for biological network alignment are reviewed in [13]. In general, network alignment may be classified as:

1) pairwise or multiple in terms of the number of networks aligned,
2) local or global in terms of the number of nodes simultaneously aligned,
3) divergence/duplication evolution, neighboring topology in PPI networks, and functional categories in MIPS [14] (GO, KO, etc.) in terms of the type of guided models.

Table 1 summarizes the classification of different network alignment tools. Note that no multiple alignment results are provided in this paper.

Alignment of protein-protein interaction networks went through three major generations. In the first generation, often called pair-wise alignment, conserved pathways/complexes between two species were indentified. PathBlast [6] is one of the pioneering works in this category. In the second generation, called multiple alignment, tools such as NetworkBlast [3], which extends PathBlast, aimed to align multiple networks and introduce a probability model for interactions. MaWish [9] focuses on divergence/duplication model guided by the evolution. Graemlin 1.0 [5] introduces an integration probability model to predict the interactions and, unlike NetworkBlast, have aligned more than three networks. Aforementioned tools are also called local alignment because their search algorithms for the conserved regions start from small local regions and then greedily expand [10]. Later, IsoRank [4] introduced the global alignment concept by adopting Google PageRank algorithm idea to network alignment. Comparing to global alignment, which finds global maximal matches across species, local alignment may just find local maximal results.

The third generation alignment tools, such as Graemlin 2.0 [8] and NetworkBLAST-M [5], are trying to improve both accuracy and speed. Previous tools such as NetworkBLAST-M are also called progressive alignment with possibly exponential representation of every set of potential orthologous proteins [10], which makes them slow and memory inefficient. The computational cost of Graemlin 2.0 is claimed to be linear with the number of PPIs in all the species.

Table 1. A summary of different network alignment tools.

Tools	local	global	pairwise	multiple	guided model
PathBLAST (PB)	×		×		evolution
NetworkBLAST(NB)	×			×	evolution
NetworkBlast-M (NM)	×			×	evolution
MaWish (MW)	×		×		duplication/ divergence
Graemlin 1.0 (Gr1)	×			×	evolution
Graemlin 2.0 (Gr2)		×		×	evolution, duplication.
Isorank (ISO)		×	×		evolution
Isorank-M (ISM)		×		×	evolution

As reported in Graemlin 2.0, which may be the only one to compare results of all tools against KO groups, the specificity of NetworkBLAST, Graemlin 1.0, MaWish, IsoRank and Graemlin 2.0 varies from species to species with the average accuracy of 42%, 53%, 57%, 70%, and 81%, respectively.

2. Our Algorithm: HopeMap

Our algorithm, called HopeMap, can be described as follows (Figure 1 shows the five-step flow of HopeMap):

Step 1: Obtain and preprocess the PPI network data from PPI network databases, such as DIP and SNDB. Find all protein pairs that are interacting with each other in a species.

Step 2: Find highly similar protein sequences across the species. Use homolog clustering to identify homolog groups across different species based on all-versus-all BLAST scores or ortholog annotations. Existing tools, such as KO groups and INPARANOID [15] can be used for this purpose. Besides sequence similarity (homology), all the genes in the same KO group are assumed to be functionally similar. Once homolog groups are identified across the species, a network alignment graph is built based on these groups. The nodes in the graph represent sets of proteins, ideally one from each species, in the same homolog group, and edges represent conserved protein-protein interactions across the compared species. One way of adding the edges between two node pairs (a_1,b_1) and (a_2,b_2) is when both (a_1,a_2) and (b_1,b_2) are directly interacting with each other in their corresponding PPI networks. Other rules for adding edges can be incorporated, such as those introduced in NetworkBLAST.

Step 3: Identify conserved protein interaction regions in the alignment graph. The major algorithm is based on strongly connected-components (clusters) in the alignment graph. A strongly connected component of a graph is a maximal set of vertices in which each vertex is reachable from another. The basic idea behind the connected-components approach is to use depth-first search. Figure 2 depicts a pseudocode for the connected-components algorithm.

Step 4: Score the identified clusters. For each cluster, *C*, our scoring function combines genomic similarity score, interaction conservation, and functional coherence. It is a normalized function with values from the [0, 1] interval so that it is convenient to compare the scores across different clusters. The scoring function of a cluster *C* is defined as:

$$Score(C) = w_1 S(C) + w_2 I(C) + w_3 F(C), \quad (1)$$

where *S(C)* is the sequence similarity score or the average confidence of homolog nodes in the cluster *C*, *I(C)* is the interaction conservation coefficient of
cluster *C*, and *F(C)* is the functional coherence score of cluster *C*; (w_1,w_2,w_3) is

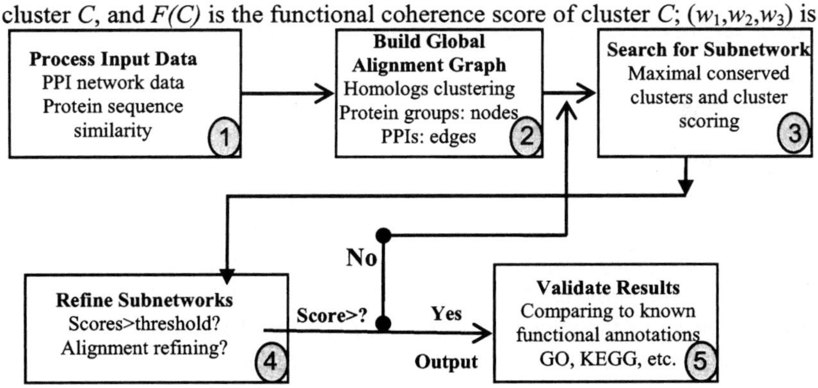

Figure 1. The HopeMap network alignment.

the corresponding weight coefficient and one-third for each is set as the default.

<u>*Similarity scoring S(C)—node scoring:*</u> We find highly similar proteins across species by identifying close homologues first. For each node (one protein from each species) of cluster *C* in the alignment graph, we find its

confidence score based on BLAST score or ortholog confidence score. Equation (2) is used for the similarity score of a cluster *C*:

```
DFS(G): V[G] is the set of all vertices in G
(1)      For each vertex u in V[G]
(2)      Do color [u]:=white  //unvisited
(3)         Pi[u]:=null; Time:=0
(4)   For each vertex u in V[G]
(5)      Do if color[u]==white
(6)       then DFS-Visit(u)
DFS-Visit(u)
(7)    color[u] :=gray;  time:= time+1;  d[u]:=time
(8)    for each v in Adj[u]  // explore u's neighbors
(9)       Do if color[v]==white
(10)         then {Pi[v]:=u;   DFS-Visit(v)}
(11)   color[u]:=black;  f[u]:=time<-time+1
Strongly-connected-components (G)
(12)   call DFS(G) to compute f[u] for each vertex u;  compute G^T, transpose of G
(13)   call DFS(G^T), consider the vertices in decreasing order of f[u] in the main loop of
       DFS
(14)   output the vertices of each tree in DFS forest formed
```

Figure 2. Pseudocode for finding strongly-connected components

$$S(C) = (\sum_{k=1}^{k=|C|} s(k)) / |C|, \qquad (2)$$

where *s(k)* is the similarity (or ortholog) confidence score of node *k* in cluster *C*. For simplicity, the confidence score of a node in the alignment graph is set to 1, if the proteins in the node have the same known ortholog annotation, such as KO group.

Interaction conservation scoring I(C)—node and edge scoring: The conserved interactions are the number of edges connecting all nodes in the identified cluster of the alignment graph. *I(C)* is set to the portion of the direct interactions conserved in the clusters. *I(C)* is formally defined as:

$$I(C) = \frac{i(C)}{|C|(|C|-1)/2}, \qquad (3)$$

where *i(C)* is the total number of conserved interactions (defined above), i.e. the total number of edges in cluster *C*, and $|C|(|C|-1)/2$ is a cliquishness measure of the cluster *C*.

Functional coherence scoring F(C)—node and edge scoring: Functional annotations, such as GO terms, complexes in MIPS [14], or pathways in KEGG can be used as functional coherence score of a cluster. We use the intersection over the union of the number of GO biological process terms covered in a cluster of a local species as *F(C):*

$$F(C) = \frac{Intersection(\text{GO process terms in } C)}{Union\ (\text{GO process terms in } C)} \tag{4}$$

The larger the *F(C)* is, the higher the significance of the cluster *C* in terms of the coherence of GO biological process terms.

Statistical significance assessment: To measure the statistical significance of the scoring functions, for each cluster (or ortholog pair), we randomly sample *N* clusters of the same size and compute the corresponding scores. Then, we find an empirical *P*-value of each cluster using the methods introduced in [16, 17]. Typically, the empirical *P*-value can be estimated as $P=(R+1)/(N+1)$, where *N* is the total number of random samples and *R* is the number of those samples that produce a test statistic greater than or equal to the value for the actual data. Finally, the scoring function can be formulated as follows:

$$Score(C) = w_1 \frac{R_{S(C)}+1}{N_{S(C)}+1} + w_2 \frac{R_{I(C)}+1}{N_{I(C)}+1} + w_3 \frac{R_{F(C)}+1}{N_{F(C)}+1}, \tag{5}$$

where $N_{I(C)}$ is the number of random sample clusters, $R_{I(C)}$ is the number of clusters that have the score values greater than or equal to *I(C)*, ($N_{F(C)}$, $R_{F(C)}$) are defined similarly for *F(C)*. For simplicity, one-third is set as default for the weight coefficient of each of the three functions. For *S(C),* which is based on a homologue node, we find the number of nodes that belongs to the same homologue groups (e.g., KO groups) as the $R_{S(C)}$ and the total possible nodes across all species as $N_{S(C)}$. Notice that the three parts in the scoring function (5) can be used jointly or independently.

Step 5: Validate the results. Since our homolog groups are based on known KO annotation groups, we currently evaluate functional coherence of the identified local clusters in each species using gene ontology (GO). This is part of local alignment and refinement. To this end, GO termFinder [18] is used, which computes the empirical enrichment *P*-value and corrected values for multiple testing using the false discovery rate procedure. Similar to NetworkBLAST-M's way of measuring the results' specificity in terms of GO biological process terms enrichment, the percent of process-enriched clusters in each species is computed. The number of clusters with *P*-value below 0.05 is used as the sensitivity metric. Other known functional annotations, such as pathways or complexes enrichment in KEGG or MIPS, respectively, may also be used for validation.

After the fourth step, local alignment in each species can be iteratively applied to improve the cluster score, if necessary. If an identified cluster has the score less than a threshold (e.g., 0.5), then HopeMap uses GO biological process terms enriched in each cluster as the indicator to keep or remove the cluster. To be more specific, we use the intersection over union (I/U) of GO biological process terms enriched in each cluster as an indicator to keep or remove the cluster. If the value of I/U is zero, we remove the cluster, otherwise, we keep it. Only two iterations were needed for convergence in the Analysis Results section.

The major step (Step 3) in HopeMap is based on identification of strongly connected-components in a graph. Its computational cost is linear with the number of nodes and edges in the alignment graph. All the other steps have linear cost with the input size. To the best of our knowledge, this is the fastest algorithm to identify maximal conserved patterns in the alignment graph.

3. Analysis Results

3.1. Input Data

Table 2 provides a summary of all the PPI networks analyzed in this paper. For GO terms, NetworkBLAST has higher specificity compared to other tools, as reported in NetworkBLAST-M. Therefore, we compare our results mainly against NetworkBLAST results for GO terms enrichment. For KO groups' enrichment, Graemlin 2.0 is reported to have the highest specificity and sensitivity, so we compare our results with Graemlin 2.0.

Table 2. The PPI networks analyzed in this paper.

Species (tax id, short name)	# proteins	#PPIs	Source
E.coli K12 (83333, eco)	4121	216426	SNDB[19]
S.typhimurium (99287,stm)	4239	94609	SNDB
C.crescentus (190650,ccr)	3365	40524	SNDB
S.cerevisiae (4932, sce)	4738	15417	DIP [18]
D.melanogaster (7227, dme)	7165	23484	DIP [18]

3.2. Alignment Results

The comparison of KO groups' enrichment against other tools is provided in Table 3. We provide specificity and sensitivity comparison in terms of KO groups for different global alignment tools (results for other alignment tools are also available in Graemlin 2.0). Specificity and sensitivity in terms of KO groups are introduced in Graemlin 2.0. In short, an equivalent class is defined as correct if all protein members in the class are in the same KO group. The fraction of equivalence classes that are correct is denoted as C_{eq}, while the fraction of nodes that are in correct equivalence classes is C_{node}. C_{or} stands for the total number of nodes in correct equivalence classes and T_{ot} is the total number of equivalence classes with k species. (C_{eq}, C_{node}) and (C_{or}, T_{ot}) are used for measuring specificity and sensitivity, respectively. Other results but HM are from the original publication of Gr2, where Gr2 stands for Graemlin 2.0, ISO for Isorank, and HM for HopeMap. Since HopeMap uses KO groups for homologue clustering, it has higher specificity and sensitivity than the other tools in Table 3.

In Table 4, the results for NetworkBLAST (NB), 117 clusters with the node size larger than one, were obtained from the supplementary material of [2] using NB and INPARANOID ortholog groups. The results of HM1 were obtained using HopeMap on the same data from NetworkBLAST after adding edges to the alignment graph in a way similar to NetworkBlast. The results of HM-ko were obtained using KO groups as ortholog groups and no additional interactions being added to original PPI networks from DIP; the edges were added between node pairs (sce_1, dme_1) and (sce_2, dme_2) only when there were edges between both sce pairs (sce_1, sce_2) and dme pairs (dme_1, fly_2). NetworkBLAST used different techniques to add edges in the alignment graph; therefore, its total number of conserved regions is larger than for HopeMap. Table 4 shows that the specificity and sensitivity for GO terms enrichment by HopeMap are comparable to NetworkBLAST's, while HopeMap is simpler to use and is faster. Using KO groups in HopeMap improves the specificity. The range for cluster size in HM-ko is from 2 to 6, in NB is from 2 to 9, in NB-ko is from 4 to 6, and in HM1 is 2, 5, 7, and 50. HopeMap discovers more unique genes than NetworkBLAST, while keeping comparable specificities for GO terms' and KO groups' enrichment.

Table 3. Specificity and sensitivity comparison in terms of KO groups.

	eco/stm				eco/ccr				sce/dme			
Tools	C_{eq}	C_{node}	C_{or}	T_{ot}	C_{eq}	C_{node}	C_{or}	T_{ot}	C_{eq}	C_{node}	C_{or}	T_{ot}
NB	0.8	0.45	457	1016	0.78	0.5	346	697	0.39	0.14	43	306
ISO	0.9	0.91	2026		0.65	0.65	1014		0.63	0.63	534	
Gr2	0.96	0.96	2024		0.78	0.78	1012		0.73	0.73	637	
HM	1.00	1.00	2159	3151	1.00	1.00	1061	1365	1.00	1.00	768	1664

Table 4. Comparison of NetworkBLAST and HopeMap for Yeast/Fly.

Methods	Spec. (%) GO	Spec. (%) KO	#GO enriched	#conserved regions	#unique genes
sce (NB)	94.87	N/A	67	117	348
dme (NB)	84.62	N/A	62	117	256
sce (HM1)	98.73	N/A	65	79	1645
dme (HM1)	78.48	N/A	46	79	1913
sce (NB-ko)	100	100	9	9	34
dme (NB-ko)	100	100	8	9	27
sce (HM-ko)	100	100	24	26	747
dme(HM-ko)	92.31	100	24	26	753

Table 5. Specificity and sensitivity results for eco/stm and eco/ccr.

Species	Spec. (%) GO	Spec. (%) KO	#GO enrich.	#conserved regions	# unique genes
eco (eco/stm) HM	100	100	49	58	2085
stm (eco/stm) HM	96.55	100	46	58	2183
eco (eco/ccr) HM	95.24	100	37	42	1069
ccr (eco/stm) HM	90.48	100	31	42	1138
eco (eco/stm) NB	100	100	267	309	657
stm (eco/stm) NB	100	100	252	309	654
eco (eco/ccr) NB	100	100	34	39	108
ccr (eco/ccr) NB	100	100	38	39	100

In Table 5, we use the PPI network data from SNDB for two organismal pairs, eco/stm and eco/ccr. The interaction probability above 0.5 was set as the cut-off. The total number of conserved regions only included those sizes that are larger than two for eco/stm, while regions with the size larger than one were included for eco/ccr. The values for cluster size in eco/stm for HM are 3, 12, 14, 20, 22, 35, 97, 177, and 280, while the cluster sizes in eco/ccr for HM are 2, 3, 4, 5, 6, and 8; NB-ko has cluster size range between 4 and 15. HopeMap discovers more unique genes than NetworkBLAST does, while keeping comparable specificities for GO terms' and KO groups' enrichment.

4. Conclusion

Based on genome similarity across different species, interactome conservations and functional coherence, we developed a pairwise network alignment tool, called HopeMap, to improve the speed, accuracy and generality of the alignment.

HopeMap is fast; it is linear in terms of the number of nodes and edges in the alignment graph. Our results show that HopeMap has specificity and sensitivity as comparable with the existing best-performing tools. Specifically, in terms of GO terms' enrichment, HopeMap performs comparably with NetworkBLAST, and HopeMap has higher specificity and sensitivity in terms of KO groups' enrichment than the other tools. Our scoring system is generic, and the main algorithm is parameter-free. HopeMap is also extensible to multiple network alignment.

Acknowledgments

The authors would like to thank NetworkBLAST and NetworkBLAST-M team, Prof. Roded Sharan and Maxim Kalaev et al. as well as the Graemlin team (Jason Flannick et al.) from Stanford University for sharing the datasets and helpful communication.

5. References

1. DIP, http://dip.doe-mbi.ucla.edu/.
2. SNDB, *Stanford Network Database (SNDB),* ftp://ftp-networks.stanford.edu/pub/graemlin_nets.
3. Sharan, R., S. Suthram, R.M. Kelley, T. Kuhn, S. McCuine, P. Uetz, T. Sittler, R.M. Karp, and T. Ideker, *Conserved patterns of protein interaction in multiple species.* Proc Natl Acad Sci U S A, 2005. **102**(6): p. 1974-9.
4. Singh, R., J. Xu, and B. Berger, *Global alignment of multiple protein interaction networks.* Pac Symp Biocomput, 2008: p. 303-14.
5. Flannick, J., A. Novak, B.S. Srinivasan, H.H. McAdams, and S. Batzoglou, *Graemlin: general and robust alignment of multiple large interaction networks.* Genome Res, 2006. **16**(9): p. 1169-81.
6. Kelley, B.P., R. Sharan, R.M. Karp, T. Sittler, D.E. Root, B.R. Stockwell, and T. Ideker, *Conserved pathways within bacteria and yeast as revealed by global protein network alignment.* Proc Natl Acad Sci U S A, 2003. **100**(20): p. 11394-9.
7. Matthews, L.R., P. Vaglio, J. Reboul, H. Ge, B.P. Davis, J. Garrels, S. Vincent, and M. Vidal, *Identification of potential interaction networks using sequence-based searches for conserved protein-protein interactions or "interologs".* Genome Res, 2001. **11**(12): p. 2120-6.
8. Flannick, J., A. Novak, C.B. Dol, B.S. Srinivasan, and S. Batzoglou. *Automatic Parameter Learning for Multiple Network Alignment.* in *RECOMB*. 2008.
9. Koyuturk, M., Y. Kim, U. Topkara, S. Subramaniam, W. Szpankowski, and A. Grama, *Pairwise alignment of protein interaction networks.* J Comput Biol, 2006. **13**(2): p. 182-99.

10. Kalaev, M., V. Bafna, and R. Sharan. *Fast and Accurate Alignment of Multiple Protein Networks*. in *RECOMB*. 2008.
11. Ashburner, M., C.A. Ball, J.A. Blake, D. Botstein, H. Butler, J.M. Cherry, A.P. Davis, K. Dolinski, S.S. Dwight, J.T. Eppig, M.A. Harris, D.P. Hill, L. Issel-Tarver, A. Kasarskis, S. Lewis, J.C. Matese, J.E. Richardson, M. Ringwald, G.M. Rubin, and G. Sherlock, *Gene ontology: tool for the unification of biology. The Gene Ontology Consortium.* Nat Genet, 2000. **25**(1): p. 25-9.
12. Kanehisa, M. and S. Goto, *KEGG: kyoto encyclopedia of genes and genomes.* Nucleic Acids Res, 2000. **28**(1): p. 27-30.
13. Sharan, R. and T. Ideker, *Modeling cellular machinery through biological network comparison.* Nat Biotechnol, 2006. **24**(4): p. 427-33.
14. Mewes, H.W., C. Amid, R. Arnold, D. Frishman, U. Guldener, G. Mannhaupt, M. Munsterkotter, P. Pagel, N. Strack, V. Stumpflen, J. Warfsmann, and A. Ruepp, *MIPS: analysis and annotation of proteins from whole genomes.* Nucleic Acids Res, 2004. **32**(Database issue): p. D41-4.
15. Remm, M., C.E. Storm, and E.L. Sonnhammer, *Automatic clustering of orthologs and in-paralogs from pairwise species comparisons.* J Mol Biol, 2001. **314**(5): p. 1041-52.
16. North, B.V., D. Curtis, and P.C. Sham, *A Note on the Calculation of Empirical P values from Monte Carlo Procedures.* Am. J. Hum. Genet., 2002. **71**: p. 439-441.
17. Zhang, B., B.H. Park, T. Karpinets, and N.F. Samatova, *From pull-down data to protein interaction networks and complexes with biological relevance.* Bioinformatics, 2008. **24**(7): p. 979-86.
18. GOtermFinder, http://go.princeton.edu/cgi-bin/GOTermFinder.

COMPUTATIONAL CHALLENGES IN COMPARATIVE GENOMICS: GENE FAMILY EVOLUTION, ANCESTRAL GENOME RECONSTRUCTION, AND APPLICATIONS TO HUMAN GENETICS

BERNARD M. E. MORET*, WEBB MILLER, PAVEL A. PEVZNER, AND DAVID SANKOFF

*Laboratory for Computational Biology and Bioinformatics
EPFL (Swiss Federal Institute of Technology)
EPFL-IC-LCBB INJ 230, Station 14
CH-1015 Lausanne, Switzerland
E-mail: bernard.moret@epfl.ch

Comparative methods have long been a mainstay of biology, particularly evolutionary biology; they are also at the core of medical research based on animal models of human physiology. They find their most challenging and most fitting application, however, in the study of whole genomes, as they are the main tools through which we can make sense of the billions of base-pairs forming the sequence of animal and other genomes. Comparing whole genomes, which is necessarily done through computational methods due to the size of the genomes, has given rise to the research area known as comparative genomics.

Comparative genomics is the tool of choice for identifying genes in both well studied and newly sequenced genomes; for studying the acquisition of virulence or drug resistance in pathogens; for tracking down gene complexes responsible for inheritable diseases or susceptibilities; and for engineering desirable new traits in crops; and for studying many forms of cancers. More generally, comparative genomics is the tool of choice to elucidate how the genetic blueprint translates into specific functions and how that blueprint evolves in populations and into various species.

Comparative genomics uses not just whole-genome sequences, but also dense single-nucleotide polymorphism (SNP) maps, genetic maps, and sequences of individual genes, but it is characterized by its emphasis on a whole-genome approach. Its computational methods include combinatorial optimization, machine learning, and data mining, while much work has also been devoted to visualization of its findings—witness, for example, the many spectacular full-color figures illustrating the correspondences between the human and mouse genomes.

The focus of our session is on computational models and algorithms. The five papers included in our session (selected from a total of 17 submissions) all exemplify the genome-wide approach of the area. Most use optimization approaches to infer evolutionary events or, equivalently, to annotate regions of the genome according to their evolutionary history.

In their paper on "Identifying parent-daughter relationships among duplicated genes," Han and Hahn tackle the evolution of gene families. When two organisms share a gene family, sorting out orthologs and various paralogs is a prerequisite for a comparative analysis. Many orthologous relationships may exist, yet perhaps the most crucial is that between the "parents" of the families. The authors present a model in which the length of shared syntenic regions is a hidden variable in an HMM model and use an EM algorithm to estimate the parameters and identify parent-daughter relationships among gene family members. They then apply their method to a collection of gene families from six mammalian genomes and use the inferred parental relationships to determine the direction of gene duplication events.

In "A parsimony approach to analysis of human segmental duplications," Kahn and Raphael tackle the two-step model of segmental duplication. These duplications are widespread in the human genome and quite complex, with many appearing to consist of a mosaic of nested duplications. The two-step model allows for such nesting, albeit to just one level. The authors give an integer linear programming formulation of the problem of explaining the data at hand with a minimum number of duplications across the two steps and apply it to the human genome. The nested formulation provides a natural tree relationship and thus a first cut at reconstructing the history of the duplication events.

In "Simultaneous history reconstruction for complex gene clusters in multiple species," Zhang, Song, Hsu, and Miller combine the two previous topics and examine the gene clusters that arise from segmental duplications. These genes form multiple families, yet their history is much more difficult to tease out than those of genes duplicated in simple tandem duplications or through retrotransposition; moreover, the presence of these large and complex duplications creates severe difficulties for sequencing. Using comparative genomics provides some leverage for the problem, leverage that the authors demonstrate how to use.

In "Inferencing genome-wide mosaic structure," Zhang, Wang, McMillan, Villena, and Threadgill take up the crucial problem of inferring the recombination history of a population of genomes. Repeated meiotic recombination gives rise to a mosaic structure in the genome, with each haplotype block (mosaic piece) traceable to a distinct ancestor from the neighboring blocks. The authors propose a graph model for the formation of these mosaics and provide a dynamic programming algorithm to infer a mosaic with the smallest number of pieces given a population of genomes, showing the results of its application to genome-wide SNP data on mice.

In "An exact solver for the DCJ median problem," Zhang, Arndt, and Tang address the issue of genomic rearrangements from the point of view of phylogenetic reconstruction. While the median of three genomes is an abstract concept, it has become a mainstay of ancestral genome reconstruction; unfortunately, under almost any rearrangement model, computing a median is NP-hard. In this paper, the authors show how to speed up such computations so as to enable application of the technique to nontrivial genomes.

We are very pleased to feature such work at this PSB'09 session and want to take this opportunity to thank attendees, presenters, all submitting authors, and the referees, who together made it possible.

IDENTIFYING PARENT-DAUGHTER RELATIONSHIPS AMONG DUPLICATED GENES[1]

MIRA V. HAN AND MATTHEW W. HAHN

Department of Biology and School of Informatics, Indiana University, Bloomington, IN 47405 USA

In this paper we use the length of the shared synteny between genes to identify "parent" orthologs among multiple lineage specific duplicated genes. Genes in the region around each duplicated paralog are compared with the genes flanking an outgroup ortholog to estimate the probability of observing homologs in syntenic vs. non-syntenic regions. The length of the shared synteny is introduced as a hidden variable and is estimated using Expectation-Maximization for each lineage specific paralog. Assuming that the original, parental gene will preserve the longest synteny with the outgroup gene, and that any daughter genes will have a shorter syntenic block, we are able to determine parent-daughter relationships. We apply this method to lineage specific duplications in the human genome, and show that we are able to determine the direction and size of the duplication events that have created hundreds of genes.

1. Introduction

Gene families are groups of genes with high sequence similarity that are derived from a common ancestor. The relationships between members of a gene family can be classified into orthology and paralogy based on their evolutionary history. Orthologs are pairs of genes that are diverged from a common ancestor by speciation, while paralogs are pairs related through duplication events [1]. Paralogs can further be divided into out-paralogs and in-paralogs [2], generally defined as relationships between paralogs either between or within species, respectively (Fig. 1a). In-paralogs are considered to be co-orthologous to a single-copy gene in an outgroup, as they are all related to this gene by a speciation event (Fig. 1a).

We can define yet another relationship based on the direction of the duplication event. We call the original copy the parent (or "primary" ortholog) and any derived copy the daughter (or "secondary" ortholog). Ortholog-paralog relationships are different from parent-daughter relationships, and identifying orthologs does not necessarily determine the direction of the duplication event. The parent-daughter relationship can be determined unambiguously only when

[1] This work is supported by grant DBI-0543586 from the U.S. National Science Foundation.

you have a gene from an outgroup species that can be used to identify the ancestral position of the gene, assuming that the outgroup gene has maintained synteny (Fig. 1b). Identifying the parent-daughter relationships among lineage specific duplicates (i.e. in-paralogs) is the problem addressed in this paper. Lineage specific duplications are interesting because they are a potential source of lineage specific phenotypes [3]. Finding the parent-daughter relationships among lineage specific duplicates can provide information on the evolutionary forces governing the origin and maintenance of gene duplicates (e.g. [4])

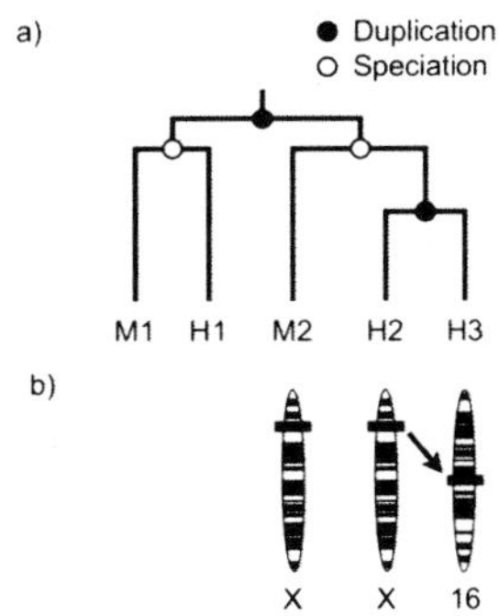

Figure 1. a) Examples of multiple types of homologous relationships. Genes M1 and H1 are orthologs, while H2 and H3 are in-paralogs of each other and are both co-orthologs of gene M2. Genes H1 and M2 are out-paralogs. b) Determining parent-daughter relationships. Genes M2 and H2 both reside on the X chromosome of their respective species, while gene H3 is found on chromosome 16. Assuming that gene H2 is in the ancestral location, H2 is the parental paralog ("primary" ortholog to M2) and H3 is the daughter paralog ("secondary" ortholog to M2).

Many computational methods have been introduced for distinguishing orthologs from out-paralogs based on synteny [5, 6], but none have addressed the problem of distinguishing ancestral and derived loci until the recent study by Jiang et al. [7]. The conceptual approach used in the current paper is similar to that introduced in this previous study [7], that is we assume that the parental duplicate shares greater synteny with the outgroup gene. The primary concern is thus how to identify the blocks of shared synteny, which is also a well studied problem [8, 9]. Many methods have also been developed to accommodate rearrangements within blocks, the general idea being to cluster or chain the pair-wise alignments by allowing a certain level of gaps between the alignments [10, 11].

In this paper we use a novel probabilistic approach to defining synteny. Instead of chaining nucleotide alignments by defining appropriate gap penalties, we allow randomness in the sharing of gene content with a probability that is estimated from the data. This approach therefore takes into account duplications, deletions, and micro-rearrangements within the syntenic block, and obviates the need to *a priori* determine gap penalties. We define pairs of genes as the experimental unit, and present a simple model based on the presence/absence of common genes in the flanking regions to estimate the lengths of duplicated blocks. The probabilities of observing genes in common in homologous vs. non-homologous regions are estimated as parameters using maximum likelihood, and the lengths of the homologous regions are introduced as hidden variables. By comparing the lengths of shared synteny, this method allows us to identify parent-daughter relationships among paralogs.

2. Methods

2.1. *Model*

We consider M families with lineage specific expansions in a species S (in this case, human). Each family m has g_m number of genes from species S and o_m number of genes from the outgroup species S_{out} (in this case, macaque). The g_m genes are restricted to duplicates whose distances between any pair are less than twice the time since speciation with the outgroup. This restriction guarantees that the duplicated genes truly are lineage specific. The outgroup genes are found by traversing the gene family tree to the closest speciation node above the lineage specific duplication and collecting all genes from the outgroup species under that node. Sometimes there is more than one outgroup gene due to lineage specific expansion within the outgroup lineage. We can therefore make $g_m \cdot o_m$ pairs between species S and S_{out} for each family. Each pair is the unit of experiment. The final dataset has a total of N experimental units consisting of pairs of genes from human and macaque where,

$$N = \sum_{m}^{M} g_m \cdot o_m$$

Note that in the end we will assign each gene to only one category (parent or daughter) by a simple rule that if any pair containing g_i is designated as a parent, then the gene g_i is a parent. For each gene we collected the gene order data that spans $\pm L$ megabases (MB) from each gene. Gene order is a series of numbers encoding the genes flanking our gene of interest. Each gene order has a different number of genes depending on the gene density of the region, and the total length of the contig available in the region. We assume that each flanking gene has been assigned to a gene family through some alignment and clustering procedure. The gene order is then encoded by family IDs, so that the same numbers correspond to homologous genes in the sequence (an example is shown in Fig. 2). This numbering scheme precludes the need for one-to-one ortholog assignment of flanking genes, and allows for duplication and rearrangement of homologous flanking genes. Each pair of focal genes has a corresponding pair of gene order data, and we will use the comparison of gene orders to extract features of synteny.

2.2. *Formulation*

To find the in-paralog that shares greater synteny with the outgroup gene, we must first mathematically define shared synteny. In this paper we define shared synteny as the higher probability of co-occurrence of homologous genes in a continuous genomic region of each species. There are two variables in this definition, the "probability of co-occurrence" and the "length of the region harboring the higher probability." In our simple model, we assume that the probability of co-occurrence of a homologous gene between two genomes is different between regions of shared synteny (p_{syn}) and regions outside shared syteny (p_{nonsyn}), but is equal within each class for all pairs. By definition p_{syn} is larger than p_{nonsyn}. The actual value will be estimated as the maximum-likelihood estimate using all pairs of gene order comparisons as data. Note that co-occurrence means homologs occur in both species within the flanking region, but does not consider the order of the genes. The length of the syntenic region has larger variance depending on the type and size of duplications and rearrangements that happen in a genome, and can be quite different from event to event. It is therefore inappropriate to model the length as one random variable for the whole dataset. We will introduce two variables l_len_i and r_len_i for each pair i ($i = 1..N$). l_len_i denotes the length of the syntenic region to the left of gene g_i, r_len_i denotes the length to the right.

If we assume that the co-occurrence of a homologous gene in the region follows a Bernoulli distribution with p as the probability of co-occurrence, then the probability of observing a certain gene order of species S for pair x_i can be expressed as,

$$P(x_i \mid \theta) = \prod_{p\_left<j<p\_right} p_{syn}{}^{m_j}(1-p_{syn})^{1-m_j} \cdot \prod_{j<p\_left \vee j>p\_right} p_{nonsyn}{}^{m_j}(1-p_{nonsyn})^{1-m_j},\ j = 1..G_i$$

$\theta:\ p_{syn}, p_{nonsyn}, l\_len_i, r\_len_i$

p_{syn} : probability of co-occurrence of homologs in a syntenic region

p_{nonsyn} : probability of co-occurrence of homologs in a non-syntenic region

l_len_i : length of the syntenic region to the left of the focal gene g_i in S

r_len_i : length of the syntenic region to the right of the focal gene g_i in S

$$m_j : \begin{cases} 1 & \text{if gene } j \text{ in species } S \text{ occurs in gene order of species } S_{out} \\ 0 & \text{otherwise} \end{cases}$$

G_i : length of the total gene order of S

$pos(x_i)$: index of gene g_i

$p\_left = pos(x_i) - l\_len_i$: index of the left border of the syntenic region

$p\_right = pos(x_i) + r\_len_i$: index of the right border of the syntenic region

Note that there is no explicit variable denoting the length of the synteny for the outgroup genome. This is to limit the dimension of the search space. When looking for matches we will consider the whole gene order of S_{out}. But since a match is a pairwise observation, the length variables for species S implicitly determines the length of the synteny in the outgroup S_{out}.

The log-likelihood of the whole data X can be expressed as the sum of the likelihood of all x_i's.

$$\ln L = \sum_{i=1}^{N}\left(s_i \ln p_{syn} + (l_i - s_i)\ln(1-p_{syn}) + r_i \ln p_{nonsyn} + (G_i - l_i - r_i)\ln(1-p_{nonsyn})\right)$$

$l_i = l\_len_i + r\_len_i + 1$: total length of the syntenic region of gene order x_i

s_i : total number of hits in the syntenic region of gene order x_i

r_i : total number of hits in the non-syntenic region of gene order x_i

The maximum likelihood estimate of p_{syn} and p_{nonsyn} are,

$$\hat{p}_{syn} = \underset{p_{syn}}{\operatorname{argmax}}\left(\ln p_{syn}\sum_{i=1}^{N} s_i + \ln(1-p_{syn})\sum_{i=1}^{N}(l_i - s_i)\right)$$

$$\hat{p}_{nonsyn} = \underset{p_{nonsyn}}{\operatorname{argmax}}\left(\ln p_{nonsyn}\sum_{i=1}^{N} r_i + \ln(1-p_{nonsyn})\sum_{i=1}^{N}(G_i - l_i - r_i)\right)$$

Implicit in the MLE estimate are the hidden variables l_len_i and r_len_i for each x_i. We use Expectation-Maximization to calculate the likelihood according to the inferred expectation of each hidden variable. The software that implements the model, PRIUS (Parental Relationship Inference Using Synteny) can be found at http://www.bio.indiana.edu/~hahnlab/Software.html.

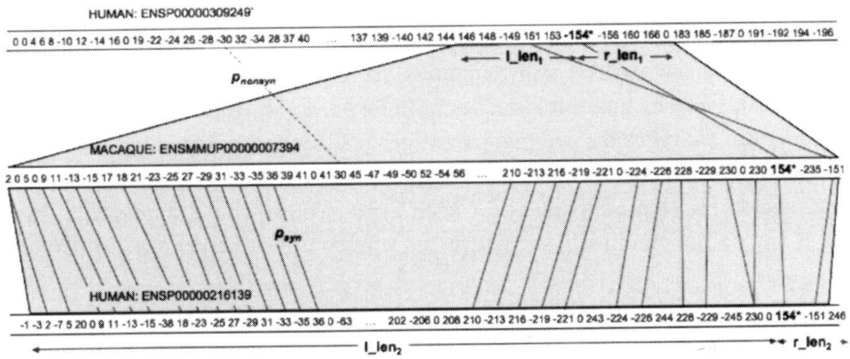

Figure 2. Description of the model. Two lineage specific human paralogs, denoted in bold, are paired with their macaque ortholog (center). The gene order of each region is encoded by the family IDs for each flanking gene, with the same family ID indicating a match; positive or negative values indicate the strand for each gene, though the strand is irrelevant to the match or mismatch. The length of the syntenic region estimated for each paralog is shown (shadowed), as are the variables representing the probabilities of finding homologous genes in syntenic and non-syntenic regions.

2.3. *Clustering and Assignment*

Once we have the length of the shared synteny for each pair of genes, we can cluster the genes of species S into parent and daughter based on the degree of shared synteny with the outgroup genome. Both the synteny length and the ratio of the synteny length to the total length are considered in distinguishing the two populations. Because the distribution of the log-transformed length of synteny shows better separation than the absolute length, the clustering used two variables: the log-transformed synteny length (V_{loglen}=log l_i), and the ratio of the shared synteny length to the total gene-order length ($V_{ratio}=l_i/G_i$). Total gene-order length varies because there are a variable number of genes contained within the regions being compared for any pair of genes. If the total gene-order length is less than 30 genes, we consider the case uninformative and exclude it. We assume that parents have longer shared synteny with the outgroup than daughters, and should therefore have both higher absolute synteny lengths and higher ratios of synteny length to gene-order length.

We cluster the genes into two groups, parents or daughters, using a simple 2-means clustering on the two variables. Since the two variables use different scales, we first scale and center them. To do this we carry out a principal component analysis, and run the clustering on the two principal components. For all data points we calculate the difference between the distances to the two cluster means, such that points close to one center has high differences. If the difference of distances is smaller than 2 S.D. from the mean, the data point is labeled ambiguous. Since tandem duplications all share the same synteny, we cannot distinguish parents and daughters among tandem duplicates with this method. This means that there can be multiple genes assigned as 'parent' due to tandem duplication of the original parent gene. Likewise, tandem duplicates of a daughter gene are also assigned as 'daughter'. The clustering of each (o_i, g_i) pair is translated into the assignment for each gene g_i following a simple rule, such that if any pair containing g_i is clustered as a parent, then the gene g_i is a parent.

3. Results

We used lineage specific duplications from the human genome for our analysis, with macaque as the outgroup species. The dataset of gene families constructed from six mammalian genomes, each with a phylogenetic tree, is described in Hahn et al. [3]. To find human specific duplicates, we collected the genes under human specific duplication nodes from the reconciled tree. The outgroup genes were collected by selecting all macaque genes that are sibling to the duplicated node. We paired each human gene and each macaque gene within a family to assemble the dataset. For each gene, we downloaded the ±10 MB flanking region from Ensembl and extracted the gene order by encoding the genes according to the gene family IDs defined in Ensembl. We chose the length of the flanking region (±10 MB) to be long enough to contain duplications of all sizes and to have an adequate number of genes outside the duplicated segment so that we can confidently assign duplication breakpoints. The sizes of duplications in primates are mostly less than 300 KB and up to about 1 MB at most [12].

In total, we found 337 families with human specific duplications, containing 871 human genes and 385 macaque genes, and we constructed 1075 pairs with these genes. We ran the EM algorithm twice, once with initial values of p_{syn}=0.99 and p_{nonsyn}=0.01 and once with p_{syn}=0.51 and p_{nonsyn}=0.49. Both runs converged within 10 iterations. The estimated probability of finding a match in a syntenic region was p_{syn} = 0.829, and the probability of finding a match in a non-syntenic region was p_{nonsyn}= 0.050 (note that the two values do not have to add up to one). The average for the estimated lengths of shared synteny was 141 genes (SD=154). The distribution of the lengths showed two separate peaks, indicating that the population is a mixture of two distributions. (Fig. 3a)

We clustered the pairs using the variables V_{loglen} and V_{ratio}, as described above. To see how these two variables contribute to variance in the data, we conducted a principal components analysis. Almost 97% of the variance was explained by the equal contribution of both variables after they were scaled and centered. We carried out a 2-means clustering with the two principal components and obtained one cluster with low syntenic lengths and low synteny ratios (syntenic length/total length) and one with high syntenic lengths and high ratios. We were able to confidently assign 425 pairs as daughters and 603 pairs as parents. We classified 25 and 22 pairs that clustered with the daughters and parents, respectively, as ambiguous. (Fig. 3b) Using the clustered human-macaque pairs, we assigned each human gene as either a parent or daughter. This resulted in 826 unambiguous calls for the human proteins, 493 of them parents and 333 of them daughters.

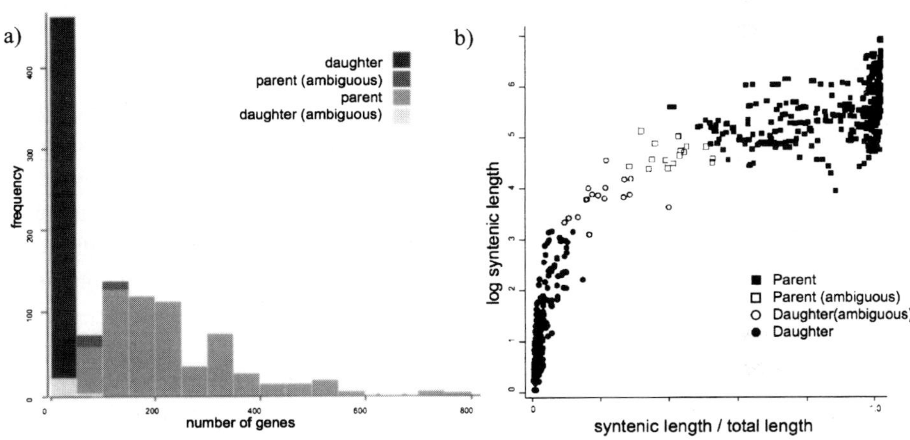

Figure 3. a) Distribution of estimated lengths of shared synteny for all pairs, measured in the number of genes in syntenic regions. b) Groups obtained by 2-means clustering on the principle components of V_{loglen} and V_{ratio} . Parent clusters have longer synteny and a higher ratio of synteny to total length. Daughter clusters have short synteny and small ratio of synteny to total length. 47 pairs have ambiguous assignment to either the parent or daughter cluster.

Based on the type and number of assigned genes, we classified the families into four classes (Table 1): 1) Families can have daughters only, likely because the original parent copy has been lost. 2) Families can have parents only, where no clear daughter can be identified because duplicate genes have been created by tandem duplication. 3) Families can have a single parent and a single daughter gene. 4) Families may contain multiple copies of both parent and daughter genes due to tandem duplication of either paralog.

Table 1. Counts of human genes in different classes based on parent-daughter assignments and the type and number of genes in a family. Ambiguous assignments are excluded from results. Counts of families are in parentheses.

Family Class	Daughter	Parent	Total
Daughter Only	201 (79)		201 (79)
Parent Only		357 (142)	357 (142)
Parent-Daughter (1:1)	85 (85)	85 (85)	170 (85)
Parent-Daughter (>1)	47 (24)	51 (24)	98 (24)
Total	333 (188)	493 (251)	826 (330)

We verified parent-daughter assignments by checking the chromosomal location of the genes. As expected, we found that all the human duplicates in the parent-only class were on the same chromosome, with two exceptions. The first case has a gene on each of chromosomes 14 and 15. These chromosomes are both homologous to the macaque chromosome 7, due to a chromosomal fission event along the human lineage [13]. The second case has one gene on chromosome 17 and four genes on chromosome X. It appears that the human gene located on chromosome 17 was misassembled with the X in our dataset, and has since been updated in the newest human genome assembly. In the parent-daughter 1:1 class, all but six pairs were on different chromosomes, as expected. The six pairs of genes that were on the same chromosome were all at least 19 MB and up to 91 MB apart, so that there were striking differences in the degree of shared synteny. The other classes do not have obvious predictions to measure against.

4. Discussion

In order to uncover the directionality of gene duplication events, we have introduced a probabilistic method for estimating shared syntenic blocks between genes. Using this method we are able to classify lineage specific paralogs as either parent or daughter copies, and thus to infer the direction of the duplication event that created the new gene. The directionality of duplications is revealing for a number of reasons. One of the most important distinctions between parents and daughters is that, though they are both technically orthologous with single-copy genes in the outgroup species, only the parent shares the same genomic environment with the outgroup (thus the term "primary ortholog") and the daughter gene is likely to experience a new genomic environment (thus "secondary ortholog"). The difference in the local environment may range from the presence of different regulatory signals to their epigenomic properties, such as different levels of histone modifications [15]. Under the model for the evolution of novel functions in gene duplicates introduced by Ohno [14], it seems probable that these functions are most likely to arise in the daughter genes because they are less likely to be able to maintain the complete ancestral function in their new location. Data on the prevalence of adaptive natural selection among paralogs (Han et al. *in review*) seems to bear this prediction out.

Because duplication via retrotransposition results in daughter copies that do not have any introns, previous research has been able to infer the directionality of these duplication events [4, 16]. While our method is able to identify parent-daughter relationships due to any mechanism of duplication, we can compare our results to these previous studies. In particular, Emerson et al. [16] found an excess of interchromosomal gene duplication involving the X chromosome. We therefore asked whether the same pattern could be found among all duplicates. We found no excess of parents or daughters on the X chromosome, though we did find an excess of tandem duplications on the X chromosome. This discordance with previous reports may be because we are looking at much younger genes than previous studies, namely only those duplicated since the human-macaque split.

Identification of parent-daughter relationships also provides insight into the molecular mechanisms of gene duplication. For instance, our results show that we can confidently identify at least 215 genes created by tandem duplication in the human lineage (357 parental genes in 142 parent-only families), vs. at least 164 genes created by duplication into a different location (79 daughter-only families + 85 parent-daughter pairs; Table 1). This high proportion of dispersed duplication is consistent with previous results [7]. Our results also provide information about the size of these interchromosomal duplications, with some duplications copying just one gene (largely via retrotransposition) and a number copying blocks up to 22 genes long between chromosomes. While the parent copies had an average synteny of 243 genes long, the daughter copies had an average duplication length of 4.5 genes. Approximately 57% of the duplications were less than 3 genes long, with the frequencies decreasing exponentially. We identified 23 unambiguous retrotransposition events—multiple exons in one copy and a single exon in the other—that could be used to verify our results. Of these, 8 out of 23 parents were misclassified into daughter-only classes due to a lack of synteny, 4 daughters were misclassified into parent-only classes because they were retrotransposed into a proximal location, and 11 out of 23 cases were correctly classified into parent-daughter relationships. Our method seems to suffer from low sensitivity when synteny with the outgroup is short due to disruptions in either lineage. The accuracy could be improved by incorporating additional variables such as exon number or a gene's rank of synteny among all family members. For instance, if we observe another member of a family assigned as parent with higher confidence, this information can add support to the assignment of an ambiguous gene as the daughter.

The major limitation of our approach is the method's dependence on shared synteny between species. Increased divergence will more likely disrupt the synteny between the ancestral regions, causing more families to be misclassified as daughter-only classes. We are planning to explore the effect of divergence with different species pairs in the future, as well as the possibility that incorporating multiple species could alleviate this problem. A major advantage of our approach is that we are able to estimate the probability of finding homologous genes in homologous genomic regions directly from the data. Considering that this value is dependent on both the time since divergence of two species and the rate of genomic rearrangement (including insertion, deletion, and inversion, among other factors), our results show that between homologous regions of the human and macaque genome there is an ~83% probability of finding homologous genes. While this represents an average probability across the genome, more complex models could incorporate regional variation. Our model for defining synteny can be applied to a number of problems in comparative genomics—such as distinguishing between orthologs and out-paralogs (Figure 1a)—and is not necessarily associated with the biological problem addressed here. However, direct comparison with alternative methods that use alignment of whole chromosomal DNA [11] will need to await further study.

References

1. Fitch, W.M., *Distinguishing homologous from analogous proteins.* Systematic Zoology, 1970. **19**(2): p. 99-113.
2. Sonnhammer, E.L.L. and E.V. Koonin, *Orthology, paralogy and proposed classification for paralog subtypes.* Trends in Genetics, 2002. **18**(12): p. 619-620.
3. Hahn, M.W., J.P. Demuth, and S.-G. Han, *Accelerated rate of gene gain and loss in primates.* Genetics, 2007. **177**: p. 1941-1949.

4. Betran, E., K. Thornton, and M. Long, *Retroposed New Genes Out of the X in Drosophila.* Genome Res., 2002: p. 1854-1859.
5. Fu, Z., et al., *MSOAR: A high-throughput ortholog assignment system based on genome rearrangement.* Journal of Computational Biology, 2007. **14**(9): p. 1160-1175.
6. Cannon, S. and N. Young, *OrthoParaMap: Distinguishing orthologs from paralogs by integrating comparative genome data and gene phylogenies.* BMC Bioinformatics, 2003. **4**(1): p. 35.
7. Jiang, Z., et al., *Ancestral reconstruction of segmental duplications reveals punctuated cores of human genome evolution.* Nat Genet, 2007. **39**(11): p. 1361-1368.
8. Nadeau, J.H. and B.A. Taylor, *Lengths of chromosomal segments conserved since divergence of man and mouse.* Proceedings of the National Academy of Sciences of the United States of America, 1984. **81**(3): p. 814-818.
9. Sankoff, D., V. Ferretti, and J.H. Nadeau, *Conserved segment identification*, in *Proceedings of the first annual international conference on Computational molecular biology*. 1997, ACM: Santa Fe, New Mexico, United States.
10. Pevzner, P. and G. Tesler, *Genome Rearrangements in Mammalian Evolution: Lessons From Human and Mouse Genomes.* Genome Res., 2003. **13**(1): p. 37-45.
11. Kent, W.J., et al., *Evolution's cauldron: Duplication, deletion, and rearrangement in the mouse and human genomes.* Proceedings of the National Academy of Sciences of the United States of America, 2003. **100**(20): p. 11484-11489.
12. Bailey, J. and E.E. Eichler, *Primate segmental duplications: crucibles of evolution, diversity and disease*, in *Nat Rev Genet.* 2006.
13. Rhesus Macaque Genome Sequencing and Analysis Consortium, *Evolutionary and biomedical insights from the Rhesus Macaque genome.* Science, 2007. **316**(5822): p. 222-234.
14. Ohno, S., *Evolution by gene duplication.* 1970, Berlin: Springer-Verlag.
15. Zheng, D., *Asymmetric histone modifications between the original and derived loci of human segmental duplications.* Genome Biology, 2008. **9**(7): p. R105.
16. Emerson, J.J., et al., *Extensive gene traffic on the mammalian X chromosome.* Science, 2004. **303**(5657): p. 537-540.

A PARSIMONY APPROACH TO ANALYSIS OF HUMAN SEGMENTAL DUPLICATIONS

CRYSTAL L. KAHN* and BENJAMIN J. RAPHAEL†

Box 1910, Brown University
Department of Computer Science & Center for Computational Molecular Biology
Providence, RI 02912
U.S.A.
E-mail: {clkahn,braphael}@cs.brown.edu

Segmental duplications are abundant in the human genome, but their evolutionary history is not well-understood. The mystery surrounding them is due in part to their complex organization; many segmental duplications are mosaic patterns of smaller repeated segments, or duplicons. A two-step model of duplication has been proposed to explain these mosaic patterns. In this model, duplicons are copied and aggregated into primary duplication blocks that subsequently seed secondary duplications. Here, we formalize the problem of computing a duplication scenario that is consistent with the two-step model. We first describe a dynamic programming algorithm to compute the duplication distance between two strings. We then use this distance as the cost function in an integer linear program to obtain the most parsimonious duplication scenario. We apply our method to derive putative ancestral relationships between segmental duplications in the human genome.

1. Introduction

Mammalian genomes consist of many repetitive sequences. Many of these are insertions of (retro)transposons and are present in many copies, but longer segmental duplications, or low-copy repeats, are also common. Current estimates suggest that approximately 5% of the human genome consists of segmental duplications > 1 kb in length with $\geq 90\%$ sequence identity between copies[1]. Moreover, the pattern of segmental duplications in primate genomes is unusual: many of these duplications are interchromosomal[1] and are recent alterations in the human genome, occurring after the separation of the Old World monkey and great ape lineages. It is estimated that

*Work supported by a National Science Foundation Graduate Research Fellowship.
†Work supported by a Career Award at the Scientific Interface from the Burroughs Wellcome Fund.

segmental duplications account for a significant fraction of the differences between humans and primate genomes, and it has been shown that some human segmental duplications contain novel gene families and genes under strong positive selection. In addition, segmental duplications are implicated in disease-causing rearrangements[2] and copy-number polymorphisms in human populations[1]. Thus, reconstructing the evolutionary history of segmental duplications is essential for understanding primate evolution and for ancestral genome reconstruction[3].

Despite the prominence of segmental duplications, both their evolutionary history and the mechanisms by which large segments of the genome are duplicated and transposed to other genomic loci are poorly understood. Studies of the human genome sequence revealed the surprising fact that many segmental duplications are complex mosaics of smaller duplicated pieces[4]. A **two-step model** of segmental duplication (reviewed in[1]) has been proposed to explain these mosaic patterns, but the convoluted nature of overlapping, interleaved duplicated material in the genome make segmental duplications refractory to traditional sequence analysis. Recently, Jiang El Ab.[5] examined the ancestral relationships between human segmental duplications, and identified "clades" of segmental duplications that share an abundance of repeated subsequences. However, their approach ignored the order and orientation of these repeated subsequences within the segmental duplications, and thus did not explicitly explain the mosaic organization of segmental duplications. While algorithms have been developed for various aspects of duplication analysis – such as the evolution of gene clusters[6,7,8], the analysis of whole-genome duplications[13,14], and the computation of reversal distance[9,10,11] and reconciliation of gene trees and species trees[12] in the presence of orthologous and paralogous genes – none of these methods are directly applicable to the analysis of the organization of segmental duplications. The problems of determining the evolutionary history of human segmental duplications and of validating the two-step model have not yet received a rigorous algorithmic treatment.

In this paper, we consider the problem of constructing a duplication scenario that is consistent with the two-step model of duplication, and that minimizes the total number of duplication operations. We formulate this problem as an integer linear program that is equivalent to the facility location problem, a classic problem in operations research. This formulation requires a measure of the minimum number of duplications to build a target string from a source string. We generalize the duplication distance algorithm introduced in [15] to compute the distance when both the source and target strings contain repeated characters, a requirement for applications to real genomic data. We apply our methods to the duplication blocks derived in [5] and discover a two-step duplication scenario in which 13 seed duplication blocks are first constructed and then duplicated to create 416 secondary duplication blocks.

2. Methods

Informally, given an ancestral genome that is devoid of duplications and a present-day genome rife with duplicated material, the problem is to compute a sequence of duplication events by which the ancestral genome is transformed into the present-day genome. In particular, we are interested in computing the simplest or *most-parsimonious* sequence of duplication events that accomplishes this transformation. In the next section, we describe a model of evolution that is consistent with the two-step model of duplication. Then in Section 2.2, we formulate the problem of computing the most-parsimonious duplication scenario as an integer linear program. Finally, in Section 2.3 we describe an algorithm to compute the duplication distance between a pair of signed strings, a distance used to define the cost function in the integer linear program. This algorithm generalizes the algorithm presented in[15], and an implementation is available upon request.

2.1. *Problem Formulation*

As described above, the segments of the present-day genome that contain duplicated material, hereafter *duplication blocks*, contain complex mosaic patterns of smaller segments, hereafter *duplicons*, that appear in multiplicity across the genome. We model both the ancestral and present-day genomes as signed strings of duplicons. We assume the present-day genome, which has incurred segmental duplications, is a superstring of the ancestral genome, and the duplication blocks are substrings of the present-day genome. See Figure 1.

In the two-step model of duplication, duplicons are copied from their ancestral loci and aggregated into larger, contiguous segments or *seed duplication blocks* during the first duplication phase. In the second phase, substrings of both the seed blocks and the ancestral genome are copied and then reinserted into the genome at disparate locations, creating *secondary duplication blocks*. We say that a seed duplication block *seeds* a secondary duplication block in the second phase if substrings of the seed block are used in the construction of the secondary block. See Figure 2a. Note that a seed block may seed multiple secondary duplication blocks but some seed

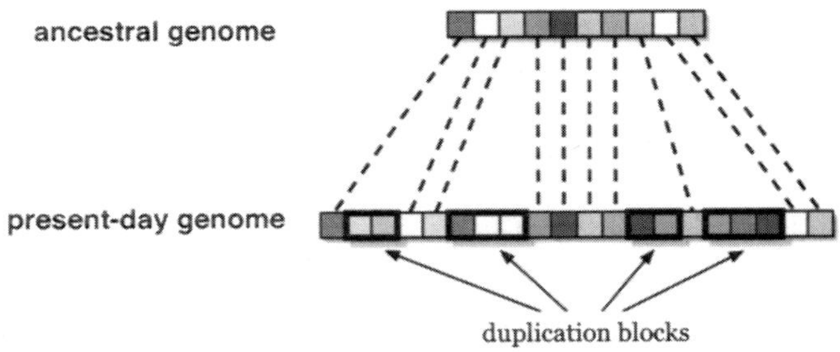

Figure 1. The present-day genome is a superstring of the ancestral genome. The duplicated material comprises *duplication blocks* which are maximal contiguous substrings of the present-day genome that were not part of the ancestral genome.

blocks may not seed any secondary blocks.

We make four simplifying assumptions about the two-step model of duplication:

(1) The ancestral genome contains exactly one copy of every duplicon.
(2) No other type of rearrangement operations – such as inversions of deletions – occur.
(3) The seed blocks are a subset of the duplication blocks observed in the present-day genome.
(4) Each secondary duplication block is seeded by exactly one seed duplication block.

Under these assumptions, we can summarize a two-step duplication scenario in a *two-step duplication tree.*

Definition 2.1. Given ancestral and present-day genomes, a *two-step duplication tree* is a tree of height three where the root is the ancestral genome and the descendants are the duplication blocks. Nodes at depth one (i.e. the children of the root) are the seed blocks created in the first phase of duplication, while nodes at depth two (i.e. children of seed blocks) are the secondary duplication blocks constructed from substrings of one seed block and of the ancestral genome. See Figure 2b.

For a given pair of ancestral and present-day genomes, a most-parsimonious two-step duplication tree is that which defines a partition of the duplication blocks into seed duplication blocks and secondary duplication blocks and defines the ancestral relationships between seed and secondary blocks such

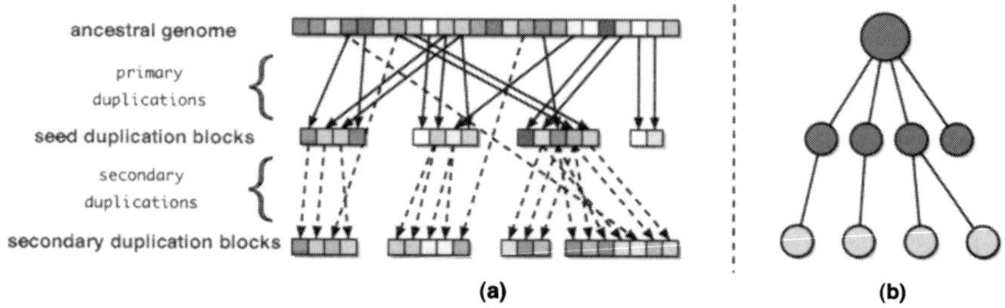

Figure 2. **(a)** The two-step model of duplication. Solid arrows indicate duplicons copied during the first phase of duplication. Dashed arrows indicate duplicons copied during the second phase of duplication. **(b)** The corresponding two-step duplication tree.

that the total number of duplication events needed to construct first the seed blocks and then the secondary blocks is minimum.

We define a duplication event as one *duplicate operation*:

Definition 2.2. A **duplicate operation,** $\delta_{s,t,p}(X)$, copies a substring $x_s \dots x_t$ of a source string X and pastes it into another string Z at position p. Specifically, if $X = x_1 \dots x_m$ and $Z = z_1 \dots z_n$, then $Z \circ \delta_{s,t,p}(X) = z_1 \dots z_{p-1} x_s \dots x_t z_p \dots z_n$.

In[15], we introduced a distance measure from a source string to a target string, called *duplication distance.*

Definition 2.3. The *duplication distance* from a source string X to a target string Y, denoted $D_X(Y)$, is the number of duplicate operations in the simplest (i.e. shortest) sequence of duplicate operations that transforms an initially empty string into the target Y by repeatedly inserting substrings of X.

The total duplication distance for a two-step duplication tree is the sum of the number of duplicate operations needed to build all the duplication blocks. We express the number of duplicate operations needed to build a seed block B_i from the ancestral genome $\mathcal{G}$ as $D_{\mathcal{G}}(B_i)$. Secondary duplication blocks are built from substrings of both its parent seed block and the ancestral genome. Thus, we express the number of duplicate operations needed to build a secondary block B_j from its parent seed block B_i and $\mathcal{G}$ as $D_{B_i \circ \mathcal{G}}(B_j)$, where $B_i \circ \mathcal{G}$ denotes the concatenation[a] of the strings B_i

[a]We insert a "dummy character" between B_i and $\mathcal{G}$ in the concatenate to avoid copying

and $\mathcal{G}$.

We now state our problem.

`Problem:` Computing a Most-Parsimonious Two-Step Duplication Tree.
`Input:` The ancestral genome $\mathcal{G}$ and a set duplication blocks $B_1, \ldots, B_N$ from the present-day genome.
`Output:` The two-step duplication tree (Definition 2.1) with minimum total duplication distance.

2.2. *Computing the Two-Step Duplication Tree*

In this section, we show how to formulate the problem of constructing a most-parsimonious two-step duplication tree as an integer linear program (ILP).

A two-step duplication tree for a given ancestral genome and a set of duplication blocks is defined by a labeling of each of the N duplication blocks as either seed blocks or as secondary blocks. In addition to this labeling, we must also define for each secondary block which seed duplication block seeded it, i.e. which seed block is its parent in the tree.

A most-parsimonious two-step duplication tree is a solution of the following integer linear program.

$$\min_{U,V} \left[\sum_{i=1}^{N} (u_i \times D_{\mathcal{G}}(B_i)) + \sum_{i=1}^{N} \sum_{j=1}^{N} \left(v_{ij} \times D_{B_j \circ \mathcal{G}}(B_i) \right) \right] \quad (1)$$

such that

$$\sum_j v_{ij} = 1 \text{ for all } i \quad (2)$$

$$v_{ij} - u_j \geq 0 \text{ for all } i, j \quad (3)$$

$$u_i \in \{0, 1\} \text{ and } v_{ij} \in \{0, 1\} \quad (4)$$

The binary variables $U = [u_1, \ldots, u_N]$ and binary matrix $V = [v_{ij}]_{i,j=1}^{N}$ describe the topology of the duplication tree. The binary variable u_i indicates whether a duplication block B_i is labeled as a seed block and thus defines an edge in the tree from the root $\mathcal{G}$ to B_i. The binary variable v_{ij} indicates that secondary duplication block B_i is seeded by seed block B_j and thus block B_i is a child of B_j in the tree.

We note that this program is equivalent to a special case of the facility location problem, a classic NP-hard combinatorial optimization problem.

---

substrings across the boundary.

The input to the facility location problem is a set of customers and a set of potential facility sites. For each site, there is a cost associated with opening a facility, and for each site-customer pair, there is a cost associated with supplying that customer from a facility at that site. The objective is to minimize the total cost of opening facilities and supplying customers such that every customer is supplied by exactly one open facility. In the context of the two-step duplication tree, each duplication block is both a customer to be supplied and the site of a potential facility. Opening a facility at site B_i corresponds to classifying B_i as a seed duplication block. Supplying customer B_j from facility B_i corresponds to classifying B_j as a secondary block that is constructed from substrings of seed block B_i and the ancestral genome $\mathcal{G}$.

2.3. *Duplication Distance Algorithm*

In this section, we present a dynamic programming algorithm for computing duplication distance from signed source string X to signed target string Y. This algorithm generalizes the $O(|Y|^4)$ algorithm introduced in[15] for the special case where the source string X is *non-ambiguous* meaning that it contains at most one instance of each character. Here, we present an algorithm whose complexity is $O(|Y|^3|X|)$ for the general case where both X and Y are ambiguous. We assume that all the characters that appear in the target string also appear at least once in the source string.

Intuitively, if signed strings X and Y are "close" in duplication distance, they share common subsequences and thus there is duplication scenario that builds Y by repeated insertions of substrings of X (see Definitions 2.2 and 2.3).

We define some notation. Given a signed string $Y = y_1 y_2 \dots y_n$, let $Y_{s,t} = y_s y_{s+1} \dots y_t$ be the substring ranging between indices s and t, inclusive, for $1 \leq s \leq t \leq n$. Given strings X and Y, let $\mathcal{I}_{X,Y}(i) = \{j | x_j = y_i\}$. Finally, let $d(Y_{s,t}) = D_X(Y_{s,t})$ be the number of duplicate operations required to build the substring $Y_{s,t}$.

First, note that we can compute the duplication distance for disjoint substrings of the target Y independently, allowing us to recursively divide the target string into subproblems. Suppose we wish to compute the duplication distance for a substring $Y_{s,t}$ of Y. There are two possible cases to consider: (i) y_s was duplicated by itself as a substring of X of length one, or (ii) y_s was copied into Y as part of a larger substring of X. In the first case, we proceed recursively on $Y_{s+1,t}$. In the second case, we note that because X may be ambiguous, we must consider the possibility that the substring of X that was copied to produce the character at index s in Y might have started at any of the indices in $\mathcal{I}_{X,Y}(s)$. If a substring of X beginning at index i and having length greater than one is duplicated and inserted into Y, the character x_{i+1} must appear in Y at some index $j \in \mathcal{I}_{Y_{s+1,t},X}(i+1)$. Note, that although x_i was copied into Y at index s, it need not be the case that x_{i+1} appear at index $s+1$ in Y because subsequent duplicate operations may have inserted substrings into Y in between the characters x_i and x_{i+1}.

This observation leads to the following theorem.

Theorem 2.1. *Given a string $Y_{s,t}$ and indices $s \leq t$, $d(Y_{s,t})$ satisfies the following recurrence.*

$$d(Y_{s,t}) = \begin{cases} 1 & \text{if } s = t \text{ (base case)}, \\ \min_{i \in \mathcal{I}_{X,Y}(s)} d_i(Y_{s,t}) & \text{otherwise}, \end{cases} \tag{5}$$

where

$$\begin{aligned} d_i(Y_{s,t}) &= \min\{a_{s,t}, b_{s,t}\}, \\ a_{s,t} &= 1 + d(Y_{s+1,t}), \\ b_{s,t} &= \min_{j \in \mathcal{I}_{Y_{s+1,t},X}(i+1)} \left(d(Y_{s+1,j-1}) + d_{i+1}(Y_{j,t})\right). \end{aligned}$$

Intuitively, $d_i(Y_{s,t})$ represents the number of duplicate operations needed to build $Y_{s,t}$ given that the character y_s resulted from the copying of a substring of X beginning at index i. The value of $d_i(Y_{s,t})$ can be expressed as the minimum of two values corresponding to the two possible cases described above. And the value $d(Y_{s,t})$ can then be computed by considering all possible indices i such that y_s is the same character as x_i.

The running time of the recurrence is bounded by the time to compute $d_i(Y_{s,t})$ for all possible values of i, s and t. For each substring $Y_{s,t}$, we compute $d_i(Y_{s,t})$ for each index $i \in \mathcal{I}_{X,Y}(s)$. This is bounded by $O(|X|)$ for all values of i, s and t. Because there are $O(|Y|^2)$ substrings $Y_{s,t}$, we compute $d_i(Y_{s,t})$ a total of $O(|Y|^2|X|)$ times. Finally, since the time to compute $d_i(Y_{s,t})$ for fixed i, s, and t is bounded by $|\mathcal{I}_{Y_{s+1,t},X}(s+1)| \in O(|Y|)$, the total running time is $O(|Y|^3|X|)$.

Note that it is straightforward to extend our algorithm to allow duplicate reversal operations on signed strings, i.e. duplicate operations in which the copied substring of X is inverted before being inserted into Y. We can accomplish by by defining $d(Y_{s,t})$ as the minimum of $d(Y_{s,t})$ and $d(-Y_{s,t})$ for all values of s and t, where $-Y_{s,t}$ denotes the reversal of $Y_{s,t}$.

3. Results

We implemented our two-step duplication tree method to analyze the ancestry of segmental duplications in the human genome. We used data from[5] who identified 4,692 ancestral duplicons that appeared in complex mosaic arrangements within 437 duplication blocks in the human reference genome (hg17, May 2004). We enumerated the duplicons according to their ancestral location in the genome and aligned each duplication block to all duplicons using Nucmer[16] thereby representing each duplication block as a signed string in the alphabet of integers between -4692 and $+4692$. This process yielded a total of 429 duplication blocks, as 8 blocks could not be aligned to duplicons. We defined the ancestral genome $\mathcal{G}$ to be the string of duplicons in the order of their ancestral loci, i.e. $+1+2\cdots+4692$. We inserted dummy characters between successive duplicons whose ancestral loci were greater than 10kb apart.

For each ordered pair of duplication blocks B_i, B_j, we computed the distance $D_{B_i \circ \mathcal{G}}(B_j)$ using the algorithm presented in Section 2.3. We also computed, for every duplication block B_i, the distance $D_{\mathcal{G}}(B_i)$ from $\mathcal{G}$ to that block. Using these distances, we solved the ILP defined in (1) using CPLEX. Given the solution to the ILP, we labeled all blocks B_i such that the binary variables $u_i = 1$ as seed blocks. We labeled the remaining blocks as secondary blocks. For any pair of blocks B_i, B_j, if the binary variable $v_{ij} = 1$, we designated secondary block B_i to be a child of seed block B_j. The corresponding two-step duplication tree is shown in Figure 3.

This tree has 13 seed blocks, with one block (chr11:49122753-50286686) having 301 children. This duplication block is long, consisting of 2922 ancestral duplicons, but is not the longest in physical distance. Interestingly, this block is located near the centromere of chromosome 11, consistent with the hypothesis that duplicon seeding occurs in pericentromeric regions[1]. However, the large number of children of this block is likely an artifact that results from our restriction that secondary blocks are descended from exactly one seed block. A more parsimonious duplication scenario might have been that this block was itself formed by duplication from several other seed blocks.

4. Discussion and Future Directions

We formulated the problem of computing the most-parsimonious duplication scenario consistent with the two-step model, and showed how to solve this problem as an integer linear program that is equivalent to the facil-

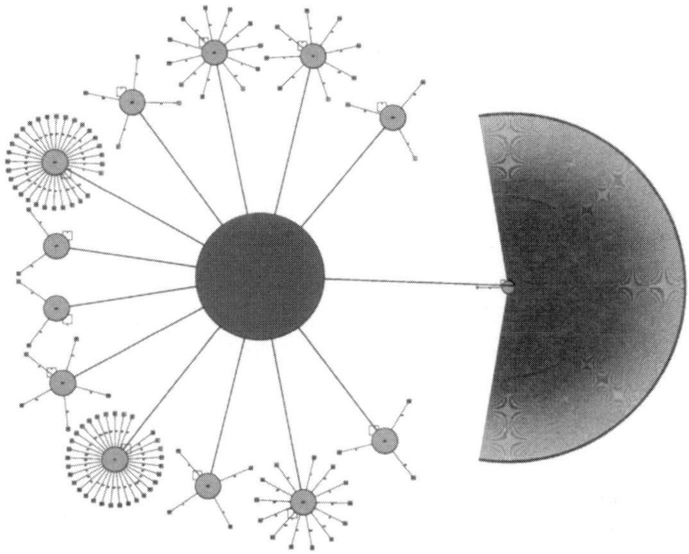

Figure 3. Most-parsimonious two-step duplication tree. The large node is the root and represents the ancestral genome. The thirteen children of the root are seed blocks that were constructed during the first step of duplication and are represented as medium-sized circles. Children of seed blocks were constructed during the second step and are represented as small squares. Starting with the seed block with the most children and proceeding clock-wise around the root, the seed blocks are: chr11:49122753-50286686, chr9:87954745-87985307, chr22:19790229-20121932, chr9:38462158-38620390, chr7:127698302-127893857, chr9:42280547-42512326, chr17:31799963-31889184, chr10:30678896-30730597, chr17:18219669-18692134, chr9:94148626-94280597, chr7:74410377-74802698, chr16:88650921-88822254, and chr17:34433993-34453309.

ity location problem. To our knowledge, this work is the first attempt to construct the ancestry of segmental duplications using parsimony methods. Our two-step duplication tree provides a set of putative seed blocks that are useful for further studies of the history of segmental duplications in the human genome.

Jiang et al.[5] used a different approach to analyze the ancestral relationships of segmental duplications. They represented each duplication block as a binary vector indicating the presence or absence of each duplicon and performed hierarchical clustering on these binary vectors. This analysis ignores the order and orientation of duplicons within a duplication block. We compared our 13 seed groups of duplication blocks (consisting of a seed block and its children) to the 24 "clades" of closely-related duplication blocks reported in Jiang et al. We found that some of our groups were enriched for blocks from a single clade. For example, two of the four members of seed group chr17:34433993-34453309 are shared with the 21 members of Clade M1 defined in[5], ($p = 0.017$ by hypergeometric test). However, in many instances the clusters produced by the two methods differed. To investigate further, we searched for instances where duplicon content and duplication distance diverged. For example, duplication block B_1 = chr7:74410377-74802698 is a seed block and is the parent of secondary block B_2 = chr16:5067675-5156097 in our analysis (Figure 3). In fact, the duplication distance $D_{B_1 \circ \mathcal{G}}(B_2) = 113$ is significantly lower than the average duplication distance to B_2 over all duplication blocks. However, the duplicon content of these two duplication blocks is sufficiently different so that the analysis in[5] placed them into different clades: chr7-2 and M1, respectively. This occurred because although these two duplication blocks did not share a large percentage of their duplicons, surprisingly they did share a few long, conserved subsequences of duplicons that appeared in the same order in both duplication blocks. The presence of conserved subsequences of duplicons suggests that deriving evolutionary relationships from duplicon content alone might be misleading in some cases. Conversely, we also found examples of duplication blocks that, while very similar in duplicon content, were "far" in terms of duplication distance indicating that the duplicons were not organized into the same order in both blocks. Duplication blocks B_3 = chr11:3365455-3580141 and B_4 = chr16:5229432-5285786 were both placed into clade M1 by[5], but were children of seed blocks chr11:49122753-50286686 and chr7:127698302-127893857, respectively, in our analysis.

Our method relied on several simplifying assumptions. These included: (1) duplicate operations were the only genome rearrangement events that occurred in constructing segmental duplications; (2) seed blocks that were constructed in the first phase of duplication still exist in an unaltered state in the present-day genome; (3) each secondary duplication block was descended from exactly one seed block. A more comprehensive analysis of segmental duplications will require the removal of some or all of these restrictions. In addition, we assumed that the work of identifying and delimiting the ancestral duplicons was already completed (e.g. by[5]). Incorporating duplication models into the annotation of duplication blocks themselves is an interesting avenue for future study. Finally, a similar two-step model involving extrachromosomal intermediates[17] has been proposed[18] to explain the extensive duplications of genomic segments in cancer genomes that also display a complex architecture[19]. It would be interesting to extend our methods to infer the sequence of duplications that occur in a cancer genome.

References

1. J. Bailey and E. Eichler, *Nat. Rev. Genet.* **7**, 552 (2006).
2. P. Stankiewicz and J. Lupski, *Trends Genet.* **18**, 74 (2002).
3. M. Blanchette, E. Green, W. Miller and D. Haussler, *Genome Res.* **14**, 2412 (2004).
4. J. Bailey, Z. Gu, R. Clark, K. Reinert, R. Samonte, S. Schwartz, M. Adams, E. Myers, P. Li and E. Eichler, *Science* **297**, 1003 (2002).
5. Z. Jiang, H. Tang, M. Ventura, M. F. Cardone, T. Marques-Bonet, X. She, P. A. Pevzner and E. E. Eichler, *Nature Genetics* **39**, 1361 (2007).
6. Y. Zhang, G. Song, T. Vinar, E. D. Green, A. C. Siepel and W. Miller, Reconstructing the evolutionary history of complex human gene clusters, in *RECOMB*, 2008.
7. O. Elemento, O. Gascuel and M.-P. Lefranc, *Mol Biol Evol* **19**, 278 (2002).
8. M. Lajoie, D. Bertrand, N. El-Mabrouk and O. Gascuel, *J. Comp. Bio.* **14**, 462 (2007).
9. N. El-Mabrouk, *J. Comput. Syst. Sci.* **65**, 442 (2002).
10. X. Chen, J. Zheng, Z. Fu, P. Nan, Y. Zhong, S. Lonardi and T. Jiang, *IEEE/ACM Trans. Comput. Biol. Bioinformatics* **2**, 302 (2005).
11. M. Marron, K. Swenson and B. Moret, Genomic distances under deletions and insertions (2003).
12. L. Arvestad, A.-C. Berglund, J. Lagergren and B. Sennblad, Gene tree reconstruction and orthology analysis based on an integrated model for duplications and sequence evolution, in *RECOMB '04: Proceedings of the eighth annual international conference on Research in computational molecular biology*, (ACM, New York, NY, USA, 2004).
13. N. El-Mabrouk and D. Sankoff, *SIAM J. Comput.* **32**, 754 (2003).
14. M. A. Alekseyev and P. A. Pevzner, *SIAM J. Comput.* **36**, 1748 (2007).
15. C. Kahn and B. Raphael, *Bioinformatics* **24**, i133 (2008).
16. S. Kurtz, A. Phillippy, A. Delcher, M. Smoot, M. Shumway, C. Antonescu and S. Salzberg, *Genome Biol.* **5** (2004).
17. B. E. Windle and G. M. Wahl, *Mutat Res* **276**, 199 (1992).
18. B. Raphael and P. Pevzner, *Bioinformatics* **20 Suppl 1**, i265 (2004).
19. B. Raphael, S. Volik, P. Yu, C. Wu, G. Huang, E. Linardopoulou, B. Trask, F. Waldman, J. Costello, K. Pienta, G. Mills, K. Bajsarowicz, Y. Kobayashi, S. Shivaranjani, P. Paris, Q. Tao, S. Aerni, R. Brown, A. Bashir, J. Gray, J. Cheng, P. de Jong, M. Nefedov, T. Ried, H. Padilla-Nash and C. Collins, *Genome Biol.* **9**, p. R59 (2008).

AN EXACT SOLVER FOR THE DCJ MEDIAN PROBLEM

MENG ZHANG

College of Computer Science and Technology, Jilin University, China
Email: zhangmeng@jlu.edu.cn

WILLIAM ARNDT AND JIJUN TANG

Dept. of Computer Science and Engineering, Univ. of South Carolina, USA
Email: {arndtw, jtang}@cse.sc.edu

The "double-cut-and-join" (DCJ) model of genome rearrangement proposed by Yancopoulos et al. uses the single DCJ operation to account for all genome rearrangement events. Given three signed permutations, the DCJ median problem is to find a fourth permutation that minimizes the sum of the pairwise DCJ distances between it and the three others. In this paper, we present a branch-and-bound method that provides accurate solution to the multichromosomal DCJ median problems. We conduct extensive simulations and the results show that the DCJ median solver performs better than other median solvers for most of the test cases. These experiments also suggest that DCJ model is more suitable for real datasets where both reversals and transpositions occur.

1. Introduction

Once a genome has been annotated to the point where gene homologs can be identified, each gene family can be assigned a unique integer and each chromosome represented by an permutation of signed integers, where the sign indicates the strand. Rearrangement of genes under reversal (also called inversion), transposition, and other operations such as translocation, fusion and fission are known to be an important evolutionary mechanism [4] and have attracted great interests from phylogenists, evolutionary biologists and comparative genomicists.

Yancopoulos et al. [12] proposed a "universal" double-cut-and-join (DCJ) operation to account for all rearrangement events. Although there is no direct biological evidence for DCJ operations, these operations are very attractive because it provides a simpler and unifying model for genome rearrangement [2].

One important problem in genome rearrangement analysis is to find the median of three genomes, that is, finding a fourth genome that minimizes the sum of the pairwise genomic distances between it and the three given genomes [7]. This problem is important since it provides a maximum parsimony solution to the smallest binary tree, thus can be used as the basis for more complex methods. The median problem is NP-hard for when DCJ distance is used [9]. Since phylogenetic reconstruction based on reconstructing ancestral states may need to compute such medians repeatedly, fast approximations or heuristics are often needed, although exact methods have done well for small genomes [6,8]. With more and more whole genome information available, it becomes very important to develop accurate median solvers for these multichromosomal genomes.

In this paper, we present an exact solver for the DCJ median problem and show our experimental results on various combinations of probabilities of reversals and transpositions. Our tests suggest that this solver is efficient and accurate, and provides better median solutions for most of the cases.

2. Background and Notions

2.1. *Genome Rearrangements*

We assume a reference set of n genes $\{1, 2, \cdots, n\}$ through which a genome can be represented by an *ordering* of these genes. Gene i is assigned with an orientation that is either positive, written i, or negative, written $-i$. Specifically, we regard a multichromosomal genome as a set $A = A(1), \ldots, A(N_c)$ of N_c chromosomes partitioning genes $1, \ldots, n$; where $A(i) = \langle A(i)_1, \ldots, A(i)_{n_i} \rangle$ is the sequence of signed genes in the ith chromosome. In this paper, we also assume that each gene i occurs exactly once in the genome, and the chromosomes are undirected [10], i.e. the flip of chromosomes is regarded as equivalent.

Let G be the genome with signed ordering of $1, 2, \cdots, n$. An *reversal* between indices i and j ($i \leq j$), produces the genome with linear ordering $1, 2, \cdots, i-1, -j, -(j-1), \cdots, -i, j+1, \cdots, n$. A *transposition* on genome G acts on three indices i, j, k, with $i \leq j$ and $k \notin [i, j]$, picking up the interval $i, i+1, \cdots, j$ and inserting it immediately after k. Thus genome G is replaced by: $1, \cdots, i-1, j+1, \cdots, k, i, i+1, \cdots, j, k+1, \cdots, n$ (assume $k > j$). A *transversion* is a transposition followed by an inversion of the transposed subsequence.

There are additional events for multiple-chromosome genomes, such as *translocation* (the end of one chromosome is broken and attached to the end of another chromosome), *fission* (one chromosome splits and becomes two) and *fusion* (two chromosomes combine to become one).

Two genes i and j are said to be *adjacent* in genome A if i is immediately followed by j, or if $-j$ is immediately followed by $-i$. Given genomes A and B, a *breakpoint* is defined as an ordered pair of genes (i, j) such that i and j are adjacent in A but not in B.

2.2. *Genomic Distance and Median Problem*

We define the *edit distance* as the minimum number of events required to transform one genome into the other. Hannenhalli and Pevzner developed an elegant theory for unichromosomal genomes and provided a polynomial-time algorithm to compute the *reversal distance* (and the corresponding sequence of events). They also extends the algorithms to handle multichromosomes [5]. Yancopoulos et al. [12] proposed a universal double-cut-and-join (DCJ) operation that accounts for events such as reversals, translocations, fissions and fusions.

The median problem on three genomes is to find a single genome that minimizes the sum of pairwise distances between itself and each of the three given genomes. The median problems are generally named after the distance they use. For example, the *reversal median problem* uses reversal distances and the *DCJ median problem* uses DCJ distances. Many solvers have been proposed for the reversal median problem. Among them, the one proposed by Caprara [3] is the most accurate. As a branch-and-bound algorithm, Caprara's solver enumerates all possible solutions and tests them edge by edge. Very recently, Adam and Sankoff [1] presented a heuristic inspired by MGR for the DCJ median problem. Experiments on various biological data suggest it is at least as good as MGR. Xu and Sankoff [11] developed a decomposition theory based on multiple breakpoint graphs, enabling the possibility of very fast and exact algorithms for the DCJ median problem in case of circular genomes. Our new DCJ median solver presented in this paper is inspired by Caprara's solver, but adjusted to handle both DCJ edit distance and the complexities introduced by multiple linear and circular chromosomes.

3. Adjacency Graph for Undirected Genome

In this section we review the notations of the Adjacency Graph [2]. A gene is an oriented sequence of DNA that starts with a *head* and ends with a *tail*. We call these the *extremities* of the gene. The tail of a gene a is denoted by a^t, and its head is denoted by a^h. An adjacency of two consecutive genes

a and b, depending on their respective orientation, can be of the following four different types: $\{a^h, b^t\}, \{a^h, b^h\}, \{a^t, b^t\}, \{a^t, b^h\}$. An extremity that is not adjacent to any other gene is called a *telomere*, represented by a singleton set $\{a^h\}$ or $\{a^t\}$.

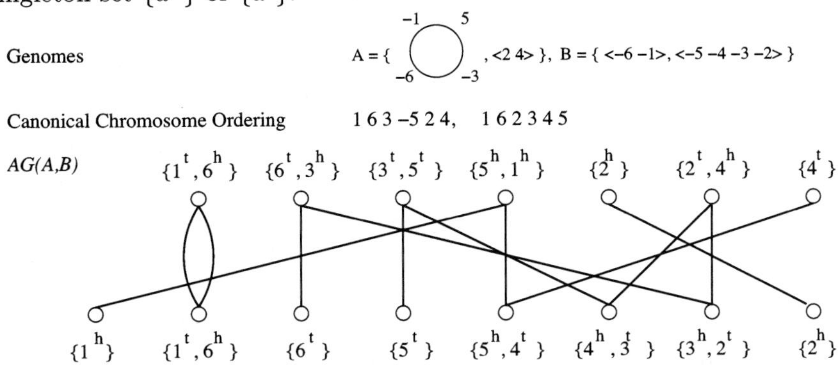

Figure 1. In $AG(A, B)$, $n = 6$, $c(A, B) = 1, |AB| = 2$, DCJ distance of A, B is 4.

We define the adjacency graph $AG(A, B)$ as a graph whose vertices are the adjacencies and telomeres of genomes A and B. For each vertices $u \in A$ and $v \in B$ there are $|u \cap v|$ edges between u and v. The adjacency graph is a union of *paths* and *cycles*. Paths of odd length (odd paths), connect telomeres of different genomes, and paths of even length (even paths), connect telomeres of the same genome. An example of adjacency graph is shown in Fig 1.

The double-cut-and-join (DCJ) operation cuts the chromosome in two places and joins the four ends of the cut in a new way. The DCJ distance can be computed using the following theorem:

Theorem 3.1. [2] *Let A and B be two genomes defined on the same set of n genes. Denote the DCJ distance for two genomes A and B by* $d_{\mathrm{DCJ}}(A, B)$, *then we have* $d_{\mathrm{DCJ}}(A, B) = n - (C + I/2)$ *where C is the number of cycles and I is the number of odd paths in the adjacency graph* $AG(A, B)$.

As a metric distance, the DCJ distance satisfies the triangle inequality:

Lemma 3.1. *Given three genomes A,B,C,* $d_{\mathrm{DCJ}}(A, C)) + d_{\mathrm{DCJ}}(C, B) \geq d_{\mathrm{DCJ}}(A, B)$.

4. Branch-and-bound algorithm

In this section, we describe a branch-and-bound algorithm for the multi-chromosomal DCJ median problem.

4.1. *Basic Notions*

We formally define the DCJ Median Problem (DMP) here. For a genome G, let $Q := \{1,2,3\}$, $\gamma(G) = \sum_{k\in Q} d_{\text{DCJ}}(G, G_k)$. We define DMP as the following: given undirected genomes G_1, G_2, G_3, find a genome G such that $\gamma(G)$ is minimized. The genome obtained after a DCJ operation may contain circular chromosome; thus we should allow circular chromosomes in the input genomes. In this paper, we also consider the solution genomes with circular chromosomes.

The following lower bound on DCJ median value is based on the fact that the DCJ distance satisfies the triangle inequality.

Lemma 4.1. *Given three undirected genomes G_1, G_2, G_3. Let γ^* be the optimal DCJ median value of these genomes,*

$$\gamma^* \geq \frac{d_{\text{DCJ}}(G_1, G_2) + d_{\text{DCJ}}(G_2, G_3) + d_{\text{DCJ}}(G_3, G_1)}{2}. \tag{1}$$

We introduce the following operation in our algorithm. Given a genome A and a pair of extremities $\{p, q\}$. Let $\{p, x\}$, $\{q, y\}$ be adjacencies or telomeres of A. $DCJ(\{p,q\}, A)$ denotes the DCJ operation on A that cuts $\{p, x\}$ and $\{q, y\}$, then connects p with q and x with y. For input genomes G_1, G_2, G_3, we use $DCJ(\{p,q\})$ to denote the set $\{DCJ(\{p,q\}, G_1), DCJ(\{p,q\}, G_2), DCJ(\{p,q\}, G_3)\}$. By applying $DCJ(\{p,q\})$ to the input genomes, we get G_1', G_2', G_3'.

For a set of genomes and $DCJ(\{p,q\})$, the following lemma holds.

Lemma 4.2. *Given three undirected genomes G_1, G_2, G_3 and a pair of extremities $\{p,q\}$, an optimal DMP solution containing the adjacency $\{p,q\}$ is also an optimal solution of the set of genomes obtained by applying $DCJ(\{p,q\})$ to the input genomes.*

As a direct consequence of Lemma 4.1 and 4.2, a new lower bound on DCJ median can be computed from G_1', G_2', G_3'.

Lemma 4.3. *Given three undirected genomes G_1, G_2, G_3 and a pair of extremities $\{p,q\}$. Let γ^* be the optimal DCJ median value of these genomes, G_1', G_2', G_3' are the resulting genomes by applying $DCJ(\{p,q\})$ to G_1, G_2, G_3. If $\{p,q\}$ is in an optimal DMP solution, then*

$$\begin{aligned}\gamma^* \geq\ & d_{\text{DCJ}}(G_1, G_1') + d_{\text{DCJ}}(G_2, G_2') + d_{\text{DCJ}}(G_3, G_3') \\ &+ \frac{d_{\text{DCJ}}(G_1', G_2') + d_{\text{DCJ}}(G_2', G_3') + d_{\text{DCJ}}(G_3', G_1')}{2}.\end{aligned} \tag{2}$$

Proof. Let S be an optimal DMP solution that contains $\{p,q\}$. If $\{p,q\} \in G_i, i \in \{1,2,3\}$, then any cycle or path in $AG(S,G'_i)$ corresponds to a unique cycle or path in $AG(S,G_i)$. Otherwise $AG(S,G'_i)$ has the cycle from $\{p,q\}$ to $\{p,q\}$, and any cycle or path in $AG(S,G'_i)$ corresponds to a unique cycle or path in $AG(S,G_i)$. In the former case, $G_i = G'_i$, $d_{\mathrm{DCJ}}(G_i,S) = d_{\mathrm{DCJ}}(G'_i,S)$. In the latter case, $d_{\mathrm{DCJ}}(G_i,G'_i) = 1$ and $d_{\mathrm{DCJ}}(S,G_i) = d_{\mathrm{DCJ}}(S,G'_i) + 1$. Therefore, $\gamma^* = \sum_{k=1}^{3} d_{\mathrm{DCJ}}(S,G_k) = \sum_{k=1}^{3} d_{\mathrm{DCJ}}(G'_k,G_k) + \sum_{k=1}^{3} d_{\mathrm{DCJ}}(S,G'_k)$. By lemma 4.1, $\sum_{k=1}^{3} d_{\mathrm{DCJ}}(S,G'_k) \geq \frac{d_{\mathrm{DCJ}}(G'_1,G'_2)+d_{\mathrm{DCJ}}(G'_2,G'_3)+d_{\mathrm{DCJ}}(G'_3,G'_1)}{2}$. Thus the lemma is proved. □

4.2. *The Branch-and-Bound Algorithm*

By Lemma 4.2 and Lemma 4.3, we extend the branch-and-bound algorithm in [13] to deal with both circular and linear chromosomes. In each step, it either selects a gene (negative or positive) as an end or an inner gene of a chromosome of the solution (median) genome. In terms of the DCJ model, the former operation fixes a telomere in the solution genome and the latter one fixes an adjacency in the solution genome, these are called *telomere fixing* and *adjacency fixing* respectively. In the fixing of an adjacency, say α, $DCJ(\alpha)$ is applied to each input genome and a lower bound on the γ value of the solution is computed.

In the beginning, the algorithm enumerates all partial genomes by first fixing, in turn, all genes $\pm 1, \ldots, \pm n$ as the first gene, say f, in a chromosome of the solution. Since this chromosome is not completed yet, we call this chromosome an *opening chromosome.* Recursively, let the latest fixed gene be a, we proceed the enumeration by fixing in the solution gene b for all genes not fixed so far or fixing a as an end gene of the current opening chromosome. In the former case, the adjacency $\{a^t, b^h\}$ is fixed in the solution. In the latter case, the telomere $\{a^t\}$ or the adjacency $\{a^t, f^h\}$ is fixed and the current opening chromosome is closed as a linear or circular chromosome. A closing of a chromosome is followed by an opening of a new chromosome except that a complete genome is available.

In adjacency fixing, the operation $DCJ(\{p,q\})$ is applied to the input genomes and the lower bound of all DMP solutions containing all adjacencies fixed so far up to $\{p,q\}$ in the median genomes can be derived according to Lemma 4.3. If it is greater than the lower bound $\{p,q\}$ is not an adjacency in the best solution and can be pruned. Otherwise, $\{p, q\}$ is fixed in the current solution as an adjacency.

After adjacency fixing, the number of common adjacencies of the intermediate genomes increases by one. In other words, the three genomes move simultaneously by DCJ operations. In the case that $\{p, q\}$ is already an adjacency in one of the intermediate genomes, say G, the $DCJ(\{p, q\}, G)$ will not change G, therefore G will not move in this step.

The algorithm also generates three series of DCJ operations that transform G_1, G_2, G_3 to the median genome accordingly. Each of the three DCJ series is one of the shortest DCJ series from each of the input genomes to the median genome.

The lower bound can be computed very efficiently. For the current fixed adjacencies set $S_l = \{\alpha_1, \alpha_2, \ldots, \alpha_l\}$, let G_k^i, $k \in \{1, 2, 3\}$, be the genome after applying the operations $DCJ(\alpha_1, G_k), DCJ(\alpha_2, G_k), \ldots, DCJ(\alpha_i, G_k)$ to G_k, $G_k^0 = G_k$. We have that $\sum_{i=1}^{l} |\{k : \alpha_i \in G_k^{i-1}\}|$ equals the number of cycles in $AG(S_l, G_1), AG(S_l, G_2)$ and $AG(S_l, G_3)$. Since a DCJ operation alters at most two cycles or paths in the adjacency graph for two genomes, the new lower bound can be easily computed from the old one, avoiding the need to compute from scratch.

It's worth mentioning that the treatment of the telomere fixing plays an important role in reducing the complexity of the branch-and-bound algorithm. In adjacency fixing, we can find three unique DCJ operations such that Lemma 4.2 holds. However, in telomere fixing, it is difficult to find such DCJ operations with the property like Lemma 4.2. Other than treating the adjacency fixing and telomere fixing in a uniform way, we do not change any genome in telomere fixing and the number of paths in the adjacency graph of a solution and each input genome can be computed very efficiently as described above.The complexity is therefore avoided.

We also extend the efficient genome enumeration method in [13] to enumerate and test the genomes containing both linear and circular chromosomes at most once, which saves considerable computing time.

Let the linear chromosome $X = \langle X_1, \ldots, X_M \rangle$. The *canonical flipping* of X is $\langle -X_M, \ldots, -X_j, \ldots, -X_1 \rangle$, if $|X_M| < |X_1|$, otherwise, X itself. For a circular chromosome $Y = \langle Y_1, \ldots, Y_l, Y_m, Y_r, \ldots, Y_M \rangle$, let $|Y_m|$ be the minimal gene in Y. The *canonical flipping* of Y is $\langle Y_m, \ldots, Y_l \rangle$, if $Y_m > 0$; Otherwise, $\langle -Y_m, \ldots, -Y_r \rangle$. After canonical flipping of each chromosome, we order the chromosomes by their first genes in an increasing order. Obviously, each genome has a unique canonical ordering, and it can be uniquely represented by the canonical ordering along with the markers of open and closed genes and an indicator of linear or circular chromosome.

We proceed the branch-and-bound algorithm in a depth first order, which can be viewed as the depth-first traversal of the trie of all canonical chromosome ordering of genomes, as illustrated in Fig 2. With this scheme we can perform the lower bound test after each adjacency fixing as described in section 4.2.

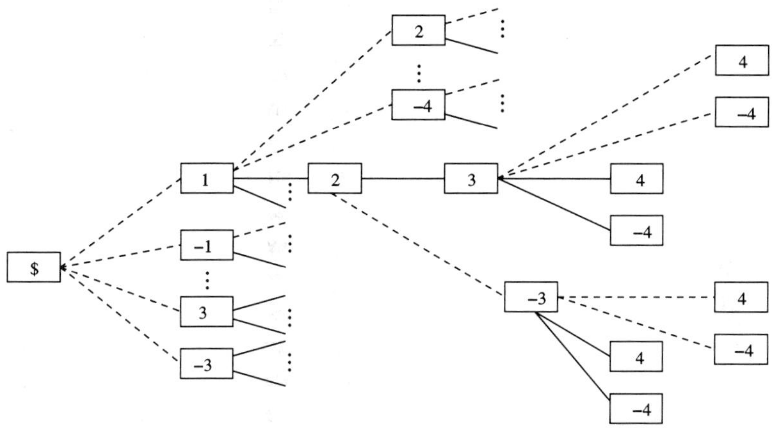

Figure 2. A sketch of the trie of canonical chromosome ordering of all genomes on gene 1..4. Solid edges correspond to adjacencies and each dotted edge corresponds to telomeres. The branch-and-bound algorithm can be viewed as the depth-first traversal of the trie. When lower bound test is failed, it backtracks and tries other genes. We assume that all genomes start at a virtual gene $.

In our implementation, the initial lower bound LD is computed from the initial input genomes (Eq. 1) based on triangular inequalities. The branch-and-bound algorithm starts searching for a DMP solution of target value $L = LD$ and if a solution of value L is found, it is optimal and we stop, otherwise there is no solution of value LD. The algorithm then restarts with an increased target value $L = LD + 1$, and so on. The search stops as soon as we find a solution with the target value.

5. Experimental Results

We have implemented the algorithm and conducted simulations to assess its performance and compare its results against other solvers.

5.1. *Performance of the New Solver*

We test the performance of our new solver on genomes with $100 \sim 200$ genes and $2 \sim 8$ chromosomes. We create each dataset by first generating a tree topology with three leaves and assigning each edge with different length. We assign a genome G_0 to the root, then evolve the signed permutation down the tree, applying along each edge a number of operations that equals the assigned edge length. We test a large range of evolutionary rates: letting r denote the expected number of evolutionary events along an edge of the model tree, we used values of r in the range of 4 to 32. The actual number of events along each edge is sampled from a uniform distribution on the set $\{r/2, \ldots, 3r/2\}$. For each combination of parameters, we generate 10 datasets and average the results.

Table 1 shows the speed of the new solver on 200 genes (results of 100 genes are similar). From this table, we find that the number of chromosomes has very large impact on the speed and genomes with one or two chromosomes can be analyzed quickly, while genomes with 8 chromosomes are much more difficult to compute than unichromosomal genomes.

Table 1. Running time of 200 genes, 2 $\sim$ 8 chromosomes. For each combination of parameters, average running time is shown (in seconds) if all ten datasets finished in 1 hour; otherwise the percentage of finished datasets is shown.

	r=8	r=16	r=24	r=32	r=40	r=48
1 chromosome	< 1	< 1	< 1	< 1	1.1	70%
2 chromosomes	< 1	< 1	< 1	40.1	81.4	60%
4 chromosomes	< 1	26.1	142.2	421.4	40%	20%
8 chromosomes	62.0	2352.0	7691.6	20%	0%	0%

5.2. *Comparing with Other Median Solvers*

In this experimental study, each genome has 100 genes. We test each dataset with the following four methods: `MGR`, the most cited method for genome rearrangement analysis; Caprara's reversal median solver, the most accurate for the reversal median problem; breakpoint median solver; and the new DCJ median solver presented in this paper. We only use unichromosomal genomes in this study, because no multichromosomal breakpoint median solver is available. Comparing DCJ medians with breakpoint medians is interesting because the breakpoint model is independent of any particular rearrangement mechanism, hence can be viewed as a neutral model for genome rearrangements.

In this study, we used values of r in the range of 4 to 32. To test the robustness of the DCJ model, we use five combinations of events: 1) reversals are the only events; 2) 75% events are reversals, 25% are transpositions; 3) reversals and transpositions are equally likely; 4) 75% events are transpositions, 25% are reversals; 5) all three events (reversals, transpositions and transversions) are equally likely; 6) all events are transpositions.

Since the true ancestor is known in our simulations, we use the breakpoint distance between the inferred median and the true ancestor to assess the accuracy of a median solver. For each combination of parameters, we generate 10 datasets, and show the averaged results in Fig. 3 to Fig. 5.

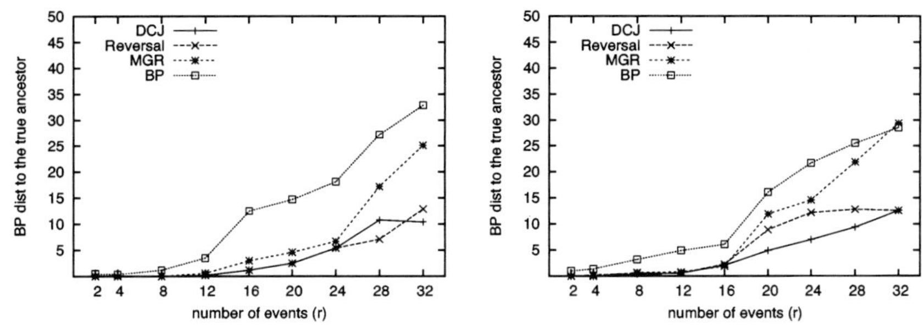

Figure 3. Breakpoint distance from the inferred median to the true ancestor when reversals are the only events (left), and when 75% of the events are reversals, 25% are transpositions (right).

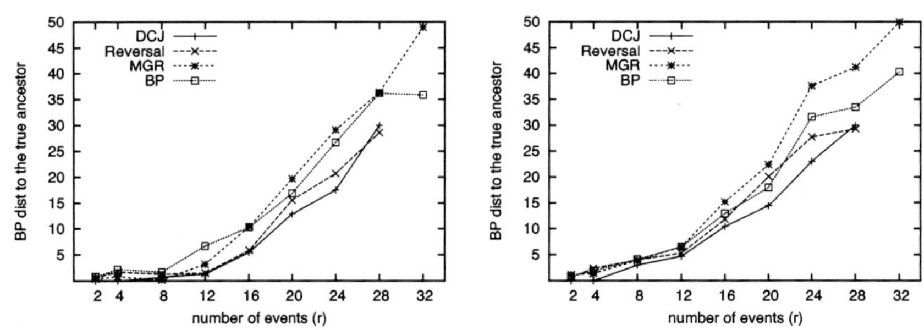

Figure 4. Breakpoint distance from the inferred median to the true ancestor when 50% of the events are reversals, 50% are transpositions (left), and when 25% of the events are reversals, 75% are transpositions (right). DCJ and reversal median solvers cannot finish all datasets for $r = 32$.

From these figures, we find that the DCJ median solver outperforms the other solvers for most of the test cases. Even for the case that favors the reversal median, i.e. when reversals are the only events, using the DCJ median solver has performance equals to the Caprara's solver for $r < 24$, and has better performance when $r = 32$. With the increasing probabilities

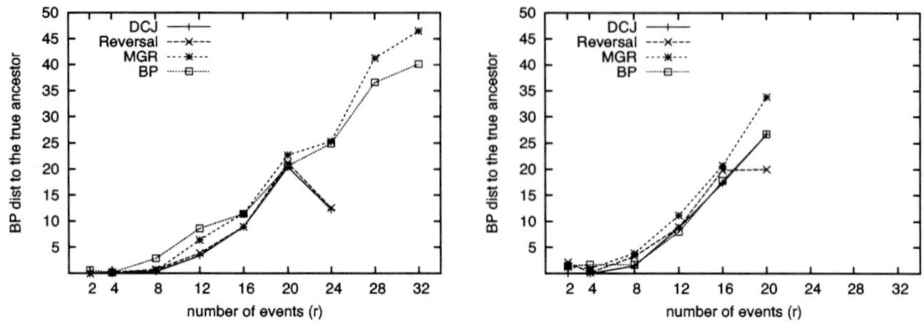

Figure 5. Breakpoint distance from the inferred median to the true ancestor when when all three events (reversals, transpositions and transversions) are equally likely (left), and when all events are transpositions (right). All median solvers cannot finish all datasets for $r > 24$.

of transpositions, the performance gap widens and the DCJ median solver has the best performance for almost all test cases. Although both DCJ and breakpoint models do not deal with actual biological events, the better performance of the DCJ median solver suggests that it is a better choice for real data where both reversals and transpositions occur.

6. Conclusions and Future Work

In this paper we present a new branch-and-bound method for the DCJ median problem. Our extensive experiments show that this method is more accurate than the existing methods. However, this solver is still primitive and further improvements of efficiency are needed. This solver (as almost all other solvers) requires equal gene content which limits the range of data we can analyze. To improve this solver, we need to develop new distance lower bounds so that complex events such as gene losses and duplications can be handled. Recent developments on breakpoint graph decomposition have shown potential and we plan to include some of these new results to further reduce the search space and improve the speed.

7. Acknowledgments

WA and JT were supported by US National Institutes of Health (NIH grant number R01 GM078991). All experiments were conducted on a 128-core shared memory computer supported by US National Science Foundation grant (NSF grant number CNS 0708391).

References

1. Z. Adam and D. Sankoff. The abcs of mgr with dcj. *Evolutionary Bioinformatics*, 4:69–74, 2008.
2. A. Bergeron, J. Mixtacki, and J. Stoye. A unifying view of genome rearrangements. In *Proc. 6th Int'l Workshop Algs. in Bioinformatics (WABI'06)*, number 4175 in Lecture Notes in Computer Science, pages 163–173, 2006.
3. A. Caprara. On the practical solution of the reversal median problem. In *Proc. 1st Int'l Workshop Algs. in Bioinformatics (WABI'01)*, volume 2149 of *Lecture Notes in Computer Science*, pages 238–251, 2001.
4. S.R. Downie and J.D. Palmer. Use of chloroplast DNA rearrangements in reconstructing plant phylogeny. In P. Soltis, D. Soltis, and J.J. Doyle, editors, *Plant Molecular Systematics*, pages 14–35. Chapman and Hall, 1992.
5. S. Hannenhalli and P.A. Pevzner. Transforming mice into men (polynomial algorithm for genomic distance problems). In *Proc. 36th Ann. IEEE Symp. Foundations of Comput. Sci. (FOCS'95)*, pages 581–592, 1995.
6. B.M.E. Moret, A.C. Siepel, J. Tang, and T. Liu. Inversion medians outperform breakpoint medians in phylogeny reconstruction from gene-order data. In *Proc. 2nd Int'l Workshop Algs. in Bioinformatics (WABI'02)*, volume 2452 of *Lecture Notes in Computer Science*, pages 521–536, 2002.
7. D. Sankoff and M. Blanchette. The median problem for breakpoints in comparative genomics. In *Proc. 3rd Int'l Conf. Computing and Combinatorics (COCOON'97)*, volume 1276 of *Lecture Notes in Computer Science*, pages 251–264. Springer Verlag, Berlin, 1997.
8. A.C. Siepel and B.M.E. Moret. Finding an optimal inversion median: Experimental results. In *Proc. 1st Int'l Workshop Algs. in Bioinformatics (WABI'01)*, volume 2149 of *Lecture Notes in Computer Science*, pages 189–203. Springer Verlag, Berlin, 2001.
9. E. Tannier, C. Zheng, and D. Sankoff. Multichromosomal genome median and halving problems. In *Proc. 8th Int'l Workshop Algs. in Bioinformatics (WABI'08)*, volume 5251 of *Lecture Notes in Computer Science*, pages 1–13, 2008.
10. G. Tesler. Efficient algorithms for multichromosomal genome rearrangements. *J. Comput. Syst. Sci.*, 63(5):587–609, 2002.
11. W. Xu and D. Sankoff. Decompositions of multiple breakpoint graphs and rapid exact solutions to the median problem. In *Proc. 8th Int'l Workshop Algs. in Bioinformatics (WABI'08)*, volume 5251 of *Lecture Notes in Computer Science*, pages 25–37, 2008.
12. S. Yancopoulos, O. Attie, and R. Friedberg. Efficient sorting of genomic permutations by translocation, inversion and block interchange. *Bioinformatics*, 21(16):3340–3346, 2005.
13. M. Zhang, W. Arndt, and J. Tang. A branch-and-bound method for the multichromosomal reversal median problem. In *Proc. 8th Int'l Workshop Algs. in Bioinformatics (WABI'08)*, volume 5251 of *Lecture Notes in Computer Science*, pages 14–24, 2008.

INFERRING GENOME-WIDE MOSAIC STRUCTURE

QI ZHANG, WEI WANG, LEONARD MCMILLAN,
FERNANDO PARDO-MANUEL DE VILLENA, DAVID THREADGILL
University of North Carolina at Chapel Hill

Genetic recombination plays two essential biological roles. It ensures the fidelity of the transmission of genetic information from one generation to the next and it generates new combinations of genetic variants. Therefore, recombination is a critical process in shaping arrangement of polymorphisms within populations. "Recombination breakpoints" in a given set of genomes from individuals in a population divide the genome into haplotype blocks, resulting in a mosaic structure on the genome. In this paper, we study the Minimum Mosaic Problem: given a set of genome sequences from individuals within a population, compute a mosaic structure containing the minimum number of breakpoints. This mosaic structure provides a good estimation of the minimum number of recombination events (and their location) required to generate the existing haplotypes in the population. We solve this problem by finding the shortest path in a directed graph. Our algorithm's efficiency permits genome-wide analysis.

1. Introduction

Ancestral genetic recombination events play a critical role in shaping extant genomes. Characterizing the patterns of recombination (*e.g.* the recombination locations and rates), is a crucial step for reconstructing evolutionary histories, performing disease association mapping, and solving other population genetics problems.

During meiosis diploid organisms recombine two homologous genome copies on a chromosome by chromosome basis to form a haploid gamete, which contributes half of the genetic content to its offspring. This mixing of genomes leads to mosaic chromosome (haplotype) structure composed of segments from each grandparent. We refer to the boundaries between the segments of each haplotype as the recombination breakpoints in this paper. Recombination breakpoints represent locations where the crossovers have occurred, either during the generation of the haplotype itself, or in previous generations.

In this paper, we are interested in inferring the possible mosaic structure of a given set of related haplotypes. This is accomplished by finding a set of recombination breakpoints that divide the haplotypes into compatible blocks according to the Four-Gamete Test (FGT)[?]. The FGT states that, under the infinite-sites assumption[?], all pairs of polymorphisms should co-occur in only three out of their four possible configurations. Thus, when four configurations are observed in a pair of markers, it implies that either a recombination or a homoplastic event has occurred between them. We propose an efficient algorithm to solve the "Minimum Mosaic Problem", which finds the mosaic with the minimum number of breakpoints. The algorithm is suitable for genome-wide study.

2. Related Work

Many algorithms have been developed for estimating a lower bound on the minimum number of recombination events necessary to generate a given set of haplotypes. Hudson and Kaplan[?] proposed a lower bound (HK bound) computed using the FGT[?]. Their algorithm computes a minimum set of non-overlapping intervals where all pairs of SNPs in an interval are compatible according to the FGT. This number of intervals, less one, is the HK bound. Myers and Griffiths[?] proposed a tighter bound, RecMin. However, it is only computationally tractable to find the optimal RecMin in relatively small data sets. Myer and Song et al.[?] proposed a RecMin approximation algorithm known as HapBound using Integer Linear Programming (ILP). They also proposed an algorithm, SHRUB, which finds a plausible evolutional history for the given haplotypes, called an Ancestral Recombination Graph (ARG). The ARG establishes an upper-bound on the minimum number of recombinations. Different from RecMin or SHRUB, our algorithm focuses on the mosaic structure of a set of sample sequences without explicitly computing the evolutionary history, assuming that the genomic structures of the sample sequences are of more interest. However, the breakpoints on the sample sequences may reflect possible recombination events happened in the history.

In addition to estimating lower-bounds on the number of recombinations, algorithms have also been proposed for finding a set of founders which generate a given set of recombinant sequences. Ukkonen first proposed the founder set reconstruction problem [?]. A dynamic programming algorithm was developed to compute a minimum number of founders with a given set of sample haplotype sequences, where a segmentation of all the sequences in the sample set can be derived which contains the minimum number of segments. Wu and Gusfield [?] studied a slightly different problem to compute a set of K founders where a segmentation of the given sequences can be derived with a minimum number of segments. They proposed a polynomial time algorithm for genotype sample sequences with only two founders. Different from these works, we do not construct the founder sequences, or rely on the existence of a founder set for segmentation and inferring the mosaic structures on the genome.

3. Problem Formulation

Suppose that we have a set of n haplotypes over m SNPs, represented by a binary data matrix $D = [d_{ij}]_{i=1..n,j=1..m}$. Row i in D corresponds to a haplotype h_i, and column j in D corresponds to a SNP s_j. Matrix entry d_{ij} is either 0 or 1, representing the majority allele or minority allele at SNP s_j respectively. In this paper, we only consider crossover recombination events, ignoring gene conversion and homoplasy (assuming they do not have a significant role).

Over any pair of SNPs s_j and $s_{j'}$, a haplotype takes one of four possible gametes 00, 01, 10, 11 (the combination of alleles at s_j and $s_{j'}$). We denote the set of haplotypes taking 00, 01, 10, or 11 at SNP pair $(s_j, s_{j'})$ as $HapSet_{00}^{j,j'}$, $HapSet_{01}^{j,j'}$, $HapSet_{10}^{j,j'}$, $HapSet_{11}^{j,j'}$ respectively. If all four sets are nonempty, according to the FGT[?], a historical recombination event must have occurred between s_j and $s_{j'}$. In this case, we say that the SNP pair $(s_j, s_{j'})$ is incompatible. We represent a recombination breakpoint as a tuple (h_i, s_b), where the breakpoint locates on haplotype h_i between SNPs s_b and s_{b+1}. It is possible that multiple haplotypes may have breakpoints at the same location since they may "inherit" the breakpoint from a common ancestor sequence.

We define a compatible block of SNPs as a continuous set of SNPs such that any two SNPs inside the block are compatible. Two SNP blocks are incompatible with each other if there exist two incompatible SNPs, one from each block.

A complete set of breakpoints creates a haplotype mosaic structure over the set of genome sequences. A *Mosaic* M over a SNP data matrix D is defined as a set of recombination breakpoints $M = \{(h_i, s_b)\}$, $i \in [1, n], b \in [1, m]$. The set of distinct[a] locations of breakpoints s_b in M divides the entire range of SNPs $[s_1, s_m]$ into blocks that satisfy: 1) each block is a

[a]Multiple breakpoints can correspond to the same s_b

compatible block, 2) any two neighboring blocks are incompatible, and 3) any two neighboring blocks (assume the boundary is between s_b and s_{b+1}) would become compatible if the set of haplotypes that have breakpoints between s_b and s_{b+1} are excluded. In this paper, we develop an efficient algorithm for computing *Minimum Mosaic* (denoted as M_{min}) – the mosaic structure that contains the least number of breakpoints. We refer to this problem as the *Minimum Mosaic Problem.*

4. Inferring the Local Mosaic

4.1. *Maximal Intervals*

We begin by defining the concept of *maximal interval*. An *interval* $I = [s_b, s_e]$ is a set of consecutive SNPs which are compatible with each other starting from s_b and ending at s_e. We define an interval I as a *maximal interval* if and only if there is no other interval I', $I' \neq I$, $I' = [s'_b, s'_e]$, which contains I: $s'_b \leq s_b$, and $s'_e \geq s_e$. The complete set of maximal intervals can be computed in $O(mn)$ time[?], assuming that the compatibility test of any two SNPs using FGT takes $O(1)$ time[?].

4.2. *Finding Local Breakpoints*

Maximal intervals are useful for inferring the local mosaic. The set of distinct breakpoint locations s_b in a mosaic $M = \{(h_i, s_b)\}$ divide the entire SNP range $[s_1, s_m]$ into compatible blocks, where neighboring blocks are incompatible. The set of breakpoints in M is the union of the set of breakpoints on the boundary of each pair of neighboring blocks. We first consider the breakpoints on the boundary of a pair of neighboring blocks. We observe that, every pair of neighboring blocks in M fall inside a pair of overlapping or adjacent maximal intervals, as stated in the following Lemma:

Lemma 4.1. *For any pair of neighboring blocks (B_L, B_R) deduced by a mosaic, there exists a pair of overlapping or adjacent maximal intervals (I_L, I_R), where B_L completely falls inside I_L ($B_L \subseteq I_L$) but not I_R ($B_L \backslash I_R \neq \phi$), and B_R completely falls inside I_R ($B_R \subseteq I_R$) but not I_L ($B_R \backslash I_L \neq \phi$). We refer to (I_L, I_R) as (B_L, B_R)'s* ***containing interval pair****; and (B_L, B_R) as (I_L, I_R)'s* ***contained block pair*** *(Fig. 1).*

Proof.

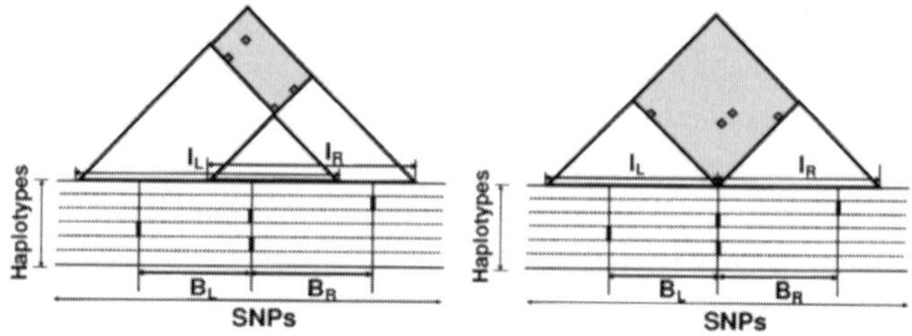

(a) I_L *and* I_R *are overlapping* (b) I_L *and* I_R *are adjacent*

Figure 1. Neighboring blocks B_L, B_R fall inside overlapping/adjacent maximal intervals I_L, I_R respectively. The dots in the shaded region represent incompatible SNP pairs of I_L and I_R.

Details of the proof are presented in ?. □

For each pair of overlapping or adjacent intervals (I_L, I_R), there exists a set of incompatible SNP pairs $SNPPair(I_L, I_R) = \{(s_l, s_r)\}$, where $l < r$, s_l and s_r are incompatible, and $s_l \in I_L \backslash I_R$, $s_r \in I_R \backslash I_L$. For example, in Fig. 1(a) and 1(b), each dot represents an incompatible SNP pair. Let (B_L, B_R) be a contained block pair of the interval pair (I_L, I_R). We denote the incompatible SNP pairs contained in (B_L, B_R) as $SNPPair(B_L, B_R) = \{(s_l, s_r)\}$, where $l < r$, s_l and s_r are incompatible, and $s_l, s_r \in B_L \cup B_R$. Apparently, $SNPPair(B_L, B_R)$ is a subset of $SNPPair(I_L, I_R)$. The incompatible SNP pairs in $SNPPair(B_L, B_R)$ determine the minimum number of the breakpoints on the boundary of B_L and B_R, as well as the corresponding set of haplotypes having these breakpoints. Given an interval pair, several candidate block pairs may be derived, each of which corresponds to a different $SNPPair(B_L, B_R)$. Fig. 2(a)-2(d) show four different candidate block pairs derived from the same interval pair. The exact set of incompatible SNP pairs in $SNPPair(B_L, B_R)$ depends on the positions of B_L and B_R, *i.e.*, the leftmost SNP of B_L, and the rightmost SNP of B_R. Formally, we define the start range R_s, the end range R_e of a block pair (B_L, B_R) as the ranges where $SNPPair(B_L, B_R)$ remains unchanged if the leftmost SNP of B_L changes within R_s and the rightmost SNP of B_R changes within R_e. Moreover, the breakpoint range R_b of (B_L, B_R) is defined as the range where the boundary of B_L and B_R falls into. The breakpoint range is the overlapping region of I_L and I_R (if I_L and I_R overlap), or the boundary of I_L and I_R (if I_L and I_R are adjacent to each other). For example, in Fig. 2(a), $SNPPair(B_L, B_R)$ contains only one incompatible SNP pair (s_q, s_r). The start range $R_s(B_L, B_R)$ is $(s_p, s_q]$, the end range $R_e(B_L, B_R)$ is $[s_r, s_s)$, and the breakpoint range $R_b(B_L, B_R)$ is $I_L \cap I_R$. In Fig. 2(b), $SNPPair(B_L, B_R)$ contains two incompatible SNP pairs (s_q, s_r) and (s_p, s_r). The start range $R_s(B_L, B_R)$ is $[s^*, s_p]$ (s^* denotes the leftmost SNP of interval I_L), and the end range $R_e(B_L, B_R)$ is $[s_r, s_s)$, and the breakpoint range $R_b(B_L, B_R)$ is $I_L \cap I_R$.

Any contained block pair (B_L, B_R) of any overlapping/adjacent maximal interval pair (I_L, I_R) can be a possible neighboring block pair inside a mosaic M. A subset of these block pairs constitute a mosaic. Specifically, for any neighboring block pair (B_L, B_R) which is inside a Minimum Mosaic M_{min}, we have the following Lemma:

Lemma 4.2. *Let (B_L, B_R) be a neighboring block pair in a Minimum Mosaic M_{min}, and $Breakpoints(B_L, B_R)$ be the set of breakpoints on the boundary of B_L and B_R in M_{min}, and $HapSet(Breakpoints(B_L, B_R))$ be the set of haplotypes having breakpoints in $Breakpoints(B_L, B_R)$. Then $Breakpoints(B_L, B_R)$ is the smallest number of breakpoints which satisfies:*

$$\begin{array}{l} \forall (s_l, s_r) \in SNPPair(B_L, B_R) \\ (\exists g_{l,r} \in \{00, 01, 10, 11\}, HapSet(Breakpoints(B_L, B_R)) \supseteq HapSet^{l,r}_{g_{l,r}}) \end{array} \quad (1)$$

Proof. Details of the proof are presented in [?]. □

It is easy to compute $Breakpoints(B_L, B_R)$ if $SNPPair(B_L, B_R)$ only contains one pair of incompatible SNPs (as shown in Fig. 2(a)). We can choose the smallest set of $HapSet^{l,r}_{00}$, $HapSet^{l,r}_{01}$, $HapSet^{l,r}_{10}$ and $HapSet^{l,r}_{11}$ to be $Breakpoints(B_L, B_R)$. If $SNPPair(B_L, B_R)$ contains more than one pair of incompatible SNPs (as shown in Fig. 2(b), 2(c), and 2(d)), we need to find the smallest set of haplotypes which is a superset of at least one of $HapSet^{l,r}_{00}$, $HapSet^{l,r}_{01}$, $HapSet^{l,r}_{10}$ and $HapSet^{l,r}_{11}$, for each pair of incompatible SNPs (s_l, s_r). The computation complexity is $O(4^k)$, where $k = |Breakpoints(B_L, B_R)|$. In practice, k is small. Moreover, many incompatible SNP pairs are caused by a small number of SNP patterns, which enables further reduction in computation.

5. Finding Minimum Mosaic - A Graph Problem

The set of all possible block pairs $\{(B_L, B_R)\}$ of all overlapping/adjacent maximal interval pairs are the building blocks of a mosaic. We can use them and construct a graph as follows. A node nd in this graph represents the

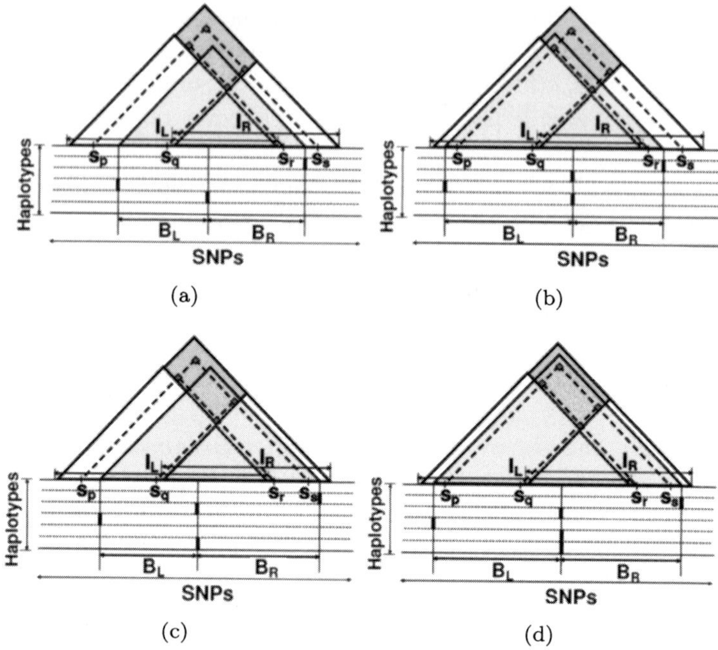

(a) (b) (c) (d)

Figure 2. Neighboring blocks B_L, B_R contain different subsets of the incompatible SNP pairs. The dots represent the incompatible SNP pairs contained in the overlapping maximal intervals I_L and I_R. The dots inside the shaded triangle are contained in the neighboring block pair B_L and B_R.

combination of three block pairs $BP_1 = (B_{L_1}, B_{R_1})$, $BP_2 = (B_{L_2}, B_{R_2})$, $BP_3 = (B_{L_3}, B_{R_3})$ that satisfy the following constraints: 1) the breakpoint range of BP_1 overlaps with the start range of BP_2: $R_b(BP_1) \cap R_s(BP_2) \neq \phi$; 2) the end range of BP_1, the breakpoint range of BP_2, and the start range of BP_3 overlap: $R_e(BP_1) \cap R_b(BP_2) \cap R_s(BP_3) \neq \phi$; 3) the end range of BP_2 overlaps with the breakpoint range of BP_3: $R_e(BP_2) \cap R_b(BP_3) \neq \phi$. As shown in Fig. 3, BP_1, BP_2, and BP_3 are the left block pair, middle block pair, and right block pair of nd, respectively. The breakpoint range of nd is the intersection of the end range of BP_1, the breakpoint range of BP_2, and the start range of BP_3: $R_b(nd) = R_e(BP_1) \cap R_b(BP_2) \cap R_s(BP_3)$. The set of breakpoints associated with nd is the same as $Breakpoints(BP_2)$, denoted as $Breakpoints(nd)$. The weight of the node is the number of breakpoints in $Breakpoints(nd)$, $weight(nd) = |Breakpoints(nd)|$.

We also create two special kinds of nodes – starting nodes and ending nodes to model the two ends of a chromosome. We first identify all

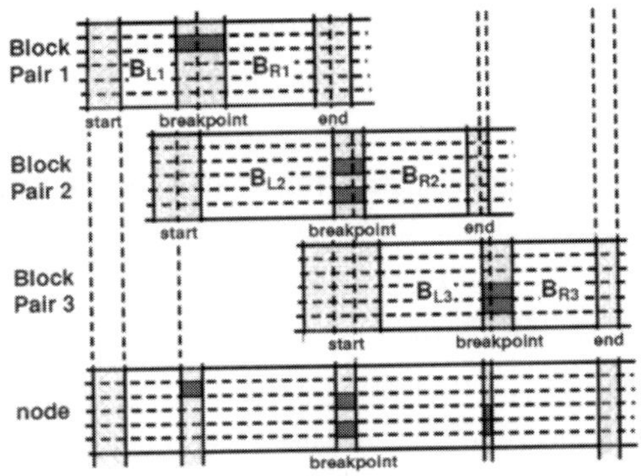

Figure 3. Three block pairs form a node. Block pair 1, 2, and 3 are the left, middle, and right block pair of the node respectively. The breakpoint range of the node is the intersection of the end range of block pair 1, the breakpoint range of block pair 2, and the start range of block pair 3. The vertical stripes correspond to the start range, breakpoint range, and end range of a block. The marked haplotypes in the stripes are the haplotypes which have breakpoints in the corresponding region.

block pairs with start range beginning from the first SNP s_1, referred to as starting block pairs. We create a starting node nd_s for every combination of a starting block pair BP_s and another block pair BP satisfying 1) the breakpoint range of BP_s overlaps with the start range of BP: $R_b(BP_s) \cap R_s(BP) \neq \phi$, and 2) the end range of BP_s overlaps with the breakpoint range of BP: $R_e(BP_s) \cap R_b(BP) \neq \phi$. BP_s is the middle block pair of the starting node nd_s, BP is the right block pair of nd_s. There is no left block pair for nd_s. The set of breakpoints associated with nd_s is the same as $Breakpoints(BP_s)$: $Breakpoints(nd_s) = Breakpoints(BP_s)$. The weight of nd_s is $weight(nd_s) = |Breakpoints(nd_s)|$. Similarly, we create a set of ending nodes $\{nd_e\}$ associated with the set of ending block pairs $\{BP_e\}$.

After generating all nodes, we connect nodes with directed edges according to the following rule. For nodes nd_1 and nd_2, if nd_1's middle block pair is the same as nd_2's left block pair and nd_1's right block pair is the same as nd_2's middle block pair, we add an edge from nd_1 to nd_2. The nodes and edges form a directed graph. A Minimum Mosaic corresponds to a shortest path from any starting node to any ending node in this graph. The weight of the path is the sum of all node weights on the path. The set of breakpoints $\{Breakpoints(nd)\}$ associated with all nodes on the shortest path is the Minimum Mosaic solution. We can use any shortest path algorithm to compute the solution. The details of the complete algorithm and the correctness proof are presented in [?].

6. Experimental Studies

Our algorithm is implemented in C++ and all experiments were performed on a machine with an Intel Core2 Duo processor of 1.60GHz and 3GB RAM.

6.1. *Kreitman's ADH Data*

The alcohol dehydrogenase (ADH) data of Kreitman[?] consists of 11 haplotypes over 43 polymorphic sites of the ADH locus of fruit fly, *Drosophila melanogaster*. The haplotypes were sampled from 5 geographically distinct populations: Washington, Florida, Africa, France, and Japan[?]. Our algorithm detected 7 breakpoints shown in Fig. 4(a). We can estimate the exact locations of 6 out of 7 breakpoints: $H_1 : (S_3, S_4)$, $H_5 : (S_3, S_4)$, $H_5 : (S_{16}, S_{17})$, $H_5 : (S_{35}, S_{36})$, $H_6 : (S_{35}, S_{36})$, $H_6 : (S_{36}, S_{37})$. For the remaining breakpoint on H_1, its location can be either (S_{12}, S_{13}), or (S_{13}, S_{14}), or (S_{14}, S_{15}), or (S_{15}, S_{16}) with equal probability.

Note that 7 is the lower and upper bounds of the minimum number of recombinations, estimated by HapBound and SHRUB, respectively[?]. Therefore, 7 is the exact number of minimum number of recombination events for the ADH data. The corresponding ARG generated by SHRUB is shown in Fig. 4(c). The breakpoints in the ARG are illustrated in a SNP matrix in Fig. 4(b). By comparing Fig. 4(a) and 4(b), we observe that almost the same set of breakpoints are inferred by our algorithm and SHRUB.

6.2. *Running Time and Scalability Analysis*

We tested the performance of our algorithm on two genome-wide SNP data sets from mouse. Both sets represent a combination of experimental and imputed genotypes[?] in two overlapping sets of laboratory inbred strains available from the Center of Genome Dynamics at the Jackson's Laboratory[?]. The 51-strain data set contains 51 inbred mouse strains with 7,870,134 SNPs[b]. The 74-strain data set contains 74 inbred mouse strains

---

[b]The 51-strain data set includes Chr 1-19 and Chr X, with 51 mouse strains: X129S1.SvImJ, X129S4.SvJae, X129X1.SvJ A.J, AKR.J, BALB.cByJ, BTBR.T....tf.J, BUB.BnJ, C3H.HeJ, C57BL.6J, C57BLKS.J, C57BR.cdJ, C57L.J, C58.J, CAST.EiJ, CBA.J, CE.J, CZECHII.EiJ, DBA.1J, DBA.2J, DDK.Pas, FVB.NJ, I.LnJ, JF1.Ms, KK.HLJ, LG.J, LP.J, MA.MyJ, MAI.Pas, MOLF.EiJ, MSM.Ms, NOD.LtJ, NON.LtJ, NZB.BlNJ, NZO.HlLtJ, NZW.LacJ, O20, PERA.EiJ, PL.J, PWD.Ph, PWK.PhJ, Qsi5, RIIIS.J, SEA.GnJ, SEG.Pas, SJL.J, SM.J, SPRET.EiJ, ST.bJ, SWR.J, and WSB.EiJ.

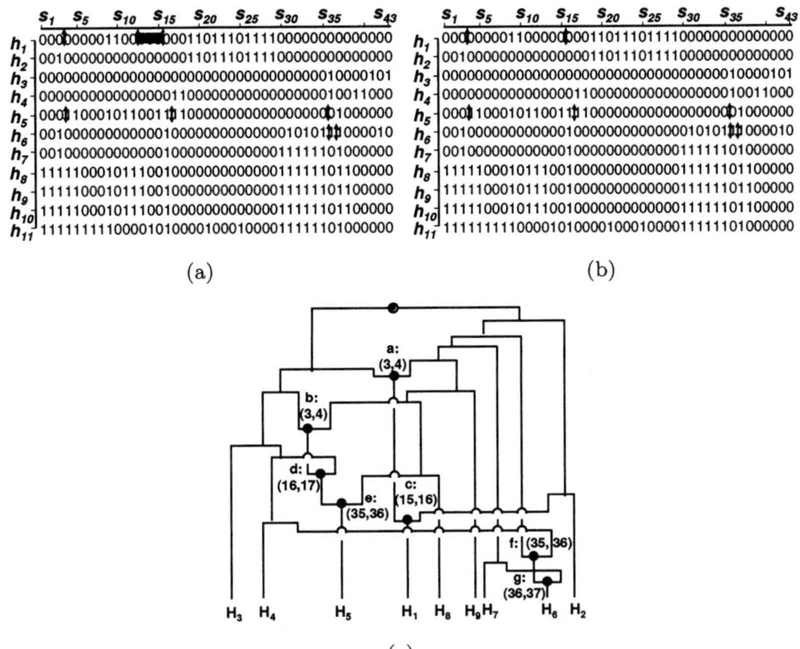

Figure 4. Comparison of Minimum Mosaic and Hapbound/SHRUB results on ADH data. (a): the Minimum Mosaic result; (b): the result inferred from the ARG in (c); (c): the ARG computed using SHRUB[7]. The bars in (a) and (b) represent the breakpoints. The dots in (c) represents the recombination events.

with 7,989,200 SNPs[c].

Fig. 5 shows the running time comparison of Hapbound and our algorithm using the first w SNPs from Chromosome 19 of both data sets where w varies from 1000 to 4000. Our algorithm is 250x - 7000x faster than Hapbound on 74-strain dataset, and 350x - 4000x faster on 51-strain dataset.

Our algorithm is efficient enough to finish on all chromosomes (Chr 1-19 and Chr X). Results from the 51-strain data set are shown in Table 1. Genome-wide, the number of breakpoints in the Minimum Mosaic varies between 15253 (Chr X) and 266006 (Chr 1), and the number of derived blocks in the Minimum Mosaic varies between 9888 (Chr X) and 68261

[c]The 74-strain dataset includes all strains in the 51-strain data set and 23 additional strains: BALB.cJ, BPH.2J, BPL.1J, BPN.3J, C57BL.10J, CALB.RK, DDY.JCLSIDSEYFRKJ, EL.SUZ_2, HTG.GOSFSN, ILS, IS.CAMRK, ISS, LEWES.EI, MOLG.DN, MRL.MpJ, NOR.LTJ, P.J, PERC.EI, RF.J, SKIVE.EI, SOD1.EI, TALLYHO.JNGJ, and ZALENDE.EiJ.

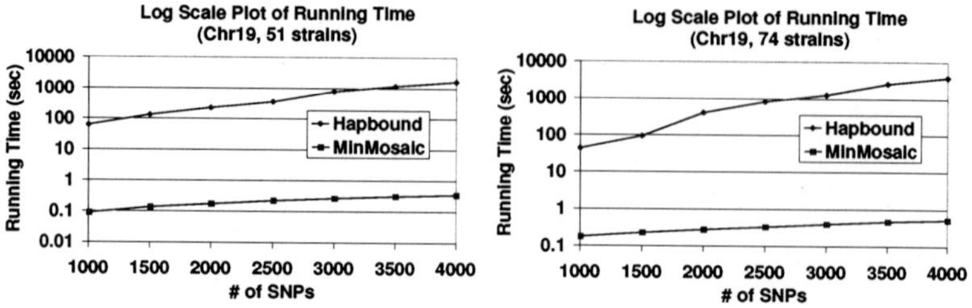

(a) 51-strain Dataset

(b) 74-strain Dataset

Figure 5. Comparison of the running times of MinMosaic and Hapbound over varying number of SNPs (in log scale). The datasets used are from Chr19 of 51-strain dataset and 74-strain dataset. The number of SNPs included varies from 1000 to 4000.

(Chr 1). The average number of breakpoints per neighboring block pair is 2.2.

Table 1. The result on genome-wide 51-strain mouse data set.

Chr	# of SNP	# of breakpoints	# of blocks	Runtime (min)
1	694809	266006	68261	6.87
2	524667	210797	47793	11.27
3	509892	113715	52487	8.90
4	476425	100702	43776	7.84
5	496888	110157	49938	33.98
6	509547	97740	49562	6.42
7	405733	94973	46884	38.83
8	444910	87659	45796	37.10
9	361571	86755	40189	3.89
10	399126	64806	35764	3.21
11	259028	65092	27575	23.52
12	396114	89243	42159	1.30
13	399930	75323	39914	3.03
14	345783	67304	34089	2.54
15	337461	78181	35776	4.08
16	305078	57257	28449	1.14
17	266421	73542	31517	0.75
18	291266	69546	31271	8.61
19	222031	49276	22839	1.46
X	223456	15253	9888	0.96

7. Conclusions

Genetic recombination during meiosis generates a mosaic structure of the genome, where each resulting haplotype consists of segments from different ancestral sequences. In this paper, we study the Minimum Mosaic model that contains a minimum number of breakpoints to generate the haplotypes present within extant populations. The resulting blocks are compatible where no recombinations can be inferred within a block according to the

FGT. We proposed a novel algorithm to compute the minimum mosaic structure of genomes. the efficiency of our algorithm allows for genome-wide analysis.

References

1. Gusfield,D. (2005) Optimal, efficient reconstruction of root-unknown phylogenetic networks with constrained and structured recombination, *Journal of Computer and System Sciences*, **70**, 381-398.
2. Hudson,R., Kaplan,N. (1985) Statistical properties of the number of the recombination events in the history of a sample of DNA sequences, *Genetics*, **111**, 147-164.
3. Kreitman,M. (1983) Nucleotide polymorphism at the alcohol dehydrogenase locus of Drosophila melanogaster, *Nature*, **304**, 412-417.
4. Moore,K., Zhang,Q, McMillan,L., Wang,W., Pardo-Manuel de Villena,F. (2008) Genome-wide compatible SNP intervals and their properties, UNC Tech Report, June 08.
5. Myers,S.R., Griffiths,R.C. (2003) Bounds on the minimum number of recombination events in a sample history, *Genetics*, **163**, 375-394.
6. Schwartz,R., Halldorson,B.V., Bafna,V., Clark,A.G., Istrail,S. (2003) Robustness of inference of haplotyp block structure, *J. Comput. Biol.*, **10**, 13-19.
7. Song,Y.S., Wu,Y, Gusfield,D. (2005) Efficient computation of close lower and upper bounds on the minimum number of recombinations in biological sequence evolution, *Bioinformatics*, **21**, i413-i422.
8. Song,Y.S., Hein,J. (2005) Constructing minimal ancestral recombination graphs, *J. Comput. Biol.*, **12(2)**, 147-169.
9. Song,Y.S., Hein,J. (2003) Parisimonious reconstruction of sequence evolution and haplotype blocks: finding the minimum number of recombination events, *Proceedings of the Third International Workshop on Algorithms in Bioinfomatics (WABI 2003)*, 287-302.
10. Szatkiewicz,J.P., Beane,G.L., Ding,Y., Hutchins,L., Pardo-Manuel de Villena,F., Churchill,G. (2008) An imputed genotype resource for the laboratory mouse, *Mamm Genome*, 19:199.
11. Ukkonen, E. (2002) Finding founder sequences from a set of recombinants, *WABI 2002*, 277-286.
12. Wang,L., Zhang,K., Zhang,L. (2001) Perfect phylogenetic networks with recombinaiton, *J. Comput. Biol.*, **8**, 69-78.
13. Wu, Y., Gusfield, D. (2007) Improved algorithms for inferring the minimum mosaic of a set of recombinants, *CPM 2007*, 150-161.
14. Zhang,K., Deng,M., Chen,T., Waterman,M.S., Sun,F. (2003) A Dynamic programming algorithm for haplotype block partitioning, *PNAS.*, **99**, 7335-7339.
15. Zhang, Q., Wang, W., McMillan, L., Villena, F.P., Threadgill, D. (2008) Inferring genome-wide mosaic structure, UNC Tech. Report (2008).
16. http://cgd.jax.org/ImputedSNPData/imputedSNPs.htm.

SIMULTANEOUS HISTORY RECONSTRUCTION FOR COMPLEX GENE CLUSTERS IN MULTIPLE SPECIES*

YU ZHANG

Department of Statistics,
326 Thomas Building, Penn State University,
University Park, PA 16802, USA
E-mail: yuzhang@stat.psu.edu

GILTAE SONG, CHIH-HAO HSU, WEBB MILLER

Center for Comparative Genomics and Bioinformatics,
506B Wartik Lab, Penn State University,
University Park, PA 16802, USA
E-mail: {gsong,chih-hao,webb}@bx.psu.edu

Genomic intervals that contain a cluster of similar genes are of extreme biological interest, but difficult to sequence and analyze. One goal for interspecies comparisons of such intervals is to reconstruct a parsimonious series of duplications, deletions, and speciation events (a putative evolutionary history) that could have created the contemporary clusters from their last common ancestor. We describe a new method for reconstructing such an evolutionary scenario for a given set of intervals from present-day genomes, based on the statistical technique of Sequential Importance Sampling. An implementation of the method is evaluated using (1) artificial datasets generated by simulating the operations of duplication, deletion, and speciation starting with featureless "ancestral" sequences, and (2) by comparing the inferred evolutionary history of the amino-acid sequences for the CYP2 gene family from human chromosome 19, chimpanzee, orangutan, rhesus macaque, and dog, as computed by a standard phylogenetic-tree reconstruction method.

1. Introduction

Repeated duplications within a cluster of similar genes provide a mechanism for rapid evolution, making those clusters particularly interesting. However, reconstructing evolutionary scenarios that include the operations of duplication and deletion has proven difficult; a number of partial solutions have been proposed [1,2,3,4], but none provide an explict evolutionary

*This work is supported by NHGRI grant HG002238.

reconstruction for multiple species under reasonably general conditions. A recent report[5] attempts to reconstruct an even more general set of operations that includes fissions/fusions of chromosomes and translocations. Under certain idealized conditions, it guarantees a most-parsimonius reconstruction, and efforts are on-going to build a practical reconstruction pipeline based on that theory.

We recently proposed an algorithm for reconstructing a hypothetical ancestral sequence and a parsimonious set of evolutionary events to explain an observed gene cluster in a given species[6]. Here we generalize our single-species algorithm to simultaneously reconstruct an evolutionary history of orthologous gene clusters in multiple species, i.e., an ordered series of duplication, deletion, and speciation events that reproduces the current orthologous gene-cluster configurations from an ancestral progenitor. We start by setting a lower bound for the percentage identity for both within-species and between-species pairwise alignments. This defines a time bound (such as 25 millions of years ago) after which the recent histories of multi-species gene clusters can be reconstructed. Our belief is that attacking this limited but critical problem of ancestral reconstruction and by employing a fundamentally different approach based on statistical methods, we may be able to compute reconstructions more economically and under more general assumptions than when using purely combinatorial methods[5].

2. Methods

Our approach takes multi-species DNA sequences of orthologous gene clusters as input. The sequences are first processed by the following pipeline: 1) use blastz[7] to construct all combinations of self-alignment and pairwise-alignment dot-plots; 2) filter out weak alignments with percentage identity less than a threshold, say 70%, that corresponds to an evolutionary separation greater than that between the most distant species pair under consideration; 3) process the dot-plots such that all local alignments satisfy the "transitive closure property", i.e., given local alignments between regions (A,B) and (B,C), there must be a local alignment between (A,C), where the regions may located in different species; the transitive closure property ensures the completeness of information; 4) chain together local alignments of similar percentage identity broken by small insertions/deletions or post-duplication insertion of interspersed repeats. For instance, Fig. 1 shows the resulting self-alignment dot-plots for five species in the CYP2 cluster.

For such preprocessed data, we propose a new algorithm to reconstruct

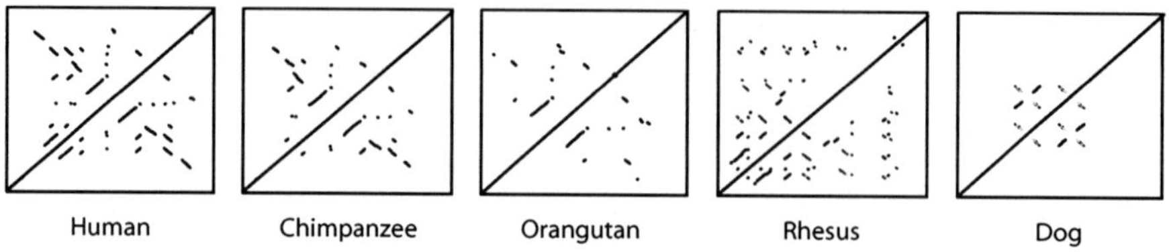

Figure 1. Self-alignment dot-plots for orthologous CYP2 clusters of cytochrome P450 genes in human, chimpanzee, orangutan, rhesus, and dog.

the detailed evolutionary history of orthologous gene clusters in multiple genomes. We consider tandem and non-tandem duplications, duplications with inversions, deletions, and speciation events. We first describe our previously developed algorithms for single-species reconstruction, then generalize the method to multiple species. For multi-species reconstruction, a speciation event corresponds to coalescence between the two sibling species backward in time. We demonstrate how to ensure coalescence in multi-species history reconstruction.

2.1. *A Basic Algorithm for Single-Species Reconstruction*

Given a preprocessed gene cluster in a single species, there is a simple combinatorial algorithm[6] that correctly reconstructs all duplication events. The underlying assumption is that the gene clusters have been exclusively generated by segmental duplications, and the alignment boundaries produced by past duplication events are not reused. The no-alignment-boundary-reuse assumption is a stronger version of the commonly used no-breakpoint-reuse assumption in genome rearrangement analysis, which is first established in Nadeau and Taylor's landmark paper [8]. Although this assumption is still in debate [9], from an inference point of view, it is needed for resolving ambiguities in duplication reconstruction.

The basic algorithm reconstructs duplication events backward in time. In each step, the algorithm finds a local alignment corresponding to a duplication event that satisfies the following criterion: when a duplication is "unwound" by removing one matched DNA segment from the sequence, the breakpoints of the corresponding event do not coincide with remaining alignment boundaries. An example is shown in Fig. 2. If several alignments satisfy the above criterion, we randomly choose one and unwind it. We further merge and extend remaining alignments after each step of reconstruction. The above procedure is repeated until all local alignments in the dot-plot are resolved. The output is thus a reconstructed history

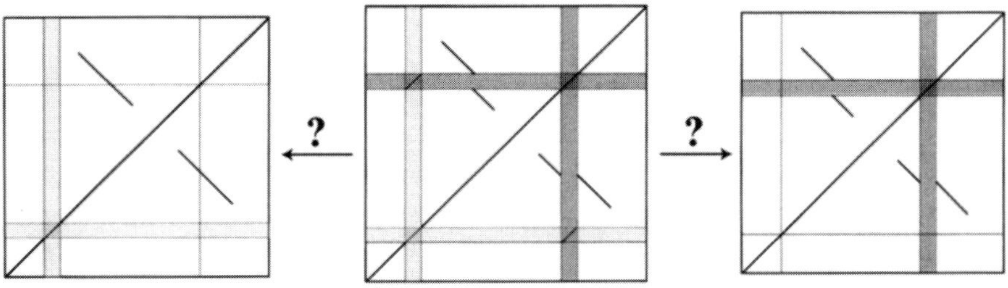

Figure 2. A gene cluster configuration is shown in the middle dot-plot. Lightly and darkly shaded segments are matched, corresponding to a duplication event. To reconstruct the duplication event, we either remove the darkly shaded segment to obtain the left dot-plot, or we remove the lightly shaded segment to obtain the right dot-plot. According to the basic algorithm, the left dot-plot is a correct reconstruction, as breakpoints of the reconstructed event do not coincide with existing alignment boundaries. The right dot-plot is incorrect because there are two breakpoints coincide with existing alignment boundaries.

consisting of a sequence of duplication events, which if unwound reduces the gene cluster sequence to a duplication-free progenitor.

Despite its simplicity, we have proven that the basic algorithm always has solutions, and all solutions are optimal[6]. In particular, there is at least one local alignment at each step of the reconstruction that satisfies the criterion. Depending on the alignments chosen in each step, different histories can be reconstructed that generate the current gene cluster sequence from some duplication-free progenitors. One of the reconstructions will correspond to the real history, while all reconstructions will contain exactly the same number of events, the same duplication sizes, yet different ancestral sequence configurations.

2.2. *An Alternative Probabilistic Solution*

The exclusive duplication assumption and the no-alignment-boundary reuse assumption are often violated in real analysis. Genomic deletions are likely to occur, and alignment boundaries may be reused due to either misalignments or rearrangement hotspots. When an alignment boundary has to be reused during reconstruction, the algorithm can no longer guarantee a correct history reconstruction or even estimate the true number of events. In addition, there are often many equivalent history reconstructions for a gene cluster, which differ by the event orders and thus produce very different progenitor sequences. To accommodate deletion events and data complications, and to infer gene cluster evolution from a large pool of plausible histories, we developed a more practical probabilistic algorithm [6].

The approach is to treat evolutionary events as random, where alignment boundaries can be reused with small probabilities. We first specify a target distribution for gene cluster histories, which defines the scope of histories that we think plausible. For example, to make inference exclusively from histories without alignment boundary reuse, the target distribution should be uniform on all such histories and 0 otherwise. In practice, we define the target distribution as $\pi(\vec{O}_T \mid X) \propto e^{-5(T+r)}$, where $\vec{O}_T = (O_1, O_2, \cdots, O_T)$ denotes a series of evolutionary events ordered backward in time, with O_i describing the type, the location, and the size of each evolutionary event, T denotes the total number of events occurred, and X denotes the gene cluster data. The exponential formula is arbitrarily chosen such that the number of events T and the number of alignment boundary reuses r are linearly penalized in the log-likelihood scale. Parameter "−5" is chosen to allow suboptimal solutions. This target distribution favors histories with fewer events and fewer alignment boundary reuses.

Directly sampling histories from the target distribution is computationally intractable. We therefore turn to sequential importance sampling (SIS)[10]. Suppose that t most recent events have been reconstructed, the SIS algorithm samples the next event O_{t+1} backward in time from a trial distribution $g_t(O_{t+1} \mid \vec{O}_t, X)$, where $\vec{O}_t$ denotes the t events. The SIS algorithm sequentially samples events from a series of trial distributions until all alignments in the dot-plot are resolved, which then produces a history reconstruction $\vec{O}_T$. We repeat the SIS algorithm to obtain many history reconstructions, and we calculate a weight for each reconstruction as $w = \pi(\vec{O}_T \mid X) / \prod_{t=0}^{T-1} g_t(O_{t+1} | \vec{O}_t, X)$. The weight is used to adjust for the sampling bias. Given m reconstructions $\vec{O}_{T_1}^{(1)}, \vec{O}_{T_2}^{(2)}, \ldots, \vec{O}_{T_m}^{(m)}$ and their weights $w_1, \ldots, w_m$, we infer evolutionary parameters using a weighted average of function $u(\vec{O}_T)$ as $E[u(\vec{O}_T)] \cong \left(\sum_{i=1}^m w_i u(\vec{O}_{T_i}^{(i)})\right) / (\sum_{i=1}^m w_i)$. Letting $u(\vec{O}_T) = T$, for example, allows us to estimate the number of events that occurred in a gene cluster evolution. The basic algorithm in the previous section is a special case of the SIS approach, with $g_t(O_{t+1} \mid \vec{O}_t, X)$ uniform on all events O_{t+1} satisfying the the basic algorithm's criterion and 0 otherwise. If the no-alignment-boundary-reuse assumption hold true, the SIS algorithm will efficiently and precisely produce the same duplication reconstructions as given by the basic algorithm. Additional details about the SIS algorithm can be found in our previous paper[6].

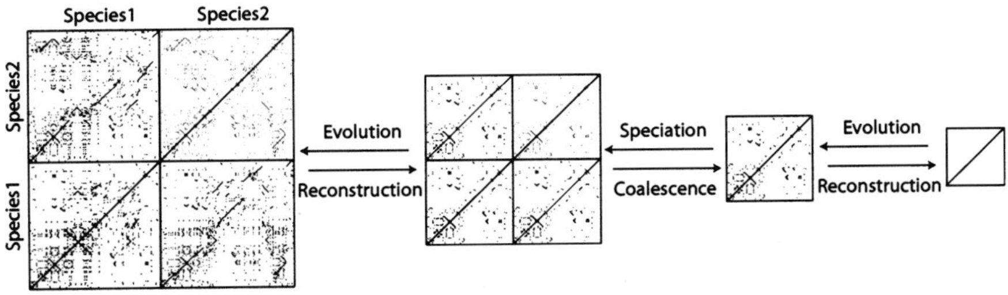

Figure 3. Two-species history reconstruction for gene clusters evolved after 150 duplications and 1 speciation. From left to right: 1) joint dot-plot of self-alignments (lower-left and upper-right quadrants) and inter-species pairwise alignments (upper-left and lower-right); 2) joint dot-plot immediately after speciation; 3) joint dot-plot immediately before speciation; 4) original duplication-free ancestor.

2.3. *Extension to Multi-Species Reconstruction*

Under certain conditions, it is easily shown that the basic algorithm for single-species history reconstruction can be directly applied to solve the multi-species history reconstruction problem. The conditions are similar to those for a single species, except that we now assume gene clusters evolve exclusively through segmental duplications in all species, and alignment boundaries are neither reused within species nor across species. The key idea is to treat speciation events as duplications that double the ancestral genomes. We first concatenate the orthologous gene clusters of all species together, then treat the joint sequence as a new gene cluster sequence for an unknown species. As a result, the single-species basic algorithm can be used to reconstruct the duplication history of the joint gene cluster. It further follows that any reconstructed duplication histories for the joint gene cluster is optimal. In practice, the SIS algorithm is used instead to infer deletion events and handle data complications. Since we know exactly how the multi-species sequences are concatenated, it is straightforward to convert the solution back to a history reconstruction for multi-species gene clusters. Our algorithm allows identification of speciation events as the entire sequence of one species being removed in reconstruction. The total number of events in a multi-species history reconstruction equals the number of duplications and deletions plus the number of speciation events. An example of two-species history reconstruction is shown in Fig. 3.

2.4. *Coalescence between Species*

If the no-alignment-boundary-reuse assumption is violated, the above multi-species reconstruction algorithm can no longer guarantee correct history reconstructions. A more challenging yet critical issue is the coalescence between species. A coalescence event is the inverse of a speciation event, where two sibling species coalesce into their common ancestor. By concatenating gene clusters sequences together, we may reconstruct some histories that are biologically impossible. For example, an event may be reconstructed as duplicating DNA segments across species, or a speciation event is reconstructed as multiple partial duplications. Those are just algorithm artifacts and must be corrected in reconstruction. Intuitively, we should impose a constraint to reconstruction stating that no alignments in pairwise (non-self) alignment dot-plots should be resolved unless it corresponds to a double-genome speciation event.

We use ortholog information to resolve the coalescence problem. Here, orthologs refer to DNA sequences between a pair of species that evolved from a common ancestor by speciation. Using orthologs, we can learn the occurrence time of an event relative to speciation. By the definition of orthologs between two species, it is obvious that no orthologous sequences should be resolved before reconstructing the speciation event, but all non-orthologous sequences should be resolved. If we restrict the algorithm to only reconstruct events within non-orthologous regions between pairs of species, eventually two alternative situations will occur: either all non-orthologous sequences between two species are resolved and thus their sequences become identical, or no more duplication and deletion events can be reconstructed to resolve the remaining non-orthologous sequences. When either case happens, the two species should be coalesced.

Identifying orthologs in two species is a hard problem in the presence of tandem gene clusters. In each step of reconstruction, we impute orthologous sequences between pairs of species using the corresponding pairwise alignments. In particular, given a pairwise alignment dot-plot of two species, we treat all forward pairwise alignments as matches and all reverse pairwise alignments and unmatched regions as gaps and mismatches. We then calculate a rough global pairwise alignment between the two species using this information, and we treat the DNA segments matched along the best global alignment path as orthologs between the two species. We currently do not consider segmental inversions, and thus reverse pairwise alignments cannot form orthologs. Since gaps are mostly created by duplications and deletions, we only penalize mismatches but not gaps. The recursive scoring function of dynamic programming is given by

$$S[x_i^e, y_i^e] = \max_j \{S[x_j^e, y_j^e] + u(x_i^e - x_i^s) - \gamma \max(c(x_j^e, x_i^s), c(y_j^e, y_i^s))\}$$

Here, (x_i^s, y_i^s) and (x_i^e, y_i^e) denote the starting and ending positions of the ith forward alignment, respectively, while $c(a, b)$ measures the size of unmatched segments within region $[a, b)$, which are regarded as mismatches. Also, u and γ are match and mismatch weights. The maximization is taken over all alignments ending before the starting position of the ith alignment, i.e., $\forall\{j : x_j^e \leq x_i^s, y_j^e \leq y_i^s\}$. We recursively calculate the scores of all forward pairwise alignments between two species, and we do this for all pairs of species.

Having identified orthologous sequences between all pairwise gene clusters, our algorithm will reconstruct duplication and deletion events exclusively from non-orthologous regions in each species. For reconstructing duplications, only self-alignments are used, to prevent reconstructing across-species events. If a self-alignment involves a non-trivial part of non-orthologous regions (e.g. >200bp), we treat the alignment as non-orthologous. This is because different species may have evolved at the same functional regions independently, which will make ortholog identification ambiguous. By iteratively imputing orthologs and reconstruct evolutionary events, pairs of species will eventually have no more reconstructible duplication and deletion events from their non-orthologous regions. We then assume the two species to be identical and coalesce them. To coalesce two species, we first calculate a consensus sequence for the two species using their pairwise alignments with other species in the dot-plot. We then remove one species from the dot-plot and replace the other species with the consensus sequence. Note that as evolutionary events are gradually resolved using orthologs, identifying orthologs will in turn become easier.

In theory, a phylogenetic tree of multiple species is not required by our algorithm, and an unrooted tree can be estimated *a posteriori* from the data. In practice, if different species suggest contradicting events caused by incorrect ortholog identification, genome mis-assembly, alignment errors, or unknown evolutionary events, we should take advantage of known phylogenetic relationships to resolve such problems. A straightforward approach is to iteratively reconstruct the history of two closest species in the tree in a bottom-up approach.

3. Results

We evaluated our algorithm using simulated gene-cluster data for $n = 2, 3, 4$ species. The assumed trees resemble that for human, chimpanzee, orangutan, and rhesus, i.e., each additional species is an outgroup species of the existing ones. Starting from a single ancestral sequence of 500 kbp, we randomly simulated k duplication and deletion events, with frequency 98% and 2%, respectively, until the first speciation. We then doubled the current sequence in correspondence to a speciation event. Between every two adjacent speciation events, we continued to simulate k events in each lineage, and repeated the process until all n species appeared in the tree. Finally, we simulated k additional events in each species. The duplication and deletion events were generated according to distributions estimated from analysis of 165 biomedically interesting human gene clusters[6], totalling around 111 million bases. Between every adjacent pair of events in each lineage, we further introduced 0.2% mutations to diversify the sequences. When simulating $n = 4$ species with $k = 20$, for example, the sequence percentage identities between species are reduced by between 8% and 24%.

We used our algorithm, including data preprocessing, to estimate the number of duplication, deletion, and speciation events for each dataset. Given $n = 2, 3, 4$ species and $k = 5, 10, 15, 20$ events between adjacent speciations in each lineage, the true number of events for a gene cluster dataset is $kn(n+1)/2+n-1$. The results from 20 datasets under each setting are shown in Fig. 4. Our algorithm achieved very high accuracy in estimating the total number of events, with about 1% over-estimation. We further checked the number of events that occurred in downstream species after each speciation. Given m downstream species, the true number of downstream events is $k(m(m+1)/2-1)+m-2$, and our estimations are shown along the lines in Fig. 4. Again, our results indicate accurate reconstruction of both coalescence times and the order of coalescence among multiple species. In terms of computational efficiency, the current implementation of the method can reconstruct a history of 100 events within 30 seconds on a regular personal computer. Since we concatenate the genomes of multiple species into a common dot-plot, the computation time for multiple species is mainly determined by the size of the joint dot-plot.

We also evaluated our method using actual sequence data from the CYP2 gene cluster, which is known to have undergone extensive lineage-specific duplications[11]. The CYP2 genes in humans, chimpanzee, orangutan, rhesus and dog were identified by using data downloaded from

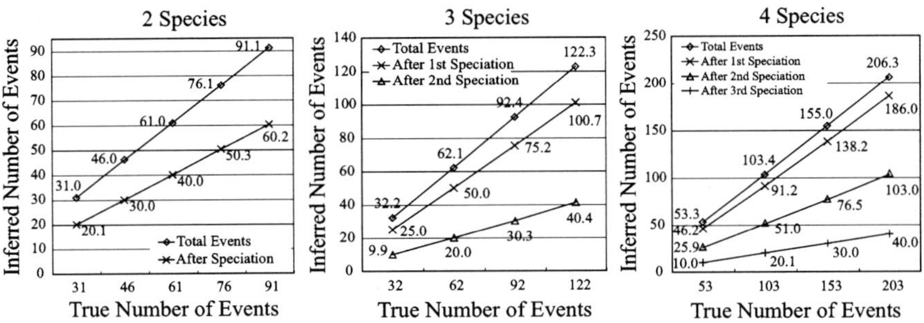

Figure 4. Reconstructed versus true number of events for simulated gene clusters. Given $n = 2, 3, 4$ species and $k = 5, 10, 15, 20$ events between adjacent speciations, the total number of events is $kn(n+1)/2 + n - 1$, as shown on x-axis. The number of events among m downstream species after each speciation is $k(m(m+1)/2 - 1) + m - 2$. All estimated numbers are shown along the lines.

the UCSC Genome Browser (http://genome.ucsc.edu) and by applying the GeneWise[15] program. The human cluster contains 8 genes including 2 pseudo-genes. We constructed a gene tree using an amino-acid-based maximum-likelihood method, as shown in Fig. 5.

We built a second gene tree based on the evolutionary history reconstructed for the CYP2 cluster by the method described in this paper, The method inferred 85 events; 6 of the duplication events involved genes. The two trees agreed perfectly. For instance, the maximum-likelihood tree (Fig. 5) and our reconstruction agree that *CYP2A6* and *CYP2A7* split after the human-chimp divergence, though our method gives the additional information that *CYP2A7* was copied and reinserted in the genome to create *CYP2A6*. The two approaches also agree that the common accestor of *CYP2A13*, c6, and o5 split from the common ancestor of *CYP2A7*, c1, and o1 after the human-rhesus split and before the human-orangutan split. This example indicates how traditional phylogenetic reconstruction methods can be used to validate our results, though our analysis is more informative since it also treats non-coding DNA and infers the source and target of a duplication event.

4. Discussion

Gene duplications are a primary mechanism of evolution[12]. Indeed, biologists have sought explanations for why so many duplicated genes are retained for long periods of time[13]. Studies of recent gene duplications have been greatly impeded by that lack of accurate sequence for gene clus-

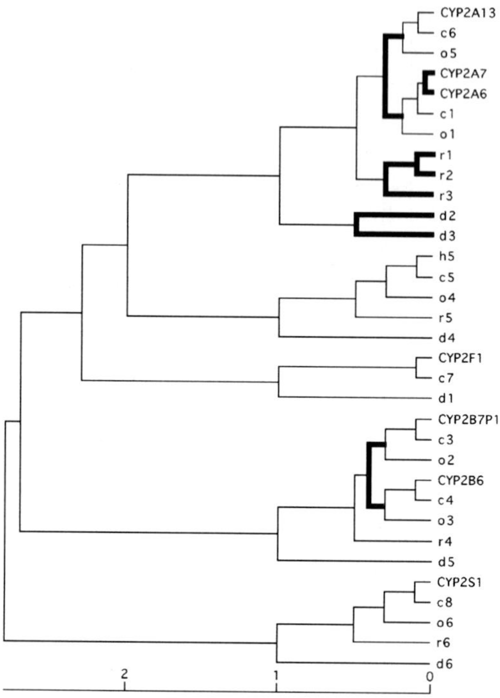

Figure 5. A maximum-likelihood Gene tree of the CYP2 gene cluster for 8 human genes, 8 chimpanzee genes (c1-c8), 6 orangutan genes (o1-o6), 6 rhesus genes (r1-r6) and 6 dog genes (d1-d6). Bold lines indicate duplication events after speciation of dog.

ters; most genomes are being sequenced by the so-called "whole-genome shotgun" approach, which has severe difficulties discriminating gene copies that exceed, say, 95% identity[14], corresponding roughly to duplication events in the last 15 million years. Fortunately, gene clusters are beginning to be sequenced in multiple primates by techniques that resolve recent duplications[16,17], and there will soon be ample data to evaluate methods for reconstructing the evolutionary history of tandem gene clusters.

This current paper is motivated by the belief that the most serious deficiency of current whole-genome alignments is inadequate handling of tandem gene clusters. Assuming that a satisfactory approach can be developed for aligning gene clusters, the solution can be "spliced into" whole-genome alignments computed by other means. However, work remains before the method explored here can be considered totally reliable. For instance, inversions not associated with a duplication event currently cause problems for our approach to identifying orthologs. Another hurdle is presented by gene-conversion events[18], which can present the appearance of an alignment with zones of differing percent identity. We anticipate that promising approaches will be developed by several groups, and that before long biologist will have free access to automatically generated alignments in gene clusters of higher quality than what is currently available.

References

1. O. Elemento, O. Gascuel and M. P. Lefranc, *Mol. Biol. Evol.* **19**, 278–288 (2002).
2. M. Lajoie, D. Bertrand, N. El-Mabrouk and O. Gascuel, *J. Comput. Biol.* **14**, 462–468 (2007).
3. Z. Jiang, H. Tang, M. Ventura, M. F. Cardone, T. Marques-Bonet, X. She, P. A. Pevzner, E. E. Eichler, *Nat. Genet.* **39**, 1361–1368 (2007).
4. J. Ma, A. Ratan, B. J. Raney, B. B. Suh, L. X. Zhang, W. Miller and D. Haussler. *J. Comput. Biol.* To appear (2008).
5. J. Ma, A. Ratan, B. J. Raney, B. B. Suh, W. Miller and D. Haussler. *Proc. Natl. Acad. Sci. USA.* To appear (2008).
6. Y. Zhang, G. Song, T. Vinar, E. D. Green, A. Siepel and W. Miller, *Proceedings of RECOMB* (2008).
7. S. Schwartz, W. J. Kent, A. Smit, Z. Zhang, R. Baertsch, R. C. Hardison, D. Haussler and W. Miller, *Genome Res.* **13**, 103–107 (2003).
8. J. H. Nadeau and B. A. Taylor,*Proc. Natl. Acad. Sci. USA* **81**, 814–818 (1984).
9. P. Pevzner and G Tesler, *Proc. Natl. Acad. Sci. USA* **100**, 7672-7677 (2003).
10. J. S. Liu, *Monte Carlo Strategies in Scientific Computing*, Springer, New York (2001).
11. H. Wang, K. M. Donley, D. S. Keeney and S. M. G. Hoffman, *Environmental Health Perspectives* **111**, 1835–1842 (2002).
12. S. Ohno, *Evolution by Gene Duplication*, Springer, Berlin (1970).
13. A. Force, M. Lynch, F. B. Pickett, A. Amores, Y. Yan and J. Postlethwait. *Genetics* **151**, 1531–1545 (1999).
14. E. D. Green, *Nat. Rev. Genet.* **2**, 573–573 (2001).
15. E. Birney, M. Clamp and R. Durbin. GeneWise and Genomewise. *Genome Res.* **14**, 988–995 (2004).
16. Rhesus Macaque Genome Sequencing and Analysis Consortium. *Science* **316**, 222–234 (2007).
17. B. Hurle, W. Swanson, NISC Comparative Sequencing Program and E. D. Green. *Genome Res.* **17**, 276–286 (2007).
18. J.-M. Chen D. N. Cooper, N. Chuzhanova, C. Ferec and G. P. Patrinos. *Nature Reviews Genetics* **8**, 762–775 (2007).

DYNAMICS OF BIOLOGICAL NETWORKS

TANYA Y. BERGER-WOLF*

Department of Computer Science, University of Illinois at Chicago
Chicago IL 60607, USA

TERESA M. PRZYTYCKA†

National Center of Biotechnology Information, NLM, NIH
Bethesda MD 20814, USA

MONA SINGH‡

Department of Computer Science, Lewis Sigler Institute for Integrative Genomics
Princeton University, Princeton NJ 08544, USA

In modern physics, all phenomena in the universe are considered to be the results of interactions between particles. Biologists are coming to the same conclusion. From genome products working together in a coordinated fashion in order to realize specific cellular functions to diseases spreading through contacts between individuals, the living world is a conflation of myriads of networks of interacting entities. To analyze this added level of complexity, new areas of research have emerged that focus on understanding the roles of interactions between genes, proteins, and other cell components, as well as between cells and even entire organisms.

* Work partially supported by NSF CAREER grant IIS-0747369 and NSF grants IIS-0705822 and IIS-0612044

† Supported by the Intramural Research Program of the National Institutes of Health and National Library of Medicine

‡ Work partially supported by NSF grant CCF-0542187, NIH grant GM076275 and NIH Center of Excellence grant P50 GM071508

Network analysis provides a unifying language to describe relations within complex systems and has played an increasingly important role in understanding biological systems. Over the past decade, computational methods have been developed to infer, analyze, and predict the structure of gene and protein networks. The majority of these approaches have focused on the topology, rather than the dynamics of these networks. Yet most biological networks change temporally, spatially and in a context-dependent manner. Therefore, in addition to a description of these networks as collections of nodes and edges, researchers have began to elucidate dynamic properties of biological networks. In molecular networks, this is frequently obtained by integrating static interactions (such as protein-protein or regulatory interactions) with time- or environment-dependent expression data, protein localization data, or other contextual information.

Understanding cellular dynamics will play a key component in efforts to reverse engineer cellular networks. Gene expression or molecular activity data (such as phophorylation state) collected in different time points or under different conditions can be used to infer direct or indirect connections between genes or gene products and, thus, to infer networks. Such reverse network engineering is increasingly more successful thanks to the accumulating availability of biological data and the development of new computational methods. Moreover, computational methods for inference and analysis of dynamic networks are currently being developed in other domains, such as social networks, and recently some of these approaches for topology reconstruction, identification of clusters, and prediction of dynamic networks are starting to be successfully applied to biological networks.

Uncovering the dynamic nature of cellular networks also has clear impacts in human health and disease, as defects in signaling and regulatory pathways are associated with many serious diseases, such as cancer. This necessitates the development of predictive computational approaches that can be used to model the underlying dynamics of signaling and regulatory networks, and may have important ramifications in drug development and discovery.

The session on Dynamics of Biological Networks brings together scientists working on various aspects of the dynamic nature of biological networks. This year, the session includes an invited talk, six contributed papers, and a panel discussion. The invited speaker, Trey Ideker, will provide his perspective on recent important directions in network and systems biology. The papers selected for presentation address a broad spectrum of problems related to network dynamics.

The first group of papers addresses questions related to the analyses of protein interaction networks. Two papers study protein interaction networks from the perspective of evolutionary dynamics. Gibson and Goldberg present a novel framework for reverse engineering the evolution of protein interaction networks of extant species using phylogenetic gene trees and protein interaction data. Colak, Hormozdiari, Schonhuth, Moser, Holman, Ester and Sahinalp utilize multiple interaction networks across large evolutionary distances to understand the formation of certain patterns in protein interaction networks. The paper by Jin, McCallen, Liu, Xiang, Almaas, and Zhou proposes a method for identifying dynamic modules in PPI networks. While temporal information is not readily available in these networks, the authors propose to infer it from the similarity of expression time series data. They then infer modules from the resulting dynamic networks by identifying connected components that are contiguous in time. The paper by Mitrofanova, Farach-Colton, and Mishra gives an efficient and robust algorithm for extracting from physical interaction networks groups of proteins that work together in the same context as protein complexes.

Gene expression data has been proven to be useful in untangling other aspects related to the dynamics of biological networks. Wang and colleagues give an algorithm for identifying post-translational modulators of transcription factor activity that they utilize to produce a first genome-wide map of the interface between signaling and transcriptional regulatory programs in human B cells. They show that the serine-threonine kinase STK38 emerges as the most pleiotropic signaling protein in this cellular context and then biochemically validate this finding by shRNA-mediated silencing of this kinase, followed by gene expression profile analysis.

Gene expression data have been also utilized in the work of Tamada *et al.* which combines protein-protein interaction networks with transcriptional networks estimated from drug-response time-course expression data to uncover autocrine pathways that are dynamically regulated by drug response.

In addition to the oral presentations two papers have been selected for presentation in the proceedings. The paper by Palisano *et al.* shows how to automatically translate a system described using ODEs into a stochastic approach featured by the BlenX language. They then apply their framework to model the yeast cell cycle. The paper of Nakamura et al. demonstrates the potential power of large-scale particle filtering for parameter estimations of *in silico* biological pathways where time-course measurements of biochemical reactions are available.

The panel will discuss questions related to the opportunities and challenges of analyzing the dynamics of biological networks gleaned from large-scale experimental data. In our tutorial, we hope to touch upon the techniques developed in the field of dynamic network analysis and contemplate how they may be relevant in biological settings. As more temporal, spatial, and contextual data become available for proteins and other cellular macromolecules, we are certain that computational methods for analyses of dynamic networks will become increasingly important. With this session, we hope to encourage computational biologists and bioinformaticians to begin to view networks from a dynamic perspective.

Acknowledgements

We are grateful to those who submitted manuscripts for consideration for inclusion in this session, and we thank the numerous reviewers for their valuable expertise and time throughout the peer review process.

DENSE GRAPHLET STATISTICS OF PROTEIN INTERACTION AND RANDOM NETWORKS

R. COLAK*, F. HORMOZDIARI*, F. MOSER *, A. SCHÖNHUTH*,
J. HOLMAN, M. ESTER AND S.C. SAHINALP+

School of Computing Science, Simon Fraser University

**Joint first authors. +Contact author: cenk@cs.sfu.ca*

Understanding evolutionary dynamics from a systemic point of view crucially depends on knowledge about how evolution affects size and structure of the organisms' functional building blocks (*modules*). It has been recently reported that statistics over sparse PPI graphlets can robustly monitor such evolutionary changes. However, there is abundant evidence that in PPI networks modules can be identified with highly interconnected (*dense*) and/or bipartite subgraphs. We count such dense graphlets in PPI networks by employing recently developed search strategies that render related inference problems tractable. We demonstrate that corresponding counting statistics differ significantly between prokaryotes and eukaryotes as well as between "real" PPI networks and scale free network emulators. We also prove that another class of emulators, the low-dimensional geometric random graphs (GRGs) cannot contain a specific type of motifs, complete bipartite graphs, which are abundant in PPI networks.

1. Introduction

On the biochemical level, life can be explained by gene products facilitating and controlling essential cellular mechanisms, thereby establishing overall cellular viability. To suitably model the structural features that underlie the complex wirings of gene products is presently at the core of computational systems biology. As further explained by the modularity paradigm, global interplay of cellular mechanisms can be decomposed into interaction of functional subunits (*functional modules*), which consist of subsets of gene products that facilitate essential functionalities by concerted actions. Therefore, beyond studying global properties of biomolecular networks (BNs) such as the degree distribution, there has been considerable interest in identifying and quantifying *local* topological properties of biomolecular networks. In particular, small subgraphs or (*graphlets*) that appear significantly more frequently in biomolecular networks than in random graphs can yield insights on cellular functionalities[10]. In order to identify dense graphlets, a variety of methods that count induced subgraphs of different types and node sizes of up to 5, 6 and 7 have been suggested[12,7,5]. In the most recent approach, an algorithm has been developed that counts all non-induced subtrees (as well as certain other "sparse" graphlets) of a PPI network[1]. Corresponding subtree statistics turned out to be robust similarity measures between PPI networks, hence, in principle, can quantify organismic changes due to *evolutionary dynamics*.

It has been argued that graphlet distributions resulting from graphlet counting algorithms can also be used to assess the suitability of random graph models for emulating the growth of and achieving the observable topological features of BNs. A thorough assessment would be highly desirable, as available models, although correctly accounting for many of the global features of BNs, can be different in terms of local features. For example, while exhibiting degree distributions similar to those of prior models (e.g. the *preferential attachment model* (PAM) which gives rise to scale-free networks), the *generalized duplication model* (GDM) is superior in terms of graphlet distributions that have recently been suggested[7]. This may be explained by that the GDM is the only model under actual consideration that is based on emulation of processes that guide BN growth, i.e. gene duplications. Apart from the GDM and the PAM, the geometric random graph model (GRGM) has recently been introduced as an intuitive alternative option. Although not being biologically motivated, PPI networks were successfully fit to GRGMs of low dimensions[6] (for definitions see Section 3).

1.1. *Motivation: Dense Biomolecular Graphlets*

It has been widely established in the biological literature[10], modules in PPI networks are most likely encoded as highly interconnected (*dense*) regions, i.e., connected subgraphs that contain relatively large numbers of edges. It is thus of interest how these regions change over evolutionary time and whether they play a significant role in how the PPI networks evolve. As a result, determining the number of specific dense graphlets and analyzing how they vary with respect to the "complexity" of their respective organism is a problem of significant interest. Unfortunately, counting dense graphlets, even approximately, is a highly non-trivial problem; for general graphs it is known to be intractable. In fact, the simpler problem of determining whether an input graph G includes one or more cliques of size k, or any graphlet with density $1 - O(1/n)$, is NP-hard. Furthermore, the number of any specific dense graphlet M in a dense graph G would be exponential with the size of M.

Fortunately it is well known that PPI networks are typically very sparse, with average degree of 7 or less. For such networks, a novel data mining tool for determining whether a given network G includes a dense graphlet M has recently been described[4]. This tool is based on a pruning technique derived from combinatorial observations on dense subgraphs[4]. Although, in the worst case, the running time of this pruning technique is exponential with the size of the dense graphlet it is looking for, it is of considerable interest whether it can be generalized to count a given dense graphlet in a PPI network of interest. Once such statistics are obtained for all dense graphlets one can consider a number of interesting questions.

(i) Are prominent PPI network generation models, in particular GRGM (i-a) and GDM (i-b) in accordance with statistics over such dense graphlets? If not, it is highly unlikely that they sufficiently account for central cellular functionalities.

(ii) Can we use dense graphlet distributions of various organisms, in order to quantify changes in organism complexity due to evolutionary dynamics? In general, it would be interesting to find out whether evolutionary trends towards more complex organisms can be monitored this way. If one accepts that PPI networks have evolved in a duplication oriented procedure, then one might be able to suggest that more evolutionary complex species might have more complex graphlets.

As a last point, we noticed that in networks generated by low-dimensional GRGMs, complete bipartite graphs of the types $K_{n,m}$ where $n, m \geq 3$ do not/can not exist. However, there is evidence that induced bipartite graphs abundantly occur in the PPI networks of E.coli[3] and others. Moreover, there is evidence that complete bipartite graphs in PPI networks are related to "parallel" functional modules which increase cellular flexibility and robustness.Furthermore, complete bipartite graphs of four nodes (known as **bi-fan**) are the main building blocks of dense overlap regulons (a regulon is the set of genes regulated by given transcription factor) [2,9]. Bi-fan generalizations to larger patterns with row of inputs and row of outputs (bipartite graphs of larger size) are also abundant in PPI networks [2]. Therefore, we put particular emphasis on bipartite graphlet statistics.

1.2. *Contributions*

Counting graphlets (induced or non-induced form) with more than 7 nodes has been a challenging computational task. Recent advances allow improvement for sparse, non-induced graphlets: it is now possible to count trees and bounded treewidth graphlets (with very small treewidth)[1]. In this paper we generalize these results to all graphlets of density ≥ 0.85, up to a node size of 14 (for definitions see Section 2), as well as all complete bipartite graphlets of density ≥ 0.55 up to a node size of 10. The reason that we limited the results to graphlets of density ≥ 0.85 is that for lower densities the number of these occurrences seems to be extremely high and the algorithm took alot of time to give results. Based on these counts we present graphlet statistics for (1) two of the best studied PPI networks: Yeast (eukaryotic) and E.Coli (prokaryotic) as well as (2) for random networks that are obtained through GDM and the GRGM models, whose parameters are set to emulate the growth of these PPI networks. Our main observation is that for the two organisms considered, the motif statistics were in accordance with their complexity: the Yeast network, in comparison to the E.coli network, contained (relatively) many more denser graphlets than sparser graphlets. We also observe that dense motif statistics of the Yeast network was highly divergent from that of the GDM whose parameters were set to emulate its growth as closely as possible. Finally, we prove that low-dimensional GRGM cannot generate complete, bipartite graphs - which abundantly occur in the PPI networks we considered. These observations may help resolve the inconclusive discussion on the suitability of random network generators in emulating the evolution of PPI networks.

2. Inference of Densely Connected and Bipartite Subgraphs

An undirected graph $G = (V, E)$ is said to be a *supergraph* of G' if G' is an induced graphletof G. As usual, a graph is defined to be *bipartite* if $V = V_1 \dot{\cup} V_2$ such that $E \cap ((V_1 \times V_1) \cup (V_2 \times V_2)) = \emptyset$ and it is moreover called *complete* if $E = (V_1 \times V_2) \cup (V_2 \times V_1)$. A $K_{n,m}$ is a complete, bipartite graph such that $|V_1| = n, |V_2| = m$.

We define the *density* $d(G)$ of a graph G as the number of edges divided by the number of possible edges $d(G) = \frac{|E|}{\binom{|V|}{2}} = \frac{2|E|}{|V|(|V|-1)}$. G is said to be α-*dense* if $d(G) \geq \alpha$ and *dense* in general if $\alpha = 0.5$. As usual, G is said to be *connected* if there is a path between any pair of nodes in G. G is said to be *densely connected* if it is dense and connected.

Given a biomolecular network $G = (V, E)$ the driving question is to count all its induced densely connected and complete bipartite subgraphs. Note that inference of all densely connected subgraphs is $\mathcal{NP}$-hard which can be shown by a straightforward reduction from the max-clique problem which is $\mathcal{NP}$-complete[8]. Therefore, one has to screen all $2^{|V|}$ subgraphs in the worst case. Help comes from the following insights:

(i) For $\alpha \geq 0.5$, in each α-dense graph of node size n there is an induced α-dense graphletof size $n - 1$
(ii) Each induced graphlet of a complete bipartite graph is a complete bipartite graph.

While proving (ii) is straightforward, (i) is based on some more subtle arguments that have been presented elsewhere[4]. As a consequence of a combination of $(i), (ii)$ one can employ a search strategy starting with induced 2-node subgraphs and iteratively not considering supergraphs that contradict (i) and/or (ii).

Thanks to the peculiarities of PPI networks (sparseness, scale-freeness), by employing a search strategy that incorporates this core idea, the inference problem becomes tractable.

3. Models for PPI Network Emulation

3.1. *The Generalized Duplication Model*

The duplication model grows iteratively in discrete time steps. It starts with an arbitrary connected network $G(t_0)$, of node size t_0. In iteration $t > t_0$, one node, denoted as v_t, is added to $G(t-1)$, the network resulting from iteration $t-1$, as follows:

A node $w \in G(t-1)$ is picked uniformly at random and then "duplicated" by creating a new node v_t that is connected to all neighbors of w but not to w itself. Subsequently,

(1) Edges (u, v_t) (where u is a neighbor of w) are deleted with probability p,

(2) Edges (u, v_t) are created with probability $r/(t-1)$ where u runs through all nodes in the network. Resulting parallel edges are merged if necessary.

In this process, p and r are fixed parameters.

3.2. *Geometric Random Graphs*

A geometric random graph (GRG) is created by drawing points uniformly at random from some restricted area (e.g. the hypercube) as nodes and, given some fixed threshold r, connecting two nodes v_i, v_j by an edge if $||v_i - v_j|| < r$ where $||.||$ is the Euclidean norm. The dimension of the hypercube is referred to as the dimension of the graph. GRGs have been suggested as a model for emulating PPI networks[12,6]. Although geometric random graphs can successfully capture a couple of PPI network features we would like to outline that GRGs cannot generate induced, complete bipartite subgraphs in the following. However, such subgraphs are crucial ingredients of PPI networks[3].

Theorem 3.1. *A 2- resp. 3-dimensional GRG does not contain a $K_{2,3}$ resp. $K_{3,3}$ as an induced subgraph.*

As we will demonstrate (see the results Section 4) that PPI networks contain significant amounts of such induced bipartite subgraphs in the following, this will rule out low-dimensional GRGs as a model that is suited to capture important local topological features of PPI networks. Moreover, as inspired by the theorem's proof, we conjecture that there is no $K_{n,2}$ in $(n-1)$-dimensional and no $K_{n,3}$ in n-dimensional GRGs. We are currently close to finishing respective results for

$n = 4$ and, indeed, we couldn't observe $K_{4,2}$'s resp. $K_{4,3}$'s in the 3- resp. 4-dimensional GRGs we generated.

3.3. *Notations*

In the following, for $X, Y \in \mathbb{R}^d$ let

$$\overline{XY} := \{v \in \mathbb{R}^d : v = \lambda X + (1-\lambda)Y : \lambda \in [0,1]\}$$

be the line segment that connects X and Y and

$$\overrightarrow{XY} := \{v \in \mathbb{R}^d : v = X + \lambda(Y - X), \lambda \geq 0\}$$

be the half ray that leaves X in direction of Y. For $X \in \mathbb{R}^d$ let

$$U_r(X), S_r(X), B_r(X) := \{v \in \mathbb{R}^d : ||v - X|| \, \{<,=,\leq\} \, r\}$$

be the open ball, the (hyper)sphere and the closed ball with radius r around X. Note that for $d = 2$, $S_r(X)$ is just a circle around X with radius r.

3.4. *Proof of theorem 3.1*

We prepare the proof with two essential lemmata.

Lemma 3.1. *Let $P, A, B, C \in \mathbb{R}^2$ such that*

$$||P - A||, ||P - B||, ||P - C|| < r \leq ||A - B||, ||A - C||, ||B - C|| \qquad (1)$$

Then, for $s \geq \max\{||P - A||, ||P - B||, ||P - C||\}$, it holds that

$$U_r(A) \cap U_r(B) \cap U_r(C) \subset U_s(P). \qquad (2)$$

Proof. Obviously, if A, B, C are located on one line, $U := U_r(A) \cap U_r(B) \cap U_r(C)$ is empty such that there is nothing to prove. Therefore, we can assume that the affine dimension of A, B, C is 2. In combination with some elementary geometric arguments, this implies that $\overrightarrow{PA}, \overrightarrow{PB}, \overrightarrow{PC}$ divide $\mathbb{R}^2$ into three sections such that the section S_{XY} which is bounded by $\overrightarrow{PX}, \overrightarrow{PY}$ does not contain Z where $X, Y, Z \in \{A, B, C\}$ (X, Y, Z pairwise different). Hence

$$\mathbb{R}^2 = S_{AB} \cup S_{AC} \cup S_{BC} \quad \text{and} \quad A \notin S_{BC}, B \notin S_{AC}, C \notin S_{AB}. \qquad (3)$$

We will show that

$$U_r(A) \cap S_{BC}, \; U_r(B) \cap S_{AC}, \; U_r(C) \cap S_{AB} \; \subset \; U_s(P) \qquad (4)$$

which yields

$$U \overset{(3)}{=} U \cap (S_{AB} \cup S_{AC} \cup S_{BC}) = (U \cap S_{AB}) \cup (U \cap S_{AC}) \cup (U \cap S_{BC})$$
$$\subset (U_r(C) \cap S_{AB}) \cup (U_r(B) \cap S_{AC}) \cup (U_r(A) \cap S_{BC}) \overset{(4)}{\subset} U_s(P)$$

and therefore establishes (2).

In order to show (4), we can restrict our attention to showing $U_r(C) \cap S_{AB} \subset U_s(P)$ as the remaining two cases follow from arguments that are completely analogous. If we can show that

$$S_r(C) \cap S_{AB} \subset U_s(P). \tag{5}$$

we will be done with the proof by the following argumentation. Let $D \in U_r(C) \cap S_{AB}$. Consider $\overrightarrow{PD}$. By definition of S_{AB} as a section we have

$$\overrightarrow{PD} \subset S_{AB}. \tag{6}$$

Note that every ray starting within $U_r(C)$ will finally hit the boundary of $U_r(C)$ which is $S_r(C)$. Therefore, while traveling from $D \in U_r(C)$ on $\overrightarrow{PD}$ away from P, one will hit $S_r(C)$. Denote by E the point of that intersection. We compute

$$E \in S_r(C) \cap \overrightarrow{PD} \overset{(6)}{\subset} S_r(C) \cap S_{AB} \overset{(5)}{\subset} U_s(P).$$

However, by construction, D is at least as close to P as E, so $D \in U_s(P)$ with which we have completed the proof.

In order to finally show (5) observe that, because of (1), $P \in U_r(C)$, that is, P is inside of $S_r(C)$, whereas A, B are not inside. Therefore, $S_r(C)$ intersects $\overrightarrow{PA}$ resp. $\overrightarrow{PB}$ between P and, at the latest (maybe just there), A resp. B. Therefore, by definition of s,

$$S_r(C) \cap \overrightarrow{PA},\ S_r(C) \cap \overrightarrow{PB}\ \subset\ B_s(P). \tag{7}$$

However, for $F \in S_r(C) \cap \overrightarrow{PC}$ we have that, as all $P, C, F \in \overrightarrow{PC}$ (*),

$$||F - P|| \overset{(*)}{=} ||P - C|| + ||C - F|| = ||P - C|| + r > r \geq s$$

where $||P - C|| > r$ follows from (i) which implies $P \neq C$ whereas the last inequation follows from the definition of s. This translates to

$$S_r(C) \cap \overrightarrow{PC}\ \not\subset\ B_s(C) \tag{8}$$

Note that two non-concentric circles (here: $S_r(C), S_s(P)$) can intersect at most two points. In our case, these two points establish the transitions of $S_r(C)$ from being inside $B_s(P)$ to being outside $B_s(P)$. Combining this insight with (7,8) implies that, within S_{AB}, $S_r(C)$ is contained in $B_s(P)$ which establishes (5). □

In the following we will write R_x, R_y, R_z for the coordinates of $R = (R_x, R_y, R_z) \in \mathbb{R}^3$.

Lemma 3.2. *Let $P, Q, A, B, C \in \mathbb{R}^3$ such that*

$$A_z = B_z = C_z = 0 \quad \textit{and} \quad P_z \geq Q_z \geq 0 \tag{9}$$

$$||P - A||, ||P - B||, ||P - C||, ||Q - A||, ||Q - B||, ||Q - C|| < r \tag{10}$$

$$||A - B||, ||A - C||, ||B - C|| \geq r. \tag{11}$$

Then, for $s \geq \max\{||P - A||, ||P - B||, ||P - C||\}$, it holds that

$$Q \in U_s(P). \tag{12}$$

Proof. Intuitively speaking, this lemma is trying to show that if three points in two-dimensional space are in proximity of two other points, then these two points have to be proximate to each other.

Let $P' = (P_x, P_y, 0), Q' = (Q_x, Q_y, 0)$ be the projection of P and Q onto the $x - y$-space. Combining (9) and (10) implies that also

$$||P' - A||, ||P' - B||, ||P' - C||, ||Q' - A||, ||Q' - B||, ||Q' - C|| < r. \tag{13}$$

Hence P', A, B, C satisfy the conditions of lemma 3.1 and applying the lemma yields

$$||P' - Q'|| \leq \max\{||P' - A||, ||P' - B||, ||P' - C||\}. \tag{14}$$

Wlog. let A such that $||P' - Q'|| \leq ||P' - A||$. We compute

$$\begin{aligned} ||P - Q|| &= \sqrt{(P_x - Q_x)^2 + (P_y - Q_y)^2 + (P_z - Q_z)^2} \\ &= \sqrt{||P' - Q'||^2 + (P_z - Q_z)^2} \overset{(14)}{\leq} \sqrt{||P' - A||^2 + (P_z - Q_z)^2} \\ &\overset{(ii)}{\leq} \sqrt{||P' - A||^2 + P_z^2} = ||P - A||. \end{aligned}$$

□

We are now in position to prove theorem 3.1.

Proof. [Th. 3.1] We assume the contrary in both cases.

For the first part let $A, B, C \in \mathbb{R}^2$ and $P, Q \in \mathbb{R}^2$ be the partitions of the sampled $K_{3,2}$. By definition of a $K_{3,2}$ in a GRG, P, A, B, C satisfy (1) in lemma 3.1 where r is the threshold of the 2-dimensional GRG. Applying lemma 3.1 yields $||P - Q|| < s \leq r$ which is a contradiction to having an edge between P and Q.

For the second part let $P, Q, R \in \mathbb{R}^3$ and $A, B, C \in \mathbb{R}^3$ be the two partitions of the sampled $K_{3,3}$. Observe that any three points in $\mathbb{R}^3$ lie on a plane. By necessarily rotating all points around the origin (which is an orthogonal transformation hence norm preserving) we can assume wlog. that A, B, C lie in the $x-y$-space of $\mathbb{R}^3$. By necessarily reflecting P, Q, R at the $x-y$-space which, again, is a norm preserving transformation, we can moreover assume that, for two of the P, Q, R, the z-coordinates are non-negative. Wlog. $P_z \geq Q_z \geq 0$. By definition of a $K_{3,3}$ in a 3-dimensional GRG, this establishes (9,10,11) of lemma 3.2 where r is the threshold of the 3-dimensional GRG. Applying lemma 3.2 to P, Q, A, B, C yields $Q \in U_s(P)$ which yields $||P - Q|| < s \leq r$. This is a contradiction to that there is no edge between P and Q. □

4. Results

We present our results on dense graphlet statistics for both Yeast and E.coli PPI networks for graphlet densities in the range $[0.85, 1.00]$ and number of nodes in the range $[7, 14]$.[a] We also present similar dense motif statistics for random graphs generated by both 3-dimensional GRGM (unit cube) and GDM which are set to generate identical number of nodes and edges to the Yeast PPI network[13] [b], i.e. $|V| \approx 4900$ and $|E| \approx 17000$.[c] Due to limitations on computational resources, we were not able to count graphlets of larger sizes or smaller density. In our experiments we have used the network available from DIP [13] for Yeast and E.coli. Yeast PPI network has around 4900 nodes and 17000 edges, and E.coli network has around 1441 nodes and 5871 edges. By considering the distribution (fraction) of dense graphlets (between range $[0.85, 1.00]$) for each species, we are trying to eliminate the effect caused by difference in average degree between these two species.

In Figure 1 we depict for each $n \in \{7, \ldots, 12\}$ (the number of nodes in the graphlet), how the proportion of all graphlets with k edges among all dense graphlets (with density ≥ 0.85) vary with respect to k - for all four PPI networks considered. Note that we give only proportional distributions here as the Yeast and the E.coli networks have different number of edges and nodes and thus it is not meaningful to compare absolute figures.

It is possible to make a number of observations on Figure 1:
(1a) The fractional distribution of dense graphlets in GRGM is significantly different from that of the other networks: for example, it contains no graphlets with density ≥ 0.85 for $n = 12, 13$ and 14 and no graphlets with density ≥ 0.89 (i.e.

[a] We remind the reader that the density of a graphlet with n nodes is the ratio of the number of edges in the graphlet and $n(n-1)/2$, the maximum number of edges possible between the nodes of the graphlet; a density of 1 indicates a clique.
[b] The DIP release date for Yeast and E.coli PPI networks is July 7, 2007.
[c] This is possible by setting the radius of the GRGM to an appropriate value and setting $p = 0.365$ and $r = 0.12$ in the GDM as per[7,1].

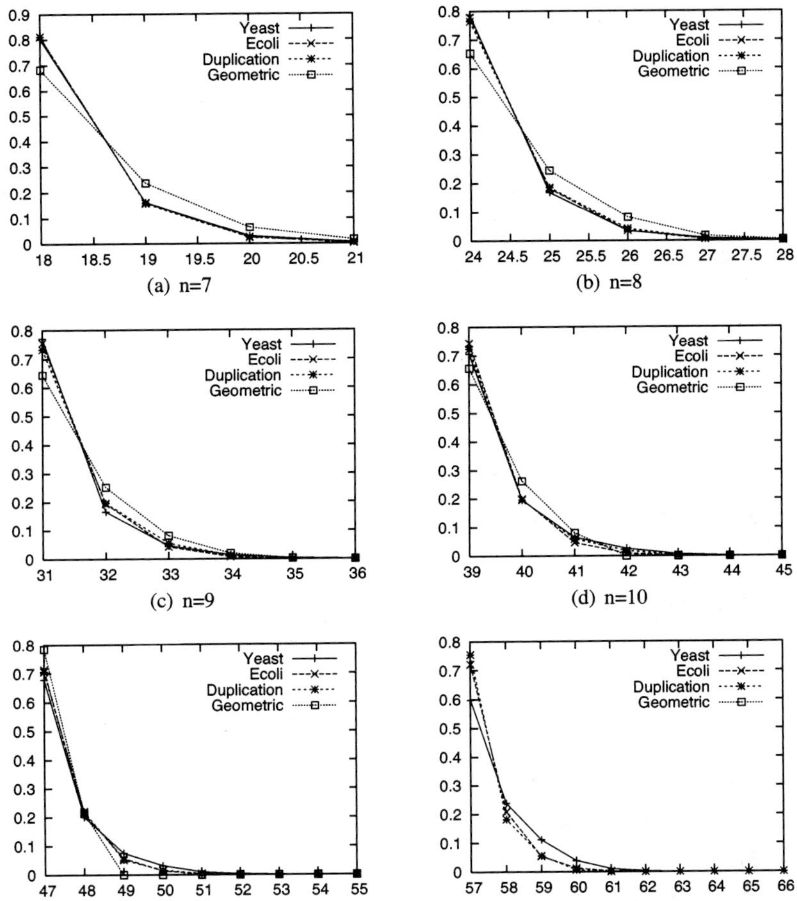

(a) n=7 (b) n=8 (c) n=9 (d) n=10 (e) n=11 (f) n=12

Figure 1. Here we demonstrate for each $n \in \{7, 8, 9, 10, 11, 12\}$ the fraction of graphlets in each PPI network which have k edges, among all graphlets with density ≥ 0.85. The specific PPI networks are depicted with the following colors: Yeast (Red), E.coli (Green), GDM (Blue) and GRGM (Purple). For example, plot (a) indicates that, the graphlets in the E.coli PPI network (Green) with 7 nodes and 18 edges, cumulatively, form 80% of all of its dense graphlets with density ≥ 0.85.

with 49 edges or more) for $n = 11$. This, together with the proof that we provide in Section 3.4, this observation seems to support a negative answer to the question (i-a) in Section 1.1.

(1b) The fractional distribution of GDM largely agrees with the two PPI networks for $n \leq 12$. However, GDM fails to generate any dense network (with density ≥ 0.85) with $n = 13$ or 14.

(2) The fractional distribution (of dense graphlets) in Yeast and E.coli are quite similar for smaller values of n. However for $n = 11$ and especially for $n = 12$, the fraction of graphlets in Yeast increase with density. For example, for $n = 12$, the fraction of graphlets with 59 edges is ~ 0.1 in Yeast whereas it is ~ 0.05 in E.coli. This observation seems to provide a positive answer to question (ii) in Section 1.1 (although we believe much more study should be done regarding this question).

In Figure 2, we depict how the total number of dense graphlets (with density ≥ 0.85) with n nodes ($n = \{3, \ldots, 14\}$) vary as a function of n - for the Yeast PPI network as well as the specific networks generated by GRGM and GDM whose parameters were set to emulate the Yeast network.

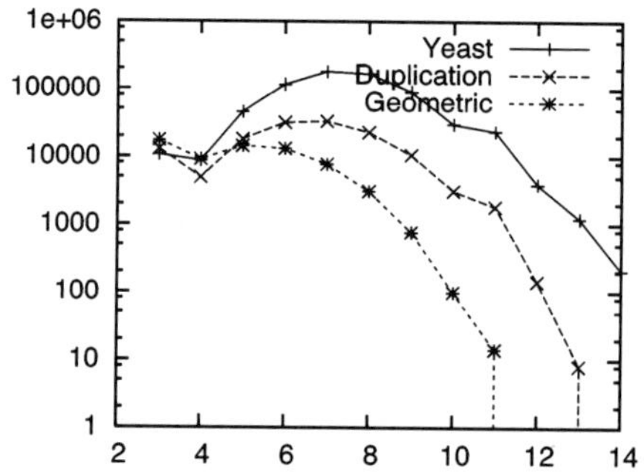

Figure 2. Total number of dense subgraphs (with density ≥ 0.85) with n nodes - in the Yeast PPI network as well as networks generated by GRGM and GDM - whose parameters are set to emulate the Yeast PPI network. The specific colors used are GRGM (Blue), GDM (Green) and Yeast (Red).

As can be seen in Figure 2, there is a wide gap between the total numbers of dense graphlets in the the Yeast PPI network and the random networks generated by GDM and GRGM. Although the number of dense graphlets for $n = 6$ is consistent with an earlier study [7], the figure 2 shows substantially difference for $n > 6$ between GDM (or GRGM) and Yeast PPI network, especially for $n \geq 8$, where there is a 7-fold (or 50-fold) difference respectively. More drastically GRGM includes no dense graphlets with $n = 12$ nodes and GDM includes no dense graphlets with $n = 14$ nodes. Once again, Figure 2 seems to support a negative answer to question (i-a) in Section 1.1: i.e. GRGM is not suitable for emulating the growth of PPI networks. The figure also seems to imply (somewhat less strongly) a negative answer to question (i-b), i.e., the GDM used in this paper does not capture the dense graphlet distribution of the Yeast PPI network. We leave the possibility of how (or if) we can modify the seed network or the GDM itself so as to better capture the distribution of denser graphlets of size bigger than 6 (see [7]) via a duplication oriented model to future study. Our final results are on the fractional distributions of all $K_{n,n}$'s and all $K_{n,n-1}$'s (which are all 0.55-dense complete bipartite graphs) up to $n = 5$ in each of the PPI networks we considered.

Bipartite Graph	Ecoli	Yeast	Duplication	Geometric
$K_{2,3}$	2685054	498844	337218	153
$K_{3,3}$	2188868	376186	23311	0
$K_{3,4}$	11103153	1677626	21623	0
$K_{4,4}$	5155489	852301	519	0
$K_{4,5}$	13561155	2077675	129	0
$K_{5,5}$	1125496+	659614	2	0

In the above table, it can be seen that in the E.Coli and the Yeast PPI networks, complete bipartite graphlets are abundant. However, as can be deduced from theorem 3.1, the GRGM cannot generate $K_{n,m}$'s for $n, m \geq 3$. Our experiments confirm this finding: there are no $K_{n,n}$'s and no $K_{n,n+1}$'s in the GRGM network for $n \geq 3$.

References

1. N. Alon, P. Dao, I. Hajirasouliha, F. Hormozdiari and S.C. Sahinalp, *Proceedings of the ISMB 2008*, (2008).
2. U. Alon, *An Introduction to Systems Biology: Design Principles of Biological Circuits* (2006).
3. K. Baskerville and M. Paczuski, *Physical Review E* **74**, 051903 (2006).
4. R. Colak, F. Moser, A. Rafiey and M. Ester, *Tech Report*, Simon Fraser University, (2008).
5. J. Grochow and M. Kellis, *Proc. RECOMB 2008*, 92, (2008).
6. D.J. Higham, M. Rasajski and N. Przulj, *Bioinformatics* **24**, 1093 (2008).
7. F. Hormozdiari, P. Berenbrink, N. Przulj and S.C. Sahinalp, *PLoS Computational Biology* **3**, e118 (2007).
8. R.M. Karp, in *Complexity of Computer Computations, Plenum Press*, 85, 1972. *Theoretical Computer Science* **369**, 234 (2006).
9. M. Middendrof, E. Ziv, and C. Wiggins, *PNAS* **102**, 9 (2005).
10. R. Milo, S. Shen-Orr, S. Itzkovski, N. Kashtan, D. Chklovskii and U. Alon, *Science* **298**, 824 (2002).
11. ME. Newman, SH. Strogatz, DJ. Watts, *Phys Rev E Stat Nonlin Soft Matter Phys.* (2001).
12. N. Przulj, D.G Corneil and I. Jurisica, *Bioinformatics* **20**, 3508 (2004).
13. L. Salwinski, C.S. Miller, A.J. Simith, F.K. Pettit, J.U. Bowie and D. Eisenberg, *Nucleic Acid Research* **32**, Database issue:D449-52 (2004).

REVERSE ENGINEERING THE EVOLUTION OF PROTEIN INTERACTION NETWORKS

TODD A. GIBSON[1] AND DEBRA S. GOLDBERG[1,2]

[1] *Computational Bioscience Program*
University of Colorado Denver, CO, USA

[2] *Computer Science Department*
University of Colorado at Boulder, CO, USA

Protein interaction network analyses have moved beyond simple topological observations to functional and evolutionary inferences based on the construction of putative ancestral networks. Evolutionary studies of protein interaction networks are generally derived from network comparisons, are limited in scope, or are theoretic dynamic models that aren't contextualized to an organism's specific genes. A biologically faithful network evolution reconstruction which ties evolution of the network itself to the actual genes of an organism would help fill in the evolutionary gaps between the gene network "snapshots" of evolution we have from different species today. Here we present a novel framework for reverse engineering the evolution of protein interaction networks of extant species using phylogenetic gene trees and protein interaction data. We applied the framework to Saccharomyces cerevisiae data and present topological trends in the evolutionary lineage of yeast.

1. Introduction

Among the tasks utilizing protein interaction networks (PIN) are inferences of how the network itself evolved. With the availability of large-scale protein interaction datasets[1-6], the properties of these networks have been analyzed and theoretical models of evolving networks have been produced which incorporate some aspect of biological evolution in an effort to reproduce properties observed in experimentally-derived networks[7,8].

Evolutionary PIN analysis has been widely explored in network comparisons which are based on the premise that disparate organisms share a common ancestor[9]. Such comparisons generally elucidated protein function and predicted interactions. Protein interaction evolution has also been studied through structures of protein complexes[10-12]. These methods have identified a number of important properties of protein interaction evolution including the role self-interacting proteins play in forming complexes of paralogous proteins and the development of homologous protein complexes through a stepwise progression of duplications of complex constituents rather than via a single monolithic duplication event. Presser, et al.[13] studied PIN evolution using a probabilistic model of motifs formed by paralogous protein pairs born of the Whole Genome Duplication (WGD) event in Saccharomyces cerevisiae's evolutionary history. With the model they found that self-interactions were prevalent in the motifs formed by pre-WGD protein interactions. This study provides a probabilistic view of pre- and post-WGD motifs. A few studies have provided an evolving view of organism-contextualized protein interaction networks. Notably Vernon, et al.[14] reconstructed the interaction evolution of MADS domain MIKC-type proteins, and Pinney, et al.[15] reconstructed the bZIP family of transcription factors.

Here we extend the ability to analyze evolutionary trajectories with a novel framework which incorporates interaction data and phylogenetic gene trees into an evolving view of the protein interaction networks of species sharing the gene trees' phylogeny. The framework meets three criteria required to measure the evolutionary trajectory of some aspect of protein interaction network evolution:

(1) The reconstruction evolves. It includes a modern network, an ancestral network, and transitions between them.
(2) The evolving PIN is associated with a specific organism. That is, each vertex in the evolving network can be traced to an extant gene.
(3) The PIN encompasses the entire interactome.

Existing studies fail to meet all three criteria. Theoretical models are decontextualized–each vertex in the model is simply a generic protein. These models have no power to elucidate organismal evolution beyond identifying broad evolutionary processes which produce network characteristics consistent with those of empirical networks. Comparative studies allow functional and topological inferences, but are less informative of the evolution of the interaction network itself. The structural and stochastic studies have identified important factors in interaction evolution, but do not incorporate them into the larger, network-level evolutionary context. Those studies which have formed organism-specific, multi-step evolutionary views of protein interaction networks have been isolated to individual protein families.

In the following sections we introduce the framework, present some initial findings having applied it to the Saccharomyces cerevisiae lineage, and discuss possible future research the framework readily accommodates.

2. Evolutionary considerations

Gene duplication is readily accepted as a primary mechanism for generating organismal complexity. Two evolutionary mechanisms have been proposed for the fate of gene duplicates: neofunctionalization and subfunctionalization. Neofunctionalization posits that the functional redundancy intrinsic to initially identical gene duplicates releases one duplicate from selective pressure. While under neutral selection one of the duplicates can accumulate random mutations and potentially acquire novel and beneficial functions[16]. Subfunctionalization states that both gene duplicates acquire mutations resulting in each duplicate assuming a complementary subset of the ancestral gene's original functions[17]. Gene duplication and subsequent neofunctionalization and subfunctionalization have straightforward analogs in models of protein interaction network (PIN) evolution. With proteins as nodes, edges between proteins represent physical interactions and serve as an indication of protein function. Proteins with identical sets of interacting partners are presumed to have identical functions. Gene duplication is modeled by copying a protein vertex in the network along with its interactions. Neofunctionalization and subfunctionalization are modeled by the gain and loss of interactions respectively.

The high false negative rate endemic of interaction data sets makes it difficult to distinguish neofunctionalization from subfunctionalization in PINs, even with the availability of PIN data from several related species. We have recently found that the established ubiquity of neofunctionalization was based on three independent flaws: a bias against observing homodimers in both yeast two-hybrid (Y2H) and affinity-purification mass-spectrometry (AP-MS) high-throughput interaction assays, a failure to consider gene duplications of interacting partners subsequent to the duplication event under analysis, and theoretical models unable to produce clustering coefficients found in empirical protein interaction networks and biologically untenable parameter requirements[18]. In the absence of compelling evidence of ubiquitous *de novo* interaction formation, we include only the well-established ubiquity of subfunctionalization[17,19,20] in our framework.

The evolutionary dynamics integrated into the evolutionary framework assume that interactions between paralogs arose via the duplication of a self-interacting protein. If either paralog is homomeric, or if a paralogous interaction exists between them, then the ancestral protein is assumed to

(A) (B)

Figure 1. Duplication of self-interacting proteins. (A) A self-interacting protein. (B) When the self-interacting protein duplicates, a paralogous interaction is formed. Either or both of the two self-interactions, or the paralogous interaction may be lost in evolution, but any one of the three interactions is sufficient information to reconstitute the original ancestral gene with a homomeric interaction.

be self-interacting (see Figure 1).

3. Methods

We now introduce our framework for reverse engineering the evolution of an organism's protein interaction network. To assist in the exposition, we refer the reader to Figure 2 which graphically represents an example of the entire framework.

3.1. *Framework inputs*

As the figure shows, there are three inputs (highlighted in yellow) the framework requires. The first is a phylogenetic tree of the species subject to the evolutionary reconstruction. Second are gene family trees reconciled against the phylogenetic tree: trees describing each gene family's speciations, duplications, and losses (not shown in figure). Figure 2 has been color-coded to identify the portion of the phylogeny each phase of the reconstruction is associated with (i.e., extant species, common ancestor, last common ancestor).

The third input is protein interaction data for the extant species present in the phylogenetic tree. Semantically, all of the genes identified in the gene trees comprise the nodes of a single large network. The edges of the network may be drawn from the interaction data of more than one species. The example in Figure 2 shows the genes for all three species, and interaction data for two of the three species.

3.2. *Evolutionary event identification*

Prior to reverse-engineering the network evolution,t his phase simple processes information found from the previously-published inputs. Each speciation, duplication, and loss event in the gene trees are associated with a node of the phylogenetic tree. In Figure 2, the evolutionary events identified from the gene trees are shown within the soft-cornered box running down the center of the figure and are color-coded to the phylogenetic node they are associated with.

3.3. *Reverse engineering*

The protein interaction network input is combined with the evolutionary events generated in the previous step to reverse engineer the network evolution. The phylogenetic tree drives the ordering of the process. The extant species leaves of the phylogenetic tree are processed first, followed by the common ancestor nodes. The tree is iteratively "rolled up" until the last common ancestor is reached.

With each phylogenetic node, the evolutionary events associated with the node are applied to the protein interaction network. The evolutionary events are reversed with respect to time. Note that each evolutionary event identified previously describes an action on a single gene. A gene loss is the loss of a single gene. A duplication event produces two genes from one. Similarly, a gene speciation event produces two genes (one for each species) from a single (common ancestor's) gene. During reverse engineering, the opposite action is taken. A gene loss becomes a gene gain. A duplication or speciation event joins two separate genes into a single ancestral gene.

Paralogs which interact with the same neighbor are assumed to be preserving a redundant ancestral function (see Figure 3). An interaction present in only one paralog is presumed to have been genetically silenced in the other paralog after duplication. As evolutionary events are reverse engineered, the ancestor of each gene duplication acquires interactions with all of the neighbors its paralogous progeny pair interact with (Figure 3B).

Running down the left side of Figure 2 is a complete example of reverse engineering network evolution. Starting with the extant species and iterating through the phylogenetic tree, speciation and duplication events are reversed to join nodes from the protein interaction network together[a]. As proteins in the network are joined, the merged protein interacts with the union of the set of neighbors each protein interacted with individually. Later when the evolutionary re-creation is rolled forward and the merged protein is again separated into two separate proteins, it will be necessary to reproduce the sets of neighbors the separate proteins interact with. Therefore as proteins are merged, the neighbors not included in each separate

[a]There are no gene loss evolutionary events in the example.

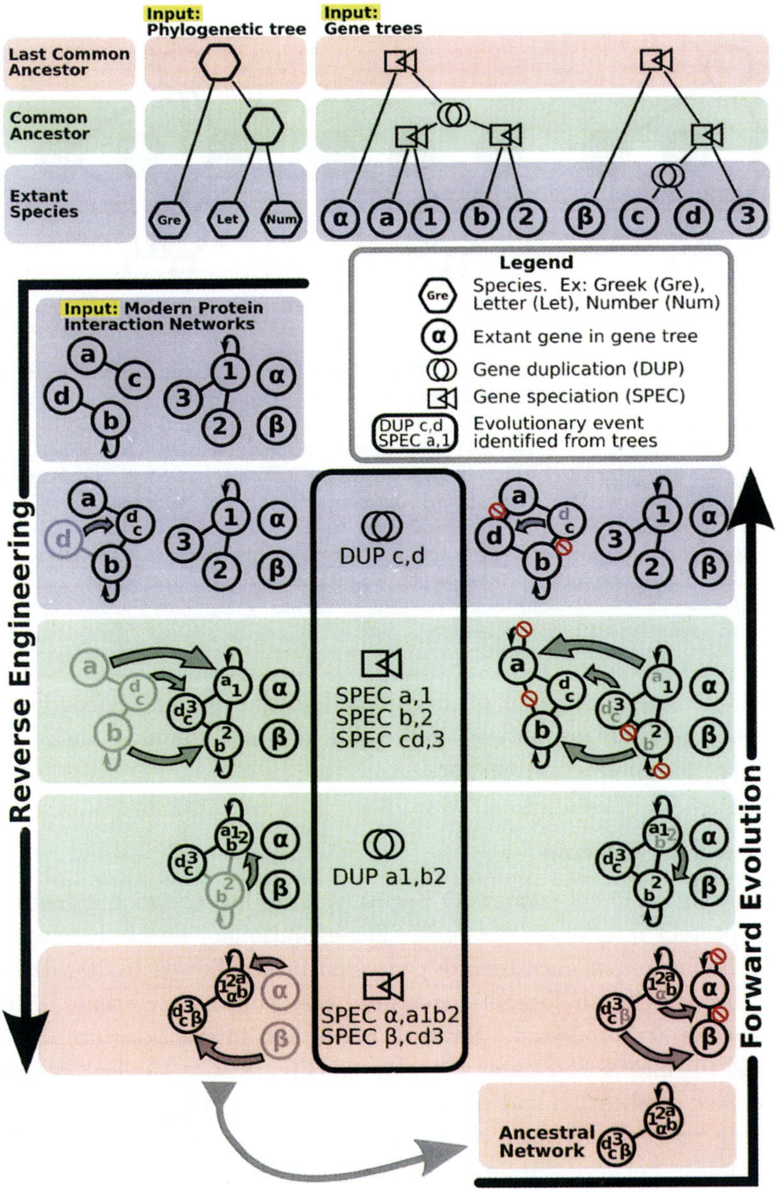

Figure 2. The reverse-engineering and forward evolutionary re-creation framework.

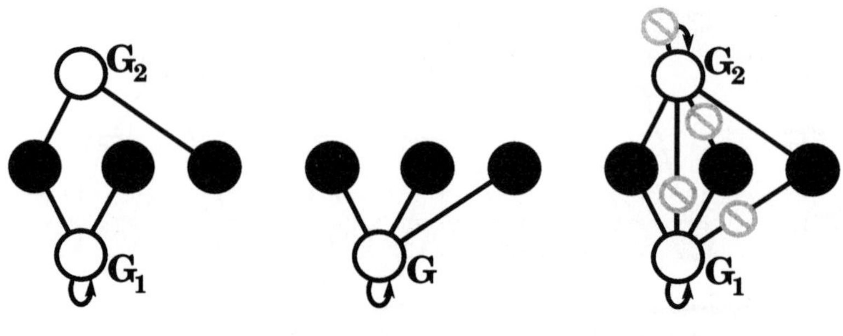

Figure 3. Reverse engineering and forward evolutionary re-creation. For this example assume a duplication event to duplicate gene G into G_1 and G_2 was generated during event identification (Section 3.2). (A) The extant protein interaction network. (B) During reverse engineering, G_1, G_2, and their interactions are collapsed into a single gene G. At this time the interactions to be lost after duplication are recorded. (C) Forward evolutionary re-creation. After the gene and its interactions are duplicated (including a paralogous interaction born of the self-interaction), the interaction losses recorded during reverse engineering are processed. Note that speciation differs from gene duplication in that a speciating self-interacting protein never produces a paralogous interaction between species duplicates. Forward evolutionary re-creation naturally segments the last common ancestor into species-specific components.

protein's interacting neighbor set is recorded. Once reverse-engineering is completed, the resulting network represents the last common ancestor's putative protein interaction network.

3.4. *Forward evolution*

Forward evolution in the framework begins with the last common ancestor's putative protein interaction network. The evolutionary events associated with the last common ancestor are processed first, followed by iteratively moving through the phylogenetic tree until the evolutionary events of the extant species are processed. Network nodes and interactions are duplicated, and the interaction losses are removed (Figure 3C). The lost interactions for each duplication and speciation event are recorded during reverse engineering process. Because redundant interactions have been found to diverge rapidly between paralogous genes[19,20], redundant interactions are removed immediately after the duplication event (i.e., prior to the next speciation event in the phylogenetic tree.

Figure 2 shows a complete example of the forward evolutionary re-creation running up the right side of the figure. The end of the forward evolutionary re-creation recapitulates the protein interaction network used at the beginning of the reverse-engineering process.

4. Results

We implemented our framework in Python and ran it on a standard PC running Linux. For the protein interaction framework input we combined multi-validated protein interactions[21] and high confidence interactions culled from a genome-wide in vivo screen using a protein-fragment complementation assay (PCA)[22]. Both data sets are Saccharomyces cerevisiae interactions

The phylogenetic tree and gene tree inputs were drawn from a study which generated 25,408 gene trees reconciled against a phylogeny of 19 species of Ascomycota fungi[23] [b].

The Evolutionary Event Identification phase generated 103,091 gene loss events, 7,711 gene duplication events, and 84,167 speciation events among the 19 species. The protein interaction network included 117,286 individual genes spread across 19 yeast species. Protein interaction input included only Saccharomyces cerevisiae so the vertices in the protein interaction network associated with the other species were not connected. The Saccharomyces cerevisiae vertices contain 5,780 genes, 4,052 of them in the largest component, and 12,341 edges, of which 12,241 are contained in the largest component. From this network, and the evolutionary events created previously, the evolving network was reconstituted.

The evolutionary reconstruction requires no computationally difficult algorithms. With the stated inputs and implementation the entire reconstruction runs in under half an hour.

Figure 4 illustrates the effect the forward evolutionary re-creation has on the evolutionary trajectories of the average degree and clustering coefficient. The average degree does not include self-interactions in the count. Self-interactions are absent from the Vázquez, et al. model, and omitting them from degree calculations is consistent with other protein interaction network analyses.

The plot has been highlighted between timesteps 5 and 6, indicating the time period within which the whole-genome duplication event occurred during Saccharomyces cerevisiae's evolution[24]. The curves are flat beyond timestep 6 indicating an absence of further duplication or loss events.

[b]We utilized an updated and expanded data set acquired from Ref. 23's companion Web site.

Figure 4A shows that the average degree decreases during evolution. Intuitively we might expect that as network vertices become less well-connected with their neighbors, they would become less clustered as well. However, Figure 4B indicates that the clustering coefficient increases.

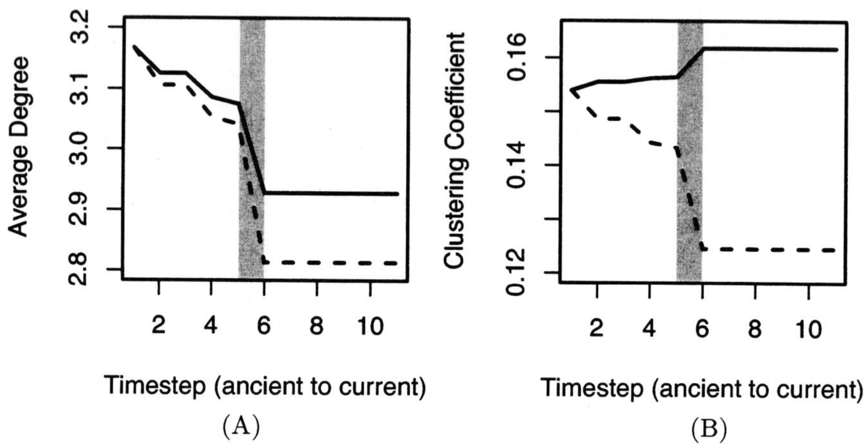

Figure 4. Basic topological properties of the evolving protein interaction network. The solid line is the evolutionary re-creation of Saccharomyces cerevisiae. The dashed line is the Vázquez, et al[4].model of protein interaction network evolution. The grey bar between timepoints 5 and 6 represent the evolutionary period within which the whole-genome duplication event occurred. (A) Average degree of the large component. (B) Clustering coefficient of the large component.

We next tracked the same topological values in an implementation of the Vázquez, et al.[8] theoretical model. Vázquez, et al. was selected because it reflects the principles of gene duplication, subfunctionalization, and homomeric duplication central to our framework. It is also appropriate as the model generates high clusterings consistent with empirical networks. Other theoretical models, in particular that of Solé, et al.[7] are unable to attain high clusterings and do not specifically address evolution of self-interactions[18].

To eliminate biases introduced by seed graphs[25], the theoretical model was seeded with the ancestral network which begins the forward evolutionary re-creation. The topological values of the theoretical model were measured each time the number of proteins in the model equaled that of the evolutionary re-creation at the same timepoint.

The Vázquez, et al. model takes two parameters: p, the probability of adding an edge between protein duplicates (i.e., probability of duplicating a self-interacting protein), and q, the probability of a redundant edge being lost from either the progeny or progenitor genes due to subfunctionalization. The parameter values were selected based on the number of paralogous interactions which survived duplication of a self-interacting protein in the evolutionary re-creation itself ($p = 0.98, q = 0.046$). The theoretical model was run 1,000 times and the mean of their topological values plotted.

Figure 4A shows that the theoretical model produces a similar reduction in the average degree as in the evolutionary re-creation. However, as Figure 4B shows, the clustering coefficient decreases with the addition of new genes. The parameter values published by Vázquez, et al.[8] ($p = 0.7, q = 0.1$) produced similar curves (data not shown).

5. Discussion

The evolutionary framework provides a novel, dynamic view of an organism's protein interaction network. Previous efforts have identified the influence of evolution on topologically-relevant factors (e.g., self-interacting proteins), and the few which have measured the evolution of protein interaction networks across several evolutionary periods have been isolated to small subnetworks. Despite the dynamic nature of theoretical models, theoretical model validation has commonly involved post hoc analysis of the generated network. A comparison of the evolutionary trajectories of Vázquez, et al.[8] and the yeast evolutionary re-creation reveal that while in both models the proteins on average interact with fewer neighbors over time, the yeast evolution generates higher clustering while the clustering is reduced in the theoretical model. This is of particular interest due to the high clustering levels found in empirically-derived protein interaction networks[19,7,26-28]. These results suggest that surviving gene duplicates are not distributed randomly throughout the interactome.

The framework may also provide some inferential power to species for which interaction data is not available but are nonetheless represented in the re-creation by phylogenetic gene trees. Under these conditions the species' genes are represented in the network as zero-degree vertices. As the gene duplication, speciation, and loss events are reverse-engineered, the ancestors of these genes interact with other ancestral genes based on interaction data available from other species. For example, the ancestral genes α and β in Figure 2 acquire interactions based on interaction data of other species.

As the evolutionary re-creation moves forward in evolutionary time, these genes speciate, duplicate and ultimately lose all ancestral interactions to return to their non-interacting empirical state. To construct initial inferences on genes or entire species which are lacking interaction data, interaction losses associated with gene duplication and speciation events can be selectively suppressed. For species in the phylogenetic tree without interaction data, this amounts to creating a first-pass putative interaction network based on homologous data sets. Though the inferred network surely contains false positives, the candidate set of interactions have novel derivation. Specifically, the putative network is derived directly from the last common ancestor rather than solely, for example, sequence similarity with homologous genes[29]. Putative interactions produced from the re-creation are a product of the gene duplication and loss events. This effectively combines the homology-based prediction of interactions[29] with parameters derived from networks of species included in the phylogeny input. Indeed, subsequent development of the framework should include a comparison with homology-based predictions.

Another area for subsequent exploration is the framework's assumption that all interactions arise through subfunctionalization. Although it has been suggested that the de novo acquisition of interactions is not a common occurrence[18], de novo interactions can not be completely ruled out. One possibility is the development of an an error model to handle this uncertainty.

As with all network analyses, the utility of the framework relies in part on the quality of the interaction data. Generally, high-confidence data sets eliminate false-positives at the expense of reduced coverage[30]. However, as disparate protein interaction datasets from different species are combined into common ancestors, ancestral networks benefit from wider coverage than the individual data sets provide.

The framework presented here reinvigorates the study of network evolution. Differences in topological properties, development of motifs, and the development of functional modules are just a few of areas that may now be analyzed in the context of their evolutionary trajectories. The framework provides new opportunities to analyze both the evolutionary trajectory of a single species as well as processes through which network features diverge between different species. As phylogenetic gene trees come available for a wider range of species, and additional interaction sets are published, evolutionary network re-creations contextualized to specific species will become increasingly valuable.

Acknowledgments

Funding. This material is based upon work supported by the National Science Foundation under Grant MCB 0630250.

References

1. Uetz P, et al. (2000) A comprehensive analysis of protein-protein interactions in saccharomyces cerevisiae. *Nature* 403:623–627.
2. Ito T, et al. (2001) A comprehensive two-hybrid analysis to explore the yeast protein interactome. *Proc Natl Acad Sci U S A* 98:4569–4574.
3. Gavin A C, et al. (2002) Functional organization of the yeast proteome by systematic analysis of protein complexes. *Nature* 415:141–147.
4. Gavin A C, et al. (2006) Proteome survey reveals modularity of the yeast cell machinery. *Nature* 440:631–636.
5. Krogan N J, et al. (2006) Global landscape of protein complexes in the yeast Saccharomyces cerevisiae. *Nature* 440:637–643.
6. Reguly T, et al. (2006) Comprehensive curation and analysis of global interaction networks in Saccharomyces cerevisiae. *J Biol* 5:11.
7. Solé R V, Pastor-Satorras R, Smith E, Kepler T B (2002) A model of large-scale proteome evolution. *Advances in Complex Systems* 5:43.
8. Vázquez A, Flammini A, Maritan A, Vespignani A (2003) Modeling of protein interaction networks. *ComPlexUs* 1:38–44.
9. Sharan R, Ideker T (2006) Modeling cellular machinery through biological network comparison. *Nat Biotechnol* 24:427–433.
10. Pereira-Leal J, Levy E, Kamp C, Teichmann S (2007) Evolution of protein complexes by duplication of homomeric interactions. *Genome Biol* 8:R51.
11. Pereira-Leal J B, Levy E D, Teichmann S A (2006) The origins and evolution of functional modules: lessons from protein complexes. *Philos Trans R Soc Lond B Biol Sci* 361:507–517.
12. Pereira-Leal J B, Teichmann S A (2005) Novel specificities emerge by stepwise duplication of functional modules. *Genome Res* 15:552–559.
13. Presser A, Elowitz M B, Kellis M, Kishony R (2008) The evolutionary dynamics of the saccharomyces cerevisiae protein interaction network after duplication. *Proc Natl Acad Sci U S A* 105:950–954.
14. Veron A S, Kaufmann K, Bornberg-Bauer E (2007) Evidence of interaction network evolution by whole-genome duplications: A case study in mads-box proteins. *Mol Biol Evol* 24:670–678.
15. Pinney J W, Amoutzias G D, Rattray M, Robertson D L (2007) Reconstruction of ancestral protein interaction networks for the bzip transcription factors. *Proc Natl Acad Sci U S A* 104:20449–20453.
16. Ohno S (1970) *Evolution by Gene Duplication* (Springer, New York).
17. Force A, et al. (1999) Preservation of duplicate genes by complementary, degenerative mutations. *Genetics* 151:1531–1545.

18. Gibson T, Goldberg D (2008) Questioning the ubiquity of neofunctionalization. PLoS Computation Biology. In Revision.
19. Wagner A (2001) The yeast protein interaction network evolves rapidly and contains few redundant duplicate genes. *Mol Biol Evol* 18:1283–1292.
20. He X, Zhang J (2005) Rapid subfunctionalization accompanied by prolonged and substantial neofunctionalization in duplicate gene evolution. *Genetics* 169:1157–1164.
21. Batada N N, et al. (2007) Still stratus not altocumulus: further evidence against the date/party hub distinction. *PLoS Biol* 5:e154.
22. Tarassov K, et al. (2008) An in vivo map of the yeast protein interactome. *Science* 320:1465–1470.
23. Wapinski I, Pfeffer A, Friedman N, Regev A (2007) Natural history and evolutionary principles of gene duplication in fungi. *Nature* 449:54–61.
24. Kellis M, Birren B W, Lander E S (2004) Proof and evolutionary analysis of ancient genome duplication in the yeast saccharomyces cerevisiae. *Nature* 428:617–624.
25. Hormozdiari F, Berenbrink P, Pržulj N, Sahinalp S C (2007) Not all scale-free networks are born equal: The role of the seed graph in PPI network evolution. *PLoS Comput Biol* 3:e118.
26. Goldberg D S, Roth F P (2003) Assessing experimentally derived interactions in a small world. *Proc Natl Acad Sci U S A* 100:4372–4376.
27. Yook S H, Oltvai Z N, Barabási A L (2004) Functional and topological characterization of protein interaction networks. *Proteomics* 4:928–942.
28. Han J D J, Dupuy D, Bertin N, Cusick M E, Vidal M (2005) Effect of sampling on topology predictions of protein-protein interaction networks. *Nat Biotechnol* 23:839–844.
29. Yu H, et al. (2004) Annotation transfer between genomes: protein-protein interologs and protein-DNA regulogs. *Genome Res* 14:1107–1118.
30. von Mering C, et al. (2002) Comparative assessment of large-scale data sets of protein-protein interactions. *Nature* 417:399–403.

IDENTIFYING DYNAMIC NETWORK MODULES WITH TEMPORAL AND SPATIAL CONSTRAINTS

RUOMING JIN [1,*], SCOTT MCCALLEN [1], CHUN-CHI LIU [2], YANG XIANG [1], EIVIND ALMAAS [3], XIANGHONG JASMINE ZHOU [2*]

[1] *Department of Computer Science, Kent State University, Kent, OH, USA*

[2] *Program in Molecular and Computational Biology, University of Southern California, Los Angeles, CA, USA*

[3] *Bioscience and Biotechnology Division, Lawrence Livermore National Laboratory, Livermore, CA, USA*

Despite the rapid accumulation of systems-level biological data, understanding the dynamic nature of cellular activity remains a difficult task. The reason is that most biological data are static, or only correspond to snapshots of cellular activity. In this study, we explicitly attempt to detangle the temporal complexity of biological networks by using compilations of time-series gene expression profiling data. We define a dynamic network module to be a set of proteins satisfying two conditions: (1) they form a connected component in the protein-protein interaction (PPI) network; and (2) their expression profiles form certain structures in the temporal domain. We develop an efficient mining algorithm to discover dynamic modules in a temporal network. Using yeast as a model system, we demonstrate that the majority of the identified dynamic modules are functionally homogeneous. Additionally, many of them provide insight into the sequential ordering of molecular events in cellular systems. Finally, we note that the applicability of our algorithm is not limited to the study of PPI networks, instead it is generally applicable to the combination of any type of network and time-series data.

1. Introduction

Cellular systems are highly dynamic and responsive to cues from the environment. Cellular function and response patterns to external stimuli are regulated by a complex web of diverse molecular interactions, such as protein-protein interactions, protein-DNA interactions, and metabolic processes. Despite the availability of large-scale biological network data[23,15,7,8], gaining a system-level understanding of the *dynamic* nature of cellular activity remains a difficult and, until recently, a much overlooked task. Since there typically is little direct information available on the temporal dynamics of these network interactions, the majority of molecular-interaction network modeling and analysis has been solely focused on static properties. For example, a protein-protein interaction (PPI) network obtained from yeast two-hybrid experiments can be viewed as a comprehensive graph of the edges (interactions) that eventually may occur under the set of tested conditions. However, it is not guaranteed that two adjacent edges would ever occur *simultaneously* or even close in time, and thus, that identified network modules and motifs will correspond to functionally relevant units. While a series of studies

*to whom correspondence should be addressed

have been aimed at identifying modules in PPI networks [17,2,19], the networks in their analysis have all been regarded as static.

In this study, we attempt to detangle the *temporal* complexity of biological networks by identifying dynamic modules, consisting of a set of proteins that coexist both spatially and temporally. We will use the PPI network as the spatial constraints. We estimate temporal characteristics of a network component by using compilations of time-series gene expression profiling data, since accurate temporal parameters are not yet available for PPI systems. A similar approach has been employed to study the dynamic protein complex formation during the cell cycle [3,12]. In order to measure the similarity between two time-series gene-expression profiles [1,18], we have implemented time-warping dynamic programming. A significant advantage of this method is that it identifies the best local alignment of segments in two time-series profiles. Consequently, it can identify time-shifted and local similarity patterns which are often overlooked by computing the Pearson's correlation of two entire profiles. We represent the local alignment of two profiles as a range pair (or interval pair), denoted as $[s1, e1]$:$[s2, e2]$, where $s1$ and $s2$ are the starting time points, and $e1$ and $e2$ are the end time points of the alignment. Since two proteins may interact over different time segments in a long time-series and they may show varying level of cooperativity in those time segments, we may also consider suboptimal alignments of two profiles. In this case, the temporal expression similarity of two interacted proteins can be represented as a list of range pairs. Given a PPI network, we add the local alignment information from the time-series gene expression data, and refer to the network as a *temporal network*.

On a temporal network, we can discover how two proteins' activities correlate with each other over time. More importantly, for multiple proteins we can identify when a dynamic module is activated in the PPI network. We define a dynamic network module to be a set of proteins which satisfy two conditions: (1) they form a connected component in the PPI network; and (2) their expression profiles form certain structures (see below) in the temporal domain. An example dynamic module satisfying these constraints could consist of all proteins, for which time-series local alignments overlap over a common time period. That is, all protein interactions in this example module are synchronized to perform their functions. Another interesting example corresponds to the case of activity cascades. Here, a dynamic module is triggered by one or a small number of proteins that activate their network neighbors. These in turn, may also propagate the signal further through neighbor activation. Given this, we use a directed edge to represent how the activity of a protein may invoke that of another protein, and the direction is determined by the range-pair of their time series alignment, i.e. the beginning aligned time point of the triggering protein is likely to be earlier than that of the triggered protein.

Although many graph algorithms are available to perform network analysis, including those recent work [21,22,4,14] for mining temporal and evolutionary networks, to our knowledge, no algorithm has been developed to identify network patterns while accounting for the interval-based temporal constraints. In this study, we take the first step in this direction by developing efficient mining algorithms to discover dynamic modules in a temporal biological network. We have performed detailed experimental testing using yeast as a model system. In addition, many of those modules provide insight into the sequential order of molecular events in cellular systems. Furthermore, a majority of the discovered modules are not densely connected, yet they comprise functional units. Such modules are very difficult to identify based on the PPI network alone. We are well aware of the limitation of microarray data in measuring protein activities, and the incompleteness of protein-protein interaction data. While using those data as an example, we want to emphasize that our algorithm is generally applicable to combining any types of network and time-series data.

2. Problem Definition

We represent the PPI network as a graph $G = (V, E)$, where proteins correspond to the vertex set V, and the protein-protein interactions are recorded as the edge set E. Each vertex (protein) is associated with a gene expression time-series, formally $T_i = (x_1^i, x_2^i, \cdots, x_n^i)$.

We use the time-warping dynamic programming algorithm [18] to align two time-series expression profiles, and use a range pair to represent the resulting best local alignment for each interacted protein pair. The range pair for protein pair $e = (v_1, v_2) \in E$ is denoted as $r(e) = ([i, i'] : [j, j'])$, where i, j are the starting time points of the aligned interval pair of an interacting protein pair, and i', j' are the end points. We denote $r(e)[v_1] = [i, i']$ and $r(e)[v_2] = [j, j']$. From the PPI network we eliminate those edges (the interacted protein pairs) that do not have any local similar range pairs, and term the resulting range-pair network as the *temporal network*.

We also assign directions to the edges in the temporal network. Such assignment is based on the observation that the range pairs in the temporal network can provide the sequential ordering of activities between edge-connected vertices. Specifically, if one interval in the range pair appears earlier than the other one, it could be possible that the activity of the vertex (protein) associated with the first interval induces the activity of the other vertex. That is, the sequential order obtained could facilitate causal inference, although causal inference requires much more information and is beyond the scope of this work. In the following, we will use a directed edge to represent the sequential order between two vertices in the temporal network. Specifically, let $[t_{A1}, t_{A2}] : [t_{B1}, t_{B2}]$ be the interval pair between protein A and B. If $t_{A1} - t_{B1} \geq k$, we build a directed edge from protein A to B; if $|t_{A1} - t_{B1}| < k$, we build an undirected edge between A and B, where k serves as a smoothing parameter against noise.

As previously mentioned, the module discovery approach in this study is bound by both *spatial* and *temporal* constraints. Intuitively, the spatial constraints refer to the physical interactions between proteins, and the temporal constraints corresponds to the closeness of range pairs, which can be formalized based on the *second order subgraph*.

Definition 1. (Second Order Graph) The second order graph of the temporal biological network $G = (V, E)$ is a graph $G' = (V', E')$, where V' records each edge e in $E(G)$ as a unique vertex v'_e, and for any two vertices v'_{e1} and v'_{e2} in $V(G')$ there exists an edge in $E' = E(G')$ between them if and only if 1) the two edges e_1 and e_2 share a common vertex y, i.e., $e_1 = (x, y)$ and $e_2 = (y, z)$, or vice versa; and 2) the temporal constraint: $|r(e_1)[y] \cap r(e_2)[y]| \geq d$.

To facilitate our discussion, we refer to a vertex in the second order graph as an *edge-vertex*. Note that, in the second order graph, two edge-vertices with a common vertex y (in the original temporal network G) are neighbors if the two intervals of y specified by the corresponding interval-pairs in G overlap an interval with length no less than d. Further, for a subgraph G_s of G ($G_s \subseteq G$), its corresponding second order graph, G'_s, is a subgraph of G' ($G'_s \subseteq G'$).

Formally, we define the *dynamic module* as follows.

Definition 2. (Dynamic Module) Given a temporal network $G = (V, E)$ and its second order network $G' = (V', E')$, a dynamic module is a subgraph G_s of G such that 1) G_s is connected (we refer to this condition as the spatial constraint); and 2) its corresponding second order subgraph G'_s is also connected (we refer to this condition as the temporal constraint).

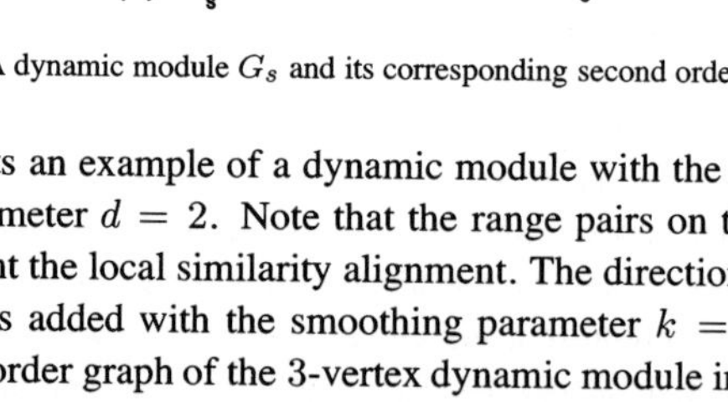

Figure 1. A dynamic module G_s and its corresponding second order graph G'_s

Figure 1(a) lists an example of a dynamic module with the choice of temporal constraint parameter $d = 2$. Note that the range pairs on the outside of the subgraphs represent the local similarity alignment. The direction of edges for the dynamic module is added with the smoothing parameter $k = 2$. Figure 1 (b) shows the second order graph of the 3-vertex dynamic module in Figure 1(a).

However, to enumerate all dynamic modules is computationally very expensive. In this paper, in order to reduce the enumeration complexity, we enumerate only the *edge-maximal dynamic modules*.

Definition 3. (Edge-Maximal Dynamic Module) Given a temporal network $G = (V, E)$ and its second order network $G' = (V', E')$, an edge-maximal dynamic module is a dynamic module $G_s \subseteq G$, and there is no any other dynamic module $G_{s'} \subseteq G$, such that $V(G_s) = V(G_{s'})$ and $E(G_s) \subset E(G_{s'})$.

3. Algorithm for Edge-Maximal Dynamic Module Discovery

The definition of dynamic modules requires to find the subgraph G_s which is connected (the spatial constraint) and its corresponding second order subgraph G'_s is also connected (the temporal constraint). An efficient enumeration algorithm thus should aggressively employ these two constraints to prune the search space. In particular, these two constraints have an interesting relationship, which will form the basis of our enumeration algorithm.

Lemma 1. *Let G_s and G'_s be the corresponding subgraph and second order subgraph. If G'_s is connected (the temporal constraint holds), then its corresponding subgraph G_s is also connected (the spatial constraint is true). However, if G_s is connected, its second order subgraph may not be connected.*

For simplicity, the proof is omitted. What this lemma suggests is that these two constraints are not *symmetric*, and one of the constraints can be easily satisfied without explicit tests. This lemma also relates to an important property, the *anti-monotone* property, which is the key for many efficient mining and enumeration algorithms.

Lemma 2. *Let G_s and its corresponding second order subgraph G'_s form a dynamic module. All the connected subgraphs of the second order subgraph G'_s always form dynamic modules.*

From the enumeration purpose, G'_s has the follow properties. Let $G'_s \cup \{v'\}$ be a new connected second order subgraph by adding a new vertex v' and associated edges to G'_s. Then, the new subgraph corresponding to the new second order subgraph will always form a *dynamic module*. However, this property does not hold for G_s. Adding a new vertex v and associated edges to G_s may produce a new second order subgraph which is not connected. Thus, if we try to enumerate all the (edge-maximal) dynamic modules from the first order graph, we do not have the anti-monotone property.

Given this, we will try to enumerate edge-maximal dynamic module directly by utilizing the second order graph. Simply speaking, we will try to recursively enumerate all the connected subsets of edge-vertices in the second order graph, and each of such subsets corresponds to a unique dynamic module. However, the issue is that how we can identify the edge-maximal ones efficiently. Clearly it is too computationally expensive to first enumerate *all* dynamic modules and then identify the edge-maximal ones. Note that the number of dynamic modules can be exponentially larger than the number of edge-maximal modules.

In the following, we introduce a novel approach which uncovers all the edge-maximal dynamic modules by coupling the original temporal network and its second order graph. It directly generates edge-maximal dynamic modules without any redundancy. Algorithm 1 provides the backbone of our approach. Conceptually, this algorithm enumerates all the edge-maximal dynamic modules in two stages. In the first stage, it tries to find a new dynamic module, which corresponds to a connected edge-vertex set in the second order graph. Then in the second

Algorithm 1 $DyModSearch(Set\ P, Set\ N, Set\ Ex)$

Parameter: P {the set of edge-vertices in G' currently in the module}
Parameter: N {the set of edge-vertices in G' that can join to the current module}
Parameter: Ex {the set of edge-vertices in G' that have been previously expanded}

1: **for each** $n \in N$ **do**
2: $P' \leftarrow Reach(n, G'_s[cover(P \cup \{n\})])$ {the edge-vertices that can be reached from vertex n in the second order subgraph $G'_s[cover(P \cup \{n\})]$}
3: **if** $(P' \backslash P) \cap Ex = \emptyset$ **then** {no edge-vertex in $P' \backslash P$ has been visited earlier}
4: Output the maximal dynamic module $G_{P'}$ {$G_{P'}$ is the subgraph corresponding to the vertex-set P' in the second order graph}
5: **if** $|cover(P')| < T$ **then** {T: user-specified size for maximal dynamic modules}
6: $Ex' \leftarrow Ex \cup P'$
7: $N' \leftarrow (N \cup Neighbor(P' \backslash P)) \backslash Ex$ {Except for initialization ($|P'| = 1$): $N \leftarrow Neighbor(P') \backslash Ex$}
8: $DyModSearch(P', N', Ex')$
9: **end if**
10: **end if**
11: $v \leftarrow cover(\{n\}) \backslash cover(P)$ {the new vertex in the original network being introduced by the new edge n}
12: $P_v \leftarrow \{v' | v' \in (P' \backslash P) \wedge v'\ covers\ v\}$
13: $Ex \leftarrow Ex \cup P_v$ {add P_v to the exclusion set E}
14: $N \leftarrow N \backslash Ex$
15: **end for**

stage, it tries to quickly expand this dynamic module into an edge-maximal dynamic module.

Algorithm 1 performs a DFS (depth-first search) style enumeration utilizing three sets P, N and Ex. Set P records all the edge-vertices of the second order graph which represents the current edge-maximal dynamic module. Set N records all the edge-vertices that are neighbors of P, and can be appended to produce new dynamic modules. Set Ex records all the edge-vertices which can no longer be used to generate any new edge-maximal dynamic modules. At each invocation of the algorithm, we will produce an edge-maximal dynamic module, represented by set P.

To make the algorithm efficient, we need specifically address the following three issues. 1) How can we transform a dynamic module into an edge-maximal dynamic module? Specifically, how can we pick up edge-maximal dynamic modules quickly without enumerating other non-edge-maximal ones? 2) How can we avoid generating an edge-maximal dynamic module more than once? In other words, how can we ensure that each edge-maximal dynamic module is only generated once? 3) How can we aggressively prune the search space to ensure that we remove as early as possible the dynamic modules that will not grow into edge-maixmal dynamic modules?

The first issue is taken care of by Line 2: simply speaking, after we generate a new dynamic module $P \cup \{n\}$ by combining the edge-maximal module P with a new edge-vertex n, we will immediately expand it to make it edge-maximal. This can be done quickly by utilizing the following property: *each newly added edge-vertex n will cover a new vertex v in the original temporal network G*, except that the first edge-vertex in P will cover two new vertices in G. This property comes from the fact that P is an edge-maximal dynamic module. Basically, it already contains all the edges which can be connected for the existing vertices to form a dynamic module. Thus, any new dynamic module $P \cup \{n\}$ ($P \neq \emptyset$) must involve a new vertex v. However, we need to note that those additional edge-vertices which can help make the new dynamic module $P \cup \{n\}$ *edge-maximal* do not necessarily links to v directly, since they may be connected through other newly introduced edge- vertices. Given this, we will apply $cover(P \cup \{n\})$ to map the edge-vertices in set $P \cup \{n\}$ to their corresponding vertices in the original network G, and then extract its induced subgraph $G_s[cover(P \cup \{n\})]$, and eventually the corresponding second order subgraph $G'_s[cover(P \cup \{n\})]$. Finally, the connected component in this induced subgraph containing n is our desired edge-maximal dynamic module for $P \cup \{n\}$ (this is achieved by Line 2: searching all the edge-vertices that can be reached from edge-vertex n in the second order subgraph $G'[cover(P \cup \{n\})]$).

The second issue is resolved by a simple test in Line 3. Basically, if the newly added edge-vertices had been previously visited, i.e., $(P' \backslash P) \cap Ex \neq \emptyset$ (Line 3), then, we would know that the newly generated edge-maximal dynamic module had already been enumerated. In other words, we will only select the modules when $(P' \backslash P) \cap Ex = \emptyset$ (Line 4).

Finally, the third issue is addressed by exploiting the coupling between the original graph and the second order graph, and by maintaining the sets N and Ex (Lines 6-7 and Lines 11-14). The set N contains edge-vertices that could be the valid extension of the current dynamic module, and the set Ex contains the edge-vertices that are invalid for extensions. Here, we will again apply the property that each newly added edge-vertex n will cover a new vertex v in the original temporal network G (when $P \neq \emptyset$). For the newly added vertex v, we identify all of its adjacent edges in the original temporal network, and further identify the subset that are connected to P in the second-order graph (P_v, Line 12). This subset (P_v) is then appended to Ex and removed from N (Line $13 - 14$). Such coupling between the two graphs helps to narrow down the search space, and the removal of edge-vertices allows the effective pruning of the search space.

4. Results

4.1. *Data source and alignment of time-series expression data*

We collected 36066 distinct protein-protein interactions of *S. cerevisiae* from The BioGrid databases [20], the MIPS database [9], and other sources [25,10]. As is customary, self interactions representing autoregulation or protein homodimerization were not included in the analysis. We collected a total of 11 microarray gene expression time-series for *S. cerevisiae*. These datasets include from 10 to 25 time points of various temporal scales (*e.g.* minutes and hours) and measure gene-regulation under very different biological conditions, such as the cell cycle, filamentous-form growth, fermentation, carbon source perturbation, and thiolutin treatment.

To analyze these time-series, we implemented local similarity analysis [18], which is a variant of the time-warping algorithm, to identify the local alignment between two time series. Since most of the collected time-series expression datasets are short, here we only identify the best local alignment for each protein pair and do not consider suboptimal alignments. For each gene pair with an identified protein-protein interaction, we performed a normal score transformation [11] for their expression profiles before analyzing their local alignment by dynamic programming. We define the local similarity score as the maximal sum of the product of the corresponding entries of all the subsequences of the two time-series within a predefined time delay D. D can be chosen to fit to individual situations. Without losing generality, we choose $D = 5$ in our analysis. To determine the significance threshold ($P < 0.025$) of the computed local similarity score, we carried out the local similarity analysis on 1,000,000 pairs of random expression profiles with a normal distribution. We assigned an edge between all gene pairs with a known protein interaction if we also uncovered a significant local expression similarity score between the two genes.

From protein interaction data and the 11 expression time series, we obtain 11 graphs, ranging from 1872 to 13298 edges. Interestingly, the terminal genes of 32% of those edges have Pearson's correlations with P-value higher than 0.05. Consequently, the local alignments are either shifted or are short for all of these edges and, thus, can not be captured by standard correlation analysis.

4.2. *Dynamic network modules are more biologically meaningful than static modules*

We implemented Algorithm 1 and applied it to obtain edge-maximal dynamic modules with different time interval overlap parameter $d = 3, 4, 5, 6, 7, 8$ and with node size from 5 to 8 [a] This constraints collects the set of edge-maximal dynamic modules Using the cell cycle dataset (GDS2347) as an example, at $d = 8$, we identified 7574 modules of node size 5 to 8. To assess whether the temporal information improves our ability to discover biologically meaningful modules, we compared the functional homogeneity of the dynamic modules with that of the static modules determined based only on the protein interaction data. That is, for each of the edge-maximal dynamic modules, we identify the connected component of the PPI network with the same size, and compare the homogeneity of their biological functions. We used the Gene Ontology (GO) biological process annotation and defined specific functions to be those associated with GO nodes containing less than 200 genes. A module is considered functionally homogeneous if its variation in functional classification, as modeled by the hypergeometric distribution, has a $P < 10^{-6}$ (P stands for P-value). Our results showed that 38%

[a] Due to the very large number of dynamic modules within each temporal network, we apply an *overlap* constraint in Algorithm 2 to prune the edge-maximal dynamic modules which share more than k nodes with an already identified edge-maximal dynamic module in the resulting set. Here, we choose $k = 2$.

of dynamic modules are functionally homogenous, while on average (from 100 randomizations), only 27% of static modules are functionally homogenous. This highlights the usage of temporal information in uncovering biologically meaningful gene constellations that are hidden by a static network representation. In general, the functional homogeneity of modules increases with increasing d.

Interestingly, a majority of the dynamic module are very sparse. For example, 81% of the dynamic modules from the dataset GDS2347 have a connectivity γ less than 0.5, where γ is defined as $\gamma = 2m/(n(n-1))$, with m being the number of edges and n the number of nodes in a module. We have illustrated two such examples in Figure 2. The non-dense modules are very hard to extract from the static protein-protein interaction network solely based on their topology. Here, we demonstrate that the temporal expression information can facilitate the identification of likely important signals that are otherwise easily overlooked.

Dataset	# modules	GO[a]
GDS124	13729	40.96%
GDS1611-1	32532	36.43%
GDS1611-2	43024	31.21%
GDS1752-1	26306	37.14%
GDS1752-2	25412	37.68%
GDS18	26662	31.33%
GDS2318	12405	41.93%
GDS2347	7574	48.26%
GDS2350	3038	26.56%
GDS39	23983	36.17%
GDS608	2814	46.55%

[a] The percentage of modules with significant enrichment in proteins from a GO term ($P < 10^{-6}$ is required and at least 2 genes in the module are annotated with the same GO term).

4.3. *Dynamic network motifs reveal temporal events in cellular systems*

The edge-maximal dynamic modules can aid in the characterization of signal propagation and the elucidation of temporal organizational patterns in cellular activities. Note that, there is little information on the detailed sequential ordering of molecular events due to the limited power and resolution of current experimental techniques. This leaves an important opportunity for the community of computational scientists, as *in silico* predictions can complement experimental approaches to provide novel insights. In the following, we will discuss two cases from our analysis in detail, for which the identified dynamic events are supported by current knowledge.

Figure 2(a) illustrates an edge-maximal dynamic module activated during the cell cycle progression (obtained using dataset GDS2350). The module contains 5 genes, all of which are involved in the biological process of "microtubule cytoskeleton organization and biogenesis". Among those, SPC105 and SPC110 are components of the spindle pole body; NUF2 is implicated in connecting the centromere to the spindle pole body during chromosome segregation [13]; Cin8 clusters

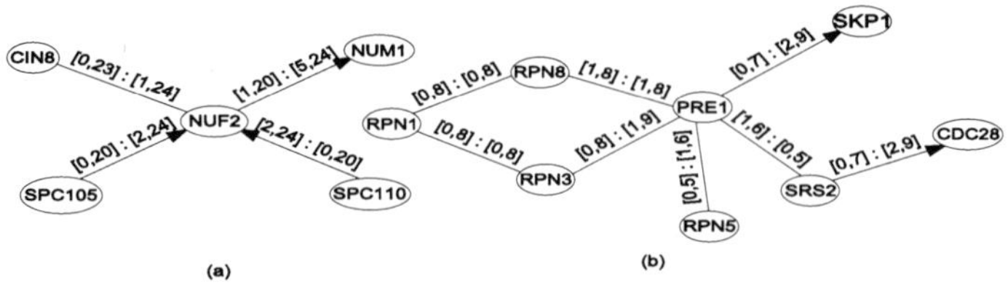

Figure 2. Two examples of edge-maximal dynamic modules.

kinetochores into their characteristic bi-lobed metaphase configuration [24]; while NUM1 is required for nuclear migration and localizes to the mother cell cortex as well as the bud tip, forming cytoplasmic microtubule capture-sites during the late anaphase [5,6]. Although all of these genes participate in microtubule dynamics, their interactions do not occur at the same time. The directed edges from SPC105/SPC110 to NUF2 indicates that the assembly of the spindle pole body takes place ahead of its interaction with the centrosome, and the directed edge from NUF2 to NUM1 implies that the connection from the centromere to the spindle pole occurs before nuclear migration. The identified edge-maximal dynamic module correctly captures the sequential order of protein-protein interactions along the time axis. It is noteworthy that all of the 4 connected gene pairs have the very low Pearson's expression correlations of -0.34, -0.08, -0.12, and 0.35. Consequently, this module can only be identified by taking the temporal characteristics into account.

Our second example, Figure 2(b), shows a dynamic module activated in the dataset GDS608 [16]. This dataset resulted from studies of the temporal expression patterns in wild-type diploid cells shifted from the typical yeast-form growth to that of a filamentous-form growth. Filamentous-form cells were collected and profiled hourly for 10 hours. The identified dynamic module contains 8 genes, in which RPN1, RPN3, RPN8, PRE1, PRE5 are involved in ubiquitin-dependent protein catabolic process. In particular, PRE5 belongs to the proteasome alpha-subunit complex, RPN3 and RPN8 are part of a proteasome regulatory particle, the lid subcomplex, while SRS, SKP1, and CDC28 are involved in the cell cycle. Most edges in this module are undirected (that is, they contain at most a time shift of 1 unit), and only two edges are directed, one pointing from PRE1 to SKP1, and the other pointing from SRS2 to CDC28. This is in agreement with the hypothesis that the ubiquitin-dependent protein degradation by the 26S proteasome is controlling the filamentous-form growth [16]. More specifically, the 26S proteasome regulates the activity of the Cdc28 kinase, which in turn controls cell-cycle progression and morphogenesis during filamentous-form growth [16].

5. Conclusions

We have presented a graph-based algorithm to identify dynamic network building-blocks by combining information from multiple temporal networks. To this end, we have designed a set of efficient algorithms to identify *dynamic* network modules. Our approach provides an alternative to traditional graph-based module finding algorithms that assume all links are simultaneously present (static). Consequently, traditional algorithms are incapable of addressing questions such as the time-ordering of link-based events.

In contrast, our approach facilitates the use of temporal information to detangle the complex wiring diagrams of cellular systems. As an example, we have applied our algorithm to a combination of microarray time-series expression data and the yeast protein-interaction network. We have demonstrated that our method uncovers functionally homogenous network modules, and more importantly, that it has the ability to identify the *sequential ordering* of molecular events taking place in biochemical systems. While the results from our approach have no direct implication on the causality of events, the sequential activity information embedded in the derived modules may serve as a starting point for causal-inference analyses.

Microarray gene-expression data and protein interaction data have their specific limitations, and as such, may not provide a measurement of all functional activities in a cell. However, we want to emphasize that as a general data integration framework, our algorithm can be applied to any type of time-series activity measurement (not limited to gene-expression data) and any type of network data (not limited to protein-interaction data). Thus, our approach provides an important new class of tools for the systems-level analysis of biological data.

6. Acknowledgments

This work of CCL and XJZ was supported by the grants NIH R01GM074163, NIH U54CA112952, NSF 0515936, and NSF 0747475. EA's work was performed under the auspices of DoE by LLNL under Contract DE-AC52-07NA27344 and funded by LDRD project 06-ERD-061.

References

1. J Aach and G M Church. Aligning gene expression time series with time warping algorithms. *Bioinformatics*, 17(6):495–508, Jun 2001. Comparative Study.
2. Dongbo Bu, Yi Zhao, Lun Cai, Hong Xue, Xiaopeng Zhu, Hongchao Lu, Jingfen Zhang, Shiwei Sun, Lunjiang Ling, Nan Zhang, Guojie Li, and Runsheng Chen. Topological structure analysis of the protein-protein interaction network in budding yeast. *Nucleic Acids Res*, 31(9):2443–2450, May 2003. Comparative Study.
3. Ulrik de Lichtenberg, Lars Juhl Jensen, Soren Brunak, and Peer Bork. Dynamic complex formation during the yeast cell cycle. *Science*, 307(5710):724–727, Feb 2005.
4. Prasanna Desikan and Jaideep Srivastava. Mining temporally evolving graphs. In *WebKDD*, 2004.
5. M Farkasovsky and H Kuntzel. Yeast Num1p associates with the mother cell cortex during S/G2 phase and affects microtubular functions. *J Cell Biol*, 131(4):1003–1014, Nov 1995.
6. M Farkasovsky and H Kuntzel. Cortical Num1p interacts with the dynein intermediate chain Pac11p and cytoplasmic microtubules in budding yeast. *J Cell Biol*, 152(2):251–262, Jan 2001.
7. D A Fell and A Wagner. The small world of metabolism. *Nat Biotechnol*, 18(11):1121–1122, Nov 2000.

8. Nabil Guelzim, Samuele Bottani, Paul Bourgine, and Francois Kepes. Topological and causal structure of the yeast transcriptional regulatory network. *Nat Genet*, 31(1):60–63, May 2002.
9. Ulrich Guldener, Martin Munsterkotter, Matthias Oesterheld, Philipp Pagel, Andreas Ruepp, Hans-Werner Mewes, and Volker Stumpflen. MPact: the MIPS protein interaction resource on yeast. *Nucleic Acids Res*, 34(Database issue):436–441, Jan 2006.
10. T Ideker, V Thorsson, J A Ranish, R Christmas, J Buhler, J K Eng, R Bumgarner, D R Goodlett, R Aebersold, and L Hood. Integrated genomic and proteomic analyses of a systematically perturbed metabolic network. *Science*, 292(5518):929–934, May 2001.
11. Ker-Chau Li. Genome-wide coexpression dynamics: theory and application. *Proc Natl Acad Sci U S A*, 99(26):16875–16880, Dec 2002.
12. Nicholas M Luscombe, M Madan Babu, Haiyuan Yu, Michael Snyder, Sarah A Teichmann, and Mark Gerstein. Genomic analysis of regulatory network dynamics reveals large topological changes. *Nature*, 431(7006):308–312, Sep 2004.
13. A Nabetani, T Koujin, C Tsutsumi, T Haraguchi, and Y Hiraoka. A conserved protein, Nuf2, is implicated in connecting the centromere to the spindle during chromosome segregation: a link between the kinetochore function and the spindle checkpoint. *Chromosoma*, 110(5):322–334, Sep 2001.
14. Joshua O'Madadhain, Jon Hutchins, and Padhraic Smyth. Prediction and ranking algorithms for event-based network data. *SIGKDD Explorations*, 7(2):23–30, 2005.
15. Christos A. Ouzounis and Peter D. Karp. Global Properties of the Metabolic Map of Escherichia coli. *Genome Res.*, 10(4):568–576, 2000.
16. Susanne Prinz, Iliana Avila-Campillo, Christine Aldridge, Ajitha Srinivasan, Krassen Dimitrov, Andrew F Siegel, and Timothy Galitski. Control of yeast filamentous-form growth by modules in an integrated molecular network. *Genome Res*, 14(3):380–390, Mar 2004. Comparative Study.
17. Alexander W. Rives and Timothy Galitski. Modular organization of cellular networks. *Proceedings of the National Academy of Sciences*, 100(3):1128–1133, 2003.
18. Quansong Ruan, Debojyoti Dutta, Michael S Schwalbach, Joshua A Steele, Jed A Fuhrman, and Fengzhu Sun. Local similarity analysis reveals unique associations among marine bacterioplankton species and environmental factors. *Bioinformatics*, 22(20):2532–2538, Oct 2006.
19. Victor Spirin and Leonid A. Mirny. Protein complexes and functional modules in molecular networks. *Proceedings of the National Academy of Sciences*, 100(21):12123–12128, 2003.
20. Chris Stark, Bobby-Joe Breitkreutz, Teresa Reguly, Lorrie Boucher, Ashton Breitkreutz, and Mike Tyers. BioGRID: a general repository for interaction datasets. *Nucleic Acids Res*, 34(Database issue):535–539, Jan 2006.
21. Jimeng Sun, Christos Faloutsos, Spiros Papadimitriou, and Philip S. Yu. Graphscope: parameter-free mining of large time-evolving graphs. In *KDD*, pages 687–696, 2007.
22. Chayant Tantipathananandh, Tanya Berger-Wolf, and David Kempe. A framework for community identification in dynamic social networks. In *KDD*, pages 717–726, 2007.
23. D. Thieffry, A. Huerta, E. Perez-Rueda, and J. Collado-Vides. From specific gene regulation to genomic networks: A global analysis of transcriptional regulation in escherichia coli, 1998.
24. Jessica D Tytell and Peter K Sorger. Analysis of kinesin motor function at budding yeast kinetochores. *J Cell Biol*, 172(6):861–874, Mar 2006.
25. Christian von Mering, Roland Krause, Berend Snel, Michael Cornell, Stephen G Oliver, Stanley Fields, and Peer Bork. Comparative assessment of large-scale data sets of protein-protein interactions. *Nature*, 417(6887):399–403, May 2002. Comparative Study.

EFFICIENT AND ROBUST PREDICTION ALGORITHMS FOR PROTEIN COMPLEXES USING GOMORY-HU TREES

A. MITROFANOVA*, M. FARACH-COLTON**, AND B. MISHRA*

**New York University, Department of Computer Science, New York, NY 10003*
***Rutgers University, Department of Computer Science, Piscataway, NJ 08855*
E-mail: antonina@cs.nyu.edu, farach@cs.rutgers.edu, mishra@nyu.edu

Two-Hybrid (Y2H) Protein-Protein interaction (PPI) data suffer from high False Positive and False Negative rates, thus making searching for protein complexes in PPI networks a challenge. To overcome these limitations, we propose an efficient approach which measures connectivity between proteins not by edges, but by edge-disjoint paths. We model the number of edge-disjoint paths as a network flow and efficiently represent it in a Gomory-Hu tree. By manipulating the tree, we are able to isolate groups of nodes sharing more edge-disjoint paths with each other than with the rest of the network, which are our putative protein complexes. We examine the performance of our algorithm with Variation of Information and Separation measures and show that it belongs to a group of techniques which are robust against increased false positive and false negative rates. We apply our approach to yeast , mouse, worm, and human Y2H PPI networks, where it shows promising results. On yeast network, we identify 38 statistically significant protein clusters, 20 of which correspond to protein complexes and 16 to functional modules.

1. Introduction

We wish to propose a new efficient and robust algorithm to infer protein complexes correctly from Y2H experiments. If the protein-protein interaction data were flaw-less and error free, then a fairly direct graph-theoretic algorithm working on graphs whose edges represent pair-wise interactions would have sufficed. The intuitively direct algorithms (e.g., clique detection, clustering or density-based methods) tend to be efficient, and work reasonably well with small number of errors that mislabel the edges falsely (both false positive and negative, errors). Our challenge is to devise more sophisticated algorithms that enjoy a comparable computational efficiency, and yet work robustly as the quality of the experimental data degrade substantially, as is common with practically all currently available PPI data. The fundamental conceptual innovation in our algorithm is to analyze structure of the graphs through their collections of edge-disjoint paths that remain relatively immune to the corrupting noises in the experiment, and yet lead to an efficient implementation through Gomory-Hu tree representations. Below, we further elaborate on these points.

Complexes of proteins are at the heart of many fundamental biological processes, including e.g. RNA metabolism, signal transduction, energy metabolism, and translation initiation. As noted, the process of efficiently purifying [5,4] protein complexes and identifying their structure and function has remained a challenge. The most common experimental techniques result in the Yeast two-hybrid protein-protein interaction networks, which encode pair-wise interactions between proteins, and thus hold the promise to yield information about large-scale phenomena such as participation in protein complexes, as examined in [3,12,7,9,13]. It has been a well-known problem that Y2H experiments suffer from high false positive (FP) and false negative (FN) rates. To overcome these limitations, one needs algorithmic approaches robust against high FP and FN rates. Thus, even when the details of protein complexes become "disguised" by FN or become intertwined with each other by FP, these algorithms could exploit the fact that proteins within complexes still remain connected by adequately many paths in the network. However, this qualitative statement requires a quantitative justification, namely, as the number of false edges (positive or negative) increases, how and when do these algorithms break down? What is the nature of the algorithmic degradation: slow and graceful, or sudden and catastrophic? What is the best algorithmic framework, in which they could be studied? Our main results are as follows:

Algorithmic Results: We devise and implement a novel algorithm based on max-flow and their representations through the classical Gomory-Hu tree data structures. We perform both theoretical and practical complexity analysis. We describe and conduct its performance and robustness analysis with respect to practical data using Meila's variational information [21] and Separation [22].

Experimental results: We consider *Saccharomyces cerevisiae* as a *model* organism for our study, since its Y2H network as well as its protein complex data are most complete. Data for protein Y2H pairwise interactions and protein complexes were taken from the BioGRID [2] and MIPS [1] databases (3930 proteins and 6219 Y2H interactions). On yeast network, we identify 38 statistically significant protein clusters, among which there are 20 protein complexes and 16 functional modules. Identified protein complexes cover 61% of all existing BioGRID/MIPS complexes, which have sufficient data coverage (or 72% of non-broken complexes, as described in Section 6).

Supplementary material: Supplementary and output data are available from http://research.rutgers.edu/~ amitrofa/predictions.html

We begin in section 2 with motivation and intuition behind our algorithm and its formal presentation in section 3. We present the robustness agains FP and FN by measuring the Variation of Information and Separation for several clustering techniques in section 4. In section 5 we discuss the criteria for statistical significance of computed clusters, and show results on yeast Y2H PPI data in section 6. Finally, we compare our results to previously published work in section 7 and conclude the paper.

2. Background

The Y2H experiments are known for high false positive and false negative rates: two adjacent proteins might not belong to the same protein complex (FP; Figure 1 A: **b**) as well as proteins from the same complex might not share an edge (FN; Figure 1 A: **a**). These phenomena raise questions about the validity of the direct statistical examination of pure Y2H networks.

With current data coverage and high FN rates, protein complexes of the Y2H PPI networks suffer from low connectivity within. Among all existing Y2H edges, only 6.14% connect protein pairs which participate in the same protein complex. In fact, there are 788 protein complexes (from BioGRID and MIPS) with at least 3 nodes. Of those, 463 do not have a single Y2H edge in the complex, 129 have only one Y2H edge, and 71 have two edges. There are only 125 complexes which contain at least three Y2H edges in the complex and can *potentially* have a minimum level of connectivity necessary to be identified by a connectivity-based computational method.

The majority of graph-based methods for extracting protein complexes look for densely connected, clique-like regions of the PPI network [9,13,12,3]. However, the problem of noise in Y2H experiments required these methods to supplement pair-wise interactions with other biological markers, as co-expression [7], functional annotation [12], small-scale immunoprecipitation [13], microarrays [19], or inter-specie data for conserved protein complexes [3,20].

If a protein complex corresponds to a clique-like subgraph in the Y2H PPI graph, then increased FP and FN rates might at least interfere with and at most preclude the search for such structures. For example, as shown in Figure 1 A: **b** high FP rate can produce areas of "false" density or increase the connectivity between complexes making them impossible to be identified in the network. Likewise, high FN rate can disguise clique-like protein complexes. However, even if proteins from a complex lose a few

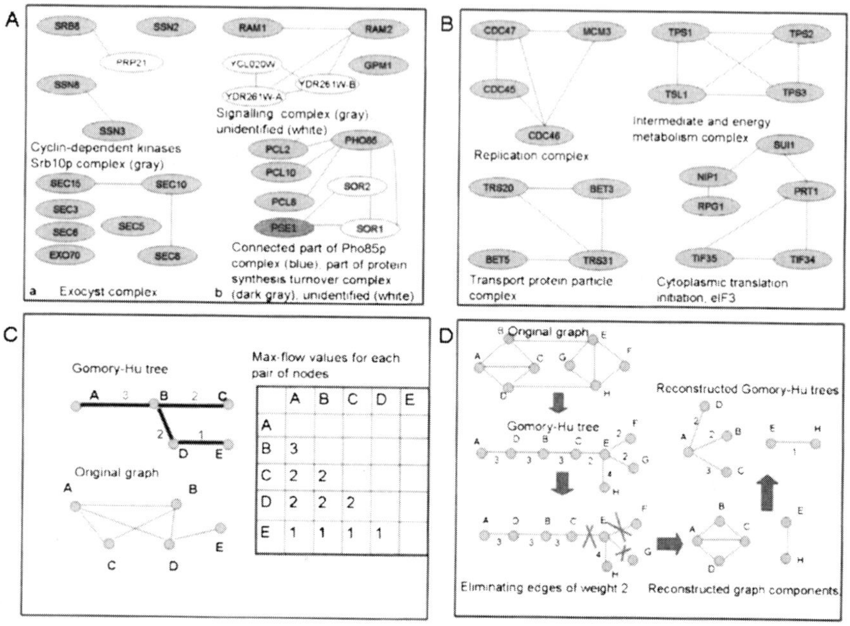

Figure 1. **A**: **(a)**Protein complexes that have low Y2H connectivity. **(b)** Protein complexes with "fused" out-of-complex proteins. **B**: Examples of protein complexes that contain 2-edge connected subgraphs. **C.** Gomory-Hu tree and its matrix representation. **D.** Cutting the small-weight edges of a Gomory-Hu tree to induce a partition on the nodes.

edges, they should still be connected by enough paths in the network.

If we take a *path* between two proteins as evidence that they are in the same complex, then the number of *edge-disjoint paths* is related to the degree of confidence we have of complex co-membership (an edge is also a path, but a path is not limited to an edge).

We examined the yeast Y2H PPI network for the number of edge-disjoint paths between protein pairs that belong to the same protein complex (in-complex) vs those that do not belong to the same protein complex (non-complex), thus covering all possible protein pairs. In Figure 2,A, we show an example of a distribution of edge-disjoint paths in each group: it is more common for non-complex group to share just one path, and in-complex group shows a clear evidence of sharing two and more paths compared to non-complex group. Overall, the fraction of protein pairs sharing one path over those sharing more than one path for the in-complex group is 1.059 and

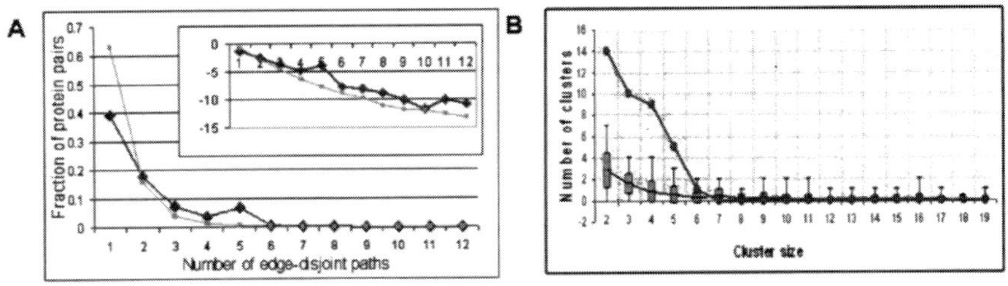

Figure 2. **A.**Distribution of edge-disjoint paths for protein pairs that belong to the same protein complex (thick dark color) vs pairs that do not belong to the same protein complex (light color). The embedded chart shows the same values in log_2 scale. **B.** Number of clusters as a function of cluster size in the whole yeast Y2H PPI network (thick dark upper line) and in the random graphs (light lower line). On the lower line: rectangles represent standard deviation, with max and min as up/down bars.

for non-complex group is 2.868, emphasizing the importance of the greater number of edge-disjoint paths for proteins from the same complex. For pairs of proteins that *do not share* an edge, the same dynamics is observed: the above is 1.081 for in-complex and 2.873 for non-complex group.

The number of edge-disjoint paths between a pair of nodes in the network (in our case unweighted and undirected) corresponds to the value of the *maximum flow* between that pair. However, there is no need to consider all $\binom{n}{2}$ node pairs in the network, since the number of edge-disjoint paths (or maximum flow) for all pairs of nodes can be calculated in only $n-1$ steps and succinctly represented in a *Gomory-Hu tree* [6], as detailed below.

3. Methods

We begin by computing a Gomory-Hu tree for each connected component of the PPI graph. A Gomory-Hu tree is a weighted tree that spans nodes of a graph such that the max-flow between any two nodes in the graph is the same as the max-flow between them in the tree. That is, the max-flow from p_α to p_β in the network has value equal to the minimum edge on the path between these nodes in the Gomory-Hu tree, as shown in Figure 1 **C**. To compute max-flow value, we use a Ford-Fulkerson method: the best known deterministic max-flow algorithm for the undirected unweighted graph is one proposed by Matula [10] and Nagamochi and Ibaraki [11] that runs in $O(|P||E|)$ steps (where $|P|$ is the number of nodes/proteins and $|E|$ is the number of edges). Thus, the time complexity of our algorithm is $O(|P|^2|E|)$.

First we remove minimum-weighted edges from the Gomory-Hu tree. Removing an edge induces a bipartition between the nodes of the tree.

Thus an edge in the Gomory-Hu tree corresponds to an edge-cut in the PPI graph. After such elimination we *recompute* a Gomory-Hu tree for each induced connected component, since the forest obtained by removing edges (of weight > 1) from the Gomory-Hu tree is no longer the Gomory-Hu trees of the partitions, as for example shown in the Figure 1 **D**. We proceed recursively, by eliminating least weighted edges and recomputing Gomory-Hu tree for each induced connected component until there are no more edges to eliminate (see full formal description in *Supplementary material*).

We call the set of nodes in each connected component of the Gomory-Hu forest a **cluster**. We eliminate singleton nodes at each phase, saying that they *disappears*. With every elimination phase, each Gomory-Hu tree becomes smaller, splitting clusters or reducing their size until each cluster disappears. Clusters found this way are then subjected to further selection according to criteria of statistical significance, as described in 5.2.

4. Robustness via Statistical Analysis

We examine the robustness of our algorithm by computing Variation of Information (VI) [21] and Separation [22] as we vary the number of FP and FN in a randomly constructed network. Since we think of protein complexes as highly connected "clique-like" structures in the network [9,13,12,3], we build our random *test graph* in the following way: we introduce complete graphs of size from 10 to 2 and singletons (following the power-law distribution: 10 graphs of size 10, 20 graphs of size 9, etc, 300 singletons), similarly to the approach described in [22]. These groups of nodes are our initial complexes. Then we delete some % of random edges (FN) and/or add edges (FP).

Separation measure is relevant to the geometric mean of sensitivity and positive predictive value, as defined in [22]. High separation values indicate bidirectional correspondence between a cluster and a complex and thus are more favorable. *Variation of Information* is another useful metric based information-theoretic criterion that measures how much information is lost or gained in going from clustering C to C'. In our case, C corresponds to the initial complexes. If we let n be the number of nodes and K be the total number of clusters, with n_k being a size of cluster C_k, then the uncertainty (or entropy) of the clustering C is defined as $H(C) = -\sum_{k=1}^{K} P(k)\log(P(k))$, where $P(k) = n_k/n$. The joint distribution that a point belongs to cluster C_k in C and to cluster $C'_{k'}$ in C' is $P(k,k') = \frac{|C_k \bigcap C'_{k'}|}{n}$. Then the mutual information between clustering C and C' is $I(C,C') = \sum_{k=1}^{K}\sum_{k'=1}^{K'} P(k,k')\log(\frac{P(k,k')}{P(k)P'(k')})$ And finally, the

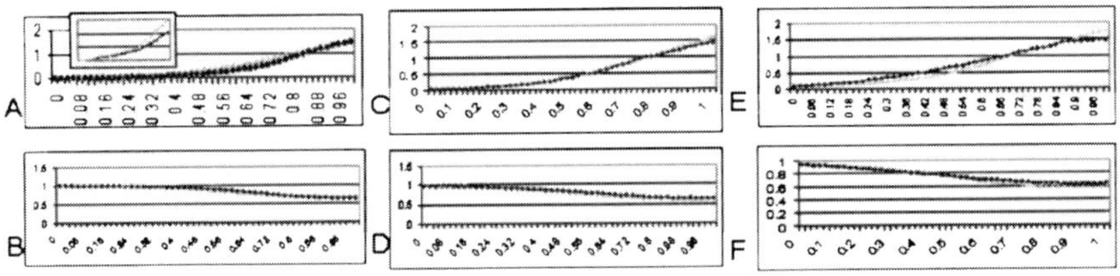

G

j	2	4	6	8	10	12	14	16	18	20
Y2H PPI network	2221	807	400	238	160	114	85	67	56	49
Protein complexes	658	428	251	132	95	62	42	35	22	17
Within-complex degrees	209	28	7	1	0	0	0	0	0	0

Figure 3. **A-F** Each curve (black for our method and gray for MC) represents the value of VI (first row) or Separation (second row) as the % of edges removed increases, averaged over 10 random runs. **(A-B)** edge removal **(C-D)** edge removal with 5% of randomly added edges. **(E-F)** edge removal with 10% of randomly added edges. **G.** Results on the training set of the yeast Y2H PPI network: threshold $d = 6, \ldots, 17$. All nodes with degree $\geq d$ are eliminated.

Variation of Information is $VI(C, C') = H(C) - I(C, C') + H(C') - I(C, C')$. Higher VI corresponds to bigger deviation from the original clustering C.

We compare our method with Markov Clustering (MC) [23], which is reported as the most robust clustering on PPI networks in [22]. Since the protein complexes in current PPI networks suffer from low connectivity, it is more important to examine the robustness of the algorithms against increasing FN rates. We present some results in Figure 3, which show that our approach is equally or more robust compared to MC when examined against increased FP (by 5% and by 10%, which are most likely to exist in Y2H PPI networks) and varying FN rates. Both methods show smooth curves toward increased FN rates.

5. Experiments

5.1. *High degree nodes*

To minimize the number of non-selective (possibly FP) interactions that would give statistically insignificant clusters, the common practice is to eliminate "excessive-degree" nodes from the Y2H PPI graph, as for example exercised in [13]. To learn the degree threshold, we select a so-called "training set" , which corresponds to about $\frac{1}{4}$ of the network (the remaining part is called a "testing set"). To choose a training set, we start with a random protein in the graph and accumulate the desired number of nodes by breadth-first search. In the learning, we eliminate nodes and the outgoing edges according to various degree thresholds d and evaluate the node/edge elimination effects by various performance measures. First, we calculate the percent coverage, P, of how many final clusters fully correspond to MIPS/BioGRID protein complexes. Additionally, we introduce a new measure of protein complex coverage, i.e., 2-edge connectedness. The graph is 2-edge connected if there are at least two edge-disjoint paths between every pair of nodes in the graph, consider some examples shown in Figure 1,B.

We found that from 125 MIPS protein complexes with at least three Y2H edges, 74 (59.2 %) are fully or partially 2-edge connected. However, these protein complexes often overlap with each other or are the subsets of each other, producing data redundancy that can negatively influence the analysis. There are 33 not overlapping non-redundant 2-edge connected protein complexes, which we use in our further statistical analysis.

We measure Q, the recall rate for 2-edge connected complexes, that are in the training set. For example, in our training set initially there exist 6 2-edge connected protein complexes and Q is the % of these 6 that we identify in each run. We show P and Q for each run in Figure 3, G : the highest Q values are observed with $d = 13-16$. Among those, $d = 13$ executes highest P, which we consider our threshold and eliminate all nodes of degree higher than 13 from our dataset (85 nodes or 2.16 % of total network nodes).

5.2. *Statistical Significance of Clusters*

As the algorithm proceeds, many clusters of different sizes are generated. The final part of our algorithm is to estimate statistical significance of computed clusters and decide which correspond to protein complexes and functional modules. To measure the statistical significance of the cluster, we need to account for the probability of finding such cluster in a random graph. To generate random graphs, we use the Maslov-Sneppen procedure [14], which shuffles the edges of the original Y2H PPI network so that the number of interactions for each protein in the network is preserved.

Size-based p-value: First, we calculate p-value for clusters of different sizes. The Figure 2,B shows enrichment in the number of clusters of sizes 2 to 6 in the original Y2H PPI graph, as compared to results on 100 random graphs. Clusters of size 7 and higher, in contrast, appear more often at random. For each cluster of size s, we calculate p-value as a probability of finding a cluster of size s at random, fit to a normal distribution. Clusters of size $2, 3, 4$ and 5 showed p-value $p < 1 \times 10^{-4}$, which we consider statistically significant. Clusters of size 6 showed $p = 0.20$ and therefore can appear at random with reasonably high probability.

Density-based p-value: We define a "cluster-network density", *CND*, as the difference between the average number of edge-disjoint paths per pair of proteins in the cluster, ED_c, and the average number of edge disjoint paths from the proteins of this cluster to the proteins in the rest of the network (ignoring proteins from different connected components), ED_r. Thus $CND = ED_c - ED_r$ reflects the difference between the connectivity inside the cluster and connectivity of this cluster with the rest of the network. Of course, all original clusters from the yeast Y2H PPI network show CND greater than 0. Here we again consider 100 random graphs generated by the Maslov-Sneppen procedure [14] and calculate p-value (fit to a normal distribution) *per cluster* produced. For each cluster of the original Y2H PPI network, p-value reflects the probability that CND at random would be greater or equal to CND in the original network. To correct for multiple hypothesis tested, we apply Bonferroni Correction (the number of hypothesis tested is equal to the number of observed original clusters). We consider those clusters with corrected p-values less than 1×10^{-4} as statistically significant. It appeared that clusters with p-values $< 10^{-4}$ do not violate the statistically significant sizes shown in Figure 2,B. We report 38 out of 56 clusters as being statistically significant according to the criteria described above. Among clusters with p-value $> 10^{-4}$, two represent protein complexes and two correspond to functional modules.

6. Results

We consider a cluster as a *match* if *all* of its proteins belong to the same protein MIPS complex (at the lowest hierarchical level). Also we use a measure of biological importance, similar to one defined in [13] – a *Functional Module*, as a group of proteins that participate in the same process in the same location, however not necessarily at the same time. In order for a cluster to be identified as a Functional module, its proteins should reside in the same cellular location and should share similar/relevant functions (Gene Ontology classification). Even stronger supporting evidence for Functional modules includes co-expression and literature co-citation (we used tables and criteria provided by [8] for pairwise log-odds scores). Since in many cases we cannot say with certainty whether proteins enter a process at different times or at the same time, clusters from this category are strong candidates for protein complex predictions.

Table 1. Final clusters of testing and training sets in the Yeast network.

Sets	$\frac{1}{4}$	$\frac{3}{4}$	$\frac{1}{4} \cup \frac{3}{4}$
Total clusters with $p < 10^{-4}$	10	28	38
Clusters that cover MIPS complexes	5	15	20
FM with co-location and (co-expression or co-citation	1	4	5
FM with co-location	4	7	11
FM with limited information	0	2	2

We present results for both the $\frac{1}{4}$ and $\frac{3}{4}$ of the network in Table 1. Among 38 clusters with p-value $< 10^{-4}$, there are 20 MIPS complexes and 18 functional modules. Five of the functional modules are supported by co-expression of participating proteins or co-citation from the literature, thus making a strongly grounded predictions for new protein complexes, as shown in *Supplementary material.* Two clusters had weaker evidence of forming a functional module primarily due to lack of information about the functional annotation or cellular location of participating proteins.

For completeness, we also studied a recall rate, which corresponds to the proportion of 33 2-edge connected complexes covered. As we select our training and testing sets, 5 2-edge connected complexes become broken, resulting in 28 2-edge connected MIPS complexes in both sets, 20 of which we identify (yielding recall rate of 72%). Additionally, we characterize the performance of our method by a parameter M, the fraction of proteins in the *matched* cluster over proteins in the 2-edge connected part of the corresponding complex: 18 out of 20 clusters show $M = 1$.

In general, among the 2-edge connected complexes, there are 17 triangles, 13 4-node, two 5-node, and one 7-node graphs. Thus, the clusters that cover 2-edge connected complexes are partially bounded by the above sizes.

We have applied our method to other species: we list 31 clusters for human, 17 for mouse, and 29 for worm, whose sizes satisfy our stringent statistical-significance criteria, as shown in *Supplementary material.*

7. Discussion

In this section, we compare our method to those previously described in the literature (the comparison is made with the best available results for each method), as shown in Table 2 and discussed below.

King et al. [12] develop the restricted neighborhood search clustering algorithm using a cost function. After generating clusters, proteins are selectively chosen from clusters using a filtering model. They identified 23 clusters matching MIPS protein complexes by 90 % of proteins in a cluster (6 of which matched by 100% – identified by "→" in the Table 2). We identify 35 new clusters not covered by their method, among which there are 17 new protein complexes.

Table 2. Precision is the number of clusters covering complexes in 100% of cluster proteins over total number of clusters. Recall is the number of 2-edge connected complexes covered by clusters over a total number of 2-edge connected complexes. Intersection reflects the overlap between complexes identified by us and other methods.

	Precision	Recall	Intersection
Our method	20/38 = 53 %	20/33 =61% → 20/28=72%	20
King et al [12]	23/30 → 6/30=20%	4/33 = 12%	4
Bader et al [9]	54/209 = 26% → 0/209=0%	0/33=0%	0
Spirin et al [13]	30/67 = 45%	18/33= 55%	16

Bader and Hogue [9] present algorithm that detects densely connected regions in PPI networks. They generate 209 clusters, 52 of which matched MIPS complexes in at least 20% of their proteins (with the highest overlap being 43%, thus bringing the number of clusters that match protein complexes by 100% of their proteins to 0, as reflected in Table 2).

Spirin and Mirny [13] look for heavily connected, clique-like groups of nodes in the network (supplemented with hypothesis-driven studies such as coprecipitation, omitted by our method). The union of *three* different clustering methods identified 30 clusters corresponding to protein complexes. We identify 4 new protein complexes and 13 new functional modules. The MCL clustering [23] has not been applied to search for protein complexes yet, but only to cluster proteins based on their sequence similarity.

An interesting min-cut clustering approach of Tarjan et al [17], which was applied to find communities in web and citation networks, introduced an artificial sink node connected to all other nodes. We plan to expand our understanding of competitive bounds for communities' sizes addressed in [17]. Another interesting approach described by Newman in [15] (later extended in [24]) describes graph decomposition based on edge betweenness, defined as the number of shortest paths which go through an edge. Hartuv et al. in [16] present an algorithm based on min cut idea, which shows an improved time complexity and generates clusters with diameter 2 (two vertices are either adjacent or share one or more common neighbors). We do not require nodes in the cluster to be adjacent or to necessarily share a neighbor; however, they can be connected by much longer edge-disjoint paths.

One promising future direction for our method would be to assign a confidence score for each Y2H interaction (i.e. conservation of the interaction across species). It is possible to define a distance-based measure between proteins and use a Diffusion Map for spectral clustering, as in [18]. However, this method is very computationally expensive and hard to scale to large datasets. We plan to explore an efficient implementation of a continuous approach of diffusion maps with discrete approach of Gomory-Hu trees.

8. Conclusions

We have presented an efficient algorithm for identifying protein complexes through manipulation of the Gomory-Hu tree of the PPI Y2H network. Our method is shown to be robust against high FP and FN rates and capable of producing clusters of high quality when compared to other approaches. Identified functional modules are strong candidates for complex predictions and constitute reliable material for experimental research.

References

1. The MIPD database: http://mips.gsf.de/
2. The BioGRID database: http://www.thebiogrid.org/
3. R. Sharan, T. Ideker, B. Kelley, R. Shamir, B. Karp, *Recomb.* 282–289 (2004).
4. A. Gavin, M. Bosche and R. Krause, *Nature.* **415**, 141–147 (2002).
5. Y. Ho, A. Gruhler and A. Heilbut, *Nature.* **415**, 180-183 (2002).
6. R. Gomory and T. Hu, *J. of SIAM.* **4**, 551-570 (1961).
7. R. Jansen, H. Yu, et al *Science.* **302(5644)**, 449-53 (2003).
8. I. Lee, S. Date, A. Adai, E. Marcotte, *Science.* **306(5701)**, 1555-1558 (2004).
9. G. Bader and C. Hogue, *BMC Bioinformatics.* bf 4(1), 2 (2003).
10. D. Matula, *FOCS.* 249–251 (1987).
11. H. Nagamochi and T. Ibaraki, *SIAM J.Disc. Math.* **5**, 54–66 (1992).
12. A. King, N. Przulj, I. Jurisica, *Bioinformatics.* **20(17)**, 3013–3020 (2004).
13. V. Spirin and L.A. Mirny,L.A, *Proc Natl Acad Sci.* **100**, 12123–12128 (2003).
14. S Maslov and K. Sneppen, *Science.* **296**, 910-913 (2002).
15. M. Newman, *Phys. Rev. E.* **74**, 036104 (2006).
16. E. Hartuv and R. Shamir, *IPL.* **76(4-6)**, 175–181 (2000).
17. G. Flake, R. Tarjan, K. Tsioutsiouliklis, *Internet Math.* **1(4)**, 385-408 (2004).
18. G. Lerman and B.E. Shakhnovich, *PNAS.* **104(27)**, 11334-11339 (2007).
19. J. Chen and B. Yuan, *Bioinformatics.* **22(18)**, 2283–2290 (2006).
20. E. Hirsh and R. Sharan, *Bioinformatics.* **23(2)**, e170-e176 (2007).
21. M. Meila. *COLT.* (2003).
22. S. Brohee and J. Van Helden, *BMC Bioinformatics.* **7:488** (2006).
23. J. Enright, S. Dongen, A. Ouzounis, *Nucl. Acids Res.* **30(7)**, 1575-84 (2002).
24. F. Luo, Y. Yang, et al *Bioinformatics.* **23(2)** 207-14 (2007).

PARAMETER ESTIMATION OF *IN SILICO* BIOLOGICAL PATHWAYS WITH PARTICLE FILTERING TOWARDS A PETASCALE COMPUTING

KAZUYUKI NAKAMURA[1,*], RYO YOSHIDA[1], MASAO NAGASAKI[2], SATORU MIYANO[2] AND TOMOYUKI HIGUCHI[1]

[1] *The Institute of Statistical Mathematics, 4-6-7 Minami-Azabu, Minato-ku, Tokyo 1068569, Japan*

[2] *The Institute of Medical Science, The University of Tokyo, 4-6-1 Shirokanedai, Minato-ku, Tokyo 1088639, Japan*

The aim of this paper is to demonstrate the potential power of large-scale particle filtering for the parameter estimations of *in silico* biological pathways where time course measurements of biochemical reactions are observable. The method of particle filtering has been a popular technique in the field of statistical science, which approximates posterior distributions of model parameters of dynamic system by using sequentially-generated Monte Carlo samples. In order to apply the particle filtering to system identifications of biological pathways, it is often needed to explore the posterior distributions which are defined over an exceedingly high-dimensional parameter space. It is then essential to use a fairly large amount of Monte Carlo samples to obtain an approximation with a high-degree of accuracy. In this paper, we address some implementation issues on large-scale particle filtering, and then, indicate the importance of large-scale computing for parameter learning of *in silico* biological pathways. We have tested the ability of the particle filtering with 10^8 Monte Carlo samples on the transcription circuit of circadian clock that contains 45 unknown kinetic parameters. The proposed approach could reveal clearly the shape of the posterior distributions over the 45 dimensional parameter space.

1. Introduction

Mathematical modeling of biological pathways has broad utility to understanding of complex networks of biochemical reactions in living cells. In systems biology, *in silico* modeling of biological pathways has proven to be

*to whom correspondence should be addressed: nakakazu@ism.ac.jp

a promising approach with successful applications.[1–4] Usually, a pathway model is comprised of a set of differential[1,5,6] or difference[2–4,7,8] equations that describe kinetic mechanism between biochemical reactants, e.g. mRNA and protein. In order to conduct simulation experiments, it is essential to find a set of model parameters and initial concentrations of reactants such that the solution fits observations. For example, one may find effective values of rate constant parameters for which time course microarray data are available.

More recently, several approaches have been proposed as a technique of estimating parameters of *in silico* pathway, e.g. genetic algorithm,[8] unscented Kalman filter,[6] and gradient methods.[5] We also have been working on this topic based on the method of particle filtering,[10–13] which has been a popular technique for the problem of system identifications in the field of statistical science and signal processing. While these existing methods could cope with the estimation of relatively small number of parameters, their ability is limited by the case that model contains much larger number of unknown parameters.

As an illustration, consider the use of particle filtering that approximates the posterior distributions of model parameters based on a set of Monte Carlo samples. The size of random numbers usually lies in 10^3 to 10^5.[13] As the number of unknown parameters becomes large, it is required to generate an exceedingly large number of Monte Carlo samples in order to cover the overall parameter space. Such a problem, often called *curse of dimensionality*, is also common in the existing methods.[5,6,8] For example, genetic algorithm explores values of parameters by alternating cross-over and mutation of initially-generated seeds for unknown parameters.[8] In its applications, it is a key to success whether or not an initial set of candidate parameters lies in the region close to the optimal values.

To address such a limitation, we adopt the direct way that uses an exceedingly large amount of Monte Carlo samples, e.g. of order 10^8 or more, for improving ability of particle filtering. In our previous works,[14,15] we applied the particle filtering with $10,000$ Monte Carlo samples to the parameter estimation of transcription network of circadian clock. Unfortunately, the number of parameters that were estimated was limited to only a few. However, as will be demonstrated, by increasing the number of samples to 10^8 particles, the ability of the particle filtering has been extended greatly so that 45 unknown parameters could be estimated with a high degree of accuracy. Particularly, the large-scale particle filtering could reveal clearly the shape of the posterior distributions over the 45 parameters. In doing so, we aim to obtain an overall view of high-dimensional parameter space of *in silico* biological pathways, e.g. multi-modality and higher-order moments.

One more attractive feature of the particle filtering is a high degree of compatibility with parallel computing. This favorable feature increases the potential of particle filtering such that we can use a larger number of parameters to deal with higher dimensional parameter space, e.g. of order 10^2, by exploiting the parallel computation schemes and computational resources which will be available in the near future.

In this paper, we first give a brief explanation about the method of the particle filtering, and then, address some computational problems that arise from the implementation of large-scale particle filtering. The development of the 10 petaflops supercomputer has been going on as a national project. It mainly targets high performance computing of life science simulators, including molecular dynamics, biomechanics and biochemical simulations. This work is a pilot study towards the estimation of large-scale *in silico* models by the aid of petascale computing.

In Section 2, we show the method of particle filtering after introducing the state space model that relates *in silico* pathway models to time course measurements. We also discuss some computational issues on the large-scale particle filtering. Section 3 shows the numerical experiments which test the applicability of the proposed approach on the analysis of transcription network of circadian clock. Finally, discussion and conclusion are given in Section 4.

2. Method

2.1. *Nonlinear State Space Model*

Suppose that we model successive change in concentration levels of biochemical reactants by a set of rate equations

$$\boldsymbol{x}_t = \boldsymbol{f}(\boldsymbol{x}_{t-1}, \boldsymbol{\theta}), \tag{1}$$

where the p dimensional vector $\boldsymbol{x}_t$ contains concentration levels of all variables, e.g. mRNAs and proteins, at time step t and $\boldsymbol{\theta}$ denotes a vector of model parameters. The state variable $\boldsymbol{x}_t$ which is assumed to be unobservable evolves during successive discrete time points from $t-1$ to t according to the p dimensional vector-valued function $\boldsymbol{f}$. For ordinary differential equations, several methods are available to be associated with (1), for example, discretization in time. [9]

To deal with an inevitable modeling error, we may rather use the stochastic difference equations, referred to as system model,

$$\boldsymbol{x}_t = \boldsymbol{f}(\boldsymbol{x}_{t-1}, \boldsymbol{\theta}) + \boldsymbol{v}_t \,, \tag{2}$$

where the noise component $\boldsymbol{v}_t$ is added to (1). In what follows, we use the general form (2) which contains (1) as a special case.

The measurement model, referred to as observation model, associates the simulation model to time course data:

$$\boldsymbol{y}_t = \boldsymbol{h}(\boldsymbol{x}_{t-1}) + \boldsymbol{w}_t \,. \tag{3}$$

Here, $\boldsymbol{h}(\cdot)$ represents the function of measurement mechanism, which defines the conversion from unobserved concentration level $\boldsymbol{x}_t$ to observed time course data $\boldsymbol{y}_t$. $\boldsymbol{w}_t$ denotes additive measurement noise. Initial state variables $\boldsymbol{x}_0$ are assumed to follow a certain distribution $p(\boldsymbol{x}_0)$. The probabilistic model consisting of (2) and (3) is referred to as the nonlinear state space model.[10–12]

Of interest is then to find values of parameters, $\boldsymbol{\theta}$, and unobserved concentrations of reactants, $\boldsymbol{x}_t$, based on time course data $\boldsymbol{y}_t, t = 1, \cdots, T$. This paper focuses on the Bayesian estimations of model parameters, which are based on the joint posterior distribution of $\boldsymbol{x}_t$, $t = 0, \cdots, T$ and $\boldsymbol{\theta}$ conditional on the observed data. However, the Bayesian estimations usually require us to approximate the posterior distributions because their closed forms are usually unavailable due to nonlinearity of models $\boldsymbol{f}$.[13,16]

2.2. *Particle Filtering*

The particle filtering is a sequential Monte Carlo technique [10–12] which approximates a joint posterior distribution of state vectors and parameters

$$p(\boldsymbol{x}_{0:T}, \boldsymbol{\theta}|\boldsymbol{y}_{1:T}) \propto p(\boldsymbol{y}_{1:T}|\boldsymbol{x}_{1:T})p(\boldsymbol{x}_{1:T}|\boldsymbol{\theta}, \boldsymbol{x}_0)p(\boldsymbol{x}_0)p(\boldsymbol{\theta}) \,, \tag{4}$$

where $p(\boldsymbol{x}_0)$ and $p(\boldsymbol{\theta})$ denote the prior distributions of initial concentrations and model parameters, respectively. Here, sets of state vectors and observations up to time t are denoted respectively by $\boldsymbol{x}_{0:t} = \{\boldsymbol{x}_0, \boldsymbol{x}_1, \ldots, \boldsymbol{x}_t\}$ and $\boldsymbol{y}_{1:t} = \{\boldsymbol{y}_1, \boldsymbol{y}_2, \ldots, \boldsymbol{y}_t\}$.

Denote N Monte Carlo samples from the conditional distribution $p(\boldsymbol{x}_t, \boldsymbol{\theta}|\boldsymbol{y}_{1:k})$ by $(\boldsymbol{x}_{t|k}^{(i)}, \boldsymbol{\theta}_k^{(i)})$, $i = 1, \cdots, N$. Then, the procedure of the particle filtering is summarized as follows:

- Set $t \leftarrow 1$. Generate N Monte Carlo samples $\{\boldsymbol{x}_{0|0}^{(i)}, \boldsymbol{\theta}_0^{(i)}\}_{i=1}^N$ from the prior distributions of initial state variables $p(\boldsymbol{x}_0)$ and parameter vector $p(\boldsymbol{\theta})$.

- ($\sharp$) Perform the following procedure:
  - (Prediction step) After drawing $\{\boldsymbol{v}_t^{(i)}\}_{i=1}^N$, generate N prediction samples of state variables $\{\boldsymbol{x}_{t|t-1}^{(i)}, \boldsymbol{\theta}_{t-1}^{(i)}\}_{i=1}^N$ according to $\boldsymbol{x}_{t|t-1}^{(i)} = \boldsymbol{f}(\boldsymbol{x}_{t-1|t-1}^{(i)}, \boldsymbol{\theta}_{t-1}^{(i)}) + \boldsymbol{v}_t^{(i)}$ for $i = 1, \cdots, N$.
  - ($\star$) (Filtering step) Execute the following procedure:
    - (i) Compute likelihood function of each particle $p(\boldsymbol{y}_t|\boldsymbol{x}_{t|t-1}^{(i)})$, $i = 1, \cdots, N$.
    - (ii) Calculate the normalizing weights $q_t^{(i)} \propto p(\boldsymbol{y}_t|\boldsymbol{x}_{t|t-1}^{(i)})$ for $i = 1, \cdots, N$ such that $\sum_{i=1}^N q_t^{(i)} = 1$.
    - (iii) Generate $\{\boldsymbol{x}_{t|t}^{(i)}, \boldsymbol{\theta}_t^{(i)}\}_{i=1}^N$ by the resampling of $\{\boldsymbol{x}_{t|t-1}^{(i)}, \boldsymbol{\theta}_{t-1}^{(i)}\}_{i=1}^N$ with the probabilities $\{q_t^{(i)}\}_{i=1}^N$.
- If t is equal to T, stop the algorithm. Otherwise, set $t \leftarrow t+1$ and go to ($\sharp$).

Here, the likelihood function $p(\boldsymbol{y}_t|\boldsymbol{x}_t)$ is defined by the probability density function, e.g. normal distribution, corresponding to the observation model (3). The final draws of $\{\boldsymbol{\theta}_T^{(i)}\}_{i=1}^N$ generated from this procedure are approximately distributed according to the augmented posterior distribution $p(\boldsymbol{x}_{0:T}, \boldsymbol{\theta}|\boldsymbol{y}_{1:T})$. Indeed, it can be proved that the empirical distributions given by N realization $\{\boldsymbol{\theta}_T^{(i)}\}_{i=1}^N$ approach to the true distribution $p(\boldsymbol{\theta}|\boldsymbol{y}_{1:T})$ with probability one as N goes to infinity.[13]

As the number of unknown parameters becomes large, the method requires to generate an exceedingly large number of Monte Carlo samples in order to cover the overall parameter space. For example, the transcription regulation model of circadian clock that will be shown in Section 3 contains 45 unknown parameters in total. Particularly, the problem of "*particle degeneracy*" is caused by a small number of Monte Carlo samples (Chapter 14 of Ref. 13). This is due to the fact that during the repeated resampling of initially generated $\{\boldsymbol{\theta}_0^{(i)}\}_{i=1}^N$ drawn from $p(\boldsymbol{\theta})$, the number of distinct values in the resampled particles can diminish drastically as the time step increases. Hence, in order to estimate a large number of parameters, it is essential to use a fairly large amount of Monte Carlo samples.

Both the time complexity and space complexity of the prediction step are linear order of the number of particles. On the other hand, the time complexity of the filtering step is $O(N \log N)$ because the resampling algorithm requires $O(\log N)$ evaluations of weight table to yield one filtered particle (see Appendix A for details). We employ this type of resampling procedure in the numerical experiment. Rather than performing the random resampling algorithms, the filtering step can be replaced by the resampling algorithm with the proportional-allocation that yields the filtered samples by increasing and decreasing the number of prediction samples proportional to their weights. In that case, the time complexity can be reduced to linear order.

Here, we briefly discuss the low coupling property of the particle filtering. In the prediction step, each predictive sample is updated independently. In the likelihood calculation (i) at the filtering step, each likelihood can be also computed independently. As a result, the tasks in each step can be divided into parallel processors. On the other hand, the processes of normalization (ii) and resampling (iii) require additional parallel computing technique in order to manage the communication among N Monte Carlo samples. In the step (ii), it is necessary that the likelihood values of N samples are collected into a root CPU. To implement a parallel computing for such interacted tasks, we can use the divide and conquer algorithm.[17] By the aid of it, time complexity of normalization and resampling can be reduced up to $O((N/M)\log M)$ where M CPU processors are available, which indicates scalability of the particle filtering in parallel computing.

3. Numerical Experiments

3.1. *Implementation*

Circadian clocks of living cells are controlled by oscillating feedback loop in the transcription circuits of clock-control genes.[2,19] The model that we consider was constructed by incorporating the regulatory mechanisms among the five mRNAs (per, cry, rev/erb, clock, bmal), the translated proteins (PER, CRY, REV/ERB, CLOCK, BMAL) and the two protein complexes (PER/CRY, CLOCK/BMAL), which were reported by Ref. 18. It consists of the 12 first-order stochastic difference equations as,

$$
\begin{aligned}
x_{1,t+1} &= x_{1,t} - k_{d_1}x_{1,t} + k_{tc_1}I(x_{5,t} < s_{1,5})I(x_{12,t} > s_{1,12}) + k_{c_1} + v_{1,t+1},\\
x_{2,t+1} &= x_{2,t} - k_{d_2}x_{2,t} + k_{tl_2}x_{1,t} - k_{b_{2,4}}x_{2,t}x_{4,t} + v_{2,t+1},\\
x_{3,t+1} &= x_{3,t} - k_{d_3}x_{3,t} + k_{tc_3}I(x_{5,t} < s_{3,5})I(x_{12,t} > s_{3,12}) + k_{c_3} + v_{3,t+1},\\
x_{4,t+1} &= x_{4,t} - k_{d_4}x_{4,t} + k_{tl_4}x_{3,t} - k_{b_{2,4}}x_{2,t}x_{4,t} + v_{4,t+1},\\
x_{5,t+1} &= x_{5,t} - k_{d_5}x_{5,t} + k_{b_{2,4}}x_{2,t}x_{4,t} + v_{5,t+1},\\
x_{6,t+1} &= x_{6,t} - k_{d_6}x_{6,t} + k_{tc_6}I(x_{5,t} < s_{6,5}) + k_{c_6} + v_{6,t+1},\\
x_{7,t+1} &= x_{7,t} - k_{d_7}x_{7,t} + k_{tl_7}x_{6,t} + v_{7,t+1},\\
x_{8,t+1} &= x_{8,t} - k_{d_8}x_{8,t} + k_{c_8} + v_{8,t+1},
\end{aligned}
$$

$$
\begin{aligned}
x_{9,t+1} &= x_{9,t} - k_{d_9}x_{9,t} + k_{tl_9}x_{8,t} - k_{b_{9,11}}x_{9,t}x_{11,t} + v_{9,t+1},\\
x_{10,t+1} &= x_{10,t} - k_{d_{10}}x_{10,t+1} + k_{tc_{10}}I(x_{5,t} > s_{10,5})I(x_{7,t} < s_{10,7})\\
&\qquad + k_{c_{10}} + v_{10,t+1},\\
x_{11,t+1} &= x_{11,t} - k_{d_{11}}x_{11,t} + k_{tl_{11}}x_{10,t} - k_{b_{9,11}}x_{9,t}x_{11,t} + v_{11,t+1},\\
x_{12,t+1} &= x_{12,t} - k_{d_{12}}x_{12,t} + k_{b_{9,11}}x_{9,t}x_{11,t} + v_{12,t+1},
\end{aligned}
\tag{5}
$$

in order corresponding to per, PER, cry, CRY, PER/CRY, rev/erb, REV/ERB, clock, CLOCK, bmal, BMAL, and CLOCK/BMAL, respectively. Here, $v_{i,t}$, $i = 1, \cdots, 12$, stand for the Gaussian noises with mean 0 and variance 2.0×10^{-7}. Of interest is then to estimate the 26 rate constant parameters $k.$, the 7 threshold parameters $s.$, and the initial concentrations of 12 reactants $x_{i,0}$, $i = 1, \cdots, 12$. Note that we did not estimate values of k_{d_8} and k_{c_8} for the reason shown below.

Ueda *et al.*[19] studied the transcription regulations of mammalian circadian clocks based on the time course gene expression data which were generated with Affymetrix mouse high-density oligonucleotide probe array. Time course data were measured at equally-spaced 12 time points during 44 hours (about two days). We extracted gene expression measurements of the five mRNAs (per, cry, rev/erb, clock, bmal) from Figure 1a and the supplementary data (`http://sirius.cdb.riken.jp/MouseLiver/MouseLiverCCG(020707).html` [19]), where the set of 12 time points were allocated to the 700 simulation time steps by the conduction of time alignment. Note that in the current model (5), there are no regulatory mechanisms which yield oscillating concentrations of clock mRNA. As a possible solution, we decided to treat the measurements of clock mRNA concentrations as an input sequence. To this end, we interpolated the observed 12 data points of clock by using spline functions, and then, the interpolated values were substituted into the corresponding state variables which remained to be fixed during the parameter estimation process.

Figure 1 shows the observed time course of gene expression profiles and the result of simulation which was generated according to (5) with the predetermined parameters without inclusion of system noises $v_{i,t}$s. This figure shows inconsistency in periods, phases and amplitude among the simulation paths and observed profiles. The objective of this numerical experiment is to capture the observed oscillating patterns of circadian rhythm by the application of the particle filtering where the currently obtained parameters have the bias in the cyclic behavior.

Before proceeding to the particle filtering, the observation noise $\boldsymbol{w}_t$ was set as the normal distribution $N(0, 0.64)$ for per mRNA and $N(0, 0.25)$

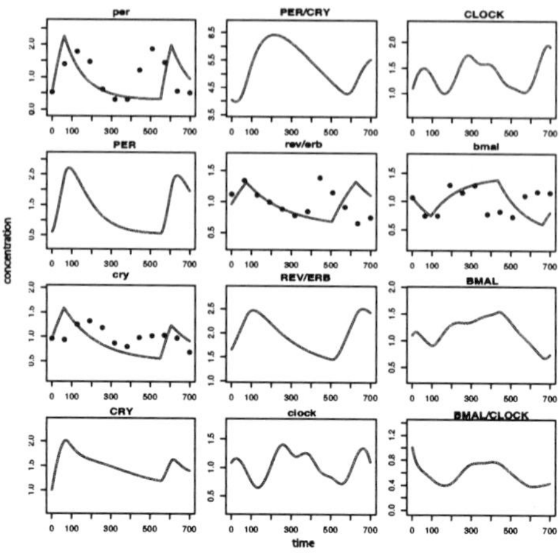

Figure 1. Result of initial simulation which was generated according to (5) with the pre-determined parameters without inclusion of system noises $v_{i,t}$s. The observed 12 data points are indicated by dots.

for the other reactants. We adopted the values that are employed in the simulation shown in Figure 1 as the prior mean of the parameter $\boldsymbol{\theta}$ and the initial state $\boldsymbol{x}_0$. Because the parameters and the states should be non-negative, the prior distributions were set to truncated normal distributions. The prior variances of initial concentration $x_{\cdot,0}$, coefficients $k_{\cdot}$ and threshold values $s_{\cdot}$ were set to 0.01, 0.001 and 0.05, respectively.

In order to realize the implementation of the particle filtering with 10^8 samples, all particles were divided into 20 sets of 5×10^6 particles, and allocated into 20 computational units. After repeating the particle filtering with 5×10^6 samples for each computational unit, which yields $\{\boldsymbol{\theta}_T^{(i,k)}\}_{i=1}^{10^6}$ for $k = 1, \cdots, 20$, we can compute the estimator by $\hat{\boldsymbol{\theta}} = \frac{1}{10^8} \sum_{k=1}^{20} \sum_{i=1}^{5\times 10^6} \boldsymbol{\theta}_T^{(i,k)}$. Due to the independence of the computational units, the integrated estimator $\hat{\boldsymbol{\theta}}$ still preserves the statistical consistency.

3.2. *Results*

Figure 2 shows the histograms of Monte Carlo samples drawn from the posterior distribution of the parameters k_{d_i} $(i = 1, \ldots, 12)$, where the sizes of Monte Carlo samples were set to 10^8 and 10^5, respectively. It is observed from the right panel of Figure 2 that the number of distinct values declines drastically for the estimation with 10^5 particles. Obviously, we would not obtain any efficient estimates by using such degenerated particles. On the

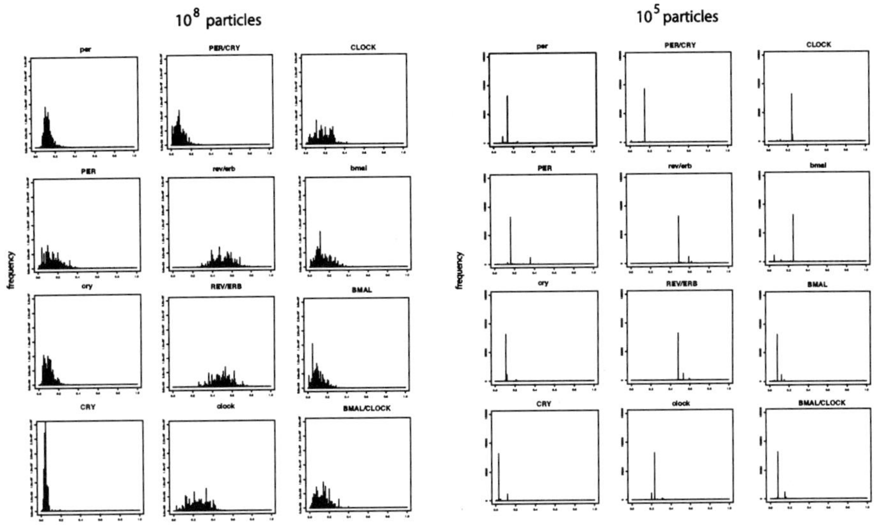

Figure 2. Approximated posterior distributions of the 12 "degradation rate" $k_{d.}$, which were generated by using 10^8 (left) and 10^5 (right) particles.

other hand, the ability of the particle filtering was improved greatly by the use of 10^8 particles as the histograms of the Monte Carlo samples reveal the shapes of posterior distributions clearly as shown in the left panel of Figure 2.

Figure 3 shows the results of simulations based on the estimated parameters and initial state variables, which were computed as the averages of 10^8 and 10^5 particles, respectively. In the case of 10^8 particles, simulation paths were generated from the estimated parameters in each of the 20 computational units, and then, the generated 20 paths were averaged in order to compute the final estimates shown in Figure 3. The observed data points are also shown in the Figure 3. The simulated concentrations corresponding to 10^5 particles preceded the observed periodicity of circadian rhythm whereas the difference in periodicity has been reduced greatly by using 10^8 particles compared with the initially-chosen parameters (prior means).

3.3. *Estimation of computational time*

We evaluate computational time for execution of large-scale particle filtering on upcoming peta-scale supercomputers.

We measured and estimated CPU times in the following three cases:

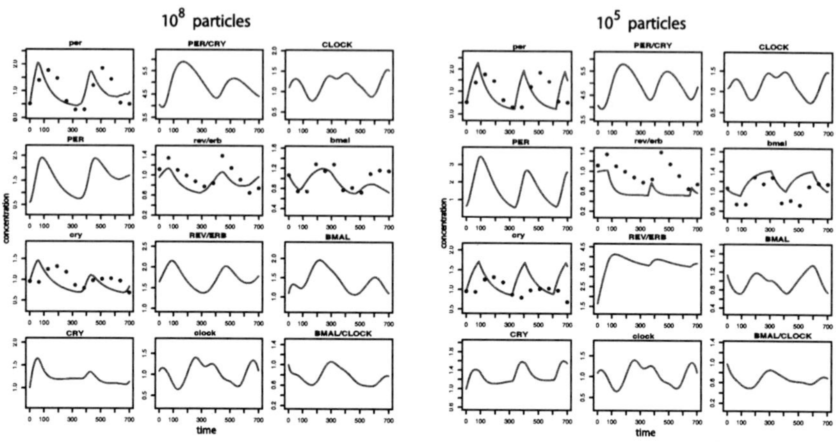

Figure 3. State transition estimated by using 10^8 (left) and 10^5 (right) particles.

(a) 10^8 particles are divided into 20 sets of 5 million particles, and then, repeat the particle filtering 20 times on Opteron 2200 (about 5 gigaflops).
(b) Particle filtering with 10^8(100 million) samples is performed on one petaflops machine.
(c) Particle filtering with 10^9(one billion) samples is performed on one petaflops machine.

For the evaluations of (b) and (c), the CPU times were estimated based on the observed CPU time of the experiment (a). The results of the tests are summarized in Table 1. The current implementation that divides 10^8 particles into 20 computational subunits requires about eight days on average, which can be reduced to 90 seconds by the aid of one petaflops computer. Even with one billion particles, the estimated CPU time can be within 20 minutes.

Table 1. Observed and estimated CPU times (second). The left column shows the measured CPU time on Opteron 2200 and 20 sets of 5 million particles. The middle and right columns show the estimated CPU times for 100 million and one billion particles.

	Opteron, 5mil. × 20	1 petaflops, 100 mil.	1 petaflops, 1 billion
Ave.	6.8×10^5	90	1.0×10^3
Min.	3.7×10^5	50	5.6×10^2
Max.	2.0×10^5	2.7×10^2	3.0×10^3

4. Conclusion

This paper has shown the promising approach based on large scale particle filtering towards exhaustive explorations for kinetic parameters of *in silico* biological pathways. The approach could offer an overall view of posterior distributions defined over the 45 unknown parameters within a reasonable computing time. To our knowledge, the use of one hundred million particles is the first attempt not only in systems biology but also in the other scientific fields. Because the idea is too obvious, such a direct and straightforward approach may have been overlooked so far.

A limitation of our approach is the sensitivity of the estimation results to the choice of the prior distributions, that is, if the prior distributions are fairly inappropriate, we would not obtain reasonable posterior estimations. In order to overcome such a difficulty, we have to develop a methodology for incorporating the existing knowledge on biochemistry and systems biology.

According to the evaluation of computational time, it has been found that we can execute the particle filtering even with one billion particles within 20 minutes by the aid of one petaflops computer that will be accessible in Japan. The results of the experiments indicate that the large scale particle filtering can be a promising approach to parameter estimation of *in silico* biological pathways in the near future.

Appendix A. Resampling Algorithm

The filtering step ($\star$) of the algorithm that has $O(N \log N)$ time complexity is achieved as follows:

- Compute likelihood function of each particle $p(\boldsymbol{y}_t|\boldsymbol{x}_{t|t-1}^{(i)})$, $i = 1, \cdots, N$. Time complexity of this step is given by $O(N)$.
- Calculate the normalizing weights $q_t^{(i)} = p(\boldsymbol{y}_t|\boldsymbol{x}_{t|t-1}^{(i)})$ for $i = 1, \cdots, N$ such that $\sum_{i=1}^{N} q_t^{(i)} = 1$. Time complexity of this step is given by $O(N)$.
- Compute cumulative weight table $\hat{q}_t^{(i)}$ for $i = 0, \cdots, N$ by $\sum_{j=1}^{i} \hat{q}_t^{(j)}$ ($\hat{q}_t^{(0)} = 0$). Time complexity is given by $O(N)$.
- For each $j = 1, \cdots, N$
  - Generate random number u from uniform distribution $U(0, 1]$. Time complexity is $O(N)$.
  - Search the number k where $\hat{q}_t^{(k-1)} < u \leq \hat{q}_t^{(k)}$ by binary search. Time complexity is $O(N \log N)$.
  - Set $(\boldsymbol{x}_{t|t}^{(j)}, \boldsymbol{\theta}_t^{(j)}) \leftarrow (\boldsymbol{x}_{t|t-1}^{(k)}, \boldsymbol{\theta}_{t-1}^{(k)})$, which requires $O(N)$.

Only the binary searching requires $O(N \log N)$. In order to deal with much larger number of particles, we can replace the binary search algorithm by the resampling with the proportional-allocation that yields the filtered samples by increasing and decreasing the number of prediction samples proportional to their weights.

References

1. K. C. Chen, L. Calzone, A. Csikasz-Nagy, F. R. Cross, B. Novak and J. J. Tyson, *Mol Biol Cell.*, **15**, 3841-3862 (2004).
2. H. Matsuno, R. Murakami, R. Yamane, N. Yamasaki, S. Fujita, H. Yoshimori and S. Miyano, *Pac Symp Biocomput.*, **8**, 152-163 (2003).
3. A. Doi, S. Fujita, H. Matsuno, M. Nagasaki and S. Miyano, *In Silico Biol.*, **4**, 271-291 (2004).
4. A. Doi, M. Nagasaki, H. Matsuno and S. Miyano, *In Silico Biol.*, **6**, 1-13 (2006).
5. K. Fujarewicz, M. Kimmel, T. Lipniacki and A. Swierniak, *IEEE/ACM Transaction on Computational Biology and Bioinformatics*, **4**, 322-335 (2007).
6. M. Quach, N. Brunel and F. d'Alché-Buc, *Bioinformatics*, **23**, 3209-3216 (2007).
7. I. Nachman, A. Regev and N. Friedman, *Bioinformatics*, **20**, i248-i256 (2004).
8. G. Koh, H. F. C. Teong, M-V ment, D. Hsu and P.S. Thiagarajan, *Bioinformatics*, **22**, e271-e280 (2006).
9. B.-E. Perrin, L. Ralaivola, A. Mazurie, S. Bottani, J. Mallet, F. d'Alché-Buc, *Bioinformatics*, **19**, ii386-ii349 (2003).
10. G. Kitagawa, *J. Computational and Graphical Stat.* **5**, 1-25 (1996).
11. N. J. Gordon, D. J. Salmond and A. F. M. Smith *IEE Proceedings-F,* **140**, 107-113 (1993).
12. T. Higuchi and G. Kitagawa, *IEICE Tras. Inf. Syst.*, **E83-D**, 36-43 (2000).
13. A. Doucet and N. De Freitas, Springer-Verlag (2001).
14. M. Nagasaki, R. Yamaguchi, R. Yoshida, S. Imoto, A.Doi, Y.Tamada, H. Matsuno, S. Miyano and T. Higuchi, *Genome Inform.* **17(1)**, 46-61 (2006).
15. S. Tasaki, M. Nagasaki, M. Oyama, H. Hata, K. Ueno, R. Yoshida, T. Higuchi, S. Sugano, S. Miyano, *Genome Inform.* **17(2)**, 226-238 (2006).
16. G. Kitagawa, *J. Amer. Stat. Assoc.*, **93**, 1203-1215 (1998).
17. Peter Pacheco, *Parallel Programming with MPI*, Morgan Kaufmann (1996).
18. Y. Fujii, Y. Okitsu, H. Matsuno, S. Miyano and S. T. Inoue, A new regulatory interactions suggested by simulations for circadian genetic control mechanism in mammals, *Genome Inform.*, available at `http://www.jsbi.org/journal/GIW0`, (2004).
19. H. R. Ueda, W. Chen, A. Adachi, H. Wakamatsu, S. Hayashi, T. Takasugi, M. Nagano, K. Nakahama, Y. Suzuki, S. Sugano, M. Iino, Y. Shigeyoshi and S. Hashimoto, *Nature*, **418**, 534-539 (2002).

FROM ODES TO LANGUAGE-BASED, EXECUTABLE MODELS OF BIOLOGICAL SYSTEMS

A. PALMISANO, I. MURA AND C. PRIAMI
CoSBi, Trento, Italy
E-mail: {palmisano, mura, priami}@cosbi.eu

Modeling in biology is mainly grounded in mathematics, and specifically on ordinary differential equations (ODE). The programming language approach is a complementary and emergent tool to analyze the dynamics of biological networks. Here we focus on BlenX showing how it is possible to easily re-use ODE models within this framework. A budding yeast cell cycle example demonstrates the advantages of using a stochastic approach. Finally, some hints are provided on how the automatically translated model can take advantage of the full power of BlenX to analyze the control mechanisms of the cell cycle machinery.

1. Introduction

Abstract models of biological systems are becoming an indispensable conceptual and computational tool for biologists. In order to be useful, a model has to allow automatic analysis, and permit the integration of new knowledge without changing too much the already built model.

Chemical kinetics has traditionally been analyzed using a mathematical formalism in which continuous variables evolve deterministically, even if molecular populations in a biological systems are integer variables that evolve stochastically [1]. There is not yet an agreement upon the characterization of the best "usable form" in modeling biology, because some aspects are better handled with the deterministic approach, while for others the stochastic one is more suitable. For example, in the literature, beside deterministic ordinary differential equations (ODE for short), we find stochastic cell cycle models built with stochastic ODE Langevine type [2], with the Gillespie method [3] and with stochasticity on transitions [4]. Moreover, as a recent paper by Nurse on Nature [5] points out, the life science field requires the development of new and more appropriate languages to describe biological systems.

After the work of Regev et al. [6], a promising trend in this direction is to use programming languages to generate executable models of biological systems. This strategy diverges from classical mathematical modeling because it is executable and not just simply solvable [7]. Execution means that we can describe the flow of control between species and reactions, i.e. not only the time, but also the causality relation among the events that constitute the history of the dynamics of the model. This interpretation is very similar to programming the behaviour of a system rather than describing only its outcome with respect to time. As a consequence, a number of process calculi have been adapted or newly developed for building biological models and performing stochastic simulations (i.e. stochastic π-calculus[8,9], BioAmbients[10], Brane Calculi[11], CCS-R[12], k-calculus[13], Bio-PEPA[14]).

The main contribution of this paper is a semi-automatic method to translate existing deterministic models written with ODEs into programs written in the stochastic modeling language BlenX [15,16]. The simple translation of general ODE terms into a stochastic BlenX model is possible because of the expressive power of the language that allows the definition of general rate functions for the transitions.

The paper is organized as follows. In the next section we concisely recall the two languages used for modeling biological systems, i.e. the mathematical (ODEs) and the computational (BlenX) one. Sect. 3 illustrates a semi-automatic procedure to translate the former into the latter and Sect. 4 applies this procedure on a biological relevant example, the cell cycle machinery of *Saccharomyces cerevisiae*. Validation results and improvements of this model are then discussed in Sect. 4.3, conclusions and future work directions close the paper in Sect. 5.

2. Modeling of Biological Systems

2.1. *The mathematical language*

The usual way in which life scientists build models starts from the organization of the known interactions of the relevant molecules into a wiring diagram [17]. There are no unambiguous rules or general conventions for building these diagrams and so the information contained in those diagrams is quite difficult to be translated automatically in different modeling languages. However, some information contained in those maps is fundamental to disambiguate between different reactions that are associated to the same mathematical model.

Once the model is represented diagrammatically, the biochemical relationships are converted into ODEs. A rate equation is written for each biochemical entity whose concentration changes over time. Those rate equations are written following different kinetic laws that take into account the known properties of the chemicals involved, possibly abstracting some unknown mechanisms.

Before doing the translation of the ODE model in BlenX, we want to make a remark, already discussed by several authors [18,14,19], about the usage of the Gillespie SSA algorithm [20] when high-level mathematical representation of non-elementary reactions are incorporated in the form of rate-dependent functions. The rate dependent functions are defining the reaction propensities of the stochastic model. Following the approach of [21] we assume that the fundamental hypothesis of Gillespie, i.e. each reaction time is a random variable following a negative exponential distribution with rate equal to the value of the propensity function, is verified for the biological system we are modeling. If this hypothesis is valid, a stochastic characterization of the reaction times as negatively distributed random variables is an accurate modeling choice, as proved by Gillespie [20]. In some studies, even considering Michaelis-Menten reactions [18], the applicability of the fundamental hypothesis was mathematically proved, and in some others dealing with the circadian rhythm this same hypothesis was experimentally verified in silico [22]. In our case, because of the presence of many non-elementary reactions, a careful validation similar to the one in [21] has been performed: here just few examples will be shown for the sake of conciseness and because the main focus of the paper is the translation procedure and not the validation process that has to be done on each single class of non-elementary reactions.

2.2. *The* BlenX *Language*

A detailed description of the BlenX language is out of the scope of this paper; here we just summarize the sub-part of the BlenX language needed for understanding the code of the presented examples. We refer the reader to [15,16] for a detailed description of the language and its modeling approach.

The basic metaphor we keep is that a *biological entity*, i.e. a component that is able to interact with other components to accomplish some biological functions, is represented by a *box* in BlenX. This box has an interface (its set of *binders*) and an internal structure that drives its behaviour (see Fig. 1).

For example, in a box modeling a protein, binders may represent *sensing* and *effecting domains*. Sensing domains are the places where the protein receives signals, effecting domains are the places that a protein uses for

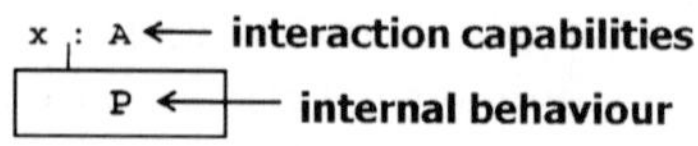

Figure 1. Boxes as abstractions of biological entities.

propagating signals, and the *internal structure* codifies the actions that transform an input signal into, for example, the activation/inactivation of a specific domain. The exchanging of signals can happen between boxes whose binders have a certain degree of affinity, which codes the strength of their interaction.

The basic primitives of the language that will be used in Sect. 3 to translate ODEs into BlenX are summarized in graphical form in Fig. 2.

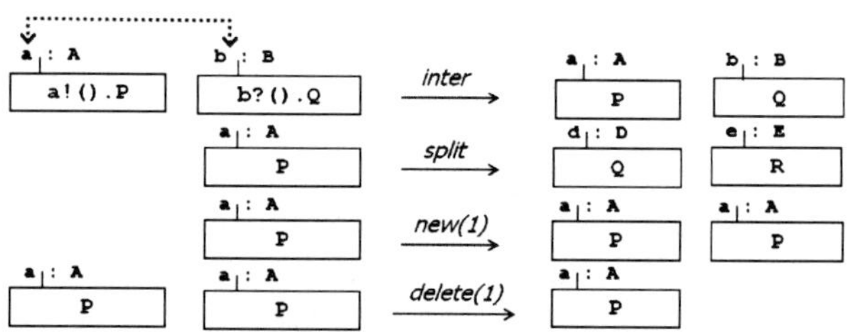

Figure 2. Intuitive behaviour of some BlenX primitives. Each row represents one of the primitives used in our translation. The first primitive codifies the interaction between two boxes, through the exchange of an input/output signal (input is in the form of b?() and the output is in the form a!()). The last three rows are the graphical representation of events of the form "when(*conditions*) *verb*", where the action *verb* is triggered when *conditions* are satisfied. The *verb* can be one of split, new or delete, that model respectively the substitution, creation, and deletion of boxes in the system (graphically depicted above). Conditions, in the models presented here, are in the form of "*entity_name : : function*", whose meaning is that the action after the condition is triggered, at rate *function*, on the entity *entity_name*.

Rate functions are associated to action and events, and those rates can be declared using real numbers that will be used as base rate for the elementary mass action law, or arbitrary functions (e.g. Hill response) that are useful when a box represents an aggregated process or when the precise mechanism of interaction between entities is not known.

3. From ODEs to BlenX models

The rationale is to use the same level of abstraction adopted in the deterministic model so that we can define an easy mapping of ODEs into BlenX.

This translation requires using few primitives of the language. Later in the paper, we propose some hints on how the user can tune the model produced by this mapping, using more powerful BlenX capabilities: this is a fundamental step toward the generation of a more effective translation of the functioning of the logic circuits that underpin biological phenomena [5].

We first modify the differential equations to get the variation of the number of molecules instead of concentrations, using a scalar constant α defined as $\alpha = (N_A 10^{-6} V)^{-1}$, where N_A is the Avogadro's number and V is the volume of the modelled system. Applying $Concs_S = \alpha \cdot MolNums_S$ conversion for each species S, we have an ODE system with the original kinetics constant, but written in terms of number of molecules. This allows us to use directly those terms as stochastic rate functions, without the need of giving to each single kinetic constant its stochastic numerical counterpart.

The second algebraic modification of the ODE system that we need is to add the species that are appearing "implicitly" in the system. If a species is present in a constant total amount but it can switch between an active and an inactive state, the time evolution of just one of the two states is usually explicitly considered, because the other (the "implicit") one can be derived from the first. In our initial simplified usage of BlenX we have to specify all the states in which a species can be, and so we need a definition also for the states not explicitly written in the initial ODEs. Hereafter, when we talk about species, we include both the original *explicit* and the added *implicit* ones.

The general mathematical description of a system of N reactant species $S_1, \ldots, S_N$ expressed in terms of number of molecules $M_1, \ldots, M_N$ and involved in $R_1, \ldots, R_R$ reactions has the following form:

$$\frac{dM_i}{dt} = \sum_{j=1}^{R_i} f_j^i(M_1, \ldots, M_N) \tag{1}$$

where $i = 1, \ldots, N$, $R_i \leq R$, and f_j^i is a rate function that can contain constants rate coefficients and/or other discrete/continuous variables dependent from time and/or species in the model (see the example of $CycB$ variable introduced in Sect. 4.1). So the complete definition of the system above has to be coupled with the set $Var_1, \ldots, Var_T$ of mathematical expressions for the variation of each discrete/continuous variable.

The translation from an ODE system written with the above conventions to BlenX is summarized by the following pseudocode:

Input: the set (1) of ODEs;
Output: the definition of the BlenX model and the rate functions.

Method:

(1) expand all the possible factors of the ODEs, so that all the equations are written as summations of positive/negative terms;

(2) for(i in $[1..N]$)
    for(j in $[1..R_i]$)
        define function(f_i_j) = ABS(mathematical_expression(f^i_j))

(3) for (i in $[1..T]$)
        define variable(var_i) = mathematical_expression(Var_i)

(4) declare all the constants contained in the previous definitions

(5) for(i in $[1..N]$)   declare empty-box(i)

(6) for(i in $[1..N]$)
    for(j in $[1..R_i]$)
    switch(kind of f^i_j reaction):
    case 'synthesis':   add when (i : : f_i_j) new(1);
    case 'degradation':   add when (i : : f_i_j) delete(1);
    case 'changing from i to k':   add when (i : : f_i_j) split(Nil, k);
    case 'changing from k to i':   add when (k : : f_i_j) split(Nil, i);

(7) remove all copies but one of the replicated events that have been created for reactions appearing in multiple ODEs

(8) set the initial state of the model.

Note that the choice on step (6) has to be done manually, looking at the wiring diagram of the system: this is the only step that cannot be done automatically, because of the *inverse problem* discussed in the conclusions of the paper.

In the next section, following [21], the whole process is applied to a budding yeast cell cycle model [23] in order to show that the translation is straightforward and that the stochastic simulations are, from one side, consistent with the results obtained by the solution of the ODE system, but from the other side, the stochastic model matches some characteristics that cannot be found with the deterministic one.

4. Cell Cycle Machinery of Saccharomyces cerevisiae

4.1. *Deterministic Model of Budding Yeast Cell Cycle*

The budding yeast is a well-studied and understood example of how the cell cycle can be controlled by the activity of only a few cyclin-dependent protein kinases. We will focus hereafter on the biochemical machinery that

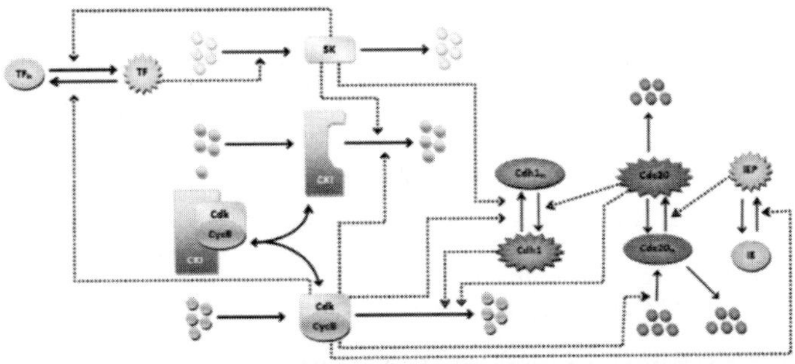

Figure 3. Graphical representation of cell cycle engine. Solid lines link reactants to products, dashed lines represent the mediation effect that some species have on reactions.

controls Cdks activity that has been modeled with ODEs by Novak and Tyson (see [23], pag. 270), and that is depicted in Fig. 3.

For the sake of conciseness, we do not report here the 8 equations composing the model in [23], but only the ODEs of two species that contain all the basic features needed for the translation in BlenX of the model.

$$\frac{d|CDC20\_IN|}{dt} = \underbrace{k5p/\alpha}_{synthesis} + \underbrace{\frac{(k5s/\alpha)*(\alpha*m*CycB)^n}{(J5)^n+(\alpha*m*CycB)^n}}_{(induced)\ synthesis} - \underbrace{k6*|CDC20\_IN|}_{degradation} + \underbrace{\frac{k8*|CDC20\_A|}{J8+\alpha*|CDC20\_A|}}_{inactivation} - \underbrace{\frac{k7*\alpha*|IEP|*|CDC20\_IN|}{J7+\alpha*|CDC20\_IN|}}_{activation} \quad (2)$$

$$\frac{d|CDC20\_A|}{dt} = \frac{k7*\alpha|IEP||CDC20\_IN|}{J7+\alpha|CDC20\_IN|} - \frac{k8|CDC20\_A|}{J8+\alpha|CDC20\_A|} - k6|CDC20\_A| \quad (3)$$

$$\frac{dm}{dt} = \mu*m*(1-m/mstar) \quad (4)$$

Figure 4. Equations for the activation/inactivation of the Cdc20 protein and the rate law for the growing of the mass. $|S|$ is the number of molecules of species S and the different ks (and Js) are the deterministic kinetic parameters. The part of the model above is also referring to a discrete variable ($CycB$) that represents the activity of the dimer CyclinB/Cdk. $CycB$ is calculated with an algebraic expression which can be found in [24] with the complete set of equations and variables.

4.2. *Stochastic Model of Budding Yeast Cell Cycle*

We now apply the translation steps described in Sect. 3 to each term of the above ODEs generating the BlenX code of the model. The model can be given as input to BetaWB[a] framework to perform stochastic simulations.

The $CDC20\_IN$ species contains a positive term for the rate of its **synthesis** ($k5p/\alpha$) and a negative term for the rate of its **degradation** ($k6 * |CDC20\_IN|$). The result of their codification is in Fig. 5(a).

(a)
```
let k5p : const = 0.005; let k6 : const = 0.1;          //step 4
let alpha : const = 0.00302023;                          //step 4
let d_dtCDC20_IN_1 : function = k5p/alpha;               //step 2
let d_dtCDC20_IN_5 : function = k6*|CDC20_IN|;           //step 2
```

(b)
```
let CDC20_IN : bproc = #(a,CDC20_IN)[ nil ];             //step 5
when(CDC20_IN : : d_dtCDC20_IN_1 ) new(1);               //step 6
when(CDC20_IN : : d_dtCDC20_IN_5 ) delete(1);            //step 6
```

(c)
```
let d_dtCDC20_IN_3 : function =                          //step 2
  (k8*|CDC20_A|)/(J8+alpha*|CDC20_A|);
let d_dtCDC20_IN_4 : function =                          //step 2
  (k7*alpha*|IEP|*|CDC20_IN|)/(J7+alpha*|CDC20_IN|);
```

(d)
```
let CDC20_A : bproc = #(a,CDC20_A)[ nil ];               //step 5
when(CDC20_IN::d_dtCDC20_IN_4) split(Nil,CDC20_A);       //step 6
when(CDC20_A::d_dtCDC20_IN_4) split(Nil,CDC20_IN);       //step 6
```

Figure 5. BlenX code showing the result of the steps of the translation method on the specific equations in Fig. 4: synthesis/degradation (a) and activation/inactivation (c) rate functions, synthesis/degradation (b) and activation/inactivation mechanism (d).

The next step is the encoding of the structure of the model. We define an empty box for the species $CDC20\_IN$ and then we add a new event that represents its synthesis and a delete event that represents its degradation. The rates of the events are the functions defined in Fig. 5(a) and structure of the model is in Fig. 5(b). The $CDC20\_IN$ species contains a positive Michaelis-Menten term that is representing the **inactivation** of the $CDC20\_A$ and a negative Michaelis-Menten term that is representing the **activation** of the $CDC20\_IN$. Coding this rate terms similarly to the previous case, we obtain the code in Fig. 5(c). In the model we just have to add the box for the CDC20_A species because we have already defined the box for the inactive species. Then we add two events, one encoding the inactivation of the active Cdc20 and one encoding the activation of the

[a]BetaWB is available at http://www.cosbi.eu/

inactive Cdc20. Their rate are the ones in Fig. 5(c) and the code for this part of the model is in Fig. 5(d). All the terms of the complete ODE system can be seen as one of the previous cases, so even if they are driven by complex kinetics (as the Hill function in the synthesis of $CDC20\_IN$) they can be easily coded by a new/delete/split event.

With the rules introduced so far we encode almost the whole set of ODEs from [23]. The only one left is ODE (4), which is not representing a chemical species but the mass of the cell. It can be seen as a variable that has to be updated, following the function in equation (4), with discrete time steps. With BlenX is possible to define those kind of variables simply copying their ODE with the other rate functions, declaring it as a **var** rather than a **function**. Also cell division is an event that is not explicitly coded in the ODE system, and it halves the mass value when the concentration of active Cdk/CycB falls below an assigned threshold (0.1, as in the [23] model) after having raised above another threshold (here 0.2). In BlenX we can simply add the event "**when**(:mCycB→0.2,mCycB←0.1:)**update**(m,mass_div);", which tells the simulator to trace the state of the variable $mCycB$ and whenever this variable overcomes the 0.2 threshold and then goes back under the 0.1 threshold, the event that updates the value of the mass with its halved value is executed. With this last rule we can complete the translation of the model from ODE to BlenX.

4.3. *Validation of the stochastic model*

Following [21], we compare the results provided by the BlenX model and the deterministic one for the wild type of budding yeast.

The model simulation was performed with BetaWB and the only difference among the runs of the model is the stochastic fluctuation (Fig. 6a-6b).

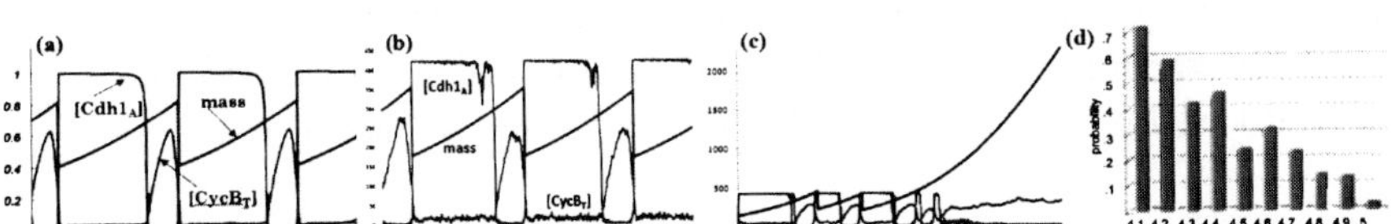

Figure 6. Deterministic (a) and stochastic (b) model results for the wild type. We plot the time courses of the cell mass, cyclin [$CycB_T$] and APC mediator [Cdh1A] concentration for a [0-400] time interval. (c) One stochastic result for the mutant with growth rate 0.0045. The cell was able to generate 2 offspring generations before losing the ability of surviving further. (d) Probability of having at least 10 generation of the mutant varying the growth rate.

Following the careful validation process of [21], we built some mutants starting from the above model: here for the sake of conciseness we show just one of them. All the results of the stochastic simulations are in agreement with the deterministic model of those mutants. In order to show the insights that can be obtained with the stochastic model, we here show results for the mutant obtained with the deletion of the destruction box of cyclin Clb2 (which reduces its degradation rates), and with the deletion of cyclin Clb5 (which reduces the overall CycB level). This model can be obtained changing just some specific parameters of the original model.

Experimental results show that this mutant is viable, but only with a decreased growth rate [25]. Therefore we reduce the rate constant of growth speed from the default value 0.005 min^{-1} to 0.0041 min^{-1}. Both the deterministic and the stochastic model results correctly show its viability. It is interesting to observe that the deterministic model is able to fit the lethality of the mutation with growth rate 0.005 and its viability with growth rate 0.0041, but cannot predict the intermediate situations: it is instead reasonable to expect a continuous transitions as the growth rate varies in the interval [0.0041-0.005], with some mutant cells that have a limited survivability for values of the growth rate inside the interval and a death probability increasing approaching the lethal situation of 0.005.

Small colonies of mutants cells with growth rate in this interval have been experimentally observed [25], so we are expecting that if a mutant is able to complete a sufficient number of cells cycles before dying, a colony may develop, even if its overall growth would be slow. We conducted an in-silico experiment to evaluate the probability that a single mutant cell would be able to generate at least 10 offspring generations before dying, varying the growth rate within the interval [0.0041-0.005] (we ran 100 simulations for each value). The results (Fig.6c-6d)) clearly show that colonies of the mutant may exist for values of the growth rate higher than the threshold which sets the upper limit for the viability of the mutant in the ODE model. We can conclude that stochastic simulations are important to check the partial viability of mutants at the border of life and death.

4.4. *Improving the structure of the model*

In this section we provide some hints on how the model structure can be improved by using the full power of BlenX. The underlying mathematical model is exactly the same as before, but using the peculiarities of this new language, as pointed out in [5], can be a fundamental step toward the generation of a more effective model that can unravel the precise mechanisms of interactions that governs biological systems.

The first modification is to use the die action for the degradation mechanism of a box, so that the death of a box is driven by its own internal behaviour and not by a global event (e.g. the independent degradation of CycBT at rate k2p). The second modification is using the communication between boxes through binders for modeling simple mass action kinetics. For example the degradation of the CycB/Cdk dimer caused by its interaction with Cdh1 (Fig. 3) can be modeled explicitly as in Fig. 4.4 and by assigning an affinity rate **k2s_stoch** of communication between CYCBT_TYPE and CDH1_TYPE binders, where **k2s_stoch** is the deterministic rate k2s converted to stochastic.

```
let CYCBT: bproc = #(x,CYCBT_TYPE)
[die(rate(k2p)) + x?().die.nil];
let CDH1 : bproc = #(y,CDH1_TYPE)
[rep y!().nil];
```

$$\frac{d|CYCBT|}{dt} = -k2p|CYCBT| + -k2s|CDH1||CYCBT| + \ldots$$

Figure 7. Code for a mass action dynamic (left) with the subpart of the deterministic model it is referring to (right).

Using this kind of mechanism on the whole set of ODEs we create a model that is more prone to be modified. As new knowledge is acquired on a specific protein, this can be added to the internal behavior of that specific box in parallel or choice with the existing coded behavior. The only drawback of this approach is that we need to have a detailed description of the interactions between the components and usually this is not the case (e.g. in the cell cycle model described above, the Hill response has been used in order to abstract a cooperative mechanism whose details are unknown).

5. Conclusions and Future Work

We presented a translation of existing ODE models into BlenX computational model. The procedure cannot be completely automatized, because it has to face the *inverse problem* of reaction kinetics [26]: more than one structurally different network of chemicals can produce the same set of ODEs. To disambiguate between those networks we need to have the knowledge of the interactions between all the species, but usually this knowledge is hidden in wiring diagrams that are not following strict or general conventions.

Our translation creates a model that can be stochastically simulated in order to answer questions that cannot be directly tackled by deterministic models. The simple mapping is possible because of the expressive power of BlenX that allows the definition of general rate functions for reactions. However the usage of the SSA algorithm with simple and complex kinetic laws leaves to the user the responsibility to validate simulation results. The example proposed here uses those different levels of abstraction and it has been validated against experimental results. It is our future goal to further analyze the mechanisms that are behind those complex kinetics in order to make more detailed considerations on the effect of the stochastic noise.

Acknowledgments

We thank A. Csikasz-Nagy and P. Lecca for the discussions on mathematical modeling, and L. Dematté and A. Romanel for bright conversations on computational approach.

References

1. D. Gillespie, *Annual Review of Physical Chemistry* **58**, 35 (2007).
2. R. Steuer, *Journal of Theoretical Biology* **228**, 293 (2004).
3. M. Sabouri-Ghomi, A. Ciliberto, S. Kar, B. Novak and J. Tyson, *J. Theor. Biol.* **250**, 209 (2007).
4. P. Lecca and C. Priami, in *Proc. of BioConcur 2003*, (ENTCS, 2007).
5. P. Nurse, *Nature* **454**.
6. A. Regev and E. Shapiro, *Nature* **419**, 353 (2002).
7. J. Fisher and T. Henzinger, *Nature Biotechnology* **25**, 1239 (2007).
8. C. Priami, *The Computer Journal* **38**, 578 (1995).
9. C. Priami, A. Regev, E. Shapiro and W. Silverman, *Inform. Process. Lett.* **80**, 25 (2001).
10. A. Regev, E. Panina, W. Silverman, L. Cardelli and E. Shapiro, *Theoretical Computer Science* **325**, 141 (2004).
11. L. Cardelli, in *Proc. of CMSB 2004*, (LNCS, 2005).
12. V. Danos and J. Krivine, in *Proc. of CONCUR 2004*, (LNCS, 2004).
13. V. Danos and C. Laneve, *Theoretical Computer Science* **325**, 69 (2004).
14. F. Ciocchetta and J. Hillston, in *Proc. of FBTC 2007*, (ENTCS, 2007).
15. L. Dematté, C. Priami and A. Romanel, *Briefings in Bioinformatics* (2008).
16. L. Dematté, C. Priami and A. Romanel, in *Proc. of SFM 2008*, (LNCS, 2008).
17. J. Sible and J. Tyson, *Methods* **41**, 238 (2007).
18. A. Arkin and C. Rao, *Journal of Chemical Physics* **118**, 4999 (2003).
19. I. Mura, Tech. Rep. TR-12-2008, CoSBi (2008).
20. D. Gillespie, *Journal of Physical Chemistry* **81**, 2340 (1977).
21. I. Mura and A. Csikasz-Nagy, *Journal of Theoretical Biology* (2008).
22. D. Gonze, J. Halloy and A. Goldbeter, *J. of Biol. Phys.* **28** (2002).
23. B. Novak and J. Tyson, *Cell Cycle Controls*, in *Computational Cell Biology*, (Springer, 2003), pp. 261–284.
24. A. Palmisano and C. Priami, Tech. Rep. TR-10-2008, CoSBi (2008).
25. F. Cross, *Developmental Cell* **4**, 741 (2003).
26. P. Érdi and J. Tóth, *Mathematical Models of Chemical Reactions* 1989.

UNRAVELING DYNAMIC ACTIVITIES OF AUTOCRINE PATHWAYS THAT CONTROL DRUG-RESPONSE TRANSCRIPTOME NETWORKS

YOSHINORI TAMADA[1*†], HIROMITSU ARAKI[2*], SEIYA IMOTO[1*], MASAO NAGASAKI[1], ATSUSHI DOI[2], YUKIKO NAKANISHI[2], YUKI TOMIYASU[2], KAORI YASUDA[2], BEN DUNMORE[3], DEBORAH SANDERS[3], SALLY HUMPHREYS[3], CRISTIN PRINT[4], D. STEPHEN CHARNOCK-JONES[3], KOUSUKE TASHIRO[5], SATORU KUHARA[5], SATORU MIYANO[1]

[1]*Human Genome Center, Institute of Medical Science, The University of Tokyo, 4-6-1 Shirokanedai, Minato-ku Tokyo, 108-8639, Japan*

[2]*Systems Pharmacology Research Institute, GNI Ltd., 3-8-33 Momochihama, Sawaraku-ku, Fukuoka, 814-0011, Japan*

[3]*Department of Obstetrics & Gynaecology, University of Cambridge, The Rosie Hospital, Robinson Way, Cambridge CB2 2SW, United Kingdom*

[4]*Department of Molecular Medicine & Pathology, School of Medical Sciences, The University of Auckland, Private Bag 92019, Auckland, New Zealand*

[5]*Graduate School of Genetic Resources Technology, Kyushu University, 6-10-1, Hakozaki, Higashi-ku, Fukuoka, 812-8581, Japan*

Some drugs affect secretion of secreted proteins (e.g. cytokines) released from target cells, but it remains unclear whether these proteins act in an autocrine manner and directly effect the cells on which the drugs act. In this study, we propose a computational method for testing a biological hypothesis: there exist autocrine signaling pathways that are dynamically regulated by drug response transcriptome networks and control them simultaneously. If such pathways are identified, they could be useful for revealing drug mode-of-action and identifying novel drug targets. By the node-set separation method proposed, dynamic structural changes can be embedded in transcriptome networks that enable us to find master-regulator genes or critical paths at each observed time. We then combine the protein-protein interaction network with the estimated dynamic transcriptome network to discover drug-affected autocrine pathways if they exist. The statistical significance (p-values) of the pathways are evaluated by the meta-analysis technique. The dynamics of the interactions between the transcriptome networks and the signaling pathways will be shown in this framework. We illustrate our strategy by an application using anti-hyperlipidemia drug, Fenofibrate. From over one million protein-protein interaction pathways, we extracted significant 23 autocrine-like pathways with the Bonferroni correction, including VEGF−NRP1−GIPC1−PRKCA−PPARα, that is one of the most significant ones and contains PPARα, a target of Fenofibrate.

*These authors contributed equally to this work.

†Corresponding author: tamada@ims.u-tokyo.ac.jp

1. Introduction

Understanding mode-of-action of drugs has received considerable attention in pharmacogenomics. Drug-response pathways at a transcriptome level are successfully predicted by cutting-edge computational techniques.[1,2] On the other hand, some drugs affect the pathways at protein level. For example, drugs affect secretion of secreted proteins (e.g. cytokines and growth factors) which are released from target cells. There is a possibility that these proteins have effects on target cells through drug-effected autocrine pathways. From the drug development viewpoint, these pathways could be useful for revealing drug mechanism of action, potentiation of drug effects and avoidance of side effects.

To validate the existence of such drug-affected autocrine pathways, we propose a novel computational method for finding signaling pathways that have the potential to regulate transcriptome networks. The method combines transcriptome networks estimated as drug-response pathways from mRNA expression data with proteome networks represented by protein-protein interactions to extract such pathways. First, we estimate a dynamic transcriptome network from drug-response time-course microarray data by dynamic Bayesian networks with nonparametric regression.[5] For this, we propose the node-set separation method that enables us to find sub-networks significantly activated at observed time points, master-regulator genes and critical paths in the drug-response pathways. We then combine protein-protein interaction (PPI) network with the estimated dynamic transcriptome network. The candidate signaling pathways that connect a ligand or a receptor to the key genes in the transcriptome network are extracted and evaluated based on statistical hypothesis testing at each observed time. Based on the computed p-values, the candidate drug-affected autocrine pathways are selected by multiplicity corrected significance level.

Comparing with the existing computational methods to find mode-of-action of drugs, the novelty of the proposed method is to consider the dynamical interactions between transcriptome and proteome networks. The methods that consider the mode-of-action of drugs only in transcriptome networks cannot be used for our purpose. Moreover, dynamic structural changes of transcriptome networks obtained by the proposed node-set separation method is a key for our purpose, but it cannot be obtained by ordinary dynamic Bayesian networks. For computational autocrine pathway identification, correlation of microarray data between known ligand-receptor pairs was used in cancer cells.[6] The proposed method can be considered as an extension of this research direction to extract dynamics of autocrine pathways from drug-response data. We would like to emphasize again that the existence of drug-affected autocrine pathways is a hypothesis and their existence has not been demonstrated. Therefore, to our knowledge, this is the first genome-wide investigation to test this hypothesis.

We illustrate the whole process of the proposed method by an application using anti-hyperlipidemia drug Fenofibrate, which is known as an agonist of PPARα. The drug effect of Fenofibrate is not only lipid lowering, but also anti-inflammatory in vascular cells.[3,4] The molecular mechanism of lipid lowering by Fenofibrate is well known. The mechanism depends on ligand dependent transcription-activity in which PPARα binds to a specific DNA element termed the PPAR response element (PPRE) of beta-oxidation enzymes and proteins implicated in the reverse cholesterol transport pathways. However, the molecular mechanism mediating trans-repression of the anti-inflammatory effects of Fenofibrate which may include post-transcriptional modification of PPARα has not been established.[3,4] Our method extracted candidate autocrine pathways including PPARα with highly statistical significance. These pathways may be involved in the trans-repression properties of PPARα. We also extracted 23 candidate drug-affected autocrine pathways from over one million protein-protein interaction pathways. Therefore, we observed candidate drug-affected autocrine pathways that support the hypothesis.

2. Methods

2.1. *Dynamic Transcriptome Network*

Figure 1 represents the overview of the proposed method. Based on drug response time-course microarray data, we estimate a dynamic transcriptome network by the dynamic Bayesian network (DBN) model with non-parametric regression.[5] However, ordinary dynamic Bayesian networks can estimate a network from time-course data, while at each observed time, different sub-networks have high activity and transmit information of external signals to other sub-networks. Therefore, we need to extend dynamic Bayesian networks to capture this feature.

The key idea of our DBN estimation, called node-set separation method, is to define the *active gene set* for each time point. That is, a gene in an active gene set is determined as a differentially expressed gene comparing with the controls. Let $\mathcal{A}_t = \{g_i : \mathrm{pv}(g_i, t) \leq \theta_t\}$ be the active gene set at time t for $t = 1, \ldots, T$, where g_i represents the ith gene, $\mathrm{pv}(g_i, t)$ is the

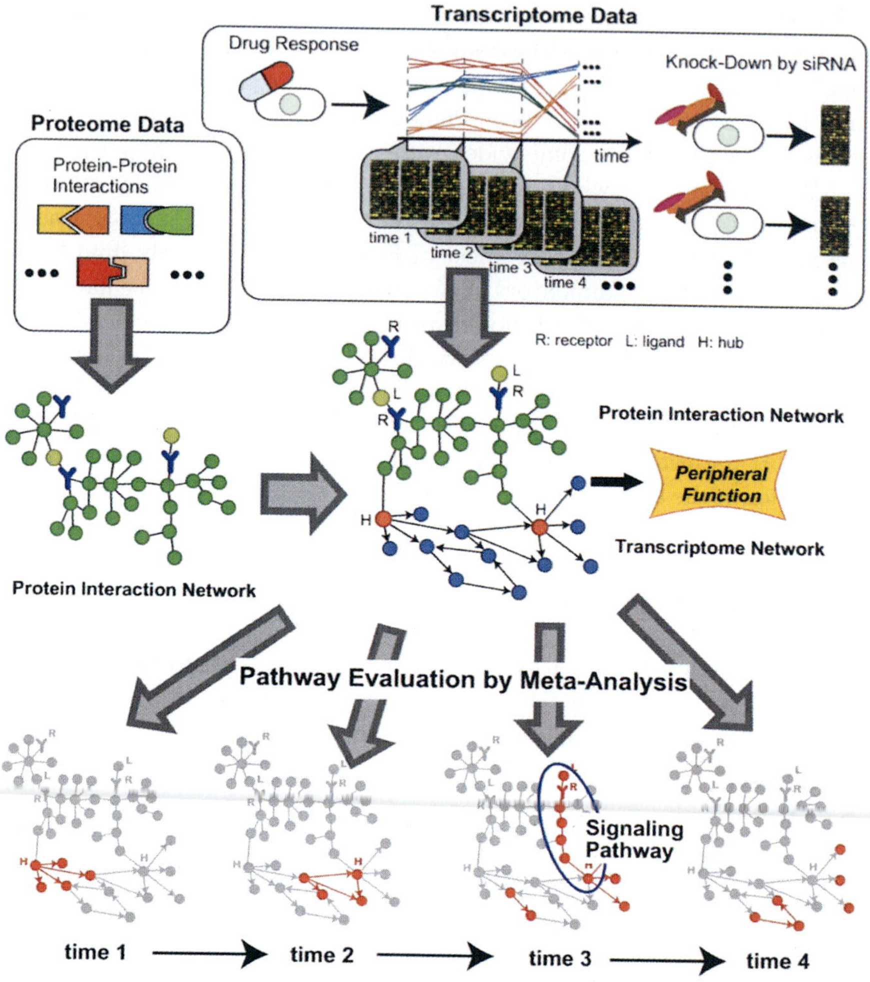

Figure 1. Overview of the proposed method.

p-value of g_i at time t, and θ_t is the threshold for time t that could be determined by using false discovery rate for example. In our case, the p-value of each gene is computed by comparing triplet expression values of the gene at a time with control four replicate expression values, i.e., expression data of non-treated cells. We then define the *node set* $\mathcal{N}_t = \mathcal{A}_{t-1} \cup \mathcal{A}_t$ for $t = 1, ..., T$, where $\mathcal{A}_0$ is the empty set.

The definition of the node set has the basis on the Markov process of the dynamic Bayesian networks, i.e., the DBN assumes the first order Markov process among time-course data; this yields the decomposition $\Pr(\boldsymbol{X}_1, ..., \boldsymbol{X}_T) = \prod_{t=1}^{T} \Pr(\boldsymbol{X}_t | \boldsymbol{X}_{t-1})$, where $\boldsymbol{X}_t$ is the expression data vector at time t and $\boldsymbol{X}_0$ is the empty set. The transcriptome network at time t, we denote G_t, is estimated for the node set $\mathcal{N}_t$ by the DBN and nonparametric regression with whole expression data $\boldsymbol{X}_1, ..., \boldsymbol{X}_T$.[5] Finally the dynamic transcriptome network is obtained by $G = G_1 \cup \cdots \cup G_T$. The advantage of this estimation procedure, i.e., using node set $\mathcal{N}_t$ separately, by comparing with other algorithms that use $\mathcal{N} = \mathcal{A}_1 \cup \cdots \cup \mathcal{A}_T$ as the node set is not only finding dynamics of transcriptome networks, but also possibility to reduce false positive edges in the network, because we can reduce the size of the gene set for each observed time efficiently; this can increase the accuracy of the structure learning.

2.2. *Signaling Pathway Extraction*

2.2.1. *PPI Paths for Candidate of Signaling Pathways*

First, we define master-regulator genes in each node set $\mathcal{N}_t$, based on the estimated G_t for $t = 1, ..., T$. In this paper, the hub genes in $\mathcal{N}_t$ are defined as the top 5% genes; the genes in $\mathcal{N}_t$ are ranked according to the numbers of their direct child-genes in G_t. We denote the set of hub genes of $\mathcal{N}_t$ as $\mathcal{H}_t$. We also focus on the direct parents of the hub genes and represent the set of parent genes of the hub genes in $\mathcal{H}_t$ as $\mathcal{P}_t$. Since the hub genes and their direct parents could control the transcription levels of many genes in $\mathcal{N}_t$, we thus define the set of master-regulator genes at time t by $\mathcal{M}_t = \mathcal{H}_t \cup \mathcal{P}_t$.

We then focus on the PPI network for exploring candidates of signaling pathways affecting master-regulator genes. On the PPI network, for $g_i \in \mathcal{M}_t$, we search receptors and ligands, denoted by r_j, that connect g_i by l or less edges, i.e., g_i connects with r_j by $l-2$ or less intermediate proteins. We denote the kth PPI path for the genes in $\mathcal{M}_t$ ending at $g_i \in \mathcal{M}_t$ as $s_{tk} = r_j - p_1 - p_2 - \cdots - g_i$ where p_1 and p_2 represent the intermediate proteins in the PPI network.

2.2.2. *P-Values for PPI Paths by Meta-Analysis*

Let $[p_i]$ represent the gene for the ith protein in the PPI network, i.e., if p_i is a protein translated from the i'th gene, we have $[p_i] = g_{i'}$. We also define $[r_j]$ in the same way. We assess the significance of s_{tk} using the p-values, $\mathrm{pv}([p'], t)$ for $p' \in s_{tk} \backslash \{g_i\}$, by statistical meta-analysis.[7] That is, we regard the p-value of each genes in s_{tk} as an evidence whether the PPI pathway s_{tk} is activated or not. We use the statistical meta-analysis method for integrating p-values of genes in s_{tk} into the p-value of s_{tk}.

The integrated p-value for s_{tk} will be computed under the null hypothesis: all p-values $\text{pv}([p'], t)$ are not significant, and the alternative hypothesis: at least one or more p-values $\text{pv}([p'], t)$ are significant. That is, if the null hypothesis is not rejected, s_{tk} seems to be not functional; otherwise if we observe the small p-value, s_{tk} is activated and is functional. For the meta-analysis, we use Fisher's inversion method to integrate p-values. We remove the p-value of g_i for the meta-analysis, because g_i was selected as a significant genes in $\mathcal{N}_t$. Therefore, it is obvious that s_{tk} is decided as significant if g_i is included in the meta-analysis calculation, and is meaningless.

Since the node set $\mathcal{N}_t$ is constructed by the active gene sets of time t and $t-1$, there are two ways to assess the significance of s_{tk} by using either p-value at time t or $t-1$. In the real data analysis, we test both cases and assess the significance of each PPI path. We determine s_{tk} is significant if and only if either $\text{pv}(s_{tk}, t) < \xi_t$ or $\text{pv}(s_{tk}, t-1) < \xi_{t-1}$ holds, where $\text{pv}(s_{tk}, t)$ is the integrated p-value of the PPI path s_{tk} with p-values at time t and ξ_t is the threshold determined by considering multiplicity of the testings. In the real data analysis, we use 1% significant level with the Bonferroni correction. Obviously, other methods for controlling multiplicity of testing, such as family-wise error rate, false discovery rate and so on, can be used for reducing false negatives. The reason why we choose the Bonferroni method is that since we use the results of statistical tests for mRNA expression data for finding the significance of protein levels, some changes of protein levels are not measured normally. Therefore, we choose the most strict correction method to achieve a conservative method.

Finally, we should indicate that if high-throughput protein expression data such as time course protein array data are available, we can replace the results of statistical testings for mRNA transcriptional data by the results from protein expression data. Our method can be applied directly to high-throughtput protein expression data immediately.

3. Discovering Fenofibrate-Affected Pathways

In order to demonstrate the capability of the proposed approach, we analyzed gene expression data of human umbilical vein endothelial cells (HUVECs) treated with the anti-hyperlipidemia drug, Fenofibrate. Fenofibrate is an agonist of the peroxisome proliferator-activated receptor α

(PPARα), which is known as a transcription factor that induces genes related to the lipid metabolism. Also, recent studies revealed that Fenofibrate has anti-inflammatory effects.[3,4] Indeed, PPARα activators positively regulate the secretion of secreted proteins.[8] Therefore, we think a PPARα activator is an optimal target (example) to evaluate our approach presented. This application aims at identifying Fenofibrate-affected autocrine pathways related to its anti-inflammatory effects, and at elucidation of unknown modes-of-action.

3.1. *Data Set*

3.1.1. *HUVEC Gene Expression Data and Transcriptome Networks*

We used CodeLink™ Human UniSet I 20K arrays for measuring drug-response time course and knockdown expression data. For the time course data, we observed 6 time points including the control, 2, 4, 6, 8 and 18 hours after treated with 25 μM Fenofibrate in 3 or 4 replicates. For the knockdown gene expression data we knocked down 400 transcripts, which are mainly transcription factors, by siRNAs. We excluded probes which have less than 90% G flags for all 400 arrays from the knockdown expression data. Missing values were imputed by LSimput.[9]

We selected genes whose SAM[10] (Significance Analysis of Microarrays) q value $\leq 5\%$ and fold change ≥ 1.5 at time t for $\mathcal{A}_t$ where $t = 1, \ldots, 5$ corresponding to 2hr, 4hr, 6hr, 8hr and 18hr, respectively. If a gene has more than one probe in the microarrays, we selected the one that has the smallest average of SAM p values for all the time points. This is required for the later steps to match each probe in the microarrays to a protein in the PPI network. Finally, the numbers of genes in $\mathcal{A}_t$ are 14, 5, 144, 129 and 370, respectively. The numbers of genes in $\mathcal{N}_t$ are 14, 19, 144, 200, and 454, respectively. The total number of unique genes in the network is 527. The online supplement[30] gives the complete list of these genes.

The transcriptome networks G_t ($t = 1, \ldots, 5$) were estimated with the prior networks which were estimated from knockdown gene expression data to incorporate transcriptome level changes which can be observed by knocking-down genes by siRNAs.[11] The reliability of the edges in the estimated networks is calculated by the bootstrap method[1] with 1 000 iterations. Edges whose bootstrap probability is less than threshold 0.05 were removed from the final transcriptome networks.

Table 1. Summary of the dynamic transcriptome networks and the extracted pathways on the PPI network ($l = 4$). "$|\mathcal{H}_t|$" is the number of hub genes, "$|\mathcal{M}_t|$" the number of hubs and their parents in G_t, and "$\# s_{tk}$" the number of all possible pathways. Columns "eval $t-1$" and "eval t" are the numbers of pathways that are statistically significant if evaluated by $\mathrm{pv}(s_{tk}, t-1)$ and $\mathrm{pv}(s_{tk}, t)$ respectively.

G_t	nodes	edges	$\|\mathcal{H}_t\|$	$\|\mathcal{M}_t\|$	$\# s_{tk}$	eval $t-1$	eval t
1 (2hr)	14	59	1	9	590 644	–	125
2 (2hr/4hr)	19	91	1	2	28 384	23	271
3 (4hr/6hr)	144	625	7	31	1 016 831	673	1 870
4 (6hr/8hr)	200	874	10	42	1 861 999	1 744	587
5 (8hr/18hr)	454	1 982	22	51	1 194 215	436	150

3.1.2. *Protein-Protein Interaction Data*

We used the integrated PPI data set publicly available in Genome Network Platform[24] (GNP) (released on May 27 2008). GNP collected the PPI data from the public PPI databases of BIND[25] (released on Jun. 25 2006), BioGRID[26] (ver. 2.0.37), HPRD[27] (rel. 7), IntAct[28] (released on Jan. 25 2008), and MINT[29] (released on Dec. 21 2007). In addition to the literature based data, it also contains their own experimental data by yeast two hybrid experiments. In total, GNP PPI data consists of 49 950 non redundant PPIs for 10 103 unique Entrez Gene IDs. By removing proteins which does not have the corresponding probes in the time course data, the final PPI network contains 42 570 edges for 9 016 proteins. We extracted 308 receptor and 149 ligand (in total 457) proteins from the PPI network, which are used as starting nodes r_j of the pathway extraction.

3.2. *Results*

Table 1 summarizes the number of nodes ($|\mathcal{N}_t|$), edges, hubs ($|\mathcal{H}_t|$), and hubs and their parents ($|\mathcal{M}_t|$) in the dynamic transcriptome network G_t. Here hub genes are defined as top 5 % genes ranked according to the number of direct child-genes in G_t. These hub and their parent genes were used as target nodes of the pathway extraction in the later step. The hub genes include cell growth related genes (SESN2), inflammatory response transcription factors (CEBPB, ANKRD1, PPARα), as well as many genes of unknown function. The complete lists of hub genes and their parents are available on the online supplement.[30]

Next we checked the number of possible pathways to determine the appropriate l (maximum distance). Table 2 shows the number of all possible

Figure 2. Extracted autocrine ligand pathways for G_3 (left), G_4 (center), and G_5 (right). The top two nodes VEGF and PDGFA are ligand genes in the network. Nodes and edges in the bottom are the transcriptome network. The middle part contains the proteins and the extracted significant PPI pathways. The large version of this figure is available on the online supplement.[30]

Table 2. The number of possible pathways s_{5k} and the final significant pathways with $\mathrm{pv}(s_{5k}, 5) < \xi_t$ with respect to the maximum distance l.

l	all	final
1	13	3
2	651	3
3	27 373	43
4	1 194 215	150
5	51 078 582	806

Table 3. The numbers of all the ALPs from ligands in $G_{t'}$ to $\mathcal{M}_t$. Column "total" represents the number of all possible pathways ($l = 4$). Columns "eval $t-1$" and "eval t" are the numbers of ALPs that are statistically significant if evaluated by $\mathrm{pv}(s_{tk}, t-1)$ and $\mathrm{pv}(s_{tk}, t)$ respectively.

$G_{t'}$ ligands	$\mathcal{M}_t$	total	eval $t-1$	eval t
2 (2hr/4hr)	3 (4hr/6hr)	437	35	126
2 (2hr/4hr)	4 (6hr/8hr)	894	160	27
3 (4hr/6hr)	4 (6hr/8hr)	1 448	177	28
2 (2hr/4hr)	5 (8hr/18hr)	533	30	27
3 (4hr/6hr)	5 (8hr/18hr)	873	23	23
4 (6hr/8hr)	5 (8hr/18hr)	873	23	23

pathways from ligands or receptors to $\mathcal{M}_5$ (the hubs and their parents in 8hr/18hr transcriptome network G_5) evaluated by p-values of 18hr Fenofibrate time course gene expression data ($\mathrm{pv}(s_{5k}, 5)$). According to this table, we decided to use $l = 4$ since it seems to be the most realistic and appropriate for the later analysis.

The number of the extracted pathways are shown in Table 1. For example, there are 1 194 215 possible pathways within distace $l = 4$ from receptors or ligands to hub and their parents in G_5. Out of these huge number of pathways, there are only 150 statistically significant pathways if their p-values were evaluated by the gene expression data observed at time $t = 5$ (18hr). The complete list of the significant pathways are avaiable on the online supplement.[30]

In order to confirm that the method can capture known pathways related to Fenofibrate, we focused on PPI paths related to PPARα, since PPARα is a target of Fenofibrate. In the dynamic transcriptome network analysis,

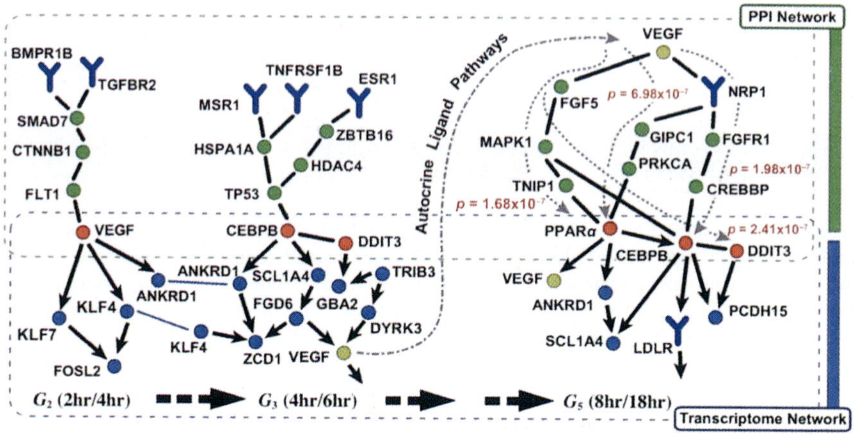

Figure 3. The top four autocrine pathways and their p-values, which are connected to G_5 (8hr/18hr transcriptome network) hub and their parent genes. Some parts of transcriptome networks and the significant pathways at the previous times are also presented to illustrate dynamical changes extracted by the proposed method.

PPARα is included in the node sets $\mathcal{N}_4$ and $\mathcal{N}_5$, i.e., PPARα was over-expressed at 8 and 18 hours. In both times, PPARα was selected as a hub gene, in G_4 PPARα has 21 children and 31 in G_5. Since we would like to investigate drug-affected autocrine pathways, we first limited the candidate PPI paths by autocrine ligand pathways (ALPs) that connect ligands included in earlier time transcriptome networks, i.e., active ligands in earlier times, to hub genes and their parent genes.

By the Bonferroni correction with 1% significance level, only 23 pathways from ligands in G_3 or G_4 to $\mathcal{M}_5$ evaluated by 8hr expression data remained as significant ALPs (Table 3). Among them, we found that the pathway including PPARα as a hub gene of the gene network has high statistical significance (the fourth highest significance). This ALP is VEGF−NRP1−GIPC1−PRKCA−PPARα. PRKCA, protein kinase C alpha, is located on the upstream of PPARα. PRKCA is one of the members of serine- and threonine-specific protein kinases and is related to phospholyration of many genes including PPARα. Protein kinase C inhibitor inactivates the phosphorylation of PPARα and induces the trans-repression activity of PPARα in hepatocytes.[14,15] Our method was able to extract this known relationship, which is related to PPARα's trans-repression, with high statistical significance. VEGF, vascular endothelial growth factor A, is also included in this pathway. VEGF is a member of the PDGF/VEGF growth factor family and is the predominant regulator of angiogeneis. It has been reported that Fenofibrate induces VEGF mRNA and prevents cell apoptotic cell death in human retinal endothelial cells (HRECs).[9] VEGF is also significantly up regulated in our microarray experiment. From this, our method suggests that the trans-repression property of Fenofibrate might be caused by PRKCA mediated thorough VEGF signaling.

We also found out that CEBPB (CCAAT/enhancer binding protein beta) is the ending node of the ALP and a hub gene of the gene network with the highest significance level. This ALP is VEGF−NRP1−FGFR1−CREBBP−CEBPB. CEBPB is a transcription factor which can bind as a homodimer or heterodimiers with the related proteins to certain DNA regulatory regions. CEBPB is important in the regulation of genes involved in immune and inflammatory responses and has been shown to bind to the IL-1 response element in the IL-6 gene, as well as to regulatory regions of several acute-phase and cytokine genes. Fenofibrate decreases CRP (C-reactive Protein) and fibrinogen which are risk factors of vascular disorders, through CEBPB.[16,17] Our method extracted the ALP including CEBPB that is a key gene for anti-inflammatory action and suggests that this ALP might also mediate the anti-inflammatory effects of Fenofibrate.

DDIT3 (DNA-damage-inducible transcript 3), which is a transcription factor of C/EBP family, is the ending node of the third highest ALP. This gene is involved in atherogenes induced by oxLDL (oxidized LDL)[18] and increases promoter activity of MCP-1.[19] PPARα increases uptake of oxLDL in bovine aortic endothelial cells (BAEC)[20] and induces atherogenesis in response to oxidized phospholipids, constituents of oxLDL, through the up-regulation of MCP-1 and IL-8 in human aortic endothelial cells (HAEC).[21] From these facts and our results, DDIT3 may partly play a role in oxLDL-induced atherogenesis through PPARα.

4. Discussion

In this paper, based on the assumption of the existence of drug-affected autocrine pathways, we presented a novel computational method capable of finding them. The results suggest that the autocrine pathways exist and have an important role in the regulation of transcriptome level networks affected by drugs. Previously, identification of autocrine signaling loops in cancer cells by microarray data was proposed,[6] but only co-expression of known ligand-receptor pairs was considered. The method proposed in this paper can be viewed as an extension of their work by adding PPI networks and dynamics of transcriptome network that are essential to extract drug-affected autocrine pathways. Moreover, unlike the transcriptome level analyses,[1,2,6] the proposed method can capture proteomic and metabolomic level signaling pathways by exploiting the PPI networks and drug-response gene expression data. By combining the dynamic transcriptome network estimated by the node-set separation method and the meta-analysis technique, the method can extract statistically significant pathways from a huge number of possible candidates. Compared to the existing methods for analyzing signaling pathways,[22] the proposed method can extract genome-wide dynamical changes of the signaling pathways that are statistically significant. Although the proposed method currently uses only gene expression data, it can be easily applied to the genome-wide proteomic time course data when they are available.

In the real data analysis, we applied the method to transcription expression data in Fenofibrate treated HUVECs. Our proposed method extracted statistically significant pathways from over a million of the possible PPI pathways. Among them, the pathway from VEGF to PPARα through PRKCA has high statistical significance. PRKCA regulates the trans-repression activity of PPAR by the phosphorylation of PPARα. Thus, we speculate that this pathway might involve the transrepression effects of Fenofibrate. Other pathways (e.g. VEGF to CEBPB) which might be related to the drug effect of Fenofibrate are also extracted by this method. The method proposed in this study has potential advantages for drug discovery and development. First, our approach can outline drug mode of action not only at the level of mRNA regulatory relationships, but also protein-protein interactions. Pathways extracted in this study can not be detected by using gene networks based only on transcriptome data. Secondly, potentiation of existing drugs has recently been recognized as an important area of anti-tumor research.[23] We think our approach will be able to detect candidate targets in silico. For example, NRP1 (Neuropilin-1) is a co-receptor for VEGF and is present in both the pathways from VEGF to PPARα and from VEGF to CEBPB. Small molecules targeting NRP1 have the potential for potentiation of the anti-inflammatory effects of Fenofibrate.

Acknowledgments

Computation time was provided by the Super Computer System, Human Genome Center, Institute of Medical Science, University of Tokyo.

References

1. S. Imoto *et al.*, *Pac. Symp. Biocomput.* **11**, 559–571 (2006)
2. D. di Bernardo *et al.*, *Nat. Biotechnol.* **23**(3), 377–383 (2005).
3. P. Lefebvre *et al.*, *J. Clin. Invest.* **116**(3), 571–580 (2006).
4. D. S. Straus and C. K. Glass, *Trends Immunol.* **28**(12), 551–558 (2007).
5. S. Kim *et al.*, *Biosystems* **75**(1-3), 57–65 (2004).
6. T. G. Graeber and D. Eisenberg, *Nat. Genet.* **29**(3), 295–300 (2001).
7. P. K. Gupta *et al.*, *Proc. 3rd ISBRA, LNBI* **4463**, 146–157 (2007).
8. L. Nilsson *et al.*, *Arterioscler. Thromb. Vasc. Biol.* **19**(6), 1577–1581 (1999).
9. T. H. Bø *et al.*, *Nucleic Acids Res.* **32**(3), e34 (2004).
10. V. G. Tusher *et al.*, *Proc. Natl. Acad. Sci. U. S. A.* **98**(9), 5116–5121 (2001).
11. M. Affara *et al.*, *Philos. Trans. R. Soc. Lond. B* **362**(1484), 1469–1487 (2007).
12. J. Kim *et al.*, *Exp. Eye Res.* **84**(5), 886–893 (2007).
13. K. A. Burns and J. P. Vanden Heuvel, *Biochim. Biophys. Acta.* **1771**(8), 952–960 (2007).
14. C. Blanquart *et al.*, *Mol. Endocrinol.* **18**(8), 1906–1918 (2004).
15. J. P. Gray *et al.*, *Biochemistry* **44**(30), 10313–10321 (2005).
16. R. Kleemann *et al.*, *Blood* **101**(2), 545–551 (2003).
17. P. Gervois *et al.*, *J. Biol. Chem.* **276**(36), 33471–33477 (2001).
18. T. Thum and J Borlak, *J. Biol. Chem.* **283**(28), 19456–19464 (2008).
19. K. Kodama *et al.*, *Am. J. Physiol. Cell Physiol.* **289**(3), C582–C590 (2005).
20. K. Hayashida *et al.*, *Biochem. Biophys. Res. Commun.* **323**(3), 1116–1123 (2004).
21. H. Lee *et al.*, *Circ. Res.* **87**(6), 516–521 (2000).
22. K. A. Janes and M. B. Yaffe, *Nat. Rev. Mol. Cell Biol.* **7**(11), 820–828 (2006).
23. S. Morgan-Lappe *et al.*, *Oncogene* **25**(9), 1340–1348 (2006).
24. GNP: `http://genomenetwork.nig.ac.jp/index_e.html`
25. BIND: `http://bind.ca/`
26. BioGRID: `http://www.thebiogrid.org/`
27. HPRD: `http://www.hprd.org/`
28. IntAct: `http://www.ebi.ac.uk/intact/`
29. MINT: `http://mint.bio.uniroma2.it/mint/`
30. Online Supplementary Information: `http://bonsai.ims.u-tokyo.ac.jp/~tamada/suppl/PSB2009/`.

DISSECTING THE INTERFACE BETWEEN SIGNALING AND TRANSCRIPTIONAL REGULATION IN HUMAN B CELLS

KAI WANG[1,2,*], MARIANO J. ALVAREZ[2], BRYGIDA C. BISIKIRSKA[2], RUNE LINDING[3], KATIA BASSO[4], RICCARDO DALLA FAVERA[4], ANDREA CALIFANO[1,2,4,†]

1. Department of Biomedical Informatics, Columbia University, New York, NY, USA

2. Joint Centers for Systems Biology, Columbia University, New York, NY, USA

3. Samuel Lunenfeld Research Institute, Mount Sinai Hospital, Toronto, Canada

4. Institute of Cancer Genetics, Columbia University, New York, NY, USA

1. Abstract

A key role of signal transduction pathways is to control transcriptional programs in the nucleus as a function of signals received by the cell via complex post-translational modification cascades. This determines cell-context specific responses to environmental stimuli. Given the difficulty of quantitating protein concentration and post-translational modifications, signaling pathway studies are still for the most part conducted one interaction at the time. Thus, genome-wide, cell-context specific dissection of signaling pathways is still an open challenge in molecular systems biology.

In this manuscript we extend the MINDy algorithm for the identification of post-translational modulators of transcription factor activity, to produce a first genome-wide map of the interface between signaling and transcriptional regulatory programs in human B cells. We show that the serine-threonine kinase STK38 emerges as the most pleiotropic signaling protein in this cellular context and we biochemically validate this finding by shRNA-mediated silencing of this kinase, followed by gene expression profile analysis. We also extensively validate the inferred interactions using protein-protein interaction databases and the kinase-substrate interaction prediction algorithm NetworKIN.

2. Introduction

A key role of signal transduction pathways is to control transcriptional programs in the nucleus as a function of signals received by the cell via complex post-translational modification cascades, thus determining the cell's response to environmental stimuli (see Figure S1 for a schematic description). Their understanding is increasingly crucial in the dissection of human disease and in the identification of therapeutic intervention targets [1], because signaling molecules (e.g., GPCR receptors or tyrosine kinases) are much more effectively targeted by small molecules than transcription factors. Unfortunately, cell-context specific dissection of signaling pathways is still an open challenge

---

* Current affiliation: Rosetta Impharmatics (Merck & Co., Inc.), Seattle, WA 98109

† email: califano@c2b2.columbia.edu

because of the inherent difficulties in the high-throughput measurement of protein concentration and post-translational modification. As a result, the dissection of signaling pathway is still, for the most part, proceeding one protein-protein interaction at a time [2].

Conversely, availability of large collections of gene expression profiles (GEP) [3] has fostered significant progress in the genome-wide dissection of transcriptional programs [4, 5]. Until recently, GEPs have not been broadly used in the dissection of post-translational interactions. Several GEP-based studies of yeast signal transduction networks have been limited to the identification of gene modules regulated by a small number of regulators, including some signaling proteins [3], or to the reconstruction of signaling pathways using known protein-protein interactions as a topological backbone [6]. In general, however, a cell-context-specific map of the interface between signaling and transcriptional regulatory programs is still an elusive target both in yeast and in higher eukaryotes.

We recently introduced the MINDy algorithm (Modulator Inference by Network Dynamics) for the genome-wide identification of post-translational modulators of transcription factor (TF) activity [7]. MINDy tests whether the conditional mutual information (CMI), $I[TF; t \mid M]$, between a transcription factor TF and a target t, as a function of a modulator M is non-constant. In that case, M is inferred as a candidate post-translational modulator of the TF. Based on this analysis, MINDy can also determine whether the modulator protein will activate or repress the TF-target interaction, resulting in either a positive or negative *mode of action* (MoA). We have biochemically validated four inferred modulators of the transcription factor MYC, including a kinase (STK38), an histone deacetylase (HDAC1), and two transcription factors (BHLHB2 and MEF2B) by shRNA mediated silencing and other biochemical assays [8]. For full details on the MINDy algorithm, its applications as well as limitations, readers are referred to [7, 8] and the Methods section 5.2. In this manuscript, we extend the MINDy algorithm to the genome-wide exploration of the interface between signaling pathways and transcriptional networks in human B cells.

3. Results

3.1. *Network components*

In this work we define the signalome as the compendium of signaling proteins (SP) annotated as protein kinases, phosphatases or cell surface receptors in the Gene Ontology (GO) [9]. The term "transfactome" is borrowed from [10] and is defined here as the compendium of proteins annotated as transcription factors (TF) in the GO. Only proteins expressed in a set of 254 GEPs from normal and tumor related human B cells were considered in the analysis (see 5.1). A total of 772 SPs and 595 TFs were selected based on these criteria, see Table 1.

Table 1. Selection of signalome and transfactome genes. # indicates the number of genes selected in each category. MF: molecular function; BP: biological process; CC: cellular component

Functional Category		#	GO Categories
Signalome	Kinases	421	Protein kinase activity (MF)
	Phosphatases	113	Phosphoprotein phosphatases (MF)
	Receptors	295	Receptor activity (MF) Cell surface receptor linked signal transduction (BP) Integral to plasma membrane (CC)
Transfactome	Transcription Factors	595	Transcription factor activity (MF)

3.2. *Signalome-transfactome interaction inference*

SP-TF interactions were inferred by assessing whether one or more TF-target interactions were modulated by the SP using the CMI test (see 5.2). The complete set of TF-targets modulated by a SP is called the SP's *regulon*, while the set of all TFs modulated by a SP is called the SP's *modulon* (Figure S3). At the genome-wide statistical significance level of 5% (see 5.2), MINDy inferred 44,349 SP-TF interactions. A summary of these results can be found in Table 2. Each SP modulates on average 29.6 TFs in human B cells, and a TF is controlled on average by 38.4 SPs. Some other interesting global properties of the signalome-transfactome interface are summarized in Figure 1 (and more are reported in the supplementary materials). These include: (a) The MoA is consistently inferred from each SP-TF-target triplet supporting a specific SP-TF interaction, even though they are independently tested. i.e., either the positive or the negative MoA is supported by the majority (95 - 100%) of the triplets. This

Transfactome \ Signalome	STK38	CDK5	BCR	CSNK1G2	PRKDC	PTK2B	CCL3	CDK2	TGFBR2	CDC25A	...	GNRHR	HRH1	PRKG1	ADRB1	GLRA1	HUS1	MPL	BMX	TACR2	DNAJC6
SMAD3	5	3	9	0	8	2	0	7	8	27	...	0	0	0	0	0	0	0	0	0	0
CREM	0	2	1	2	19	1	0	3	5	5	...	0	0	0	0	0	0	0	0	0	0
ZNF263	0	1	1	11	6	1	10	2	0	3	...	0	0	0	0	0	0	0	0	0	0
MEF2D	4	1	10	3	0	0	4	1	0	0	...	0	0	0	0	0	0	0	0	0	0
ETS1	0	16	1	5	4	4	63	0	0	1	...	0	0	0	0	0	0	0	0	0	0
PHTF1	3	22	0	7	32	0	0	27	8	0	...	0	0	0	0	0	0	0	0	0	0
NR4A1	0	2	0	8	3	0	2	18	0	28	...	0	0	0	0	0	0	0	0	0	0
ATF3	0	3	0	13	5	1	0	1	8	8	...	0	0	0	0	0	0	0	0	0	0
TAF7	12	0	1	0	3	0	7	0	0	3	...	0	0	0	0	0	0	0	0	0	0
ZNF85	0	20	0	9	28	1	0	48	11	15	...	0	0	0	0	0	0	0	0	0	0
...	...	...	...	...	...	...	...	...	...	...	...	...	...	...	...	...	...	...	...	...	...
IRF7	0	0	0	0	0	0	0	0	0	0	...	0	0	0	0	0	0	0	0	0	0
NFIC	1	0	0	0	0	0	0	0	0	0	...	0	0	0	0	0	0	0	0	0	0
SHOX	0	0	0	0	0	0	0	0	0	1	...	0	0	0	0	0	0	0	0	0	0
THRB	2	0	0	0	0	0	0	0	0	0	...	0	0	0	0	0	0	0	0	0	0
JARID1B	0	0	0	0	0	0	0	0	0	0	...	0	0	0	0	0	0	0	0	0	0
MDS1	0	0	0	0	0	0	0	0	0	0	...	0	0	0	0	0	0	0	0	0	0
FOXI1	0	0	0	0	0	0	0	0	0	0	...	0	0	0	0	0	0	0	0	0	0
HOXB1	0	0	0	1	0	0	0	0	0	0	...	0	0	0	0	0	0	0	0	0	0
NKX2-2	0	0	0	0	0	0	0	0	0	0	...	0	0	0	0	0	0	0	0	0	0
FOXJ1	0	0	0	0	0	0	0	0	0	0	...	0	0	0	0	0	0	0	0	0	0

Table 2. Summary of MINDy results in the signalome-transfactome inference. Signaling proteins are shown on the columns and sorted in decreasing order with respect to the number of TFs each SP modulates. TFs were shown on the rows and are also sorted in decreasing order with respect to the number of SPs they are under control of. Each cell indicates the number of TF-target interactions modulated by the SP.

is highly biologically consistent, since a given SP is expected to either activate or repress a TF's activity but not to do both at the same time. (b) MINDy-inferred TF-target interactions exhibits the previously observed scale-free like degree distribution [11]. However, while previously inferred regulatory networks included only static (i.e. not SP-modulated) TF-target interactions, MINDy inferred network includes conditional interactions as well.

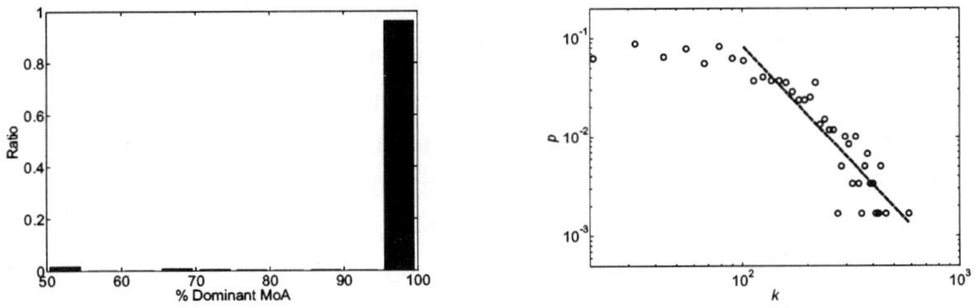

Figure 1. Properties of the signalome-transfactome network inferred by MINDy. (a) Histogram of the proportion of dominant MoA for all SP-TF interactions. Plotted on the x-axis is the percent of MINDy inferred SP-TF-*t* triplets supporting the dominant MoA for a SP-TF interaction. If all triplets support a positive MoA, for instance, the fraction would be 1; if 50% of the triples support a positive MoA and 50% a negative one, the fraction is 0.5. (b) Degree distribution of the MINDy inferred TF-targets network.

3.3. *In-silico validation*

To benchmark the accuracy of MINDy inferences, a set of gold standard SP-TF interactions in human B cells would be required. Compiling such a reference set is relatively difficult because: (a) experimentally validated protein-protein interactions in databases are still very sparse, especially for transient interactions (e.g., kinase-substrate); (b) MINDy-inferred modulators can be either direct (i.e. a physical SP-TF interaction), or pathway-mediated (i.e. SP upstream of the TF in a signaling pathway). While the former could be represented in existing databases, the latter are poorly characterized; and (c) while MINDy inferred SP-TF interactions are highly cell–context specific (to human B cells), human protein-protein interactions have been validated in highly heterogeneous or even artificial (e.g., yeast two-hybrid, Y2H) cellular contexts. To address (a) and (b) we benchmarked MINDy with the following datasets:

Protein-protein interactions (PPIDB): We collected all known human PPIs from high-quality public databases including HPRD [12], BIND [13], DIP [14] and IntAct [15]. These have been experimentally assessed, either in single biochemical assays, or by high throughput techniques such as Y2H. Compared to the 772 SPs and 595 TFs analyzed by MINDy, these datasets cover 428 SPs and 141 TFs (i.e., 13.1% of the MINDy search space).

Kinase-substrate interactions (KSIDB): We also included kinase-substrate interactions inferred by the experimentally-validated algorithm NetworKIN [16]. This algorithm utilizes information from consensus motifs on the kinase catalytic sites, substrate phosphorylation sites assessed by Mass Spectrometry, as well as cellular context and curated pathways. Such information is completely orthogonal to that used by MINDy (i.e. strictly GEP). Therefore enrichment of their common predictions can be used to assess MINDy's validity, as false positives from the two methods should not be correlated if either of them makes random predictions. Due to the limited number of kinase families for which consensus motifs are known, NeworKIN covers only 74 of 772 MINDy SPs and 240 of 595 MINDy TFs (i.e., 3.9% of the MINDy search space).

Table 3 offers a comparative view of MINDy inferred SP-TF interactions as well as those from PPIDB and KSIDB. Due to the higher coverage of GEP, MINDy covers a much larger space (~7-fold) than the other two data sources combined and may thus provide important information for SP-TF interactions that cannot be studied using other methods.

Table 3. Summary of predictions made by MINDy, PPIDB interactions and KSIDB predictions.

	MINDy	PPIDB	KSIDB	PPIDB + DSIDB
Interactions	9017	434	1105	1506
No. SPs	772	428	74	439
No. TFs	595	141	240	291
% Coverage	100%	13.1%	3.9%	15.8%
% Prediction	2.0%	0.7%	6.2%	1.2%

We first tested the hypothesis that modulator sets affecting the same TF should be more physically inter-connected than random genes (as they are more likely to cluster within signaling pathways). This was done by counting the number of PPIDB and KSIDB interactions among MINDy-inferred SPs that modulate the same TF, compared to that among the same number of SPs chosen at random. Of 595 tested TFs, 400 (67.2%) show significant enrichment among their inferred modulator SPs by Fisher's Exact Test (FET) (i.e. p-value < 0.05/595). This suggests that SPs modulating the same TF tend to cluster in physical pathways.

Next, we measured the overlap between MINDy-inferred interactions and those in PPIDB and KSIDB. Since interactions predicted by MINDy include both direct and pathway-mediated interactions, we expanded the interactions in our reference databases with additional ones for which a linear chain of PPIDB (undirected) and/or KSIDB (directed) interactions exists between a SP and a TF, as first proposed in [6]. We plotted the precision of MINDy (i.e. percent of MINDy predictions in the extended database) as a function of the MINDy p-value threshold (i.e. the threshold that controls the MINDy false positive rate). As expected (Figure 2), MINDy precision and number of predictions respectively increase and decrease as the p-value threshold becomes more stringent. Based on this plot, we selected the inflection point, corresponding to the MINDy p-value of 7.5×10^{-7}, as an optimal p-value cutoff, resulting in 9,017 inferred interactions. Of these, 3,739 (41.5%) are supported by either direct or pathway-mediated PPI interactions. The probability that such an overlap occurs by chance is $< 2.9\times10^{-89}$ by FET. A similar analysis using KEGG [17] and GenMAPP [18] annotated pathways also yielded statistically significant enrichment (results reported in the supplementary materials). These highly significant enrichments suggest that MINDy is able to recapitulate known interactions at the signalome-transfactome interface.

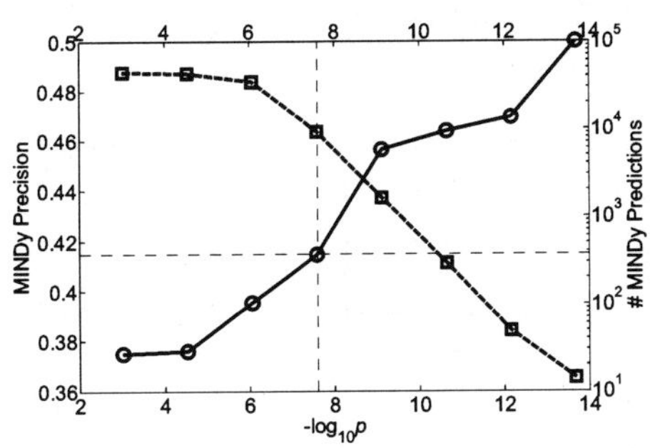

Figure 2. MINDy benchmarking using PPIDB and KSIDB interactions. X-axis shows the $-\log_{10}$ of MINDy p-value. Precision of MINDy as a function of p-value cutoffs is plotted using the solid line on the left y-axis. The dashed line plots the number of MINDy predictions at each p-value cutoff on the right y-axis. Dotted lines indicate the optimal p-value cutoff selected in the text.

3.4. *Experimental validation*

Given the exceedingly large set of MINDy prediction space, a systematic experimental validation plan is clearly impractical. Instead, we decided to validate the candidate modulator controlling the largest number of TFs. This is a serine-threonine kinase STK38 [19] that is poorly characterized in the literature, a rather surprising fact since MINDy infers it as a modulator of 303 TFs. We silenced STK38 by lentiviral-vector mediated shRNA expression in the ST486 Burkitt's line, and verified that the STK38 protein level significantly decreased at 60h after transduction (Figure 3a).

MINDy regulon analysis: We first tested whether the 1219 TF-targets that were inferred as STK38 modulated via the 303 TFs (the STK38 *regulon*) were indeed affected by STK38 silencing. Gene Set Enrichment Analysis (GSEA) [20] confirmed a very significant enrichment ($p < 10^{-4}$) of the STK38 regulon in genes that were differentially expressed after silencing (Figure 3b). Moreover, the set of 303 STK38 modulated TFs (the STK38 *modulon*) was not enriched in differentially expressed genes (Figure 3c), suggesting that the differential regulation of the STK38-regulon is not mediated by transcriptional activation/repression of the TFs in its modulon, consistent with the MINDy model which is designed to identify post-translational interactions.

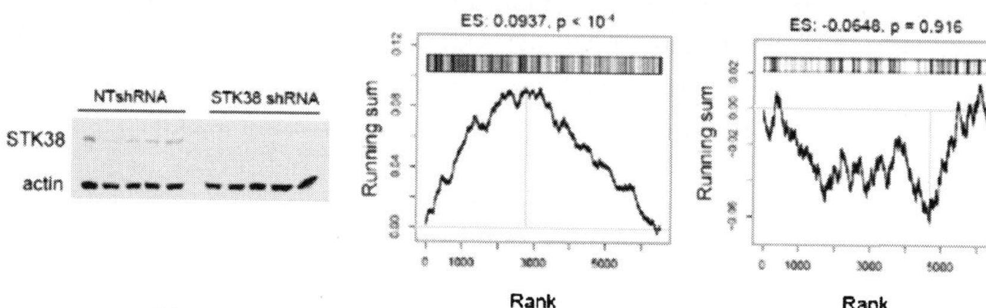

(a) (b) (c)

Figure 3. (a) STK38 and β-actin immunoblot on total cell lysates from ST486, at 60h after lentiviral vector-mediated non-target control (NTshRNA) and STK38 shRNA expression. (b) GSEA enrichment of STK38 regulon and (c) STK38 modulon among differentially expressed genes, following STK38 silencing. GSEA test is performed as in [20] using default parameters. Genes were ranked along the x-axis based on their differential expression *p*-value based on two-sample t-test. The "bar code" on top indicates the position of STK38 regulon (for panel b) and modulon (for panel c) genes in the ranked list. The black intensity of each bar is proportional to the local density of surrounding bars. False positive rate is estimated by permutation test, in which 10^4 null scores were obtained by selecting at random the same number of genes as STK38 regulon (for panel b) and modulon (for panel c).

Modulator comparison: When compared to the GSEA enrichment of the other SP regulons, the STK38 regulon scored 8th out of 772. Additionally, the most significantly enriched MINDy regulon was that of the SP having the highest overlap with the STK38 regulon (CDC2L5, with 506 common regulon genes out of 1039), further suggesting that STK38 silencing affects the MINDy-inferred STK38 regulon in a highly specific way. Indeed, based on this high overlap, we hypothesize that CDC2L5 is directly downstream of STK38 so that the higher enrichment is justified by a more specific regulon (1039 for CDC2L5 vs. 1219 for STK38). Further experimental data are required to confirm this hypothesis.

STK38-TF interaction validation: To test MINDy's ability to infer individual SP-TF interactions, we selected 257 TFs in the STK38 modulon with more than 150 STK38 modulated targets. The 150-target threshold was chosen to ensure sufficient statistical power of the GSEA test. We then tested whether the

MINDy inferred targets of each TF were enriched in differentially expressed genes after STK38 silencing. The analysis shows that 78 out of 257 target-sets (30%) are significantly enriched at 5% false positive rate, corresponding to a false discovery rate of 16% (i.e. 257×0.05/78). This is a very high percent because (a) STK38 is so pleiotropic that individual TFs may not be affected at all due to combinatorial regulation effects, and (b) modulation was not inferred by MINDy in the ST486 cell line but rather from a collection of 17 distinct B cell phenotypes (see 5.1). Hence, some TF's targets may not be affected because key co-factors, signals, or effectors are missing.

3.5. *Signalome-transfactome interaction network*

Table 4 and Figure S8 summarize the first genome-wide *in silico* map of direct and pathway mediated SP-TF interactions. The SPs are clustered based on their modulon overlaps (see 5.4). As expected, the interactions are sparse in general, but tightly clustered into modules representing functionally coherent SP-sets associated with known biological processes. This further validates MINDy's ability to characterize the modulon of arbitrary signaling genes and to annotate their functions.

#	*N*	*Annotation*	*p*
1	26	Protein biosynthesis	2.6×10^{-3}
		Cell homeostasis	2.6×10^{-3}
2	23	Cell cycle	3.9×10^{-5}
		Apoptosis	2.1×10^{-2}
3	21	Cell surface receptor linked signal transduction	2.8×10^{-4}
4	20	G1 to S cell cycle reactome	1.9×10^{-3}
5	14	Cell-cell adhesion	9.0×10^{-4}
		Cell motility	1.3×10^{-3}
6	12	G-protein coupled receptor class B	8.0×10^{-3}
		G-protein coupled receptor class C	8.0×10^{-3}
7	11	Gap junction	6.6×10^{-3}
		Integrin-mediated cell adhesion	1.4×10^{-2}
8	10	G-protein coupled receptor class A	7.3×10^{-5}

Table 4. Modules of SPs identified by clustering the signalome. *N*: module size; *p*: fisher's exact test *p*-value. Only top 8 modules with size equal to or greater than 10 are listed. Significant pathways were selected using *p*-value cutoffs specified in 5.4.

Finally, the most likely topology of the interface between SPs and TFs in human B cells, supported by significant physical evidence from both literature and novel biochemical assays, was obtained by mapping the MINDy predictions onto direct interactions in PPIDB and KSIDB, producing a hybrid network depicted in Figure 4. The connectivity of nodes in this network follows a power-law distribution (see Figure 4 inset), suggesting that the network is scale-free. MINDy-inferred signaling interactions supported by KSIDB appear to be more pleiotropic than those supported by PPIDB, perhaps due to the nature of the two evidence sources: NetworKIN makes predictions on well-studied kinases (with known consensus motif), whereas high-throughput PPI measurements in the database, e.g. Y2H, tend to have less selection bias and a high false negative rate.

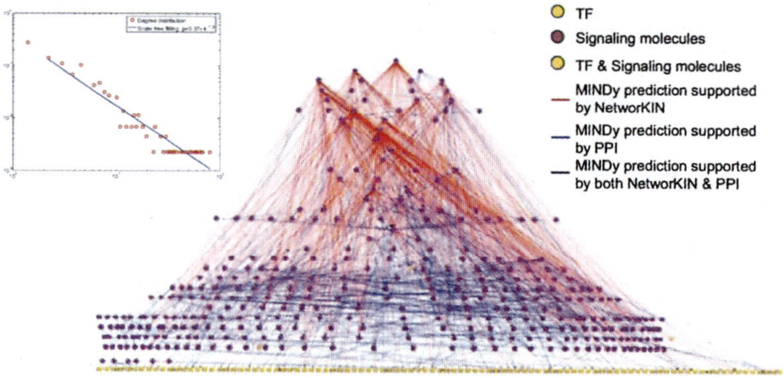

Figure 4. Visualization of the signalome-transfactome network by integrating MINDy predictions with PPIDB and KSIDB interactions. Two types of interactions are represented in the network: 1) MINDy predicted SP-TF interactions supported by PPIDB or KSIDB (i.e. modulatory interactions predicted by MINDy that have physical interaction evidence); 2) MINDy predicted SP-SP interactions supported by PPIDB or KSIDB (i.e. between modulators predicted by MINDy of the same TF, and are supported by physical interaction evidence). Depending on the source of evidence, these interactions can be either un-directed (supported by known PPIs), or directed (supported by NetworKIN, i.e. kinase→substrate). The inset figure shows the degree distribution of this network, with a linear fitting in the log-log space indicating a scale-free topology.

4. Discussion

In this manuscript we have provided both computational and experimental evidence suggesting that the MINDy algorithm can be effectively used to map physical and pathway-mediated post-translational interactions between signaling proteins and transcription factors, using only large GEP datasets. Due to lack of appropriate high-throughput technologies, such as microarray expression profiles and ChIP-Chip/Seq assays, dissection of post-translational interactions is lagging significantly behind that of their transcriptional counterpart. As a result, algorithms providing high-accuracy, cell-context specific hypotheses for biochemical validation may significantly improve our ability to elucidate post-translational processes and their effect on transcriptional networks. Specifically, we have shown that MINDy predictions are highly enriched in experimentally validated interactions and that silencing the most pleiotropic modulator, STK38, produces expression profiles that are highly consistent with the inferred STK38 modulon and regulon.

Such a network topology can be interrogated to address specific biological questions, such as (a) what are the signaling proteins that control a specific transcriptional program (i.e. TF)? (b) What are the shortest paths through which a signaling protein may affect the activity of a TF? And (c) what signaling proteins are upstream/downstream of other signaling proteins. Taken together, this represents the first genome-wide computational analysis of the interface between signaling and transcriptional networks. The combination of these results and those from *in vivo* experiments may significantly improve our understanding of the role of cellular signaling in the regulation of transcriptional programs and provide new targets for therapeutic intervention.

Lastly, another insight that can be gleaned from the signalome-transfactome network reconstructed by MINDy is the specificity of the signaling genes in terms of their ability to regulate the transcriptional response of a cell. In Table 2 shown earlier, signaling genes on columns to the left are more pleiotropic, whereas those to the right are more specific, with respect to the number of TFs (i.e. distinct transcriptional programs) they control. Similarly, TFs on top rows are controlled by broader signaling pathways than those on the bottom rows. Since signaling proteins are often selected as drug target (e.g. by small molecule compound), these results can provide guidance to the selection of intervention point that has the least side effect. Specifically, one may want to target a kinase that is very specific, so that it causes less cross-talk with other transcriptional programs not intended to be affected by the drug molecule.

5. Methods

5.1. *Gene expression profile dataset*

254 GEP were generated using the Affymetrix HG-U95Av2 GeneChip® System (~12,600 probe sets) from a collection of normal and tumor related B cell samples. Probe sets with expression mean $\mu < 50$ and standard deviation $\sigma < 0.3\mu$, were excluded as non-informative, leaving 8680 probe sets. Further details on the GEP dataset can be found in the supplementary materials.

5.2. *MINDy Analysis*

Given a triplet (TF, M, t), with ($t \neq TF$ and $t \neq M$), MINDy assesses whether the CMI, $I[TF;t \mid M]$, is constant as a function of M. Assuming that the CMI is a monotonic function of M, this can be efficiently tested by measuring $\Delta I = I\left[TF;t \mid M \in L_m^+\right] - I\left[TF;t \mid M \in L_m^-\right] \neq 0$, where L_m^+ and L_m^- represent two subsets including the 35% of the samples where M is respectively most and least expressed. The p-value corresponding to a specific ΔI is obtained by permutation tests, and Bonferroni corrected for the total number of tested modulator-target pairs. Significant triplets are further pruned if there exists a third gene, x , such that $I\left[TF;x\right] \geq I\left[TF;t\right]$ and $I\left[t;x\right] \geq I\left[TF;t\right]$ in both $L_m^{\pm}$, indicating an indirect relationship between TF and t , mediated by x , as suggested by the Data Processing Inequality [11]. Readers are encouraged to refer to the supporting online materials and [7, 8] for more details on MINDy.

For each TF in the transfactome, MINDy first identifies the set of candidate modulators among all SPs whose expression profiles are independent of that of the TF. All other genes are then tested as candidate targets t of the TF. Each TF-SP-t triplet is then analyzed by the MINDy test at a 5% statistical significance level, after asymptotic Bonferroni correction for multiple testing (i.e. using a p-value cutoff $= 0.05/\left(N_m \times N_t\right)$, where N_m is the number of candidate modulators and N_t the number of candidate target genes). The analysis was run at the Affymetrix probe set level, and duplicated probe sets mapping to same gene were merged *a posteriori*.

5.3. *Lentiviral mediated STK38 knock-down*

Human embryonic kidney 293T and Burkitt's lymphoma cell line ST486 were maintained in DMEM and IMDM, respectively. All cell culture media were supplemented with 10% FBS (Invitrogen) and antibiotics. Supernatants for the lentiviral vector containing the STK38 shRNA (TRCN0000010216, Sigma) and non-target control shRNA (SHC002, Sigma) were produced in 293T cells. 5 independent samples of ST486 cells ($2\text{x}10^6$ cells/ml) were transduced with viral supernatants for either STK38 shRNA or non-target control shRNA. Transduction was performed by centrifugation at 450xg for 2h with supernatants supplemented with 8 μg/ml polybrene. Total RNA was extracted 60h after transduction and prepared for gene expression profiling according to Affymetrix's protocol.

5.4. *Signalome clustering*

Signaling proteins are clustered based on the similarity of their modulon inferred by MINDy. Specifically, each signaling protein is associated with a vector $T_j = \{t_1, t_2, \ldots, t_N\}$ where t_i, $i = 1, 2, \ldots, N$ is the number of targets of TF i modulated by the signaling protein j, and N is the total number of TFs in the transfactome. We then performed hierarchical clustering of the signaling genes using average linkage method and Pearson correlation between these vectors as the distance metric. Signaling genes were assigned into modules when linkage stopped at 70% of the maximal linkage score. Modules consisting of more than 10 genes are subsequently queried for over-represented pathways against the background of all SPs. Only pathways with ≥5 SPs are searched, including 286 GO biological process categories and 44 KEGG/GenMAPP pathways. Enrichment is calculated using FET with p-value cutoff set to 1/286 for GO and 1/44 for KEGG/GenMAPP.

5.5. *Supporting online materials*

Supplementary materials, including lists of PPIDB and KSIDB interactions, MINDy predictions and MINDy software are available at: http://wiki.c2b2.columbia.edu/califanolab/PSB2009/.

Acknowledgements

This work is supported by the NCI (R01CA109755), the NIAID (R01AI066116), and the National Centers for Biomedical Computing NIH Roadmap Initiative (U54CA121852).

References

1. Gough, N.R., Signal Transduction Pathways as Targets for Therapeutics. 2001. p. pe1-.
2. Zhang, Y., et al., Time-resolved Mass Spectrometry of Tyrosine Phosphorylation Sites in the Epidermal Growth Factor Receptor Signaling Network Reveals Dynamic Modules. Mol Cell Proteomics, 2005. **4**(9): p. 1240-1250.
3. Roberts, C.J., et al., Signaling and circuitry of multiple MAPK pathways revealed by a matrix of global gene expression profiles. Science, 2000. **287**: p. 873-880.
4. Gardner, T.S., et al., Inferring genetic networks and identifying compound mode of action via expression profiling. Science, 2003. **301**(5629): p. 102-5.
5. Basso, K., et al., Reverse engineering of regulatory networks in human B cells. Nat Genet, 2005. **37**(4): p. 382-390.
6. Steffen, M., et al., Automated modelling of signal transduction networks. BMC Bioinformatics, 2002. **3**(34).
7. Wang, K., et al., Genome-Wide Discovery of Modulators of Transcriptional Interactions in Human B Lymphocytes. Lecture Notes in Computer Science, 2006. **3909**: p. 348 - 362.
8. Wang, K., et al., Genome-wide Identification of Post-translational Modulators of Transcription Factor Activity in Human B Cells. Submitted, 2008.
9. Ashburner, M., et al., Gene Ontology: tool for the unification of biology. Nat Genet, 2000. **25**(1): p. 25.
10. Foat, B.C., R.G. Tepper, and H.J. Bussemaker, TransfactomeDB: a resource for exploring the nucleotide sequence specificity and condition-specific regulatory activity of trans-acting factors. Nucl. Acids Res., 2007: p. gkm828.
11. Margolin, A., et al., ARACNE: An Algorithm for the Reconstruction of Gene Regulatory Networks in a Mammalian Cellular Context. BMC Bioinformatics, 2006. **7**(Suppl 1): p. S7.
12. Peri, S., et al., Development of human protein reference database as an initial platform for approaching systems biology in humans. Genome Res, 2003. **13**(10): p. 2363-71.
13. Bader, G.D., D. Betel, and C.W. Hogue, BIND: the Biomolecular Interaction Network Database. Nucleic Acids Res, 2003. **31**(1): p. 248-50.
14. Xenarios, I., et al., DIP, the Database of Interacting Proteins: a research tool for studying cellular networks of protein interactions. Nucleic Acids Res, 2002. **30**(1): p. 303-5.
15. Hermjakob, H., et al., IntAct: an open source molecular interaction database. Nucleic Acids Res, 2004. **32**(Database issue): p. D452-5.
16. Linding, R., et al., Systematic Discovery of In Vivo Phosphorylation Networks. Cell, 2007. **129**(7): p. 1415-1426.
17. Kanehisa, M., et al., From genomics to chemical genomics: new developments in KEGG. Nucleic Acids Res, 2006. **34**(Database issue): p. D354-7.
18. Dahlquist, K.D., et al., GenMAPP, a new tool for viewing and analyzing microarray data on biological pathways. Nat Genet, 2002. **31**(1): p. 19-20.
19. Tamaskovic, R., S.J. Bichsel, and B.A. Hemmings, NDR family of AGC kinases--essential regulators of the cell cycle and morphogenesis. FEBS Lett, 2003. **546**(1): p. 73-80.
20. Subramanian, A., et al., From the Cover: Gene set enrichment analysis: A knowledge-based approach for interpreting genome-wide expression profiles. PNAS, 2005. **102**(43): p. 15545-15550.

EMERGING CHALLENGES IN MULTISCALE MODELING IN BIOLOGY

ROY KERCKHOFFS
Cardiac Mechanics Research Group
Department of Bioengineering
University of California San Diego
La Jolla, CA 92093-0412

JAMES BASSINGTHWAIGHTE
National Simulation Resource
University of Washington
Seattle, WA 98195

JEFF REINBOLT
Simbios
NIH Center for Biomedical Computing
Stanford University
Palo Alto, CA 94305

PETER ARZBERGER
National Biomedical Computation Resource
University of California San Diego
La Jolla, CA 92093-0440

1. Introduction

With growing computer power, novel diagnostic and therapeutic medical technologies, coupled with an increasing knowledge of pathophysiology from gene to organ systems, it is increasingly practical to apply multi-scale patient-specific modeling based on proven disease mechanisms to guide and predict the response to therapy in many aspects of medicine. This is an exciting and relatively new approach, for which efficient methods and computational tools are of the utmost importance. This session will address multi-scale modeling activities that are focused on crossing levels from proteins, cells, tissues, organs, up to the system level.

There are several significant motivations for multi-scale modeling in physiology. First, the need to model at multiple levels – spatially and temporally – becomes clear by simply considering physiology: organisms consist of organs, which are made of different tissues; these tissues are made up of sets of specialized cells interacting with each other. Within the cells complex biochemical processes regulate cell metabolism and short and long term behavior, all taking place on the underlying protein kinetics and gene expressions. Hence, a strong interconnection exists between all levels: normal function at a given level depends on the underlying synergy of sublevels – but also on higher levels, such as feedback via neural and hormonal mechanisms.

Multi-scale patient-specific modeling is an emerging frontier in computational biology and possibly the future of medicine. Just recently, NIH posted a new Funding Opportunity Announcement "Predictive Multiscale Models of the Physiome in Health and Disease", for which the first submission round closed in January 2008. A number of workshops and panel recommendations have also recognized and addressed the importance and the difficulties in interpreting experimental results that are cross-scale in space, time and state. The Multi-Scale Modeling (MSM) consortium, along with the participation of the Interagency Modeling and Analysis Group (IMAG) of a consortium of federal agencies aims to promote the development and exchange of tools, models, data and standards for the multi-scale modeling community.

2. Session summary

2.1. *Papers*

This session includes an invited talk, four reviewed oral presentations, two additional accepted papers and a tutorial. The studies presented include work on the merging of models that operate at different time-scales, semi-automatic generation of meshes at the (sub-)cellular level, multi-scale models of blood vessel growth, an efficient multi-scale approach to simulate small to macro-molecular processes, a comparison of finite element analysis at different scales, and combining continuum and discrete models applied to large biomolecules.

Our invited speaker **Peter Hunter** will talk about multi-scale, patient-specific modeling of lower limb mechanics. Such a biophysically detailed model of the human musculoskeletal system will aid diagnostic evaluation and surgical treatment.

Sachse ***et al*** describe an experimental/image processing method to obtain input for computational models at the subcellular-cellular scale – specifically spatial information on structures such as ryanodine receptors and the sarcolemma – to build geometrical models for computational models of cardiac excitation-contraction coupling. These models are important in studying the signaling cascade that transduces transmembrane depolarization to cellular mechanics. Gaining better insight in excitation-contraction is critical to understanding the healthy and diseased heart.

Neal ***et al*** present a novel case study of the integration of two different and independent multi-scale cardiovascular system models by making use of a "semantic simulation" framework. The two models are multi-scale in space and time. One of the challenges was to accomplish the integration of the different time-scales the two models operate on. The framework makes use of established ontology for anatomy and physiology to reason on and define relationships between the two models. Multi-scale modelers will be interested in the approaches and insights obtained from the combination of these specific models, and the informatics community will be interested in the conclusions about how models should be represented.

Voelz ***et al*** present a new multi-scale method that combines all-atom molecular dynamics with coarse-grained sampling. Two levels of physiology are encompassed: the atomic scale of protein side chains and small molecules, and the scale of macromolecular complexes. Their work takes an important step towards realistic modeling of a critical biological process such as the synthesis of proteins at the ribosome exit tunnel, and seems feasible to be applied to even larger systems.

Qutub ***et al*** describe a multi-scale model of vessel growth, which encompasses five different models (blood flow, oxygen transport, growth factor distribution and signaling, cell sensing, cell movement and proliferation), which operate on different time and spatial scales. They address issues regarding the integration of these models and an interesting feature is the coupling with an experimental database, which facilitates the comparison between model and experimental results on one side, and the prescription of initial and boundary conditions on the other.

Two other papers were accepted to the session as part of the final publication of the meeting.

Cheng ***et al*** developed adaptive finite element methods to calculate electrostatic interactions and ligand binding rate constants for large biomolecules. They combined continuum modeling with discrete methods to efficiently model both diffusion and Brownian dynamics.

Eswaran ***et al*** developed micro-finite element and larger-scale continuum finite element models of the vertebral body, and assessed when micro-FE modeling might be necessary. It is a promising tool for understanding and predicting clinical fracture risk.

2.2. *Tutorial*

The tutorial is to address two interrelated topics: 1) an exemplary problem, multi-scale modeling applied to cardiac electromechanics of the failing heart and 2) generic problems associated with and the necessity for model sharing.

Cardiac electromechanics

In an integrated model of the failing heart it is necessary to integrate electrophysiological and contractile function at the intracellular level since heart failure is associated with dysregulation of intracellular excitation-contraction coupling mechanisms. But heart failure is also a tissue level problem with fibrosis and conduction abnormalities, an organ level problem (dilatation and hypertrophy), a circulatory system problem (increased plasma volume and peripheral resistance), and a systemic syndrome with shortness of breath and fatique. We and others have developed multi-scale three-dimensional models of ventricular electromechanics that integrate from molecular to cellular, tissue, organ and system scales. These models are multi-scale in that they simultaneously solve molecular and cellular systems ODE models, tissue-scale constitutive models, organ-scale finite-element models of action potential propagation and stress equilibrium, subject to dynamic boundary conditions generated by circulatory-system scale lumped-parameter models of hemodynamics and ventricular-vascular coupling. They are also functionally integrated in that they simultaneously solve electrophysiological and mechanical models (i.e. excitation-contraction coupling) and include feedback across physical scales (e.g. from baro-reflex to adrenergic receptor signaling) and between functions (e.g. mechanoelectric feedback).

Model sharing

The open sharing of models is critical to the advancement of the science. Reproducibility is ostensibly the goal of publication of scientific research, but the diligence exhibited by scientists in describing their experimental methods and procedures and approaches to analyses is not paralleled when it comes to modeling.

Rarely are models as fully and clearly defined as was accomplished by Hodgkin and Huxley (1952): every figure can be reproduced from their equations and parameters. Reasons for inadequacy in journal articles include the complexity of matching the written report to the computational code, the reluctance of journals to accept long detailed articles, and the additional labor required for putting code, manuals, descriptions, tutorials, test cases, solutions, etc, in good form, all of which are beyond the minimal effort required to produce the research result. There are also the issues of reluctance to reveal the details, like secret ingredients in a recipe, or to provide the latest and best science to one's competitors. While it is more difficult to present technically complicated models to a broad user community than it is to provide simplified conceptual models. It is more important to make the difficult ones available for workers to explore, understand and build upon. "On the shoulders of giants...". What's more, the funding agencies, scientific societies, and journal editors are beginning to require full disclosure: It won't be just "Publish or Perish" but "Disseminate usable products (models, techniques, data) or Perish! It is an oligation accompanying federal and foundation funding to make our world better.

3. Acknowledgements

The session organizers would like to thank the set of reviewers for their efforts in evaluating the papers that were submitted. The first special session on multi-scale modeling was held in 1999, with a session on Computer Modeling in Physiology: From Cell to Tissue. The second session was held last year (Multiscale Modeling and Simulation: from Molecules to Cells to Organisms). Last year we hoped that it would not be another 8 years before the richness of this area would be seen again at the Pacific Symposium on Biocomputing, so we also acknowledge the assistance of many individuals who helped support the notion of holding this third session on multi-scale modeling.

Finally, the authors wish to acknowledge the support from their individual NIH centers on computational issues in multi-scale modeling, the National Biomedical Computation Resource (NBCR, RR 08605) led from the University of California San Diego, the National NIH Center for Physics-Based Simulation of Biological Structures (Simbios) at Stanford and the National Simulation Resource (NSR) at the University of Washington.

FINITE ELEMENT ANALYSIS OF DRUG ELECTROSTATIC DIFFUSION: INHIBITION RATE STUDIES IN N1 NEURAMINIDASE

YUHUI CHENG, MICHAEL J. HOLST AND J. A. MCCAMMON*

Uinversity of California, San Diego 9500 Gilman Dr. MC 0365
La Jolla, CA 92037, USA
E-mail: ycheng@mccammon.ucsd.edu

This article describes a numerical solution of the steady-state Poisson-Boltzmann-Smoluchowski (PBS) and Poisson-Nernst-Planck (PNP) equations to study diffusion in biomolecular systems. Specifically, finite element methods have been developed to calculate electrostatic interactions and ligand binding rate constants for large biomolecules. The resulting software has been validated and applied to the wild-type and several mutated avian influenza neurominidase crystal structures. The calculated rates show very good agreement with recent experimental studies. Furthermore, these finite element methods require significantly fewer computational resources than existing particle-based Brownian dynamics methods and are robust for complicated geometries. The key finding of biological importance is that the electrostatic steering plays the important role in the drug binding process of the neurominidase.

1. Introduction

Electrostatics and Diffusion play the important role in a variety of biomolecular processes, which have been studied extensively using various biophysical, biochemical and computational methods. Computational models of diffusion have been widely studied using both discrete [1,2,3,4,5] and continuum methods [6,7,8,9,10,11]. The discrete methods concentrate on the stochastic processes based on individual particles, which include Monte Carlo[12,13,14,5], Brownian dynamics (BD) [15,16,17] and Langevin dynamics[18,19] simulations. Continuum modeling describes the diffusional processes via concentration distribution probability in lieu of stochastic dynamics of individual particles. Comparing with the discrete methods, continuum approaches needn't deal with the individual Brownian particles

*This work is supported by nih gm31749, nsf mcb-0506593 and mca93s013 (to jam). additional support from the howard hughes medical institute, the nsf supercomputer centers, the san diego supercomputing center, accelrys, inc., the w.m. keck foundation, the national biomedical computational resource and the center for theoretical biological physics is gratefully acknowledged.

and the computionational cost can be substantially less than for the discrete methods.

In the present work, we apply adaptive finite element methods to solve the PBS equation and PNP equation in several neurominidase structures. The diffusion results have been compared with those from recent experimental kinetic studies. The H5N1 avian influenza neurominidase is a highly pathogenic virus that might acquire the ability to pass readily among humans and cause a pandemic[20,21]. Our continuum modeling demonstrates that it is efficient but accurate enough to address the drug binding. To take account of the stochastic dynamics of diffusion particles, we are developing a state-in-art SMOL package by integrating the continuum modeling and Brownian dynamics together to study the drug diffusion in the neurominidase.

2. Theory and Modeling Details

Our SMOL package (http://mccammon.ucsd.edu/smol/index.html) models the diffusion of ligands relative to a target molecule, subject to a potential obtained by solving the Poisson-Boltzmann equation. It is perhaps most easily explained by initially considering motion of an ensemble of Brownian particles in a prescribed external potential $\phi(\vec{R})$($\vec{R}$ being a particle's position) under conditions of high friction, where the Smoluchowski equation applies.

2.1. *Boundaries and Initialization of the Poisson-Boltzmann and Smoluchowski Equations*

The original Poisson-Boltzmann and Smoluchowski equations have the form of continuity equations (see Fig. 1):

$$\nabla\cdot(\varepsilon\nabla\phi(\vec{R}))+\kappa^2 sinh(\phi(\vec{R}))=\sum_i q_i\delta(\vec{R}-\vec{R}_i)+\sum_j q_j p_j(\vec{R};t) \tag{1}$$

$$\frac{\partial p_j(\vec{R};t)}{\partial t}=-\vec{\nabla}\cdot\vec{J}_j(\vec{R};t) \tag{2}$$

where the particle flux is defined as:

$$\begin{aligned}\vec{J}_j(\vec{R};t)&=D_j(\vec{R})[\vec{\nabla}p_j(\vec{R};t)+\beta\vec{\nabla}\phi(\vec{R})p_j(\vec{R};t)]\\&=D_j(\vec{R})e^{-\beta q_j\phi(\vec{R})}\nabla e^{\beta q_j\phi(\vec{R})}p(\vec{R};t)\end{aligned} \tag{3}$$

Here $p_j(\vec{R};t)$ is the distribution function of the ensemble of Brownian particles, including the drug molecule, coions and counterions. $D_j(\vec{R})$ is the diffusion coefficient of the diffusion particle, $\beta = 1/k_B\mathrm{T}$ is the inverse Boltzmann energy, k_B is the Boltzmann constant, T is the temperature, and $\phi(\vec{R})$ is the potential of mean

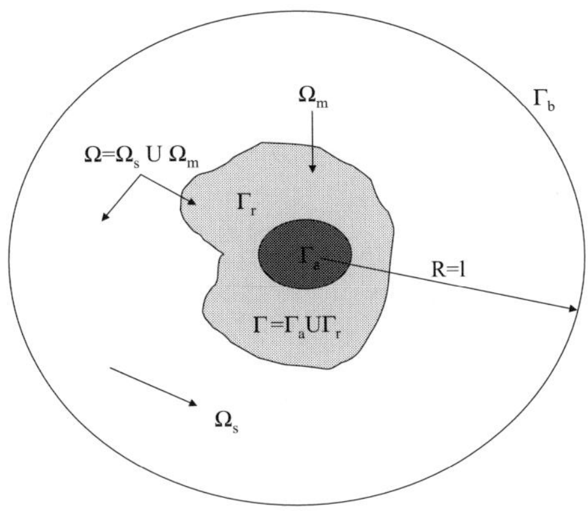

Figure 1. The illustration of the problem domains for Poisson-Boltzmann and Smoluchowski Equations. The Poisson-Boltzmann equation is solved on Ω, while the Smoluchowski equation is solved only on Ω_s. Γ represents the molecular surface, while Γ_a and Γ_r correspond the reactive and nonreactive boundaries.

force (PMF) for a diffusing particle due to solvent mediated interactions with the target molecule. For simplicity, $D_j(\vec{R})$ can be assumed to be constant. The two terms contributing to the flux have clear physical meanings. The first is due to free diffusional processes, as quantified by Fick's first law. The second contribution is due to the drift velocity - $\vec{\nabla}\phi(\vec{R})\gamma$ induced by the systematic forces - $\vec{\nabla}\phi(\vec{R})$ and friction quantified by the friction constant γ. The relation between diffusion coefficient D and friction constant γ is given by Stokes-Einstein equation: $D\beta\gamma = 1$. $\kappa = \sqrt{2n_0e^2/(\varepsilon_0 k_B T)}$, ε_0 is the vacuum permittivity and n_0 is the bulk ionic concentration.

To accurately solve the Poisson-Boltzmann equation, the potential $\phi(\vec{R})$ has been decomposed into singularized (ϕ_s) and regularized (ϕ_r) components [22]. The singularized component ϕ_s can be further decomposed into the harmonic (ϕ_h) and the singular green function G: $\phi_s = \phi_h$ + G. The ϕ_h can be obtained via solving the below equation:

$$\nabla \cdot (\nabla \phi_h(\vec{R})) = 0 \ \ on \ \ \Omega_m$$
$$\phi_h = -G \ \ on \ \ \Gamma \tag{4}$$

Where the green function G is simply the electrostatic potential induced by all the singular charges inside the biomolecule, i.e.,

$$\nabla \cdot (\varepsilon_m \nabla G(\vec{R})) = \sum_i q_i \delta(\vec{R} - \vec{R}_i) \ \ on \ \ \Omega_m \tag{5}$$

and can be directly given by $G(\vec{R}) = \sum_i \frac{q_i}{\varepsilon_m |\vec{R} - \vec{R}_i|}$

Eq. 4 can be solved using the finite element method or boundary element method. For ϕ_r , we need to solve the below equation:

$$\nabla \cdot (\varepsilon \nabla \phi_r(\vec{R})) + \kappa(\vec{R})^2 sinh(\phi_r + \phi_s) = 0 \ \ on \ \ \Omega$$
$$\phi_r = 0 \ \ on \ \ \Gamma$$
$$\varepsilon \frac{\partial \phi_r}{\partial n} = -\varepsilon_m \frac{\partial \phi_s}{\partial n} \ \ on \ \ \Gamma \tag{6}$$

Since $\kappa(\vec{R}) = 0$ on Ω_m and $\phi_s = 0$ on Ω_s, therefore Eq. 6 can be further simplified as

$$\nabla \cdot (\varepsilon \nabla \phi_r(\vec{R})) + \kappa(\vec{R})^2 sinh(\phi_r) = [\varepsilon \phi_s]_\Gamma \delta_\Gamma \tag{7}$$

Where $[\varepsilon \phi_s]_\Gamma \delta_\Gamma$ is the jump on the molecular interface.

Finally, the solution of the PB equation cab be given by:

$$\phi = \phi_r + \phi_h + G \tag{8}$$

The Smoluchowski equation (Eq. 2) can be solved to determine biomolecular diffusional encounter rates before steady state is established. Following the work of Song et al.[9,10] and Zhou et al.[23,24,25], the application of the Smoluchowski equation to this question involves the solution of Eq. 2 in a three-dimensional domain Ω, with the following boundary and initial conditions. A bulk Dirichlet condition is imposed on the outer boundary $\Gamma_b \subset \partial\Omega$,

$$p_j(\vec{R}_l;t) = p_j^{bulk}, \ \ for \ \ \vec{R}_l \in \Gamma_b, \tag{9}$$

where p_j^{bulk} denotes the bulk concentration at the outer boundary. A reactive Robin condition is implemented on the active site boundary $\Gamma_a \subset \partial\Omega$,

$$n(\vec{R}_0) \cdot \nabla p(\vec{R}_0;t) = \alpha(\vec{R}_0) p(\vec{R}_0), \ \ for \ \ \vec{R}_0 \in \Gamma_a, \tag{10}$$

providing an intrinsic reaction rate $\alpha(\vec{R}_0)$. Here, $n(\vec{R}_0)$ is the surface normal. For the diffusion-limited reaction process, such as inhibitor binding in N1, the concentration of the inhibitor at the binding site is approximately zero. Therefore, the reactive Robin condition on the inner boundary can be simplified as:

$$p_j(\vec{R}_0;t) = 0, \ \ for \ \vec{R}_0 \in \Gamma_a, \tag{11}$$

Where p_j in the inhibitor concentration. It must be noted that for coions and counterions, there are no reactive boundaries. For the non-reactive surface parts of the inner boundary $\Gamma_r \subset \partial\Omega$, a reflective Neumann condition is employed.

$$n(\vec{R}_0) \cdot jp(\vec{R}_0;t) = 0, \tag{12}$$

Finally, we set up the initial conditions as

$$p_j(\vec{R};0) = \begin{cases} p_j^{bulk} & |\vec{R}| = l \\ 0 & |\vec{R}| < l \end{cases} \tag{13}$$

The solution of the PBS equation can be obtained by sequentially solving the Eq. 1 and Eq. 2; To solve the PNP equation, we need to solve the Eq. 1 and Eq. 2 iteratively, until the potential value ϕ converges to the initial threshold set by the user.

Therefore, the diffusion-determined biomolecular reaction rate constant during the simulation time can be obtained from the flux $\vec{j}(\vec{R};t)$ by integration over the active site boundary, i.e.

$$k(t) = p_{j;bulk}^{-1} \int_{\Gamma_a} n(\vec{R}) \cdot \vec{j}(\vec{R};t) dS \tag{14}$$

Finally, the corresponding concentration distribution can be obtained by $p_j(\vec{R};t) = e^{-\beta\phi(\vec{R})} u(\vec{R};t)$.

2.2. *Adaptive Finite Element Mesh Generation*

Before generating the tetrahedral meshes for the external and internal space for the biomolecule, we first generate the surface triangular meshes for the molecule using the MSMS code (http://www.scripps.edu/ sanner/html/msms_home.html) [26]. Certainly, Although MSMS is very efficient for surface mesh generation, the quality of the surface mesh still need to improve. Fortunately, the "Adventure Project" provides the excellent surface mesh smoothing for the MSMS surface mesh (http://adventure.q.t.u-tokyo.ac.jp/). With the molecular surface mesh, Tetgen was implemented to generate adaptive finite element meshes for biomolecules (http://tetgen.berlios.de/).

In the present study, molecular dynamics studies on the wild-type N1 structure (PDB code: 2HU4) have obtained "open" and "closed" conformations by Amaro et al. (unpublished results). Adaptive meshes for the "open" and "closed" conformations have been generated respectively. To compare the difference between the wild type and the mutants, we also prepared the meshes for the His274Tyr mutant (PDB code: 3CL0 and 3CKZ) and the Asn294Ser mutant (PDB code: 3CL2). The active site of the anti-neuraminidase inhibitors has been labeled as reactive boundaries. Reactive boundaries were defined following the typical BD methods: a spherical reactive surface is defined at an arbitrary radius from the biomolecular active site [27]. Generally, the user can label the active site with several spheres and write into the input file.

3. Results and Discussion

3.1. *Validation of the PBS and PNP solver with A Spherical Test Case*

Before applying our SMOL code to a biomolecular system with complicated geometries, we first tested it with the classic spherical system [28] and compared the calculated result with the known analytical solution. For this test case, we chose a diffusing sphere with a 2 $\AA$ radius and $+1e$ charge. The receptor molecule was represented with a sphere with a 10 $\AA$ radius and -1e. The diffusion domain is another concentric sphere with a 400 $\AA$ radius, which was discretized with 141,792 tetrahedral elements. A detailed view of the surface mesh for the stationary sphere is also shown in Fig. 2(a). The diffusing particle's dimensionless bulk concentration was set to 1 mM. The concentration of coion and counterion was set at 150 mM. The dielectric constant is 2 in the receptor and 80 in the diffusion domain. Ignoring hydrodynamic interactions, the diffusion constant D is calculated as $7.8 \times 10^4 \AA^2/\mu s$ using the Stokes-Einstein equation with a hydrodynamic radius of 3.5 $\AA$, solvent viscosity of $0.891 \times 10^{-3} kg/(m \cdot s)$, and 300 K temperature.

For a unit sphere with radius r_0 and charge q in the center of the sphere, the linearized Poisson-Boltzmann equation have the below analytical solution ϕ_a:

$$\phi_a = \frac{q e^{\kappa r_0}}{\varepsilon(1+\kappa r_0)} \cdot \frac{e^{-\kappa r}}{r}, \quad r \geq R \tag{15}$$

Where R is the outer boundary.

Fig. 3 shows the calculated potential comparing with the analytical values. The relative error is less than %3.

Generally, under physiological condition, the ionic strength is approximately 0.15 M, here we solve the PBS and PNP equations with 150 mM Na^+ and Cl^- and 1 mM $+e$ charged inhibitor. If the active site of the receptor is as shown in

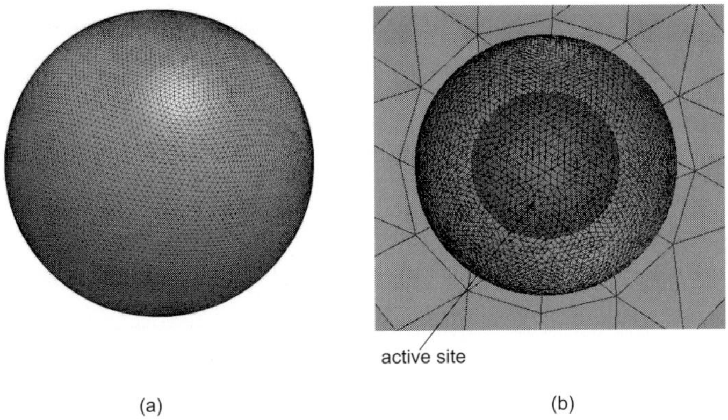

Figure 2. (a) the external diffusion domain; (b) the molecular surface and the active site.

Fig. 2(b), we can obtain the potential value ($\phi(\vec{R})$) and concentration distribution ($p_j(\vec{R};t)$) on each node in the diffusion domain. The total free energy for a given distribution $p_j(\vec{R};t)$, i.e., the quantity which develops towards a minimum during the diffusion process, is a functional defined through

$$G[p_j(\vec{R};t)] = \int_{\Omega} g(p_j(\vec{R};t))dr \tag{16}$$

where $g(p_j(\vec{R};t))$ is the free energy density connected with $p_j(\vec{R};t)$.

$$g(p_j(\vec{R};t)) = p_j(\vec{R};t)(\phi(\vec{R}) + \beta^{-1} ln \frac{p_j(\vec{R};t)}{p_j^{bulk}}) \tag{17}$$

Similarly, we define the local entropy density $s_j(\vec{R};t)$ as

$$s_j(\vec{R};t) = -k_B p_j(\vec{R};t) ln \frac{p_j(\vec{R};t)}{p_j^{bulk}} \tag{18}$$

Solving the PBS equation, we obtained $k_{on} = 9.967 \times 10^{10} M^{-1} \cdot min^{-1}$ when the steady state comes. The distribution of Na^+, Cl^- and +1e inhibitor is shown

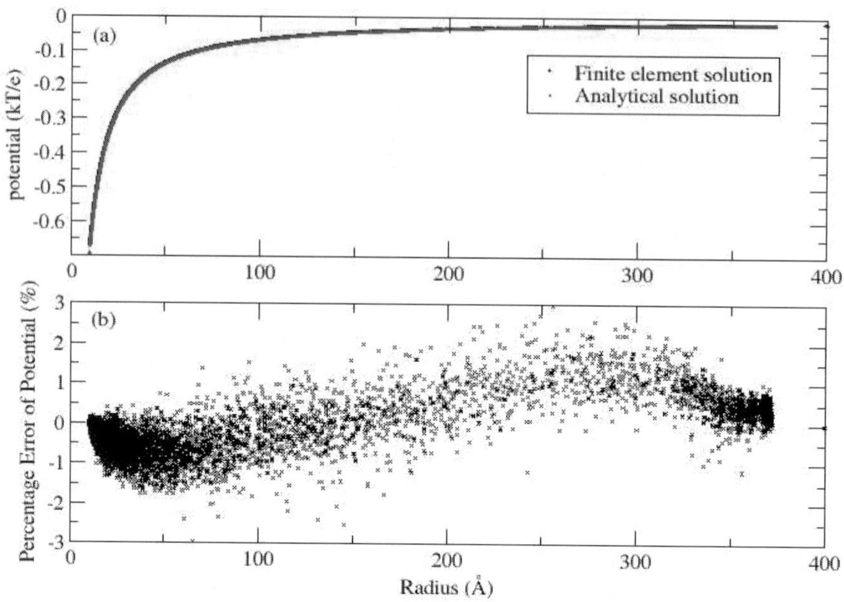

Figure 3. (a) the calculated and analytical potential in the diffusion domain; (b) the relative error for calculated potential in the diffusion domain.

in Fig.4. Furthermore, the equilibrium has not been reached due to the influx of all the ions. To accurately describe the equilibrium state, we need to iteratively solve the PB and Smoluchowski equations until the potential and diffusion rate constant reach the stable value, which is the "so-called" PNP equation. The final k_{on} is $6.517 \times 10^{10} M^{-1} \cdot min^{-1}$. It seems that coions and counterions play the role in compensating the fast consumption of the $+1e$ charged inhibitor. Comparing to the PBS model in Fig. 4, it clearly shows that there are more Na^+ ions are pumped out of the diffusion domain, while more Cl^- ions distribute near the receptor in our PNP model. The local free energy density and entropy density are much more tight and close to the receptor surface in the PNP model (Fig. 5).

3.2. *Application of the PNP solver on different neuraminidase models*

The active sites of all the neurominidase models are represented by two spheres, one is a 3 $\AA$ radius with the center on the carboxyl oxygen of Thr-225 and the other is in the gorge and around 5 $\AA$ away from the carboxyl-O of Thr-225. Finally, it turns out to be around 200 $\AA^2$ for the area of each active site. Similarly, we suppose

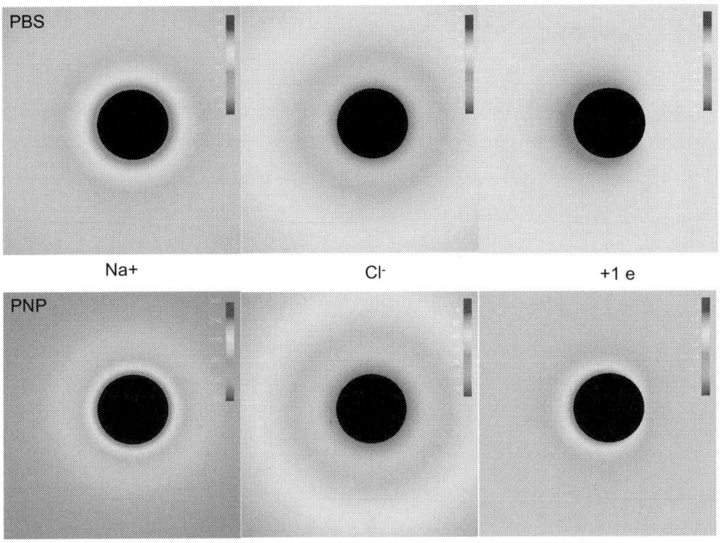

Figure 4. The concentration distribution calculated for Na^+, Cl^- and $+1e$ charged drug via the PBS and PNP models, respectively.

that there is 150 mM Na^+ and 150 mM Cl^-, 1 mM $-e$ inhibitor in the bulk. It must be noted that the potential drug inhibitor carries $-e$ charge.

First, the "open" and "closed" conformations of the wild-type neuraminidase have been modeled. When the steady state comes, the association reaction rate constant k_{on} is $2.76 \times 10^8 M^{-1} \cdot min^{-1}$ for the "open" conformation, while $3.66 \times 10^8 M^{-1} \cdot min^{-1}$ for the "closed" conformation. It turns out that the "closed" conformation have stronger binding affinity to anti-neuraminidase drugs. Experimental studies also observed the "tighter interaction with ligand" [20]. Meanwhile, we simulated three mutants 3CL0, 3CKZ and 3CL2. The calculated steady-state rate constants have been listed in Table.1. Comparing with the experimental data, our results are very consistent. Additionally, it must be noted that all the calculated rate constant is larger than experimental data. Actually, the active site of the neuraminidase is very flexible to change according to both the experimental finding and unpublished molecular dynamics simulations by Amaro et al. The size of the possible drugs, such as sialic acid, oseltmivir and zanamivir, is relatively big (at lease 30 atoms). Therefore, different from our previous acetylcholinesterase model [29], the non-electrostatic interaction, such as VdW and steric effect, and the

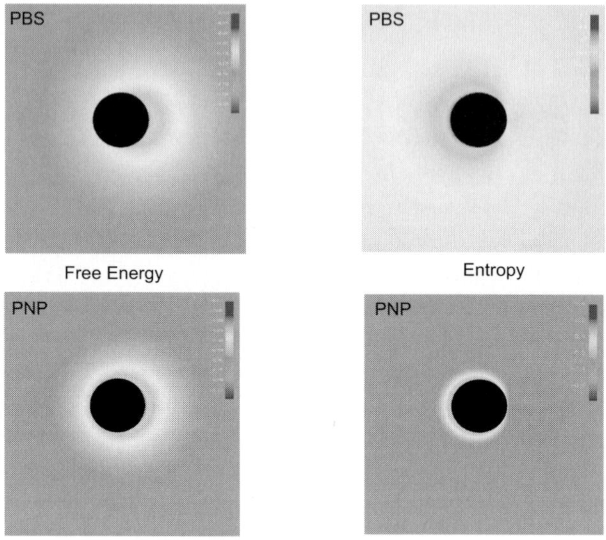

Figure 5. The local free energy and entropy densities in the PBS and PNP models.

	rate(PBS)	rate(PNP)	experimental
wild-type open	18.1	14.6	2.52
wild-type closed	23.1	19.1	2.52
3CL0	11.9	7.32	0.24
3CKZ	12.4	8.17	0.35
3CL2	16.5	11.8	1.1

steric effect, must be taken account into the simulation for accurately depicting the drug diffusion. Currently, we are trying to hybrid the Brownian dynamics and finite element method to obtain the collision reaction probability. Combining with the above diffusion coefficient, it can predict the diffusion association constant more accurately.

4. Conclusion

In this study, we describe continuum-based methods for studying electrostatic diffusion in biomolecular systems. Specifically, we present the SMOL software package, a finite element-based set of tools for solving the electrostatics and diffusion to calculate ligand binding rate constants for large biomolecules under pre-steady-state and steady-state conditions. The main improvement from the previous SMOL solver [9,10,29] can be addressed as below: first, the new regularized Poisson-Boltzmann algorithm instead of the previous APBS. It makes the solver read the potential value from the same diffusion node and avoid the previous data mapping error; second, the iteratively solving Poisson-Boltzmann and Smoluchowski equations results in a stable PNP solver; third, local free energy and entropy densities have been calculated during the diffusion process. finally, a simple mesh generator has been included for preparing tetrahedral meshes for the diffusion domain and molecular domain.

Although our above solution for the neuraminidase is qualitatively consistent with experiments, the continuum method on studying the diffusion still has two main limitations: the size of the diffusion ligand and non-electrostatic interaction are not included. Currently, we are combining the continuum modeling method and Brownian dynamics together for simulating the drug diffusion in the neuraminidase. According to Fig. 4, it must be noted that only a quite limited area close to the active site has substantial concentration/free energy density gradient. Our new method singles out the active site to implement Brownian dynamics while continuum method on other areas. The ultimate goal of this work is to develop scalable methods and theories that will allow researchers to begin to study biological macromolecules in a cellular context.

References

1. D. L. Ermak and J. A. McCammon, *J. Chem. Phys.* **69**, 1352 (1978).
2. S. H. Northrup, S. A. Allison and J. A. McCammon, *J. Chem. Phys.* **80**, 1517 (1984).
3. N. Agmon and A. L. Edelstein, *Biophys. J.* **72**, 1582 (1997).
4. R. R. Gabdoulline and R. C. Wade, *Methods* **14**, 329 (1998).
5. J. R. Stiles and T. M. Bartol, Monte Carlo methods for simulating realistic synaptic microphysiology using MCell, in *Computational Neuroscience: Realistic Modeling for Experimentalists*, ed. E. D. Schutter (CRC Press, Inc., New York, 2000) pp. 87–127.
6. J. L. Smart and J. A. McCammon, *Biophys. J.* **75**, 1679 (1998).
7. M. G. Kurnikova, R. D. Coalson, P. Graf and A. Nitzan, *Biophys. J.* **76**, 642 (1999).
8. Z. Schuss, B. Nadler and R. S. Eisenberg, *Phys. Rev. E* **6403** (2001).
9. Y. H. Song, Y. J. Zhang, C. L. Bajaj and N. A. Baker, *Biophys. J.* **87**, 1558 (2004).
10. Y. H. Song, Y. J. Zhang, T. Y. Shen, C. L. Bajaj, A. McCammon and N. A. Baker,

Biophys. J. **86**, 2017 (2004).
11. K. S. Tai, S. D. Bond, H. R. Macmillan, N. A. Baker, M. J. Holst and J. A. McCammon, *Biophys. J.* **84**, 2234 (2003).
12. H. Berry, *Biophys. J.* **83**, 1891 (2002).
13. D. Genest, *Biopolymers* **28**, 1903 (1989).
14. M. J. Saxton, *Biophys. J.* **61**, 119 (1992).
15. J. A. McCammon, *Science* **238**, 486 (1987).
16. S. H. Northrup, J. O. Boles and J. C. L. Reynolds, *Science* **241**, 67 (1988).
17. R. C. Wade, M. E. Davis, B. A. Luty, J. D. Madura and J. A. Mccammon, *Biophys. J.* **64**, 9 (1993).
18. P. Eastman and S. Doniach, *Proteins* **30**, 215 (1998).
19. L. Yeomans-Reyna and M. Medina-Noyola, *Phys Rev E Stat Nonlin Soft Matter Phys.* **64**, p. 066114 (2001).
20. R. J. Russell, L. F. Haire, D. J. Stevens, P. J. Collins, Y. P. Lin, G. M. Blackburn, A. J. Hay, S. J. Gamblin and J. J. Skehel, *Nature* **443**, 45 (2006).
21. P. J. Collins, L. F. Haire, Y. P. Lin, J. F. Liu, R. J. Russell, P. A. Walker, J. J. Skehel, S. R. Martin, A. J. Hay and S. J. Gamblin, *Nature* **453**, 1258 (2008).
22. I. Chern, J. Liu and W. Wang, *Methods Appl. Anal.* **10**, 309 (2003).
23. H. X. Zhou, *J. Chem. Phys.* **92**, 3092 (1990).
24. H. X. Zhou, S. T. Wlodek and J. A. McCammon, *Proc. Natl. Acad. Sci. U. S. A.* **95**, 9280 (1998).
25. H. X. Zhou, J. M. Briggs, S. Tara and J. A. McCammon, *Biopolymers* **45**, 355 (1998).
26. M. F. Sanner, A. J. Olson and J. Spehner, *Proc. 11th ACM Symp. Comp. Geom.* **C6-C7** (1995).
27. S. Tara, A. H. Elcock, J. M. Briggs, Z. Radić, P. Taylor and J. A. McCammon, *Biopolymers* **46**, 465 (1998).
28. E. B. Krissinel and N. Agmon, *J. Comput. Chem.* **17**, 1085 (1996).
29. Y. H. Cheng, J. K. Suen, Z. Radić, S. D. Bond, M. J. Holst and J. A. McCmmon, *Biophys. Chem.* **127**, 129 (2007).

MULTI-SCALE MODELING OF THE HUMAN VERTEBRAL BODY: COMPARISON OF MICRO-CT BASED HIGH-RESOLUTION AND CONTINUUM-LEVEL MODELS

SENTHIL K. ESWARAN
AARON J. FIELDS
PREM NAGARATHNAM
TONY M. KEAVENY

Orthopaedic Biomechanics Laboratory, Department of Mechanical Engineering, University of California, Berkeley, CA 94720, USA

The overall goal of this study was to assess the mechanistic fidelity of continuum-level finite element models of the vertebral body, which represent a promising tool for understanding and predicting clinical fracture risk. Two finite element (FE) models were generated from micro-CT scans of each of 13 T9 vertebral bodies — a micro-FE model at 60-micron resolution and a coarsened, continuum-level model at 0.96-mm resolution. Two previously-reported continuum-level modulus-density relationships for human vertebral bone were parametrically varied to investigate their effects on model fidelity using the micro-CT models as a gold standard. We found that the modulus-density relation, particularly that assigned to the peripheral bone, substantially altered the regression coefficients, but not the degree of correlation between continuum and micro-FE predictions of whole-vertebral stiffness. The major load paths through the vertebrae compared well between the continuum-level and micro-FE models (von-Mises distribution), but the distributions of minimum principal strain were notably different. We conclude that continuum-level models provide robust measures of whole-vertebral behavior, describe well the load transfer paths through the vertebra, but provide strain distributions that are markedly different than the volume-averaged micro-scale strains. Appreciation of these multi-scale differences should improve interpretation of results from these sorts of continuum models and may improve their clinical utility.

1. Introduction

Patient-specific voxel-type continuum-level finite element models of the vertebral body, that are generated automatically from clinical-resolution quantitative computed tomography (CT) scans, represent a promising alternative to standard densitometric techniques in understanding and predicting osteoporotic fracture risk [1, 2]. Due to the averaging process during clinical CT imaging, the microstructure of the trabeculae, which has a characteristic dimension on the order of 150 microns, is homogenized into a density-varying continuum of 1 mm-sized image voxels. The thin cortical shell, having a thickness of less than half a millimeter, is also averaged, and indeed there is blurring at the bone edges due to partial volume averaging effects. As a result, it is unclear if there remains a correspondence in the ability of these continuum-level models to properly capture the underlying physics of the bone failure. Understanding the relation between the behavior of these continuum-level models and more realistic micro-mechanical models that capture the microstructure is important since it may lead to identification of means to improve the fidelity of the continuum level models and improve interpretation of the outcomes of such models.

While a previous study that performed multi-scale comparison of the continuum-level and micro-FE models for the human femora provided much insight [3], differences in the specific modeling details as well as the mechanics between the femur and the vertebral body confounds the extrapolation of their results to the vertebral body. Due to difficulties in explicitly characterizing the continuum behavior of the cortical shell and associated limitations in using a generic model of the cortical shell [4-6], continuum-level models typically ignore explicit modeling of the cortical shell [1, 2, 7] and apply experimental mechanical property-density relation — developed for vertebral trabecular bone but not the cortical shell and adjacent bone — to all bone elements. The implications of using different material property-density relationships are unclear, and in particular, the validity of extrapolating these relations beyond their reported density range. Silva et al. [8] found that strain contours from continuum-level models were better indicators of experimental fracture locations than stress contours, but that study was limited to vertebral sections in which load transfer patterns through the vertebral body may have been over-simplified. Thus, there remains an incomplete understanding of ability of continuum-level models to predict stress or strain distributions within the vertebral body.

The overall goal of this study was to assess the mechanistic fidelity of continuum-level models of the whole vertebra — and some of the assumptions used in creating those models — by using a multi-scale approach to compare its predictions to that of more mechanistic micro-FE models. Our first objective was to determine the effect of different modulus-density relations on the agreement between the continuum-level and micro-FE model predictions of vertebral stiffness, and our second objective was to compare the stress and strain distributions of the multi-scale models. This study extends the use of the multi-scale modeling approach, previously applied to human femora [3], to a cohort of elderly whole vertebrae and is unique in its focus on the micromechanics of the multi-scale models.

2. Methods

After removal of posterior elements, thirteen T9 cadaveric vertebral bodies (88 ± 4 years, n=3 male, n=10 female) were micro-CT scanned at 30-μm voxel size

(SCANCO Medical AG, Brüttisellen, Switzerland). Segmented images were obtained using a global threshold value. Two finite element models were generated for each vertebra – a micro-FE model with a 60-micron element size and a coarsened, continuum-level model at 0.96 mm resolution. The micro-FE model was compartmentalized into trabecular bone, cortical shell and cortical endplates using previously validated algorithms [9, 10].

All bone tissue in the micro-FE models was assigned a Young's modulus of 18.5 GPa and Poisson's ratio of 0.3 [11]. A tissue density of 2.05 g/cm^3 [12] was used to convert the volume fraction of each continuum element into apparent density. Two different modulus-apparent density relations reported in the literature for human vertebral trabecular bone [12, 13] were modified to generate a total of four parametric variations (Table 1, Fig. 1). To study the effect of extrapolating the modulus-density relations beyond the experimental data, the two reported relations [12, 13] (Models A, B) were modified within or beyond the experimental density range (Models C, D). The number of continuum-level elements (expressed as a percentage of the total number of elements in the model) beyond the experimental density range was calculated for each vertebral body. Transversely isotropic properties were assumed for all continuum-level elements [2]. A layer of polymethylmethacrylate (PMMA) was added to the superior and inferior endplates in order to provide uniform loading conditions across all vertebrae. The resulting micro-FE models contained up to 80 million elements. Using custom code [14] and a maximum of 640 processors in parallel, a uniform compressive strain of 1% was applied in the superior-inferior direction to each model.

Table 1. Different modulus-density relations analyzed in this study.

Model	Modulus (MPa) – Apparent Density (g/cm^3) Mapping
A	$E_{zz} = 2130\rho - 97.1$ [13]
B	$E_{zz} = 4730\rho^{1.56}$ [12]
C	Model B for $\rho \leq 0.38$, Model A for $\rho > 0.38$
D	Model A for $\rho \leq 0.38$, Model B for $\rho > 0.38$

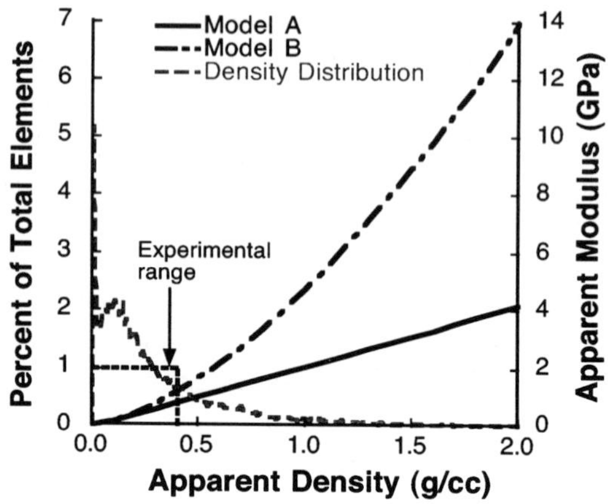

Figure 1. Modulus-density relations (Models A, B) for vertebral trabecular bone and a typical density distribution extrapolated to the entire density range. See Table 1 for equations for Models A and B.

Whole vertebral stiffnesses of the continuum-models (Models A–D) were correlated with that of the micro-FE model. The difference in the stiffness predictions between models A and B which use the modulus-density relations reported in the literature was correlated with the micro-FE stiffness in order to statistically test whether the choice of the modulus-density relation affected the slope and/or intercept of the correlation between continuum-level and micro-FE stiffness predictions. The stiffness predictions of models C and D were compared in a similar way to test whether differences in the modulus-density relations within (Model C vs. A) or beyond (Model D vs. A) the experimental density range had a greater effect on the regression coefficients.

The minimum principal strain and von-Mises stress distributions were compared between the multi-scale models. The results from the micro-FE models were averaged over each continuum-level element to facilitate this comparison (Fig. 2). Moreover, the distribution within each finite element model was compartmentalized into three groups: 0–50th percentile, 51–75th percentile, and 76th–100th percentile. Quantitative comparison of the correspondence between the continuum-level elements identified as being at high stress/strain (76th–100th percentile) and the averaged results from the micro-FE models was performed. Preliminary investigation showed that the strain distribution of the continuum-level models may largely be governed by the axial stiffness of each transverse layer, *i.e.*, $\Sigma E_i A_i$ summed over $i = 1, n$ continuum-level elements in that transverse slice. For each vertebral body and for both the continuum-level and micro-FE models, the number of elements belonging to the high strain category (76th-100th percentile) in each transverse layer was plotted against the axial stiffness of that layer. The transverse layers that included the endplates were not included as part of this regression since the curvature of the endplates confounds measurement of axial stiffness.

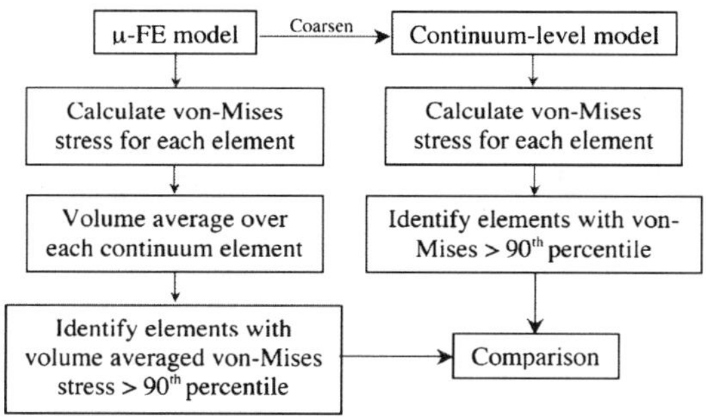

Figure 2. Flowchart describing the procedure by which regions of high-stress were compared between micro-FE and continuum-level models. A similar approach was used for minimum principal strain.

3. Results

Stiffness predictions of all parametric variations of the continuum-level model were highly correlated with micro-FE stiffness (Fig. 3, R^2=0.89–0.93). Choice of the modulus-density relation had a substantial, statistically significant effect on the regression coefficients between the predicted continuum-level FE stiffness and micro-FE stiffness but did not appreciably affect the R^2 values.

Despite the majority of elements being within the experimental density range (73.2±5.8%), differences in the modulus-density relations beyond the reported density range (Model D vs. A) had a greater effect ($p<0.0001$) on the regression with the micro-FE stiffness than did any differences associated with changes to the modulus-density relation for bone within the experimental density range (Model C vs. A). The elements with density values beyond the experimental density range were typically at the periphery of the vertebra and included the cortical shell and endplates.

While the major load paths compared well between the continuum-level and micro-FE models (as seen from the von-Mises distributions), the strain distribution in the continuum-level model was notably different from that of the micro-FE model (Fig. 4). Comparison of regions with high von-Mises stress showed that the correspondence between the continuum-level and micro-FE models was similar in the trabecular centrum and cortical shell (Fig. 5).

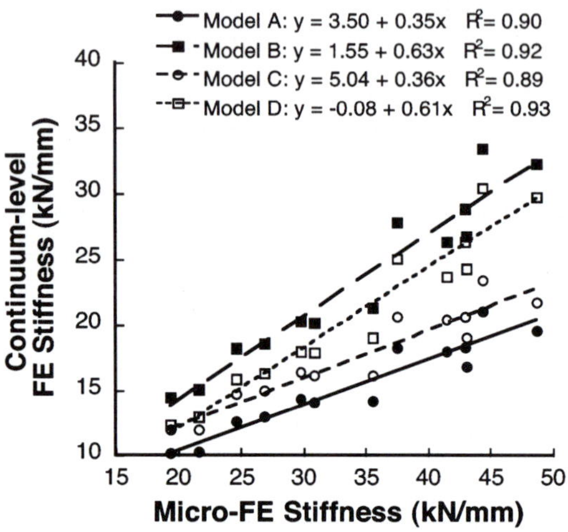

Figure 3. Stiffness predictions of all parametric variations of the continuum-level model were highly correlated with micro-FE stiffness. Material property changes to the bone beyond the experimental density range had a greater effect on the regression coefficients than changes to the bone within the experimental range. See Table 1 for descriptions for Models A-D.

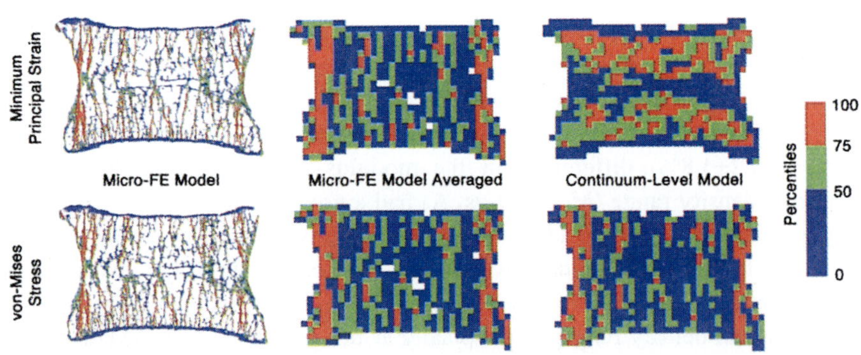

Figure 4. Typical distribution of von-Mises stress and minimum principal strain at a mid-coronal slice (0.96 mm thick) of a vertebral body. Middle and right-hand panels show volume-averaged measures. While the continuum-level models captured the major load paths for this loading mode (von-Mises distribution), the distribution of minimum principal strain of the continuum-level model was notably different compared to that of the micro-FE model.

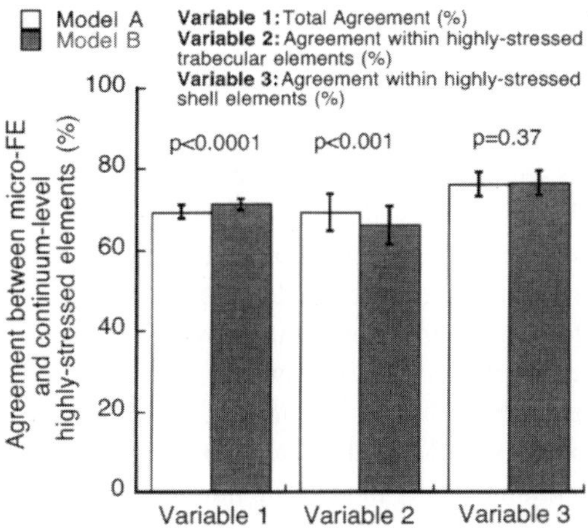

Figure 5. Agreement of von-Mises and minimum principal strain distributions (75th-100th percentiles) between the micro-FE and continuum-level models. See Table 1 for descriptions for models A and B.

In general, the transverse layers in the continuum-model that had lower axial stiffness had greater number of elements experiencing high strains whereas this trend was not observed in the micro-FE models. In the continuum model, the trend for transverse slices with lower axial stiffness to have greater number of elements in the high-strain category was observed in 11 specimens ($p<0.05$), while two specimens showed no significant correlations. On the contrary, in the micro-FE models, there was only 1 specimen that had a significant negative correlation ($p<0.05$) between axial stiffness and number of elements in the high-strain category; 5 specimens had a significant positive correlation ($p<0.05$); while 7 specimens had no significant correlation.

4. Discussion

The goal of this study was to investigate the mechanistic fidelity of the continuum-level models by comparing its predictions with that of a more detailed micro-FE model. The absolute predictions of stiffness in the continuum-level models were more sensitive to the material properties assigned to the peripheral bone as compared to the trabecular centrum. The degree of correlation between stiffness predictions of the continuum-level and micro-FE models was insensitive to the choice of the modulus-density relation, presumably because the spatial distribution of density was captured in all parametric variations. The slope of the stiffness correlation was substantially lower than an ideal slope ($Y=X$) for all parametric variations, indicating that the tissue-level modulus assumed for the micro-FE models (E=18.5 GPa) may be too high (indeed, subsequent independent tests in our laboratory have shown the effective tissue modulus of the trabecular bone to be closer to 10 GPa.) The major load paths compared well between the continuum-level and micro-FE models (von-Mises distribution) since the high-modulus continuum elements carry significant load and thus, dominate the stress distribution. The largely transverse strain bands observed in the continuum-level models were consistent with a springs-in-series analogy wherein the highest strains were observed in the most compliant layers of the vertebral body. The volume-averaged strain distribution in the micro-FE model, however, was not consistent with this analogy and was similar to that of the stress distribution of micro-FE models, indicative of an axial tissue-level stress and strain state. Taken together, these results indicate continuum-level models of human vertebrae capture the stress distributions better that the strain distributions, and that whole-vertebra metrics such as stiffness are highly correlated across the scales of modeling.

A major strength of this study was its focus on both the apparent-level and tissue-level predictions of the multi-scale models. The micro-CT models, using 60-micron-sized voxel elements, had sufficient geometric resolution to model the physics associated with bending and deformation of individual trabeculae. These models also included a detailed geometric description of the thin cortical shell. The substantial porosity between individual trabeculae is also included explicitly in these models. By contrast, the continuum models had no such porosity, the cortical shell was averaged with surrounding material, and the physics associated with deformation of individual trabeculae was absent. Coarsening of the micro-CT based vertebral body models allowed us to simulate the behavior of clinically-relevant continuum-level models and to directly relate the mechanics of the two scales. The results presented here represent the best-case performance of the continuum-level models since errors due to such factors as image streaks or low signal-to-noise ratio — both of which may be present in clinical QCT-scans — have not been simulated in this study. Parametric variations of the modulus-density relationships in the continuum models enabled us to extract the relative roles of the peripheral bone and the trabecular centrum in determining the accuracy of the continuum-level stiffness predictions.

Despite these strengths, there are a few notable limitations. First, the boundary conditions in these models were simplistic compared to physiological loading through a non-linearly and inhomogeneous elastic disc. This approach was justified at this juncture since these analyses represent a necessary first step in understanding the mechanics of the multi-scale models under controlled loading conditions without any confounding effects of the complex disc.

Second, linear analyses were performed in all models and hence, failure distribution or strength predictions were not compared between the multi-scale models. The results from the comparison of the elastic stress/strain distribution need not necessarily hold for failure distributions since there may be load redistribution around fractured trabeculae in the real bone. However, since the locations of high strain in the micro-FE model are likely representative of regions of initial failure [15], these results may well represent measures of a yield strength.

This study provides contrasting results to that of a previous study on multi-scale analysis of human femora. Verhulp *et al.* [3] found that both the stress and strain distributions of the continuum-level model compared well with that of the micro-FE model of a healthy and an osteoporotic femur during fall loading conditions. However, the micro-mechanics of the femur under fall loading conditions are much different than those for the spine under compression. Deformation mechanisms depend on the bone volume fraction, role of the cortical shell, and orientation of the trabecular bone with respect to the loading axis — all different between hip and spine. Thus, the relation between multi-scale strains is expected to depend on such factors as anatomic site and loading conditions and in that sense the results are entirely consistent.

The results from this study provide unique insight into the mechanics of the multi-scale models. Consistent with previous studies which have shown that the peripheral, dense bone plays a large role in vertebral stiffness [9, 10], our results demonstrated that absolute predictions of stiffness were more sensitive to the material properties assigned to the bone around the periphery as compared to the bone within the trabecular centrum. This finding suggests that improved characterization of the continuum-level material properties of the peripheral bone may improve apparent-level predictions of strength at the continuum-level model. Using the current modeling approach of the vertebral body at the continuum scale, our results indicate that the volume-averaged stress distribution predicted by the continuum model in the elastic range may be more accurate compared to the strain distribution. However, the discrepancy for the latter is likely due to the volume-averaging process since the results demonstrated how this process allows high strains to occur in the continuum level model in regions of very low apparent density (and consequently, in transverse layers with low axial stiffness), whereas in the micro-mechanical model the strain in such regions were zero because of the absence of bone. Interesting, regions of high apparent strain appear to agree well with directly observed regions of failure in thick sections of vertebrae loaded to failure in the cadaver laboratory [8]. Thus, it remains to be seen if — despite the disagreement between continuum-level and micro-level volume-averaged strain distributions reported here — the former remains highly correlated with directly observed failure regions in whole vertebrae. Further work is also recommended to extend these studies to more physiological type loading conditions, including load transfer through a disc with and without forward flexion.

Acknowledgments

Funding from National Institute of Health (AR49828). Computational resources from grant UCB-266 (NPACI). Finite element analyses performed on an IBM Power4 supercomputer (Datastar, San Diego Supercomputer Center). Specimen preparation by J. Buckley and micro-CT scanning by Dr. M.K. Liebschner (Rice University). Dr. Keaveny has a financial interest in O.N. Diagnostics and both he and the company may benefit from the results of this research.

References

1. Faulkner, K.G., C.E. Cann, and B.H. Hasegawa, *Effect of bone distribution on vertebral strength: assessment with patient-specific nonlinear finite element analysis.* Radiology, 1991. **179**: 669-74.
2. Crawford, R.P., C.E. Cann, and T.M. Keaveny, *Finite element models predict in vitro vertebral body compressive strength better than quantitative computed tomography.* Bone, 2003. **33**: 744-50.
3. Verhulp, E., B. van Rietbergen, and R. Huiskes, *Comparison of micro-level and continuum-level voxel models of the proximal femur.* Journal of Biomechanics, 2005. **9**: 2951-7.
4. Liebschner, M.A., et al., *Finite element modeling of the human thoracolumbar spine.* Spine, 2003. **28**: 559-65.
5. Cao, K.D., M.J. Grimm, and K.H. Yang, *Load sharing within a human lumbar vertebral body using the finite element method.* Spine, 2001. **26**: E253-60.
6. Silva, M.J., T.M. Keaveny, and W.C. Hayes, *Load sharing between the shell and centrum in the lumbar vertebral body.* Spine, 1997. **22**: 140-50.
7. Homminga, J., et al., *Osteoporosis changes the amount of vertebral trabecular bone at risk of fracture but not the vertebral load distribution.* Spine, 2001. **26**: 1555-61.
8. Silva, M.J., T.M. Keaveny, and W.C. Hayes, *Computed tomography-based finite element analysis predicts failure loads and fracture patterns for vertebral sections.* Journal of Orthopaedic Research, 1998. **16**: 300-8.
9. Eswaran, S.K., et al., *Cortical and trabecular load sharing in the human vertebral body.* J Bone Miner Res, 2006. **21**: 307-14.
10. Eswaran, S.K., et al., *The micromechanics of cortical shell removal in the human vertebral body.* Computer Methods in Applied Mechanics and Engineering. 2006. **196**: 3025-32.

11. Bevill, G., et al., *Influence of bone volume fraction and architecture on computed large-deformation failure mechanisms in human trabecular bone.* Bone, 2006. **39**: 1218-25.
12. Morgan, E.F., H.H. Bayraktar, and T.M. Keaveny, *Trabecular bone modulus-density relationships depend on anatomic site.* J Biomech, 2003. **36**: 897-904.
13. Kopperdahl, D.L., E.F. Morgan, and T.M. Keaveny, *Quantitative computed tomography estimates of the mechanical properties of human vertebral trabecular bone.* Journal of Orthopaedic Research, 2002. **20**: 801-5.
14. Adams, M.F., et al. *Ultrascalable implicit finite element analyses in solid mechanics with over a half a billion degrees of freedom.* in *ACM/IEEE Proceedings of SC2004: High Performance Networking and Computing.* 2004.
15. Eswaran, S.K., A. Gupta, and T.M. Keaveny, *Locations of bone tissue at high risk of initial failure during compressive loading of the human vertebral body.* Bone, 2007. **41**: 733-9.

ADVANCES IN SEMANTIC REPRESENTATION FOR MULTISCALE BIOSIMULATION: A CASE STUDY IN MERGING MODELS

MAXWELL LEWIS NEAL[1], JOHN H. GENNARI[1], THEO ARTS[2] & DANIEL L. COOK[3]

[1]*Biomedical & Health Informatics*, [3]*Physiology & Biophysics, University of Washington, Seattle, WA, 98195, USA*

[2]*Biophysics, University of Maastricht, The Netherlands*

As a case-study of biosimulation model integration, we describe our experiences applying the SemSim methodology to integrate independently-developed, multiscale models of cardiac circulation. In particular, we have integrated the CircAdapt model (written by T. Arts for MATLAB) of an adapting vascular segment with a cardiovascular system model (written by M. Neal for JSim). We report on three results from the model integration experience. First, models should be explicit about simulations that occur on different time scales. Second, data structures and naming conventions used to represent model variables may not translate across simulation languages. Finally, identifying the dependencies among model variables is a non-trivial task. We claim that these challenges will appear whenever researchers attempt to integrate models from others, especially when those models are written in a procedural style (using MATLAB, Fortran, etc.) rather than a declarative format (as supported by languages like SBML, CellML or JSim's MML).

1. Integrating multiscale biosimulation models

A goal of many biosimulation researchers is to develop libraries of re-usable simulation modules that can be combined in a plug-n-play manner. Ideally, biosimulation researchers should be able to retrieve a module from such a library and easily integrate it with their own models. Although this vision has been well-described[1, 2], to our knowledge there have been no published processes or methodologies by which model integration might occur.

In this paper, we* report on our experiences integrating two independently developed models for cardiovascular circulation. As we describe in greater detail in Section 2, our first goal was to make the shape of the aortic pressure waveform in our cardiovascular simulation model (CV) more canonical by raising the pulse pressure. Our second goal was to test and extend the semantic simulation (SemSim) framework[3] for biosimulation model reuse with a challenging case of cross-platform integration: integrating MATLAB code with code for the JSim environment. Our most important results and findings have to do with three critical challenges to understanding and translating *procedural*

* This research involved two independent research groups. For narrative simplicity, this paper was written from the point of view of the University of Washington authors. Contributions from the University of Maastricht author are indicated explicitly in the text.

biosimulation code, such as MATLAB, into a *declarative* environment, such as JSim[4].

1.1 SemSim modeling

We used the declarative SemSim modeling framework for our model integration task. SemSim models (called "AMO models" in [3]) are light-weight ontologies that leverage knowledge in larger reference ontologies to match concepts across biosimulation models. The SemSim framework captures the biological meaning of model components along with the mathematical/logical dependencies necessary for model simulation. We have designed this framework to facilitate model sharing and reuse by automating the processes of merging and translating legacy biosimulation models. Presently these tasks are accomplished through hand-coding—a very costly and error-prone process.

Figure 1 shows the representational schema by which a SemSim model serves as a middle layer between biosimulation code (on the right) and biomedical reference ontologies (on the left). SemSim model variables and equations are represented by SemSim *Data structure* and *Computation* classes in a *Computational model* and then mapped (via *encodes* relations) to corresponding *Physical properties* and *Property dependencies* classes in a *Physical model.* The *Physical properties* in turn are unambiguously referenced (*has_property* relation) to a *Physical entity* that references a specific class in the Foundational Model of Anatomy (FMA)[5]. *Physical property*, on the other hand, is referenced to a *Physical property* class of the Ontology of Physics for Biology (OPB [6]). For example, the physical property *Volume of blood in systemic arteries* is

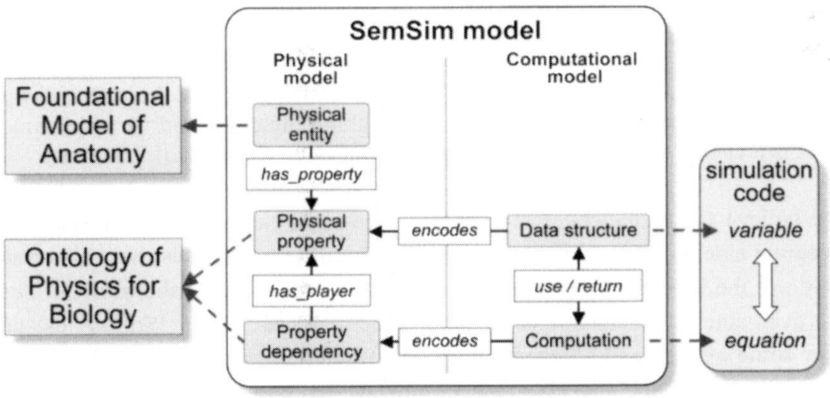

Figure 1. A diagram of the structure of a SemSim ontology. These ontologies can improve the reusability of biosimulation code (on right) by connecting variables and equations with knowledge stored in reference ontologies (on left).

defined using the class *Blood in systemic arterial tree* from the FMA and the class *Fluid volume* in the OPB. Such annotations uniquely encode and identify the biological meaning of model variables in a computable form that allows concepts to be mapped across SemSim models.

1.2 Procedural versus declarative code

Procedural programming languages, such as MATLAB and Fortran, specify the steps computers must take to reach a desired outcome. including calls to subroutines and ODE solvers. In contrast, declarative languages simply specify the problem without specifying a procedure for solution. SBML, CellML and JSim's mathematical modeling language (MML, not to be confused with MathML) are declarative languages. There are strengths and weaknesses to both styles of programming. However, our SemSim architecture for reuse is declarative. Thus, we needed to express the models in a declarative manner.

In Section 3, we identify three important challenges to integrating procedural code into the declarative SemSim architecture for reuse:

- First, the overall structure of the procedural model must be understood. Most critically, multiscale models may solve equations and update variables at more than one time scale, and these different scales must be clearly articulated and differentiated.

- Second, different languages, whether procedural or declarative, employ a variety of data structures for storing and manipulating variables. Languages such as Fortran and MATLAB can use matrices, arrays, lists, etc., whereas SBML, CellML[7] and JSim's MML have a much smaller range of variable representational choices.

- Third, the dependencies among biosimulation variables much be understood. In procedural code, this can be challenging, as the relationships among variables can be scattered throughout the code, and variable names can change as they are passed into and out of computational functions.

We believe these challenges are general—they will exist whenever researchers integrate externally developed procedural code into their models, whether or not they use the SemSim approach. Sections 3 and 4 below describe in detail how we overcame these challenges for our particular case study. We also discuss how some extensions to SemSim (section 4) will help address these challenges. However, these general problems may indicate that no model integration system can be fully automatic. Researchers should be forewarned and prepared for these complexities when integrating multiscale models.

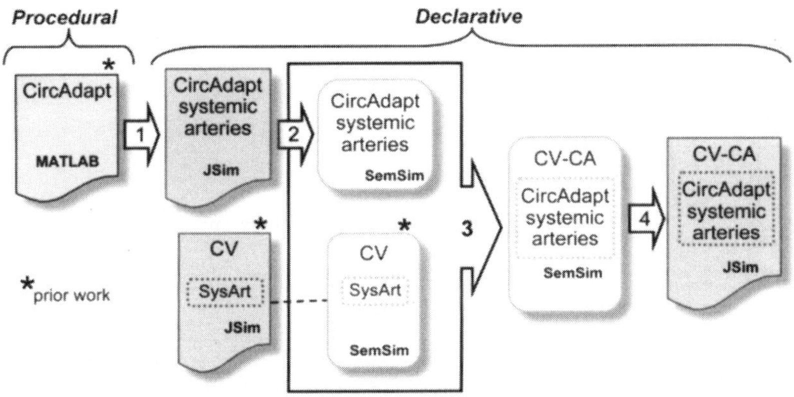

Figure 2. Overview of the 4-step process by which the CircAdapt MATLAB systemic arterial components were extracted and merged with the UW CV model. **1**: Extract relevant procedural CircAdapt code and encode in functionally equivalent JSim MML code (results shown in Figure 3). **2**: Derive a SemSim model from JSim version of CircAdapt Systemic arteries model . **3**: Use Protégé Prompt plug-in to merge SemSim models and replace original CV systemic arterial (SysArt) components with CircAdapt systemic arteries. **4**: Use SemSim Coder software to automatically generate executable JSim MML code from the merged model (results shown in Figure 4).

2. Motivation and modeling results

We encountered the CircAdapt modeling of tissue and organ level cardiovascular dynamics at PSB08[8] and recognized an opportunity to improve our own CV model and to stress-test the representational capabilities of our SemSim modeling framework. Our modeling motivation was to improve the realism of our aortic pressure curve by integrating the CircAdapt systemic arterial component which included fluid inertial effects of pulsatile blood flow.

Figure 2 is a roadmap of our model merging process as described in this paper. We needed to: (1) translate the relevant portions of the CircAdapt MATLAB code into JSim MML code, (2) create a SemSim model of the CircAdapt systemic arterial system, (3) merge the SemSim version of the CircAdapt systemic arterial component with the SemSim version of our CV model and (4) re-encode the merged SemSim model into MML for execution in JSim.

We validated our MATLAB-to-JSim integration in two ways. First, we compared the response of the native MATLAB systemic arterial pulse pressure profiles with those generated by the JSim version by driving the JSim version with aortic flow data produced by CircAdapt. Figure 3 shows that the numerical results were virtually identical and any discrepancies were probably attributable to differences in the ODE solver routines. Second, we found that incorporating the more sophisticated modeling of systemic arterial dynamics of the CircAdapt

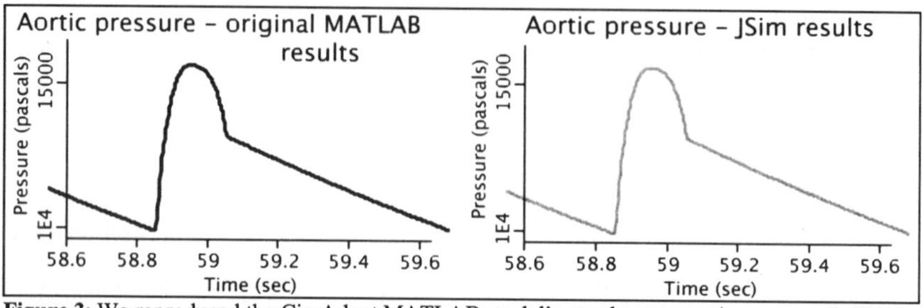

Figure 3: We reproduced the CircAdapt MATLAB model's steady state aortic pressure (left) in our JSim version of the systemic arterial component (right).

model did indeed provide a much more canonical aortic pressure waveform than our original CV model (Figure 4).

3. Translating from procedural to declarative models

We encountered three main challenges when translating the procedural code of CircAdapt's systemic arterial model to JSim's declarative MML code: 1) simulating multiple time scales, 2) resolving differences in the simulation environments' allowed data structures, and 3) translating the mathematical dependencies from the original model completely and faithfully.

3.1 Assessing computational architecture and identifying time scales

To reproduce and reuse biosimulation models, researchers must rely on several sources. Some source code may be publicly available in model repositories but in other cases, the primary source is a publication (e.g., the publication we used:[9]) which generally contains only incomplete details about model code implementation. Thus, direct contact with model authors must often be established to obtain source code and crucial clarifications about modeling choices. In this case, we had no prior collaboration with the T. Arts group, and given the geography, communication was limited to a handful of emails. We believe this

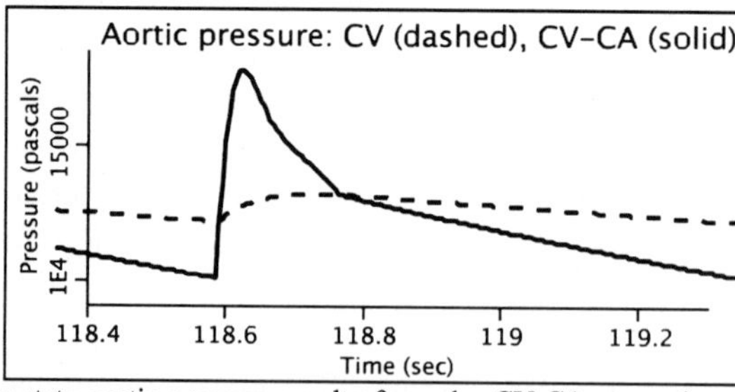

Figure 4: Steady state aortic pressure results from the CV-CA merged model (solid) showing improved pulse pressure over the original CV model results (dashed).

will be common for many reuse scenarios, and adds to the challenge of understanding and integrating code developed by others.

We first assessed the CircAdapt MATLAB model's computational structure which consists of 14 separate MATLAB files and a README documentation file. Model execution is under the control of the CircAdaptMain.m file which calls a MATLAB ODE solver (ode23) to calculate circulatory volume and flow state variables for a single heart beat. At the end of each heart beat, CircAdapt discretely updates several model variables to simulate the slower process of genetically-controlled blood vessel adaptation to hemodynamic variables. The adapted variables are then used to restart the ODE solver for the next heart beat repeatedly until a user-specified end time. Therefore, the CircAdapt model operates on two different time scales: a faster one for simulating within-beat fluid dynamics and a slower one for genetic adaptations of blood vessel geometry. This aspect of the model was not immediately apparent until we assessed the procedural code in detail. In section 3.3, we detail the workarounds in our JSim code to handle integrations on dual time scales.

Given our experiences, we believe that it is important for model annotations to explicitly describe the time scales upon which models operate. This will be especially salient when models are translated from a procedural to a declarative language, since procedural languages offer more control over how the model is solved. We recognize, however, that these annotations are only a first step in solving the more general and more difficult computational problem of efficient and accurate numerical integration across multiple time-scales. Our current focus is on identifying temporal *semantics*, a first and necessary step before devising efficient mathematical methods for simulating dynamics on multiple time-scales.

3.2 Identifying data structures and physiological variables

Our modeling goal was to include only the dynamics of the systemic arterial tree of the CircAdapt model. Thus, we needed to identify and extract only those variables relevant to that portion of the CircAdapt model. CircAdapt uses a single, globally-accessible MATLAB tree-like data structure called *Par* that represents physiological variables (blood pressures, flows, volumes, etc.) as leaves. This structure allows the developer to group related variables together as sub-trees, so that they can be passed together into various functions. For example the *Par.TubeLArt* subtree includes variables for the systemic arterial vessels whereas the *Par.TubeLArt.Adapt* subtree contains variables for the adaptive calculations. Thus, for example, the variable representing volume of blood in the systemic arteries is accessed via *Par.TubeLArt.V*, and the corresponding pressure via *Par.TubeLArt.p*. Some information about this data structure was supplied by T. Arts in a README file.

JSim's MML does not support the complex data type used for *Par*. In general, whenever translating across simulation platforms and languages, there will be some level of mismatch among supported data types. Because we were using only a single component of the CircAdapt model, we had little use for the grouping advantages this data structure provides. Thus, we simply flattened the list of variables and parameters while retaining CircAdapt's naming conventions: variables like *Par.TubeLArt.V* became *Par_TubeLArt_V*

In general, it may be that some information will be lost during translation from complex data structures to simpler ones. Although we could flatten the CircAdapt *Par* tree structure, multi-dimensional matrices cannot be so easily accommodated. Model authors should be explicit about their data structure choices, and researchers that wish to reuse model components should be aware of potential problems translating such structures.

3.3 Assessing variable dependencies

Assessing variable dependencies is critical to understanding a model: which variables depend on time vs. which variables are fixed in the model, which variables are inputs vs. those that are outputs, and, critically, how variables within the model depend upon one another. Once we understood the overall CircAdapt architecture, data structures and dependencies we created a corresponding list of parameter and variable declarations in the MML model that included the role players in CircAdapt relevant for reproducing the systemic arterial dynamics (such as *Par.TubeLArt.V*). The next task was to determine the mathematical dependencies among these variables.

As described above, some variables, like systemic arterial volume, are solved continuously within the MATLAB ode23 call, but other variables depend on processing the results from the previous heartbeat and are updated discretely following each heart cycle. Reproducing the mathematical dependencies present in the CircAdapt ode23 call was straightforward within JSim. However, to update the slower adaptation variables at the end of each heart cycle we used MML "realState" variables, which can be updated discretely. Additionally, some of these discrete variables in CircAdapt were updated using mean values of pressure and flow data from the previous heartbeat. Because MATLAB allows more procedural control over how model data is produced, we could not directly replicate these computations in MML. We instead wrote a new set of equations conducive to the JSim simulation engine to compute these mean values.

Some time-continuous variables are computed piecewise throughout the original CircAdapt code execution, and these separated dependencies had to be merged together in the MML version. In general, dependencies among variables in procedural code can be scattered and possibly buried deep within nested function calls. Tracking such nested dependencies was difficult, time-consuming and depended on direct communication with the CircAdapt author (T. Arts).

Communication with T. Arts was important throughout the translation process. These communications helped the team understand the structure, inherent assumptions and computational implementation of the CircAdapt model much more readily. Specifically, Dr. Arts: 1) identified those model codewords that are global constants versus those that update discretely versus those that update continuously, 2) identified the units used for the model constants and variables, 3) clarified that the model was designed to conserve blood volume, and 4) identified appropriate CircAdapt output data for validating the JSim results.

4. SemSim model-merging and code generation

In this section we describe our use of the SemSim architecture to merge the CircAdapt systemic arterial component into our CV model, and then to generate code for the combined system (steps 2-4 in Figure 2).

4.1 Creating a SemSim model of the CircAdapt systemic arteries

The first step in our merging process was to create a SemSim model of the JSim code that implements the CircAdapt systemic arterial model. Using Protégé [10] and the SemSim framework (see Figure 1), we began by creating a *Physical property class* for each variable in the CircAdapt JSim model. Thus, the JSim variable *Par_TubeLArt_V* (Volume of blood in the systemic arteries) was represented by a *Physical property* subclass *Fluid volume*, which references the *Fluid volume* class in the Ontology of Physics for Biology. We linked this *Property class* to the *Physical entity* class *Blood in the systemic arterial tree*, which references the corresponding class in the Foundational Model of Anatomy.

Next, we created classes describing the quantitative *Physical dependencies* (see Figure 1) between *Physical properties*. For example, the *Systemic arterial blood volume dependency* class has a positive flow rate role player *Blood flow through aortic valve* and a negative rate player *Blood flow through systemic arterial tree*. This *Physical dependency* describes how *Systemic arterial blood volume* is solved as the temporal integral of the fluid analog of Kirchoff's current law (change in vessel fluid volume equals vessel inflow minus outflow).

In addition to these *Physical model components* of the CircAdapt SemSim module, we completed the SemSim model by building appropriate *Datum* and *Computation* classes under the SemSim *Computational model component* class. We have developed a prototype SemSim Coder program that uses the Protégé API to access a SemSim file and automatically generate simulation code from its

Computational model component contents. In the future, this program will use platform-independent mathematical pseudo-code, but currently, it is restricted to generating MML code for JSim.

We tested the completeness of the SemSim version of the CircAdapt systemic arteries by running it through the SemSim Coder program and checking the numerical simulation results against the MML version. This process revealed several manual data entry errors, but these were easily corrected. For relatively small biosimulation models such as the CircAdapt systemic arteries, creating a SemSim model is straightforward, and it will become easier as we improve our tools.

4.2 Merging the SemSim models with Prompt

We next used the Protégé Prompt plug-in [11] to merge models and create a new circulatory system that retains most components of the CV model but replaces its original systemic arterial component with the CircAdapt systemic arteries component. Prompt can be custom-tailored; however, we simply used the off-the-shelf capabilities of this ontology merging tool.

First, we used the Prompt plug-in to copy all the classes of the previously built SemSim CV model and the new SemSim CircAdapt model into one file. Next, we had to identify and resolve the points of overlap between the CircAdapt module and the CV model. From our own knowledge we knew *a priori* which physiological properties had to be resolved: those properties shared by the two models that we as the modelers considered equivalent. Prompt automatically recognized three of these shared properties: blood flow through the aortic valve, pressure in the aorta, and heart rate. However, we also wished to replace the property *Volume of blood in systemic arteries in capillaries* in the CV model with *Volume of blood in the systemic arteries* from the CircAdapt module. For Prompt to identify this kind of resolution, it must recognize that there is a structural relationship between the participating entities in these two properties. This information was already encoded in the SemSim CV model, where *Blood in systemic arteries and capillaries → has_part → Blood in systemic arteries.* In the future we plan to implement a Prompt plug-in or standalone program as part of our SemGen package that analyzes the structural relationships between anatomical entities shared by both models before it suggests points of resolution.

Next, we hand-resolved these four points of overlap between the models. For each of the shared *Physical Properties* between the models, we picked the one we wanted to use in the merged system. For example, because we wanted to compute aortic pressure using the CircAdapt equations, we used the *Pressure in aorta* property class from the CircAdapt module, and not the one from the CV model. For each matching Property pair the values in the *Input_Role_Player_In* slot from the unused property were copied into the used property in order to preserve the dependencies specified in the unused property. Next we identified all those *Datum* classes which had unused property classes in their *Computational_Component_For* slots. All values in the *Input_Data_For* slots of these unused *Datum* classes were copied into the *Datum* classes that referred to the corresponding used property. This step preserved dependency information from the unused *Datum* classes. We then deleted all the Datum classes that referred to Properties that were not going to be used in the merged system (like *Paorta*, which referred to the unused *Pressure in aorta* property from the CV model). Then we deleted all the unused property classes, all the computation classes that had no *Output_Data* slot entries, and all the dependencies that had no *Has_Output_Player* slot entries.

To resolve the shared *Physical entity* classes of the two models, we arbitrarily chose to keep those entities from the CircAdapt model, and delete their twin entities in the CV model but only after copying all the slot values from the CV classes into their twin CircAdapt classes. This step preserved all the information from the CV *Physical entity* classes in the new merged model. To remove irrelevant *Physical entities* from the resulting merged system we deleted any *Physical entity* classes in which the *Has_Physical_Property*, *Part_of*, *Adjacent_to*, and *Continuous_with* slots were empty.

Based on our work merging these two models and others in the past, we plan to generalize this process of rectifying resolution points and create a tool within our SemGen program that better automates the steps described above.

4.3 Generating MML code from the merged SemSim model

Finally, we generated MML code from the merged CV-CA SemSim model using our SemSim Coder program. After entering the parameter values and initial conditions from the original CV and CircAdapt models, we attempted to run the merged model. However, the initial run failed because our model merging had introduced a circular dependency among variables. In particular, the equation for aortic pressure depended on the change in systemic arterial volume which depended on aortic flow which in turn depended on aortic pressure. This circular dependency was not present in either of the individual models, but rather was a feature of the new merged system. In general, this type of integration problem may be common when merging models. We argue that our SemSim approach, with its declarative specification of variables and dependencies, will make it easier to detect and resolve such circularities. In our case, we changed the equation for arterial volume to an ODE so that this variable could be set with an initial start value, and thus break the circular dependency.

After we corrected the circularity, the model was too stiff to solve numerically. Checking over our parameter values, we noticed that the value for heart rate in the merged model was incorrect, given its units. This error appeared because both the CV and CircAdapt models used the same codeword to represent heart rate (*HR*) but used different units and thus, different numerical values. In their respective parameter sets, which do not include unit designations, the numbers for *HR* are an order of magnitude apart, and when the parameters for the individual models were loaded into the merged system, *HR* was reparameterized with a value from the wrong model. To fix the problem, we simply corrected the value of the parameter in the MML code. Aside from *HR*, no changes were made to the parameters of the merged model. As presented in section 2, Figure 4 shows that aortic pressure in the CV-CA merged model has a more canonical waveform than the original CV model.

5. Discussion and Conclusions

A pervasive problem in reusing and merging biosimulation code is their inconsistent and incomplete annotations and explanations. There are ongoing efforts to standardize model annotation (e.g., [1]) in terms of units, nomenclature, etc. We contend that one of the most important aspects of every biosimulation model is its biological meaning—what physical aspects of which biological structures are being modeled and to what purpose. Thus we are developing the SemSim approach to provide a machine-readable, and thus searchable, annotation for models that are formally referenced to ontologies and constrained vocabularies of biological structure and biophysics.

Whereas we believe a declarative approach for model representation will be important for model reuse and integration, we recognize that procedural programming is widely used among modelers and offers important advantages in efficiency and customization. In our case study we illuminated three important challenges for representing procedural models in a declarative framework:

- Models may include procedures that occur on multiple time scales
- Data structures and naming conventions used to represent model variables may not translate across simulation platforms
- Dependencies among model variables may be obscured due to piecewise computation and/or variable renaming within functions

We claim that these challenges will appear whenever researchers attempt to integrate models from others, especially when those models are written in a procedural style (using MATLAB, Fortran, etc.) rather than a declarative format (as supported by languages like SBML, CellML or JSim's MML). We were able to solve these challenges in our particular case study, and our solutions will inform the further development of our SemGen software, and more broadly, our SemSim architecture. Our goals is to provide modelers with useful tools to annotate, create, and merge multiscale biosimulation models.

Given the inherent complexity of biosimulation modeling and the variety of independent modeling approaches for solving specific problems, it is no wonder that these challenges can be major impediments to model reuse and integration. A major premise of our SemSim approach is borrowed from the Java mantra: "write once, run anywhere." Thus, we propose that the hard work of translating native biosimulation code into a SemSim model can be performed once, and using more automated code re-writing methods, new models can be generated to run anywhere.

Acknowledgments

This work was partially funded by the NIH: for M. Neal: #T15 LM007442-06, and for J. Gennari & D. Cook: #R01HL087706-01.

References

1. Le Novere N, Finney A, Hucka M, Bhalla US, Campagne F, Collado-Vides J, et al. Minimum information requested in the annotation of biochemical models (MIRIAM). *Nat Biotechnol* 2005;23(12):1509-15.
2. Hunter PJ, Borg TK. Integration from proteins to organs: the Physiome Project. *Nat Rev Mol Cell Biol* 2003;4(3):237-43.
3. Gennari JH, Neal ML, Carlson BE, Cook DL. Integration of multi-scale biosimulation models via light-weight semantics. *Pac Symp Biocomput* 2008;13:414-425.
4. JSim. *The JSim Home Page at NSR*. http://physiome.org/jsim/index.html
5. Rosse C, Mejino JLV. A Reference Ontology for Bioinformatics: The Foundational Model of Anatomy. *Journal of Biomedical Informatics* 2003;36:478-500.
6. Cook DL, Mejino JLV, Neal ML, Gennari JH. Bridging Biological Ontologies and Biosimulation: The Ontology of Physics for Biology. *AMIA* 2008;(in press).
7. CellML. *CellML*. http://www.cellml.org
8. Lumens J, Delhaas T, Kirn B, Arts T. Modeling ventricular interaction: a multiscale approach from sarcomere mechanics to cardiovascular system hemodynamics. *Pac Symp Biocomput* 2008:378-89.
9. Arts T, Delhaas T, Bovendeerd P, Verbeek X, Prinzen FW. Adaptation to mechanical load determines shape and properties of heart and circulation: the CircAdapt model. *Am J Physiol Heart Circ Physiol* 2005;288(4):H1943-54.
10. Gennari JH, Musen MA, Ferguson RW, Grosso WE, Crubezy M, Eriksson H, et al. The evolution of Protege: an environment for knowledge-based systems development. *Int. J. Human–Computer Studies* 2003;58:89-123.
11. Noy NF, Musen MA. The PROMPT suite: Interactive tools for ontology merging and mapping. *Int J of Human-Computer Studies* 2003;59(6):983-1024.

INTEGRATION OF ANGIOGENESIS MODULES AT MULTIPLE SCALES: FROM MOLECULAR TO TISSUE*

AMINA A. QUTUB

Department of Biomedical Engineering, Johns Hopkins University, School of Medicine, 613 Traylor Bldg., 720 Rutland Ave, Baltimore, MD 21205 U.S.A.

GANG LIU

Department of Biomedical Engineering, Johns Hopkins University, School of Medicine, 425 Traylor Bldg., 720 Rutland Ave, Baltimore, MD 21205 U.S.A.

PRAKASH VEMPATI

Department of Biomedical Engineering, Johns Hopkins University, School of Medicine, 613 Traylor Bldg., 720 Rutland Ave, Baltimore, MD 21205 U.S.A.

ALEKSANDER S. POPEL

Department of Biomedical Engineering, Johns Hopkins University, School of Medicine, 611 Traylor Bldg., 720 Rutland Ave, Baltimore, MD 21205 U.S.A.

Multiscale modeling has emerged as a powerful approach to interpret and capitalize on the biological complexity underlying blood vessel growth. We present a multiscale model of angiogenesis that heralds the start of a large scale initiative to integrate related biological models. The goal of the integrative project is to better understand underlying biological mechanisms from the molecular level up through the organ systems level, and test new therapeutic strategies. Model methodology includes ordinary and partial differential equations, stochastic models, complex logical rules, and agent-based architectures. Current modules represent blood flow, oxygen transport, growth factor distribution and signaling, cell sensing, cell movement and cell proliferation. Challenges of integration lie in connecting modules that are diversely designed, seamlessly coordinating feedback, and representing spatial and time scales from ligand-receptor interactions and intracellular signaling, to cell-level movement and cell-matrix interactions, to vessel branching and capillary network formation, to tissue level characteristics, to organ system response. We briefly introduce the individual modules, discuss our approach to integration, present initial results from the coordination of modules, and propose solutions to some critical issues facing angiogenesis multiscale modeling and integration.

* This work is supported by NIH Grants RO1 HL79653 and R33 HL087351, and NIH 1F32HL085016-01 (A.Q.).

1. Introduction

1.1. *Mechanisms of Angiogenesis*

The top ten fatal diseases in the developed world, along with the physiological effects of exercise and wound healing, involve a shared process: angiogenesis, the growth of capillaries from preexisting vessels. Because of angiogenesis's prevalence and importance in disease progression and health maintenance, angiogenesis has been studied extensively experimentally, and more recently, computationally. In the last few years, multiscale modeling has emerged as a useful approach to interpret and capitalize on the biological complexity underlying blood vessel growth. To advance computational approaches, we present a way to integrate and coordinate multiscale biological modules of angiogenesis.

Designing the structure for model connectivity is a key part of successful integration. We approach angiogenesis as a complex, connected system of events occurring in sequence and in parallel, on multiple levels, triggered by a main stimulus, e.g., hypoxia.

In response to hypoxia, the transcription factor hypoxia-inducible factor 1 (HIF1) activates hundreds of genes [1]. These genes include vascular endothelial growth factor, VEGF. VEGF proteins stimulate chemotaxis and proliferation in endothelial cells during capillary sprouting. The balance of proangiogenic factors, e.g., VEGF, with antiangiogenic factors, e.g., endostatin, thrombospondin-1 and angiostatin, controls the extent of microvascular growth. At the onset of angiogenesis and throughout the process, cell activation, cell migration and cell proliferation are dependent on local growth factor concentrations and gradients.

Angiogenic sprouting also involves cell-matrix interactions. Matrix metalloproteinases, MMPs, are key molecules that allow an activated cell to proteolyze its surrounding extracellular matrix (ECM), form a moving sprout tip, and release ECM-bound factors as it migrates. The local microenvironment determines the fate of the growing sprout: it can anastomose and attach to adjacent vessels; it can retract; it can split and/or it can branch. As sprouts form and connect, they form a new capillary network. The new network can become capable of carrying blood and bringing oxygen to hypoxic regions.

To put the complexity of this process into perspective, Table 1 highlights a few numbers associated with the molecular level processes of angiogenesis.

Table 1. Selected molecular compounds included in the model and involved in the angiogenic process. The number of experimentally characterized species of these compounds is given.

	Main Compounds Modeled	Number of Characterized Species
Hypoxia	HIF	3
	Prolyl hydroxylases	3
Growth Factor Signaling	VEGF-A protein isoforms	7
	VEGF receptors	5
Matrix Degradation	MMPs	26
	Tissue inhibitors of MMPs (TIMPs)	4

At the molecular level, various degrees and durations of hypoxia yield different activity of HIF degradation enzymes (prolyl hydroxylases), HIF synthesis and reactive oxygen species, and hence oxygen sensitivity. At the single and multiple cell level, other ligand-receptor combinations such as notch and delta-like 4 ligand change tip cell density and capillary branching [2]. Furthermore, due to growth factor gradients, the position of a cell on a capillary sprout determines what growth factor concentrations it sees, its activation and protein expression, and its locomotive properties [3]. Parenchymal cells, precursor cells and stromal cells, as well as the extracellular matrix, constitute the nascent sprout microenvironment that influences cell signaling, growth factor production, endothelial cell (EC) movement, vessel stability and capillary permeability [4]. Finally, tissue composition and heterogeneity ultimately determine network structure, and paracrine signaling from different organs and organ systems, blood flow, inflammatory response, and lymphatics all can alter angiogenesis [4].

Not only is multiscale modeling a useful tool in determining how these factors interact, deciphering new experimental findings in angiogenesis and proposing new experiments, one could argue it is absolutely an essential tool to understand a process of such biological complexity.

In our laboratory, we have developed molecular-based models of HIF1α [5, 6], VEGF [7, 8], MMPs [9, 10], FGF2 [11], all key parts of angiogenesis. Below we introduce an approach to integrating models of angiogenesis, describe techniques used to represent the separate modules corresponding to processes in angiogenesis, discuss the challenges encountered in the process of multiscale integration, and propose solutions to address upcoming challenges.

1.2. *Modeling Approaches*

The large number of detailed existing angiogenesis models presents an opportunity to both develop increasingly sophisticated computational approaches to understanding biology, and to connect existing work together, taking the best pieces of developed models.

2. Modeling & Integration Methods

Integrating multiscale angiogenesis models presents a way to maintain and capitalize on biomedical knowledge in a quantitative form. It is a complex endeavor. Models were built to answer different biological questions and generate hypotheses that may have been very specific. As such, the models often use different algorithms, employ diverse languages, and focus on specific angiogenic factors, cell types or molecular species. They can be on different time and spatial scales.

Requirements then for integration are many. The most general form of integration must allow data exchange between models, feedback, asynchronous and synchronous running of models, and variable temporal and spatial scales. Furthermore, the integration must be designed so that the desired integrated or global biological phenomenon is best represented, while still maintaining the integrity of individual models.

With these criteria in mind, we began the development of a multiscale integrated model, with a goal of building a prototype from five existing models (called modules to distinguish them from the integrated model): a tissue geometry module, a blood flow module, an O_2 transport module, a VEGF module, and a cell module (Figure 1).

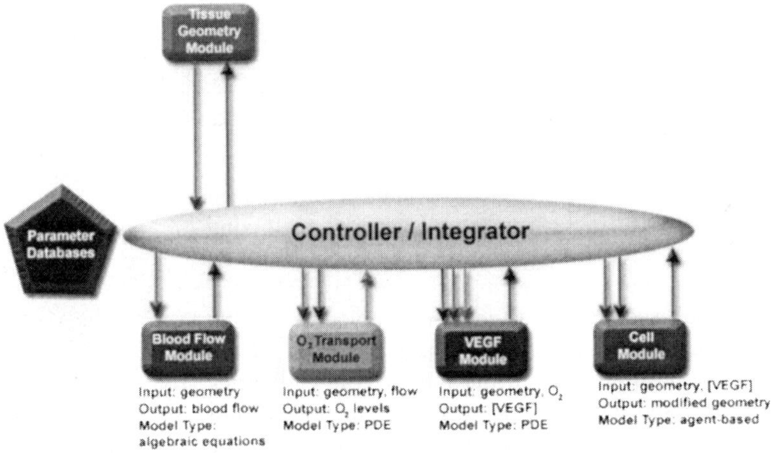

Figure 1. Schematic for the integration of five modules. Blood flow and the O_2 transport module are at the tissue level, while the VEGF module and the cell module are focused on the cellular level. Time scales range from seconds (O_2 and blood flow modules) to hours (VEGF and cell modules) to days (cell module).

To introduce the design of the integration in the context of the biology, as well as the capabilities of individual modules, we briefly describe each module, and follow-up by defining the controller.

2.1 *Geometry, Blood Flow & Oxygen Transport Modules*

A blood flow model was previously built and validated for a given geometry of microvascular network in 3D tissue, specifically applied to skeletal muscle (for a review, see [12]). The model predicts the amount of blood flow and hematocrit through capillaries in the network, by solving a set of nonlinear algebraic equations for pressure at the network nodes (bifurcations) and blood flow rate and hematocrit in the vascular segments (Figure 2). From blood flow, the relative oxygen distribution through the network is then predicted by convection-diffusion-reaction partial differential equations governing oxygen transport.

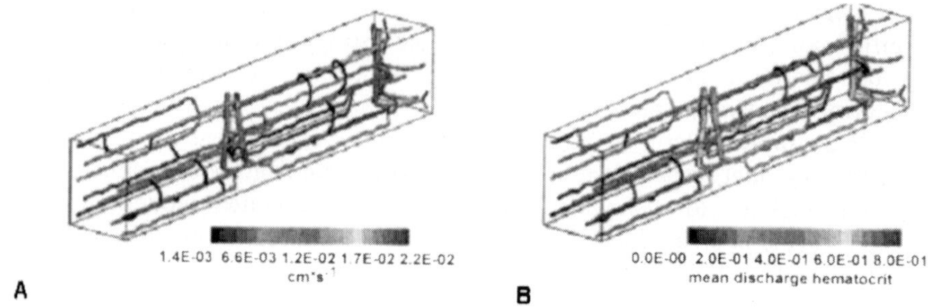

Figure 2. Velocity (A) and hematocrit (B) distributions for a representative 100 μm by 100 μm by 800 μm capillary network in rat skeletal muscle. The figure is adapted from [13], where the equation-based models representing blood flow and oxygen transport are discussed.

2.2 *VEGF Module*

The 3D model of the rat skeletal muscle introduced above was used in development of a model predicting VEGF distributions and VEGF receptor (VEGFR) occupancy in 3D tissue during exercise [14]. In the model, VEGF secretion is a function of myocyte oxygen pressure and an experimentally observed oxygen-HIF-VEGF relationship. Within the muscle, the capillaries are nonuniformly spaced, and this leads to nonuniform oxygen distribution and VEGFR expression within the tissue (Figure 3). VEGFR activation and VEGF binding is also predicted in the 3D model from a series of developed molecular-based chemical kinetic models [15].

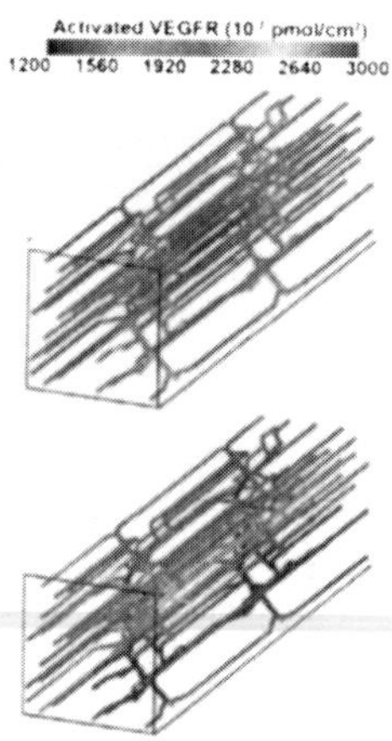

Figure 3 (right). Distribution of VEGF receptor activation along microvessels in skeletal muscle, during moderate exercise (top panel) and moderate exercise with low inspired oxygen (bottom panel).

2.3 *Cell Module*

Interaction of VEGF, MMPs, and the extracellular matrix at the molecular level trigger events at the cell level. These events include tip cell activation, endothelial cell migration, chemotaxis and haptotaxis and cell proliferation. We developed a computational model to mimic cellular sprouting at the onset of angiogenesis using agent-based methodology, governed by rules (Figure 4).

Examples of logical rules that guide cell activity in the model include migration rates as a function of local VEGF levels; or Boolean behavior for whether cellular elongation is allowed or not. Two examples of rules, in a restricted 2D case, are shown in Figure 5. The module rules based on experimental work compiled from extensive literature research, and applied to conditions that might occur in a 2D or 3D in vitro setting. Individual cell behavior (activation, elongation, migration, and proliferation) combine to produce a novel capillary network, emerging out of combinatorially complex interactions of single cells. A particular benefit of the rule or event-based modeling is the ability to easily produce in silico knockouts on multiple biological levels. Here we describe an example, an in silico molecular level knockout of the Dll4 ligand.

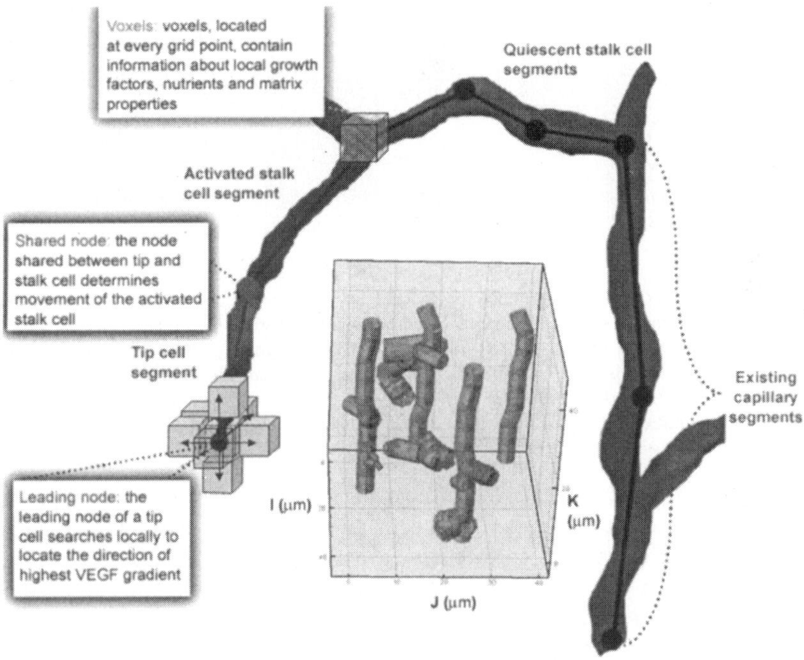

Figure 4. . Schematic of the 3D cell module. Capillaries are represented by endothelial cells. An example of a growing network with four capillaries is shown in the gray inset. Cells are divided into segments. Each segment is represented by two nodes. Currently, cell segments are modeled as cylinders specified by a length and radius (inset); an activated segment's length and radius can change during a model run. The local environment surrounding a cell is defined in each voxel of the grid. In the present model, voxels contain values for the local VEGF concentration. All cell segments have the capability of sensing what is located in voxels surrounding each of its nodes. For every timestep of the current model, this sensing is restricted to the leading node of the tip cell and the adjacent shared node.

2.3.1. Dll4 Knockouts

Delta-like ligand 4 (Dll4) is a transmembrane ligand for Notch receptors. It is a critical ligand for vascular development; haploinsufficiency of the Dll4 gene is embryonically lethal in many mouse strains [16]. Dll4 is primarily expressed in endothelial cells, and it is correlated to the local concentration of VEGF [17], as well as to VEGF receptor concentrations; a blockade of VEGF leads to a decrease of Dll4 [18], while Notch-Delta signaling downregulates VEGFR2 [19]. A Dll4 deficiency causes an increase in sprout formation but vessels appear nonproductive, with less capability of carrying blood or reducing hypoxia in surrounding tissue [18]. Overexpression of Dll4 diminishes the growth of new sprout tips.

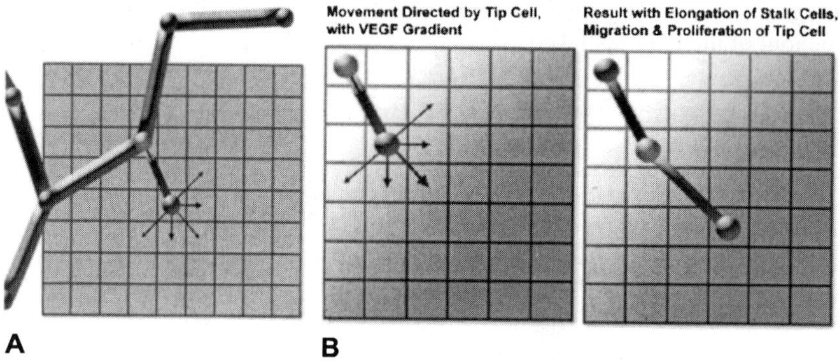

Figure 5. Illustrations of cell movement represented by rules in the cell module. The tip cell is represented by the leading node and segment; the node shared between the tip and stalk cells follows it; and the last node and segment is the adjacent stalk cell segment. Other segments and nodes represent quiescent vessels. Arrows represent direction of movement for nodes. (A) Schematic of a capillary with an activated tip cell. (B) Movement and the resulting cell segment positions when there is a VEGF gradient, showing the effects of allowed elongation and tip cell proliferation.

The cell module predicts the effects of VEGF protein concentrations and Dll4 haploinsufficiency on endothelial cells and capillary growth. In the module, cell-level behavior alone contributes to differences in capillary network formation, and the vascular networks resulting from the model appear similar in vascular density, branching and tortuosity to networks found in Dll4 experiments (Figure 6).

2.4 *The Controller: Running Models, Data Handling& Feedback*

The controller provides a platform to seamlessly integrate the modules mentioned above and simulate angiogenesis at multiple temporal and spatial scales. These modules can be, and currently are, rewritten as different Application Programming Interface (API) libraries and provide language binding for a variety of languages (such as C, C++, FORTRAN, JAVA, MATLAB). By this means, libraries can be called from different languages and shared by the community as well.

Integration revolves around the interaction of the controller and the modules, and between the controller and parameter databases or parameter text files. Parameters include biochemical parameters (such as kinetic rate constants) and physical parameters (such as vascular geometry). The parameters, currently inputted via text files, are being compiled into a database format. In this way, parameter sensitivity analysis can be performed to generate experimental hypotheses. Modules are compiled into the library files. Parameters dependent on a biological microenvironment are maintained outside of the library files, so that the modules are general, and function independent of their application to a specific tissue or cell. The controller reads the parameters from the database or text files, and invokes the modules. Each module takes input from the controller and outputs its results to the controller, which then passes the relevant data on to subsequent modules. The forward and feedback interactions between different modules are regulated through the controller.

3. Initial Output

3.1 Emerging New Vasculature Network

Results from the cell module are based on input from a sample of four initial capillaries, grown for 200 hours post stimulation by an angiogenic source, in this case elevated levels of VEGF (Figure 6). The growth of the vessels is unrestricted, and the initial space occupied by the four capillaries is a 130 μm by 130 μm cross-section, in the xy-plane. In Figure 6, the view is approximately from the top, looking down the z-plane. The concentration of VEGF is highest at the bottom of the section and lowest at the top, and we assume the VEGF levels are above the minimum required for endothelial cell activation. Activated endothelial cells are allowed anywhere on the initial capillaries, but growth is restricted to two dimensions, within a voxel movement in the z-direction. This first test of the model uses VEGF concentration and geometric input, and is coded in a manner to be easily integrated, and to write and read parameters appropriately from files.

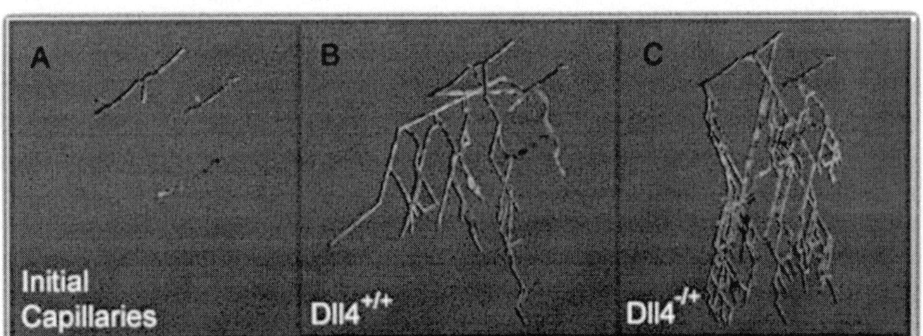

Figure 6. Snapshot of the model output directly after angiogenic stimulation (A) and for simulated hypoxia in Dll4$^{+/+}$ conditions, after 160 hrs (B) and in Dll4$^{-/+}$ conditions, after 160 hrs (C).

VEGF levels corresponding to conditions in skeletal muscle for mild to moderate hypoxia (e.g., during mild to moderate exercise), are being tested in the model to provide initial conditions for the cell module. A section of simulated skeletal muscle vasculature (a 200 μm by 200 μm by 800 μm section) is the input network. While requiring more computational memory and speed than the first tested capillary network input, we use this larger size of skeletal muscle tissue, since the corresponding vascular geometry has previously been tested as input separately into the blood flow, oxygen, and VEGF modules. In these renditions, there is no feedback. The different cell and molecular-level conditions that can be explored in the integrated model are listed in Table 2.

Table 2. Different conditions being represented and tested in skeletal muscle vasculature using the integration model.

$Dll4^{+/+}$	Low hypoxia	Elongation Allowed	Proliferation Allowed	Migration Allowed
$Dll4^{-/+}$	Moderate hypoxia	Elongation Restricted	Proliferation Restricted	Migration Restricted

Output includes new vessel network geometry, with new VEGF distribution, vessel length changes and a unique branching structure. Simulating conditions of exercise in skeletal muscle with low or moderate levels of tissue hypoxia involves changes in all parts of the angiogenic process; and we can assess the effects on local VEGF distribution surrounding tip and stalk cells, VEGFR binding, and cell proliferation and branching, as a function of local VEGF. An eventual application of this research is to test in silico molecular therapeutic intervention, and predict cellular changes and degree of angiogenic growth.

4. Discussion

4.1 Main Challenges & Current Approaches

4.1.1 Differences in Model Structure and Units

There are several approaches to tackling the issue of integrating multiscale modules written in diverse languages with different code architecture and data structures. One way to approach the "language barrier" is by compiling all of the modules into a JAVA-readable Dynamic Link Library (DLL) file format, in a Windows Operating system (or in a Shared Object Library, SO, file in a Linux platform). By this means, the controller, written in JAVA, can easily link with modules written in different languages, allowing the parameters to be passed readily between the controller and each module. The modules then are essentially black-box modular, executable files. Changes can be made to code in a single self-contained DLL or SO file shared by many modules.

This is not the most general form though. The issue arises that some of the models may need to be individually recompiled based on new equations, new parameters or a new threshold of parameters, e.g., a change in desired time-intervals. Recompiling in their original language can be a necessity for accurate results. Because of this, we are currently compiling each module in its original language, and having the controller call different compilers.

Dealing with differences in data structures comes through careful synchronization of input and output parameters from each of the modules. For us, this required some initial modification for each module in the body of the code, for writing and reading files, and for passing parameters. An example is that the cell module is defined by cellular agents comprised of nodes and segments – the location of the initial nodes and segments are derived from the geometry module which describes the initial vascular network as a series of voxel coordinates and a given radius. If endothelial cells or capillaries were defined as an amorphous shape or in 2D in one module but in 3D in another, redesign of the original code would likely be necessary. Alternately, in a more sophisticated controller, extrapolation between coordinate systems, types of data structures, and dimensions could be a feature.

4.1.2. Parameter Sensitivity, Parameter Optimization and Model Adaptation

Dealing with large number of parameters, some, if not most, unknown experimentally, comes with the process of multiscale, integrative modeling. One of the ways in which we are dealing with the large number of parameters in the modules and integrated model is by developing a database that can store parameters for different model runs. The database can be searched, and called dynamically by the controller to access input parameters for different modules as they are executed, or modules themselves can read and write directly to the database. Furthermore, part of the database will maintain experimental data, so the model results (e.g., total vessel growth over time in conditions of moderate hypoxia) can be compared immediately and directly to experimental findings. The necessary initial conditions can also be readily set to match those in experiments, by a query of the experimental protocol from the stored database.

Sophisticated, accurate, efficient means of analyzing parameter sensitivity are starting to be applied in biological models, and will likely become a requirement for progress in the field, as well as a means to help correlate model parameters with experiments. We are exploring means to do large scale parameter sensitivity rapidly, both for individual modules separately and for the combined model. Optimization will be based on the degree of experimental validation we are able to achieve for the integrated model.

Learning algorithms, too, limited in their application to the field of angiogenesis, will likely emerge as a means to keep programs updated and evolving as new data become available – furthermore, such algorithms could be employed to better portray biological adaptation to disease states and angiogenic development.

4.1.3. Software Location and Future Adaptation

Currently the described software is being developed on servers at Johns Hopkins University (Biomedical Engineering and Computer Science Departments). We will continue to develop individual modules and the multiscale integration platform. It is our goal to involve the larger community of researchers in systems biology, and provide a tool that will intelligently integrate models from laboratories around the world.

5. Conclusions

A large-scale initiative to combine existing and new modules of angiogenesis offers a way to maintain and cultivate knowledge of angiogenesis in a quantitative and dynamic format from the molecular to the cellular level and on to the tissue, organ and organism levels. By combining modules that represent each process in angiogenesis, from blood flow to oxygen transport to growth factor signaling to cell activation, movement and proliferation, we will begin to produce a comprehensive mechanistic simulation of capillary growth that can be used as a platform to test and generate experimental hypotheses. As new tools develop for the systematic validation, integration, visualization and adaptation of these models, multiscale modeling will become a key component of rigorous experimental design and therapeutic advances.

Acknowledgments

We thank Feilim Mac Gahbann, Emmanouil Karagiannis, James Ji, Carmen Kut and Michael Yang for their contributions to this research, and Mikhail Basilyan for his work on software development. We also thank laboratory members Marianne Stefanini, Kejing Chen, David Noren, Jacob Koskimaki and Florence Wu for discussions.

References

1. G. L. Semenza, *Physiology (Bethesda)*, vol. 19, pp. 176-82, 2004.
2. G. Thurston, I. Noguera-Troise, and G. D. Yancopoulos, *Nat Rev Cancer*, vol. 7, pp. 327-31, 2007.
3. A. F. Siekmann, L. Covassin, and N. D. Lawson, *Bioessays*, vol. 30, pp. 303-13, 2008.
4. R. H. Adams and K. Alitalo, *Nat Rev Mol Cell Biol*, vol. 8, pp. 464-78, 2007.
5. A. A. Qutub and A. S. Popel, *J Cell Sci*, vol. 119, pp. 3467-80, 2006.
6. A. Qutub and A. Popel, *Mol Cell Biol*, pp. in press, 2008.
7. M. O. Stefanini, F. Wu, F. Mac Gabhann, and A. S. Popel, *submitted*, 2008.
8. F. M. Gabhann and A. S. Popel, *Microcirculation*, pp. 1, 2008.

9. P. Vempati, E. Karagiannis, and A. Popel, *J Biol Chem*, pp. in revision, 2007.
10. E. D. Karagiannis and A. S. Popel, *J Theor Biol*, vol. 238, pp. 124-45, 2006.
11. R. J. Filion and A. S. Popel, *Ann Biomed Eng*, vol. 32, pp. 645-63, 2004.
12. D. Goldman, *Microcirculation*, vol. in press, 2008.
13. J. W. Ji, N. M. Tsoukias, D. Goldman, and A. S. Popel, *J Theor Biol*, vol. 241, pp. 94-108, 2006.
14. F. Mac Gabhann, J. W. Ji, and A. S. Popel, *J Appl Physiol*, vol. 102, pp. 722-34, 2007.
15. F. Mac Gabhann and A. S. Popel, *Am J Physiol Heart Circ Physiol*, vol. 286, pp. H153-64, 2004.
16. N. W. Gale, M. G. Dominguez, I. Noguera, L. Pan, V. Hughes, D. M. Valenzuela, A. J. Murphy, N. C. Adams, H. C. Lin, J. Holash, G. Thurston, and G. D. Yancopoulos, *Proc Natl Acad Sci U S A*, vol. 101, pp. 15949-54, 2004.
17. I. B. Lobov, R. A. Renard, N. Papadopoulos, N. W. Gale, G. Thurston, G. D. Yancopoulos, and S. J. Wiegand, *Proc Natl Acad Sci U S A*, 2007.
18. I. Noguera-Troise, C. Daly, N. J. Papadopoulos, S. Coetzee, P. Boland, N. W. Gale, H. C. Lin, G. D. Yancopoulos, and G. Thurston, *Nature*, vol. 444, pp. 1032-7, 2006.
19. R. C. Sainson, J. Aoto, M. N. Nakatsu, M. Holderfield, E. Conn, E. Koller, and C. C. Hughes, *Faseb J*, vol. 19, pp. 1027-9, 2005.

TOWARDS COMPUTATIONAL MODELING OF EXCITATION-CONTRACTION COUPLING IN CARDIAC MYOCYTES: RECONSTRUCTION OF STRUCTURES AND PROTEINS FROM CONFOCAL IMAGING

FRANK B. SACHSE[1,2], ELEONORA SAVIO-GALIMBERTI[1],
JOSHUA I. GOLDHABER[4], AND JOHN H. B. BRIDGE[1,2,3] *

[1]*Nora Eccles Harrison Cardiovascular Research and Training Institute,*
[2]*Bioengineering Department, and* [3]*Division of Cardiology, University of Utah,*
Salt Lake City, UT 84112, USA

[4]*David Geffen School of Medicine, University of California, Los Angeles, CA 90095, USA*

Computational models of excitation-contraction (EC) coupling in myocytes are valuable tools for studying the signaling cascade that transduces transmembrane voltage into mechanical responses. A key component of these models is the appropriate description of structures involved in EC coupling, such as the sarcolemma and ion channels. This study aims at developing an approach for spatial reconstruction of these structures. We exemplified our approach by reconstructing clusters of ryanodine receptors (RyRs) together with the sarcolemma of rabbit ventricular myocytes. The reconstructions were based on dual labeling and three-dimensional (3D) confocal imaging of segments of fixed and permeabilized myocytes lying flat or on end. The imaging led to 3D stacks of cross-sections through myocytes. Methods of digital image processing were applied to deconvolve, filter and segment these stacks. Finally, we created point meshes representing RyR distributions together with volume and surface meshes of the sarcolemma. We suggest that these meshes are suitable for computational studies of structure-function relationships in EC coupling. We propose that this approach can be extended to reconstruct other structures and proteins involved in EC coupling.

*Work supported by the Richard A. and Nora Eccles Harrison endowment, awards from the Nora Eccles Treadwell Foundation, and the National Institutes of Health research grants HL62690 and HL70828.

1. Introduction

Excitation-contraction (EC) coupling in myocytes involves a complex signaling cascade that transduces transmembrane voltage into cell contraction. In cardiac myocytes, most stages of this cascade involve calcium signaling[3]. The initial stimulus is membrane depolarization, which activates voltage-gated L-type calcium channels (LCCs) and allows for flux of extracellular calcium into the cell interior. This calcium activates sarcoplasmic calcium channels, the ryanodine receptors (RyRs), allowing release of calcium from an intracellular calcium store, the sarcoplasmic reticulum (SR)[10]. Both, calcium influx via LCCs and calcium induced calcium release (CICR), raise the cytosolic concentration of calcium. Subsequently, binding of cytosolic calcium to troponin C causes shifting of the troponin-tropomyosin complex followed by structural changes of the tropomyosin-actin configuration. Finally, these changes enable molecular interactions of actin and myosin causing cell contraction.

Electrophysiological measurements, functional and structural imaging, as well as computational modeling have been the major methods to characterize EC coupling. Several modeling approaches have been developed to describe calcium signaling underlying EC coupling. The approaches differ in the number of compartments considered, and the description of the cell anatomy and the properties of ion channels. In general in these approaches, simplified models of cell anatomy and protein distributions are applied due to difficulties to create accurate three-dimensional (3D) reconstructions from imaging data. In particular, physical limitations of imaging systems complicate the reconstruction. The limitations include small signal-to-noise ratio, resolution limits, image blurring and small contrasts between objects of interest and background. However, recent advances in imaging, protein labeling and methods for digital image processing promise to mitigate these difficulties.

In this study, we addressed the development of realistic anatomical models of ventricular myocytes for application in computational studies of EC coupling. Our focus was on describing spatial distributions of RyR clusters and the geometry of the sarcolemma including the transverse tubular system (t-system). The modeling of RyR clusters and sarcolemma was based on dual labeling with Alexa dyes and 3D scanning confocal imaging. RyRs were labeled with antibodies linked to Alexa 488. The sarcolemma associated glycocalix was labeled with wheat germ agglutinin (WGA) conjugated to Alexa 555. We imaged segments of cells lying flat on a cover slip

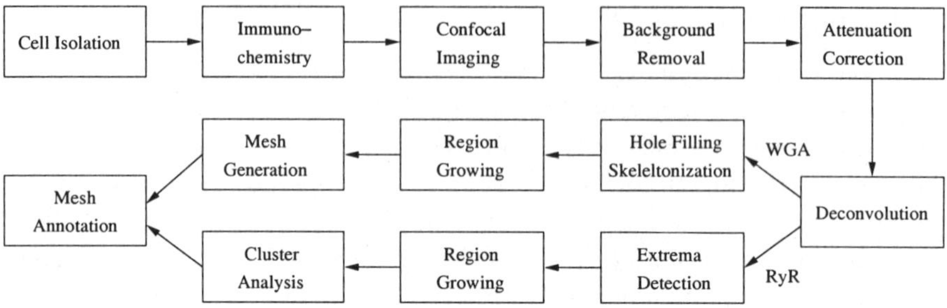

Figure 1. Processing pipeline for reconstruction of RyR cluster distributions and the sarcolemma of cardiac myocytes based on dual labeling and confocal microscopy.

or on end to take advantage of anisotropic optical properties of confocal microscopy.

Confocal imaging has several limitations, in particular anisotropic convolution and depth-dependent signal attenuation. We accounted for these limitations with 3D deconvolution and attenuation correction. Deconvolution was carried out with the Richardson-Lucy algorithm using measured point spread functions (PSFs), which characterize the optical properties of our confocal microscopic imaging system. We applied various methods of image processing to identify the intracellular space and the RyR cluster distribution. The resulting reconstructions were transformed into volume, surface and point meshes which provide a spatial domain for computational studies of EC coupling at sub-cellular level.

2. Methods

2.1. *Overview*

An overview of our approach for modeling of RyR cluster distribution and the sarcolemma is shown in Fig. 1. After isolation, labeling and imaging of myocytes, the image data were processed with various methods of digital image processing for background removal, correction of depth-dependent attenuation, and image deconvolution. Image processing methods were also applied to identify the sarcolemma in the WGA images and segment clusters in the RyR images. The image data were combined and represented with volume and surface meshes for application in computational simulations of EC coupling.

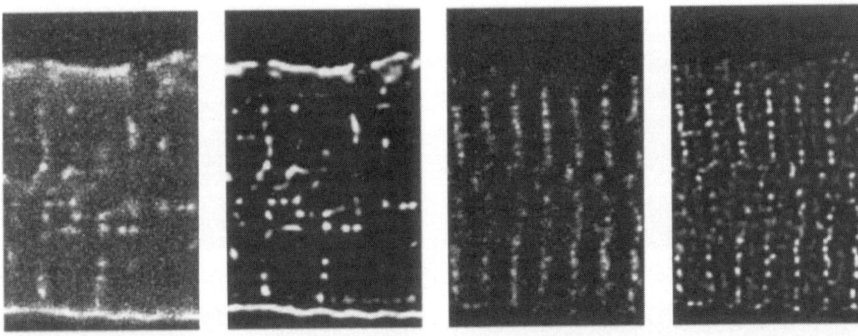

(a) (b) (c) (d)

Figure 2. Exemplary image of segment of ventricular myocyte lying flat on the cover slip. The imaging was performed in a hexahedral region with a size of 50 μm x 30 μm x 10 μm. The image stack was stored in a lattice of 248 x 150 x 51 cubic voxels. Raw intensity distributions for (a) WGA and (c) RyR labeling are shown in the central XY plane. Intensity distributions after background removal, attenuation correction and deconvolution are depicted in (b) and (c) for WGA and RyR labeling, respectively.

2.2. *Isolation, Labeling and Imaging of Cardiomyocytes*

Rabbit ventricular myocytes were obtained according to a previously described procedure[9]. Cells were fixed with 2% para formaldehyde and labeled with WGA conjugated to Alexa 555[22]. Afterwards, cells were permeabilized and RyRs were labeled with a primary antibody and a secondary antibody attached to Alexa 488. This procedure has been described previously in detail[9].

A Zeiss LSM 5 confocal microscope (Carl Zeiss, Jena, Germany) together with a 60x oil immersion objective lens (Numerical aperture: 1.4) was used for imaging. It resulted in 3D image stacks consisting of cubic voxels of size 100 nm x 100 nm x 200 nm, respectively. The data volume of each of the stacks was 25 million voxels.

The imaging of the WGA and RyR signals was performed on myocytes lying flat on the cover slip (Fig. 2). In addition, we imaged cells tilted on end (Fig. 3) using a recently developed transverse sectioning approach[8]. In short, the approach consisted of embedding cells in agar for mechanical stabilization and selection of vertically oriented cells for imaging.

2.3. *Image Pre-Processing*

Our approaches for correction of depth-dependent signal intensity attenuation and deconvolution have been detailed in our previous work [21,22].

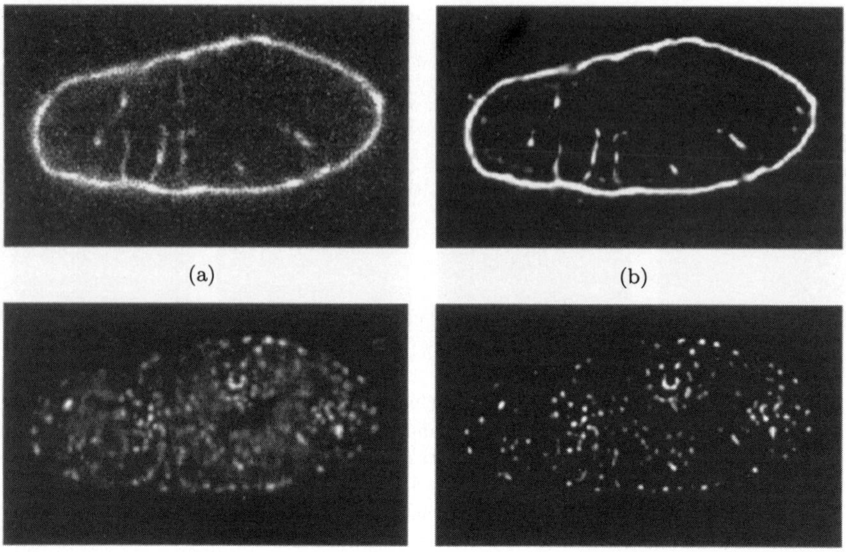

(a) (b) (c) (d)

Figure 3. Exemplary image of segment of ventricular myocyte tilted on end embedded in agar. The imaging was performed in a hexahedral region with a size of 26 μm x 42 μm x 21 μm. The image stack was stored in a lattice of 128 x 212 x 107 cubic voxels. Raw intensity distributions for (a) WGA and (c) RyR labeling are shown in the central XY plane. Intensity distributions after background removal, attenuation correction and deconvolution are depicted in (b) and (d) for WGA and RyR labeling, respectively.

Briefly, a scaling factor was calculated characterizing depth-dependent attenuation for each image slice. The factor was used for slice-wise correction. The iterative Richardson-Lucy algorithm[19] was applied with PSFs extracted from images of fluorescent beads of 100 nm in diameter residing in the first 10 μm above the cover slip.

Removal of background signal was carried out for all WGA and RyR images before deconvolution. For each image we detected the background level in regions outside the cell and without debris. The background level varied between image stacks, but was not dependent on depth. We subtracted the detected background level in each image stack.

2.4. *Processing of WGA Images*

WGA stained the glycocalix associated with the sarcolemma. Thus, WGA images served for identification of the sarcolemma. We applied a 3D thinning operator[17] on the deconvolved image stacks (Fig. 4a). Minor holes in

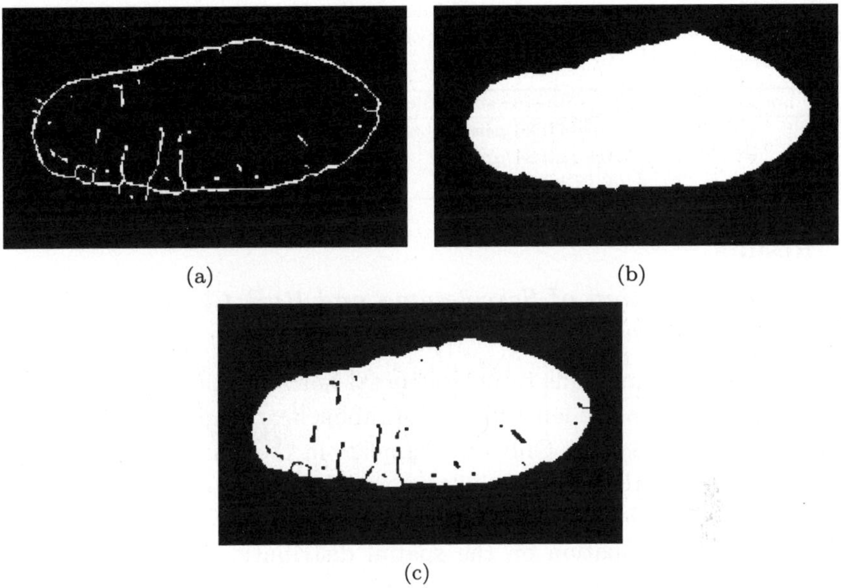

Figure 4. Segmentation of WGA images stacks. (a) A 3D thinning operator created skeletons of deconvolved images. The skeletons exhibited holes in the outer sarcolemma, which were closed before further processing. (b) Region-growing produced a mask image stack containing intracellular space and t-system. (c) The mask image stack was combined with a thresholded segmentation of the WGA image stack.

the resulting skeletons were closed either automatically[1] or by manual editing of the image data. We created a mask image by using the region-growing technique[20] in the skeletons images (Fig. 4b). The mask image contained the intracellular space and the t-system. Subsequently, the mask was applied with thresholded WGA images to create extracellular and intracellular segments (Fig. 4c), whose borders were identified with the sarcolemma.

2.5. *Processing of RyR Images*

Processing of RyR images consisted of segmentation of RyR clusters and their statistical analysis from the deconvolved RyR image stacks. The clusters were segmented with the region-growing technique. Seed points for region growing were determined by thresholding and maxima detection. We derived centers of mass and summary intensities of RyR clusters from the segments.

Table 1. Mean and standard deviation of nearest neighbor distance of RyR clusters in rabbit ventricular myocytes. The distances were calculated from segments of 3 horizontal and 3 vertical cells.

horizontal	distance (mean±stddev)	vertical	distance (mean±stddev)
#1	0.88±0.23 μm	#1	0.88±0.31 μm
#2	0.90±0.24 μm	#2	0.92±0.27 μm
#3	0.91±0.25 μm	#3	0.97±0.29 μm

3. Results

3.1. *Reconstruction of Sarcolemma and RyR Cluster Distribution*

We applied these methods to reconstruct 3 horizontal and 3 vertical cells. The cells were from the left ventricle of rabbit hearts and selected from an image library of 48 cells. Only cell segments in proximity to the cover slip were processed. Processing of cell segments in regions distal (¿20mum) to the cover slip was hindered by low signal to noise ratio.

Statistical information on the spatial distribution of RyR clusters is shown in Tab. 1. The mean nearest neighbor distance of RyR cluster is 0.91 μm. An example of our reconstruction of a vertical ventricular myocyte is shown in Fig. 5. The image dataset includes 128 x 212 x 20 voxels and describes a volume of 12.8 μm x 21.2 μm x 40 μm. This visualization shows RyR cluster associated with the outer sarcolemma and the t-system. The majority of clusters was not associated with the sarcolemma (Fig. 5c).

3.2. *Generation of Meshes for Computational Simulations*

All processing was performed with data stored in 3D lattices. The data was converted to other formats with regard to application in computational simulations. In particular, software allowed for automated generation of cubic voxel meshes of various spatial resolution. Application of a modified marching-cube algorithm on these meshes facilitated the creation of surface meshes[15]. The algorithm generated meshes of triangular elements approximating iso-intensity surfaces with sub-voxel resolution in the 3D image stacks. This approach was used for the visualizations shown in Fig. 5. Modifications of the original algorithm assured closeness of the generated surfaces and permitted sub-voxel resolution by adjusting positions of mesh nodes based on edge-wise interpolation of intensities[12].

The distribution of RyR clusters was represented with a point set. Each point is annotated with the cluster size, second order moments of the cluster, and its distance to the sarcolemma.

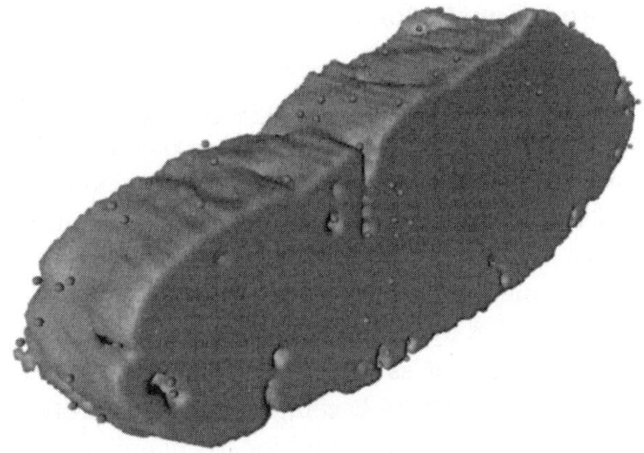

(a)

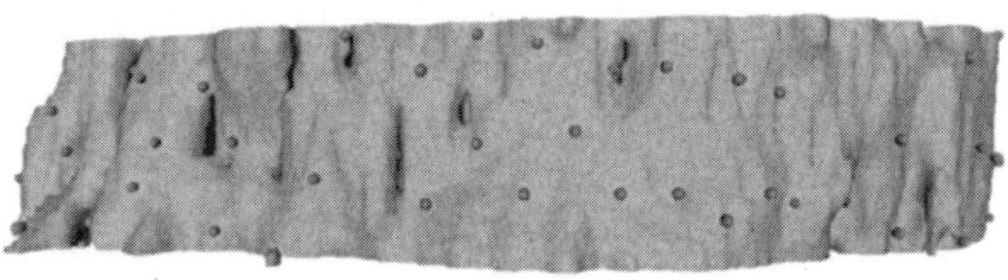

(b)

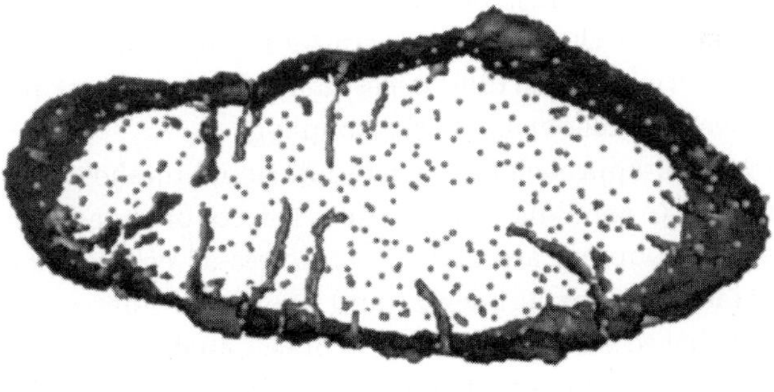

(c)

Figure 5. 3D visualization of cell fragment from different perspectives. (a) The mycoyte interior is shown overlaid with RyR clusters as spheres. (b) A number of RyR clusters was located in the outer sarcolemma. (c) The majority of RyR clusters was found in the cell interior.

4. Discussion and Conclusions

In the present study, we introduced a new approach to reconstruct distributions of RyR clusters and the sarcolemma of cardiac myocytes. The study extends our previous work on modeling of the sarcolemma based on confocal imaging of isolated living myocytes superfused with membrane impermeant dextran conjugated to fluorescein[21]. Our previous work allowed us to create volume meshes of myocytes and surface meshes of sarcolemma suitable for computational simulations. In both studies, our focus on modeling the sarcolemma was motivated by its central role as a border between the intra- and extracellular environment as well as for cell signaling. Modeling of the sarcolemma is significant because it comprises various proteins for cellular signaling such as electrical signaling and for EC coupling.

The present study describes methods to extract 3D reconstructions of the sarcolemma from images of WGA labeled myocytes. WGA has been used in previous studies for identification of the cell membrane[4]. In a previous study, we showed that dextran and WGA signals were correlated in the t-system and are therefore equally good markers[22]. In this and the present study, WGA labeling allowed us to create detailed descriptions of the t-system. More importantly, WGA labeling can be applied in combination with immuno-labeling of proteins, such as RyRs. Thus, the present approach allows modeling of the membrane together with proteins. An alternative approach would be to apply published information on protein density distributions. However, it appears difficult to evaluate the reliability and accuracy of this alternative approach.

Our estimate of nearest neighbor distance (0.91 μm) of RyR clusters in rabbit was higher than those previously reported for rat (0.66 μm) and human (0.78 μm)[27]. These differences might be significant to gain insights in species-specific phenomena of EC coupling.

The presented approach led to computational meshes with applications in modeling of EC coupling in myocytes. Effects of changes in components involved in EC coupling are difficult to assess at cellular and sub-cellular level with traditional experimental and analytical approaches. Computational studies based on realistic models of cell anatomy might give insights in these effects and thus complement the traditional approaches. Modeling of EC coupling is a very active field of research and various sorts of models have been developed, for instance common pool and local control models.

Common pool modeling use a cytosolic reservoir acting as source or sink for all sarcolemmal and sarcoplasmic calcium fluxes. The cytosolic calcium controls RyR gating. Local control models are based on an additional compartment representing a small cleft space (diad) between sarcolemma and junctional SR membrane[29]. The cleft space is sink for calcium fluxes through LCCs and RyRs. The calcium concentration in the cleft space controls RyR gating. The cleft space is connected to the cytosolic compartment. Several studies with local control models focused on description of calcium transients in the cleft space[14,7,25]. We suggest that our cell models provide insights into the validity of EC coupling models and support a hybrid of the two EC coupling models to describe ventricular myocytes.

Our methodology for reconstruction of the sarcolemma including the t-system is related to previously published approaches with rat ventricular cardiac myocytes based on confocal microscopy[26] and myocytes from guinea pig papillary muscle based on thin section electron microscopy[2]. In our studies, we profited from the largeness and low complexity of rabbit versus rat t-system. T-tubule diameter in rabbit ventricular cells was in 400±172 nm[22], which is above the resolution of the confocal imaging system.

Our methodology for reconstructing the sarcolemma together with RyR cluster distributions is new. Related approaches for dual labeling of RyRs and other proteins have been described previously for α-actinin[27], calsequestrin[23], sodium calcium exchangers[23], sodium channels[23], and L-type calcium channels[24,23,6]. Extending our dual labeling approach for labeling of the sarcolemma together with other proteins is straight forward. Of particular interest for modeling of EC coupling is information on the distribution of L-type calcium channels and their relationship to RyRs.

The presented approach can also be applied for modeling cells during development and aging as well as affected by cardiac diseases. Changes of the t-system and RyR cluster distributions have been described for diseased human[30,13] and animal ventricular myocytes[18,11,16,28].

Limitations. We and others have discussed physical limitations of confocal imaging studies and ambiguity of image processing in previous publications[5,22]. A major limitation is spatial resolution of confocal imaging. The resolution can be specified by e.g. the Rayleigh criterion and estimated from measured PSFs. From our measurements[22] we expect that t-tubules and clusters can be separated having a distance larger than 0.83 and 0.28 μm in laser beam direction and transversal to it, respectively.

Our approach necessitated myocyte isolation, fixation and membrane perforation. These procedures affect cell anatomy. We found insignificant myocyte distortion caused by fixation and perforation (data not shown).

References

1. Z. Aktouf, G. Bertrand, and L. Perroton. A 3D-hole closing algorithm. In *Proc. 6th Int Workshop Discrete Geom Comp Imag*, pages 36–47, 1996.
2. J. Amsellem, R. Delorme, C. Souchier, and C. Ojeda. Transverse-axial tubular system in guinea pig ventricular cardiomyocyte: 3D reconstruction, quantification and its possible role in K^+ accumulation-depletion phenomenon in single cells. *Biol Cell*, 85:43–54, 1995.
3. D. M. Bers. *Excitation-Contraction Coupling and Cardiac Contractile Force.* Kluwer Academic Publishers, Dordrecht, Netherlands, 1991.
4. V. P. Bhavanandan and A. W. Katlic. The interaction of wheat germ agglutinin with sialoglycoproteins. The role of sialic acid. *J Biol Chem*, 254(10):4000–8, 1979.
5. S. Bolte and F. P. Cordelieres. A guided tour into subcellular localization analysis in light microscopy. *J Microsc*, 224(3):213–32, 2006.
6. M. D. Bootman, D. R. Higazi, S. Coombes, and H. Llewelyn Roderick. Calcium signalling during excitation-contraction coupling in mammalian atrial myocytes. *J Cell Sci*, 119(19):3915–25, 2006.
7. M. B. Cannell and C. Soeller. Numerical analysis of ryanodine receptor activation by l-type channel activity in the cardiac muscle diad. *Biophys J*, 73:112–22, 1997.
8. Y. Chen-Isu, S. L. McCulle, C. W. Ward, C. Soeller, B. M. Allen, C. Rabang, M. B. Cannell, C. W. Balke, and L. T. Izu. Three-dimensional distribution of ryanodine receptor clusters in cardiac myocytes. *Biophys J*, 91(1):1–13, 2006.
9. J. M. Cordeiro, K. W. Spitzer, W. R. Giles, P. E. Ershler, M. B. Cannell, and J. H. Bridge. Location of the initiation site of calcium transients and sparks in rabbit heart Purkinje cells. *J Physiol*, 531(Pt 2):H1471–9, 2001.
10. A. Fabiato and F. Fabiato. Contractions induced by a calcium-triggered release of calcium for the sarcoplasmic reticulum of single skinned cardiac cells. *J Physiol*, 249:469–95, 1975.
11. J. He, M. W. Conklin, J. D. Foell, M. R. Wolff, R. A. Haworth, R. Coronado, and T. J. Kamp. Reduction in density of transverse tubules and L-type Ca(2+) channels in canine tachycardia-induced heart failure. *Cardiovasc Res*, 49(2):298–307, 2001.
12. W. Heiden, T. Goetze, and J. Brickmann. 'Marching-Cube'-Algorithmen zur schnellen Generierung von Isoflächen auf der Basis dreidimensionaler Datenfelder. In M. Frühauf and Martina Göbel, editors, *Visualisierung von Volumendaten*, pages 112–7. Springer, Berlin, Heidelberg, New York, 1991.
13. R. R. Kaprielian, S. Stevenson, S. M. Rothery, M. J. Cullen, and N. J. Severs. Distinct patterns of dystrophin organization in myocyte sarcolemma and transverse tubules of normal and diseased myocardium. *Circulation*, 101:2586–94, 2000.
14. G. A. Langer and A. Peskoff. Calcium concentration and movement in the diadic cleft space of the cardiac ventricular cells. *Biophys J*, 70:1169–82, 1996.
15. W. E. Lorensen and H. E. Cline. Marching cubes: A high resolution 3D

surface construction algorithm. *Computer Graphics*, 21(4):163–9, 1987.

16. W. E. Louch, H. K. Mork, J. Sexton, T. A. Stromme, P. Laake, I. Sjaastad, and T. J. Kamp. T-tubule disorganization and reduced synchrony of Ca2+ release in murine cardiomytes following myocardial infarction. *J Physiol*, 574:301–23, 2006.
17. A. Manzanera, T. M. Bernard, F. Preteux, and B. Longuet. Medial faces from a concise 3D thinning algorithm. In *Proc. IEEE Conf. Computer Vision*, pages 337–43, 1999.
18. E. Page and L. P. McCallister. Quantitative electron microscopic description of heart muscle cells. applications to normal, hypertrophied and thyroxin-stimulated hearts. *Am J Cardiol*, 31(31):172–81, 1973.
19. W. H. Richardson. Bayesian-based iterative method of image restoration. *J Opt Soc Am*, 62:55–9, 1972.
20. F. B. Sachse. *Computational Cardiology: Modeling of Anatomy, Electrophysiology, and Mechanics*, volume 2966 of *Lecture Notes in Computer Science*. Springer, Heidelberg, 2004.
21. F. B. Sachse, E. Savio-Galimberti, J. I. Goldhaber, and J. H. B. Bridge. Submicrometer anatomical models of the sarcolemma of cardiac myocytes based on confocal imaging. In *Pac Symp Biocomp*, volume 13, pages 390–401, 2008.
22. E. Savio-Galimberti, J. Frank, M. Inoue, J. I. Goldhaber, M. B. Cannell, J. H. B. Bridge, and F. B. Sachse. Novel features of the rabbit transverse tubular system revealed by quantitative analysis of three-dimensional reconstructions from confocal images. *Biophys J*, 95:2053–62, 2008.
23. D. R. L. Scriven, P. Dan, and E. W. Moore. Distribution of proteins implicated in excitation-contraction coupling in rat ventricular myocytes. *Biophys J*, 79:2682–91, 2000.
24. F. Sedarat, L. Xu, E. W. Moore, and G. F. Tibbits. Colocalization of dihydropyridine and ryanodine receptors in neonate rabbit heart using confocal microscopy. *Am J Physiol Heart Circ Physiol*, 279:H202–9, 2000.
25. C. Soeller and M. B. Cannell. Numerical simulation of local calcium movements during l-type calcium channel gating in the cardiac diad. *Biophys J*, 73:97–111, 1997.
26. C. Soeller and M. B. Cannell. Examination of the transverse tubular system in living cardiac rat myocytes by 2-photon microscopy and digital image-processing techniques. *Circ Res*, 84:266–75, 1999.
27. C. Soeller, D. Crossman, R. Gilbert, and M. B. Cannell. Analysis of ryanodine receptor clusters in rat and human cardiac myocytes. *Proc Natl Acad Sci USA*, 104(38):14958–63, 2007.
28. L. S. Song, E. A. Sobie, S. McCulle, W. J. Lederer, C. W. Balke, and H. Cheng. Orphaned ryanodine receptors in the failing heart. *Proc Natl Acad Sci USA*, 103:4305–10, 2006.
29. M. D. Stern. Theory of excitation-contraction coupling in cardiac muscle. *Biophys J*, 63:497–517, 1997.
30. C. Wong, C. Soeller, L. Burton, and M. B. Cannell. Changes in transverse-tubular system architecture in myocytes from diseased human ventricles. In *Biophys J (Annual Meeting Abstracts)*, volume 82, page a588, 2002.

A MULTISCALE APPROACH TO SAMPLING NASCENT PEPTIDE CHAINS IN THE RIBOSOMAL EXIT TUNNEL

V. A. VOELZ, P. PETRONE and V. S. PANDE*

Department of Chemistry, Stanford University,
Stanford, CA 94305, USA
**E-mail: pande@stanford.edu*
www.stanford.edu

We present a new multiscale method that combines all-atom molecular dynamics with coarse-grained sampling, towards the aim of bridging two levels of physiology: the atomic scale of protein side chains and small molecules, and the huge scale of macromolecular complexes like the ribosome. Our approach uses all-atom simulations of peptide (or other ligand) fragments to calculate local 3D spatial potentials of mean force (PMF). The individual fragment PMFs are then used as a potential for a coarse-grained chain representation of the entire molecule. Conformational space and sequence space are sampled efficiently using generalized ensemble Monte Carlo. Here, we apply this method to the study of nascent polypeptides inside the cavity of the ribosome exit tunnel. We show how the method can be used to explore the accessible conformational and sequence space of nascent polypeptide chains near the ribosome peptidyl transfer center (PTC), with the eventual aim of understanding the basis of specificity for co-translational regulation. The method has many potential applications to predicting binding specificity and design, and is sufficiently general to allow even greater separation of scales in future work.

Keywords: multi-scale modeling, molecular dynamics, ribosome exit tunnel, generalized ensemble, translation, PMF, coarse-grain, Monte Carlo.

1. Introduction

It is often the case that we wish to predictively simulate the molecular interaction of something small (a peptide or other ligand) with something very large (a macromolecular complex). Molecular dynamics (MD) simulations that capture chemical atomic detail are crucial to this goal. For instance, the effect of individual water molecules near a surface may be critical to ligand affinity, yet the computational cost of simulating large systems in atomic detail can be a practical impossibility. New multiscale methods are essential to bridging the atomic scale with the much larger scale of macromolecular complexes.

One example of such a system is the ribosomal exit tunnel. The ribosome is a complex catalyst for the synthesis of new proteins. As nascent polypeptide chains are synthesized at the ribosome peptidyl transferase site (PTC), they emerge from the ribosome only after traversing a long tunnel.[1,2]

The ribosome exit tunnel is believed to act as a discriminating gate for protein sequences. Some peptide sequences have been shown to regulate their own translation process by directly interacting with components of the exit tunnel as they are being translate although the specific details of this molecular mechanism is unclear. For example, stalling of the SecM[3–5] and tnaC[6,7] peptides requires critical residues in the nascent peptide: a terminal Pro near the PTC site and a Trp near the constriction site, between ribosomal proteins L4 and L22 which protrude into the cavity at the tunnels narrowest point. Macrolide antibiotics are also known to bind near the constriction site, arresting translation, and inducing conformational changes in L22 observed in crystal structures.[8]

What is the molecular mechanism for peptide recognition in these stalling events? All-atom computer simulation is key to understanding this. There have been previous efforts to simulate the ribosome[9] in atomic detail, yet this remains a challenge for all-atom molecular dynamics, due to its enormous size. Furthermore, studying sequence specificity requires multiple simulations to compare different sequences, and accurate free energy calculations incur even more computational cost.

Here, we present a multiscale simulation approach we call the Multi-Scale Fragment Potential (MSFP) method, that aims to overcome some of these challenges. In MSFP, the ligand of interest is divided into fragments (a peptide into amino acids, for instance). First, all-atom molecular dynamics simulation is used to calculate the affinity of each fragment along the binding surface or other molecular environment as a 3D spatial potential of mean force (PMF). The fragment potentials are then used to locally bias conformational sampling in a coarse-grained model of the full ligand (a polypeptide chain, for example), allowing for efficient sampling of conformations and fragment sequences (different orderings and combinations of fragments). If desired, this process can be repeated iteratively, using predicted sequences and conformations from the coarse-grained sampling to seed further all-atom simulations.

In this report, we present a proof-of-principle study, using a reduced alphabet of single amino acid side chain analog PMFs constructed from large-scale MD simulations of single amino acid probes sampling the entire cavity of the ribosomal exit tunnel. In principle, the MSFP method is not limited to any particular molecular system. Here we use PMF maps of the ribosomal exit tunnel as a realistic example.

For the coarse-grained sampling, we use an efficient generalized ensemble Monte Carlo technique, which we describe in the Methods section. We demonstrate, for short polypeptide chains, how MSFP can obtain accurate estimates of free energy differences between sequences and conformational states, and with and without the inuence of the ribosomal exit tunnel PMFs. We discuss the differences between MSFP and existing multiscale methods, the approximations assumed in the method, and future directions of this work.

2. Methods

2.1. *Theory*

Here we rigorously define the approximations in the multiscale fragment approach. As we will see, there is a natural decomposition to the approximation that corresponds to the additive terms in the coarse-grained potential used in the MSFP method. This theoretical treatment is useful for substantiating the the method, and is a useful starting point for making accurate approximations.

2.1.1. *The fragment approximation*

Consider the Hamiltonian $\mathcal{H}(\mathbf{x})$ for a molecular system described by coordinates $\mathbf{x} = \{\mathbf{p}, \mathbf{s}, \mathbf{e}\}$, where $\mathbf{p}$ represents a set of protein coordinates, $\mathbf{s}$ represents a set of solvent coordinates, and $\mathbf{e}$ represents a set of 'environment' coordinates, for example, the atoms of a protein receptor or other molecular surface (Figure 1).

$$\mathcal{H}(\mathbf{x}) = U_{pp}(\mathbf{p}) + U_{ps}(\mathbf{p}, \mathbf{s}) + U_{pe}(\mathbf{p}, \mathbf{e}) + U_{ss}(\mathbf{s}) + U_{ee}(\mathbf{e}) + U_{se}(\mathbf{s}, \mathbf{e})$$

Here, U_{pp} is the potential from the protein self-interactions, U_{ps} is the potential due to all protein-solvent interactions, U_{pe} is due to all protein-environment interactions, and so on. (Here, we are considering the ligand to be a polypeptide, but the treatment is general.)

Consider the partition function for this system at inverse temperature $\beta = 1/k_BT$ when we integrate out the solvent and environment coordinates to achieve a description of $Z(\mathbf{p})$, the partition function as function of the

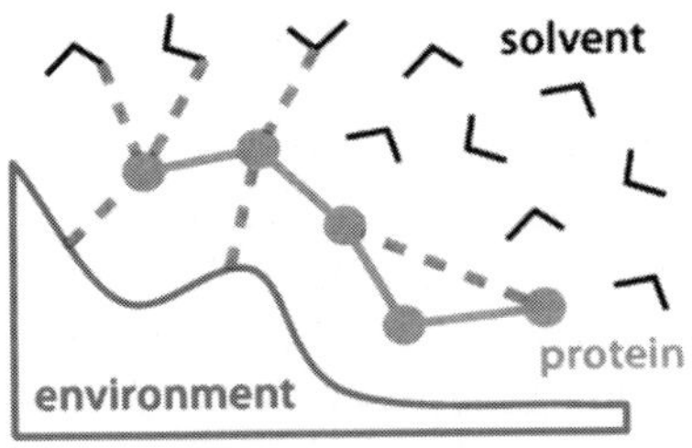

Fig. 1. A molecular system is decomposed into protein, solvent, and environment coordinates.

protein coordinates alone.

$$Z(\mathbf{p}) = \int d\mathbf{s} \int d\mathbf{e} \exp(-\beta \mathcal{H}(\mathbf{p}, \mathbf{s}, \mathbf{e})) \tag{1}$$

For the time-being, we will consider the case where the environment coordinates are fixed with respect to the protein and solvent. In this case, U_{ee} is constant and can be ignored, while U_{pp} and U_{pe} depend only on the protein coordinates, and U_{se} depends only on the solvent coordinates. This yields

$$Z(\mathbf{p}) = Z_p(\mathbf{p}) Z_s(\mathbf{p}) \tag{2}$$

where

$$Z_p(\mathbf{p}) = \exp(-\beta[U_{pp}(\mathbf{p}) + U_{pe}(\mathbf{p})]) \tag{3}$$

$$Z_s(\mathbf{p}) = \int d\mathbf{s} \exp(-\beta[U_{ps}(\mathbf{p}, \mathbf{s}) + U_{ss}(\mathbf{s}) + U_{se}(\mathbf{s})]) \tag{4}$$

Here, $Z_p(\mathbf{p})$ is the Boltzmann factor due to protein self-interactions and the environment, and $Z_s(\mathbf{p})$ can be thought of as a solvation free energy term. $Z_s(\mathbf{p})$ contributes an extra "weight" to each conformation $\mathbf{p}$ due to interactions with the solvent.

Next, we introduce the fragment approximation. Suppose we have estimates for the partition functions $Z(\mathbf{p}_i) = Z_p(\mathbf{p}_i) Z_s(\mathbf{p}_i)$ over a set of disjoint fragments $\{\mathbf{p}_1, \mathbf{p}_2, ..., \mathbf{p}_n\} = \mathbf{p}$. If we assume that each fragment is independent, then the partition function for the entire protein can be approximated as

$$Z(\mathbf{p}) \approx \prod_i Z(\mathbf{p}_i)$$

To correct this approximation for the true $Z(\mathbf{p})$, we would have to multiply by a correction factor

$$C(\mathbf{p}) = \frac{Z(\mathbf{p})}{\prod_i Z(\mathbf{p}_i)}$$

This correction factor represents the cooperativity of fragments as a function of the protein position $\mathbf{p}$. Any non-additivity in the free energies is accounted for by $F_{coop}(\mathbf{p}) = -\beta^{-1} \log C(\mathbf{p})$. Note that we can separate the protein non-additivity of fragments and the solvent non-additivity:

$$C(\mathbf{p}) = C_p(\mathbf{p}) C_s(\mathbf{p}) = \frac{Z_p(\mathbf{p})}{\prod_i Z_p(\mathbf{p}_i)} \cdot \frac{Z_s(\mathbf{p})}{\prod_i Z_s(\mathbf{p}_i)}$$

If the protein-protein interactions are represented by an additive pairwise potential (as with most all-atom forcefields), then we can further reduce the cooperativity correction factor to the following form:

$$C(\mathbf{p}) = \exp[-\beta \sum_{i \neq j} U_{pp}(\mathbf{p}_i, \mathbf{p}_j)] \frac{Z_s(\mathbf{p})}{\prod_i Z_s(\mathbf{p}_i)}$$

A similar approach, with the environment coordinates not fixed, leads to a similar expression, but with $Z_s(\mathbf{p})$ substituted with $Z_{se}(\mathbf{p}) = \int\int d\mathbf{s} \exp(-\beta[U_{ps}(\mathbf{p}, \mathbf{s}) + U_{ss}(\mathbf{s})]) d\mathbf{e} \exp(-\beta[U_{pe}(\mathbf{p}, \mathbf{e}) + U_{se}(\mathbf{s}, \mathbf{e}) + U_{ee}(\mathbf{e})])$

Now we can interpret the terms in the correction factor to have some intuitive meaning. The Boltzmann factor term on the left accounts for purely inter-residue interactions, mostly protein chain connectivity. The term on the right accounts for cooperativity (between fragments) due to solvent effects (hydrophobicity, charge screening, etc., and any mutual interaction of the solvent with the environment such as surface tension, molecular shape, etc.).

These two effects are exactly what we must add to the fragment potentials in order to construct an accurate coarse-grained model. In this paper, we use a simple dihedral potential with chain connectivity (described below) as a first approximation to a more detailed correction.

It is worth noting some useful features of the theory above. One is that it provides a framework for making successive many-body (2-, 3- and 4-body, say) corrections to the MSFP method, to systematically make corrections to a coarse-grained potential to improve the approximations. One way to do this is to simulate larger fragments (i.e. groups of smaller fragments) to quantify many-body interactions, at the cost of much more simulation. This kind of iterative approach is also a useful way to quantify the accuracy of any given coarse-grained approximation.

2.2. *MD simulation of amino acid analogs in the ribosomal exit tunnel*

The MD simulation protocol is described as in.[10] The model is built from a cut-out of the *Haloarcula marismortui* X-ray structure (pdb 1S72[11]), including the ribosome tunnel and nearby areas, explicit water and ions (Figure 3A). The inner surface of the ribosome tunnel was probed with small molecule analogs to amino acid sidechains. All-atom molecular simulations using the AMBER forcefield were performed with probe molecules exhaustively umbrella-sampled at a series of restrained grid points, which were then analyzed using WHAM,[12,13] yielding comprehensive maps of the 3D potential of mean force (PMF) for the single amino acid sidechain analogs throughout the ribosomal exit tunnel. The resolution of these maps is 1Å, in the center-of-mass coordinate of the sidechain. It is important to note that while these 3D PMF maps are constructed from small molecule sampling, they are designed to capture the interactions of solely the single amino acid side chain with the chemical environment of the tunnel, including the effect of the solvent.

2.3. *Coarse-grained MC sampling of protein chain*

A Monte Carlo scheme was used to broadly sample conformational and sequence space under the influence of a coarse-grained potential. In this scheme, the peptide chain object is an all-atom structure, but with fixed bond lengths, which is convenient for transporting conformations from the coarse-grained phase back into all-atom simulations (Figure 3B). The form of the potential is

$$U = U_{steric} + U_{trans} + \sum_{n=1}^{N} U_{dih}(x_n, s(n)) + PMF(x_n, s(n))$$

where $x_1, x_2, ...x_n$ are the coordinates, and $s(n)$ the amino acid, of each residue for chain index n. U_{steric} is a steric repulsion term calculated over all atoms in the structure as $\sum_{i \neq j} \Theta(r_{ij} - r_0)(\sigma_{ij}/r_{ij})^6$, where r_{ij} is the distance between atoms i and j, Θ is the Heaviside step function, and σ_{ij} are standard atomic radii. r_0 is set to 1.2Å. U_{trans} is a harmonic potential that allows translations about an anchor point, which we tether to the backbone carbon of the last residue in the chain (nearest to the PTC). The PMF potential is a function of the side chain centroid, and is evaluated by linear interpolation from the grid of stored values. There is one PMF grid for each residue type: ALA, ASP, ILE, LYS, TRP. U_{dih} is a backbone dihedral potential compiled from the *top500* database.[14] Note that in this model, there is no interaction of the chain with itself, except for simple sterics, respecting the distribution of dihedral angles.

Allowed move sets for the coarse-grained chain include: (i) chain growth or deletion by one residue, at the N-terminal end of the chain (ii) backbone dihedral (ϕ,ψ) rotations, (iii) random perturbations to side chain χ angles, (iv) mutation of a residues to a new amino acid, (v) translation of the entire chain.

The nascent peptide anchor location has been defined based on the crystal structure of a pre-translocational intermediate of *Haloarcula marismortui* (pdb 1KQS[15]) which includes the tRNA analog A76. We structurally aligned the pre-translocational intermediate to the ribosome structure on which we based our model (pdb 1S72[11]), and find little variation in the area of the PTC site, with especially the A and P loops structurally unchanged. It is believed that the polypeptide is attached close to atom 3OH of the tRNA analog A76.[15] We use the location of this atom as an anchor point for the growing polypeptide chain in our MSFP simulation.

2.4. *Flat-histogram generalized ensemble sampling*

For enhanced efficiency, we sample from what can be called a *generalized ensemble*.[16,17] Instead of a single potential, we sample from an ensemble of potentials, U_k, each with index k representing a binned macrostate of the chain that we wish to sample over. For this study, we have chosen states $k \equiv (s, N)$, each representing a chain of length N where the Nth residue is amino acid s.

Consider a generalized ensemble of Hamiltonians

$$\mathcal{H}_k(x) = \beta U_k(x) + g_k$$

where $\beta = 1/k_B T$, k_B is the Boltzmann constant, T is temperature, and we have perturbed each Hamiltonian $\mathcal{H}_k(x)$ by a weighting term g_k. The generalized partition function for the entire ensemble is

$$Z = \sum_k \sum_x \exp(-\mathcal{H}_k(x)) = \sum_k Z_k \exp(-g_k)$$

where Z_k is the unperturbed partition function for the ensemble indexed by k. To achieve uniform sampling across states k, we use the flat-histogram method of Wang and Landau[18] to iteratively converge upon weights $g_k = \log Z_k + C$, for some arbitrary constant C. Once converged, the free energy of state k is obtained as $-k_BT \cdot g_k$. In this algorithm, the number of times a state k is visited is stored as histogram counts W_k. Initially, all the g_k are equal. Each time a state k is visited, g_k is incremented by a small amount δ, and the histogram count W_k is increased by one. If the histogram is sufficiently flat (when all $W_k \geq \gamma\langle W_k\rangle$), δ is scaled to $\delta \leftarrow \alpha\delta$, where $0 < \alpha < 1$, and $0 < \gamma < 1$.

3. Results

3.1. *Validation of the sampling algorithm*

We simulated the sequence $(Ala)_5$, using a sterics-only (no external PMF) coarse-grained potential, to validate the performace of the sampling. Here, the sequence stays fixed, but the chain is allowed to grow or shrink by one monomer at a time. The spring constant for the C-terminus tether was set as to produce fluctuations smaller than $0.1\mathring{A}$ in the the anchored atom position. The flat-histogram method ensures that that the simulation converges to a random walk in chain length.

In Wang-Landau histogram method, there is a trade-off in the initial choice of the weight increment δ. For accuracy, δ should be kept small, to achieve good resolution in the resulting histograms, as the free energy estimates are influenced more heavily by the sampling early in the simulation. However, as δ becomes small, the rate of convergence slows. By trial and error with the ALA_5 sequence, we found that $\delta = 0.001$, $\gamma = 0.95$, and $\alpha = 0.9$, with weights updated every 100 MC steps, achieved a good balance and accuracy and speed. The free energies reach convergence within a few hours of real-time computation on a single processor of an 3 GHz Intel Xeon MacPro.

Because the only terms in the potential are dihedral preference and sterics, the only determinant of the free energies in this case is the accessible conformational volume, which should be roughly equal. After about 2 million MC steps, the computed free energies as a function of chain length converge to within $< 0.1k_BT$ from each other (Figure 2). These remaining free energy differences are due to both the accuracy resolution of the technique, and the slight differences in excluded volume for different chain lengths.

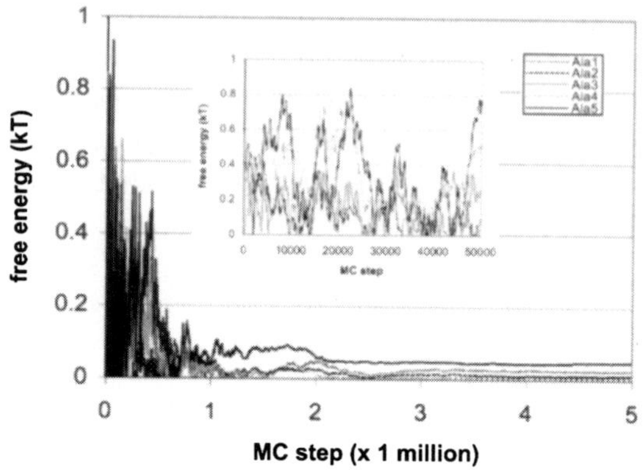

Fig. 2. Chain-growth sampling of Ala_5 shows convergence of generalized ensemble sampling to within $0.1k_BT$ the predicted values . The free energies are plotted with a reference state so that the lowest free energy is set to zero.

3.2. *Sampling conformational and sequence space of peptide pentamers*

Generalized-ensemble Monte Carlo using MSFP was performed across all sequences from the 5-letter alphabet of residues (A,D,I,K,W) and chain lengths ranging from 1 to 5 monomers, in both the absence and the presence of the computed PMF potential. The macrostates $k = (N, s)$ were used for the Wang-Landau flat-histogram algorithm. The MC parameters were those derived from our previous validation simulations above.

Pentamer sampling in the absence of the PMF (with dihedrals and sterics), produced an equilibrated trajectory, but with fluctuations in the weights slow to converge due to the number of possible macrostates. As expected, free energies were determined by the accessible conformational volume of each residue. But because the weight updating increment is not decreased until a sufficiently flat sampling histogram has been achieved, the rate-limiting step for weight convergence is the sampling transition rates to unfavorable states. In this case, mutations to the bulky tryptophan residue had the lowest acceptances.

Pentamer sampling in the presence of the PMFs showed similar convergence behavior, but also showed strong specificity for particular peptide sequences and conformations during the coarse-grained trajectories that were simulated. Most notably, the sequence ADDAA sampled a conformational basin that was enthalpically favored by about 30 kcal/mol (Figure 4).

With more sampling, and complete analysis will allow us to determine the entropic components of this binding affinity that our model predicts. For sequences predicted to have high affinity such as these, we can transport conformations back into detailed all-atom MD simulations of the ribosome, to test the predictions of MSFP. The role of water and/or entropic stabilization of peptides confined within larger macromolecules is important to many fundamental biological processes, and is an area of active research.

Some of the convergence issues we observe with the Wang-Landau algorithm can be ameliorated by augmenting the weight-updating procedure,[19] or using more sophisticated generalized ensemble methods.[16,20,21] Reference states for fragment PMFs may need to be validated and adjusted for optimal convergence.

4. Discussion

Many approaches have been developed for multiscale simulation,[22,23] as well as coarse-grained models for proteins.[24] The main goal of much of this work has been to develop accurate models that require fewer degrees of freedom, in order to speed up molecular simulations with minimal loss of accuracy in inter-particle interaction. Many of these methods rely on atomically detailed reference simulations. The Multiscale Coarse-Graining Method (MS-CG)[25,26] is an approach for generating a consistent[27,28] coarse-grained approximation to a many-body potential by using a force-matching technique from reference simulations. Inverse Monte Carlo approaches have also been used to optimize potentials for coarse-grained models using distributions calculated from atomistic reference simulations[29,30]

Two kinds of information are used to build a coarse-grained potential using the MSFP method: a potential of mean force derived from atomically detailed fragment simulations, and a potential that accounts for fragment cooperativity, mainly due to chain connectivity (although other cooperative effects are important). While large-scale fragment simulations are computationally feasible by a variety of methods,[10,31] detailed simulation of a ligand's self-interaction within its proper context (in this case, a nascent polypeptide in the unique environment of the ribosomal exit tunnel) can be prohibitively expensive. With the MSFP approach as presented here, we have instead modeled the polypeptide as a simple dihedral model with excluded volume to capture chain connectivity and sequence dependent torsional propensities. The statistical mechanical framework we present suggests a straightforward way to iteratively improve this simple chain potential to arbitrary accuracy, by using multi-body correction factors calculated from atomically detailed simulations.

In this paper, we have shown how the MSFP method can efficiently sample conformational and sequence space for polypeptides in a coarse-grained potential. Using generalized ensemble Monte Carlo sampling, we can obtain accurate estimates of free energy differences between chain macrostates. For this test problem, we have shown that some sequences are much more preferred than others, suggesting that MSFP could be a generally useful tool for ligand and/or peptide design that overcomes some of the traditional limitations of all-atom simulation.

Acknowledgments

We wish to thank M. Scott Shell, John Chodera, Xuhui Huang, the contributers of the Folding@Home distribution computing resource, and the Stanford Bio-X cluster. Authors gratefully acknowledge support from NIH and NSF, in particular Simbios (NIH U54 GM072970) and NSF FIBR (NSF EF-0623664).

References

1. N. R. Voss, M. Gerstein, T. A. Steitz and P. B. Moore, *Journal of Molecular Biology* **360**, 893 (2006).
2. A. S. Mankin, *Trends in Biochemical Sciences* **31**, 11 (2006).
3. H. Nakatogawa and K. Ito, *Cell* **108**, 629 (2002).
4. H. Muto, H. Nakatogawa and K. Ito, *Molecular Cell* **22**, 545 (2006).
5. H. Nakatogawa and K. Ito, *ChemBioChem* **5**, 48 (2004).
6. L. R. Cruz-Vera, S. Rajagopal, C. Squires and C. Yanofsky, *Molecular Cell* **19**, 333 (2005).
7. F. Gong, K. Ito, Y. Nakamura and C. Yanofsky, *Proceedings of the National Academy of Sciences* **98**, 8997 (2001).
8. T. Auerbach, A. Bashan and A. Yonath, *Trends in Biotechnology* **22**, 570 (2004).
9. K. Y. Sanbonmatsu and C.-S. Tung, *Journal of Structural Biology* **157**, 470(Mar 2007).
10. P. M. Petrone, C. D. Snow, D. Lucent and V. S. Pande, *Proceedings of the National Academy of Sciences* **in press** (2008).
11. N. Ban, P. Nissen, J. Hansen, P. B. Moore and T. A. Steitz, *Science* **289**, 905 (2000).
12. S. Kumar, D. Bouzida, R. Swendsen, P. Kollman and J. Rosenberg, *Journal of Computational Chemistry* **13**, 1011 (1992).
13. C. Bartel and M. Karplus, *Journal of Computational Chemistry* **18**, 1450 (1997).
14. S. C. Lovell, I. W. Davis, I. W.B. Arendall, P. I. W. de Bakker, J. M. Word,

M. G. Prisant, J. S. Richardson and D. C. Richardson, *Proteins: Structure, Function and Genetics* **50**, 437 (2003).
15. T. M. Schmeing, A. C. Seila, J. L. Hansen, B. Freeborn, J. K. Soukup, S. A. Scaringe, S. A. Strobel, P. B. Moore and T. A. Steitz, *Nature Structural Biology* **9**, 225 (2002).
16. S. Park, *Phys. Rev. E* **77**, p. 016709 (2008).
17. A. P. Lyubartsev, A. A. Martsinovski, S. V. Shevkunov and P. N. Vorontsov-Velyaminov, *The Journal of chemical physics* **96**, 1776 (1992).
18. F. Wang and D. P. Landau, *Physical Review E* **64**, p. 056101 (2001).
19. M. S. Shell, P. G. Debenedetti and A. Z. Pangiotopoulos, *Proteins: Structure, Function and Genetics* **62**, 232 (2006).
20. S. Park, D. L. Ensign and V. S. Pande, *Physical Review E* **74**, p. 066703 (2006).
21. X. Huang, G. R. Bowman and V. S. Pande, *Journal of Chemical Physics* **128**, p. 205106 (2008).
22. G. S. Ayton, W. G. Noid and G. A. Voth, *Current Opinion in Structural Biology* **17**, 192 (2007).
23. M. Praprotnik, L. D. Site and K. Kremer, *Annual Reviews of Physical Chemistry* **59**, 545 (2008).
24. V. Tozzini, *Current Opinion in Structural Biology* **15**, 144 (2005).
25. S. Izvekov and G. A. Voth, *Journal of Physical Chemistry B* **109**, p. 2005 (2005).
26. W. Noid, J. Chu, G. Ayton and G. Voth, *J. Phys. Chem. B* **111**, 4116(Jan 2007).
27. W. Noid, J. Chu, G. Ayton, V. Krishna and S. Izvekov, *J Chem Phys* **128**, p. 244114(Jan 2008).
28. W. Noid, P. Liu, Y. Wang, J. Chu and G. Ayton, *J Chem Phys* **128**, p. 244115(Jan 2008).
29. A. Lyubartsev and A. Laaksonen, *Phys. Rev. E* **52**, 3730(Jan 1995).
30. A. Lyubartsev, *European Biophysics Journal* **35**, 53 (2005).
31. M. Clark, F. Guarnieri, I. Shkurko and J. Wiseman, *J Chem Inf Model* **46**, 231(Jan 2006).

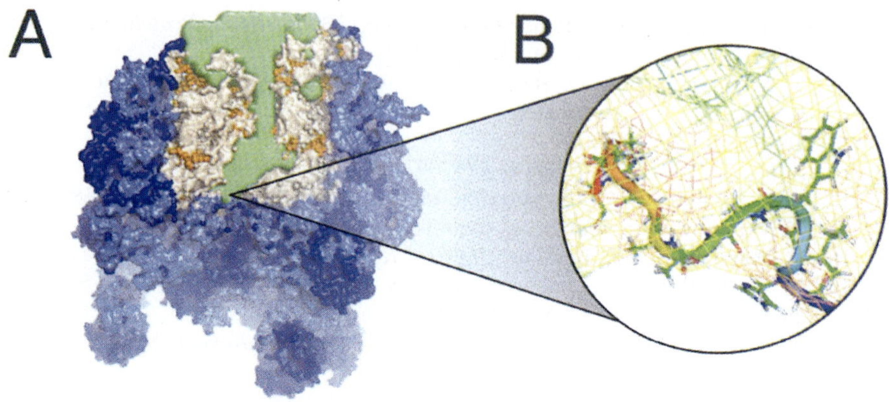

Fig. 3. (A) The cutout of the ribosome used for molecular simulation. In green is the accessible volume of the exit tunnel. (B) A close-up view of a conformation sampled by the coarse-grained chain potential using MSFP.

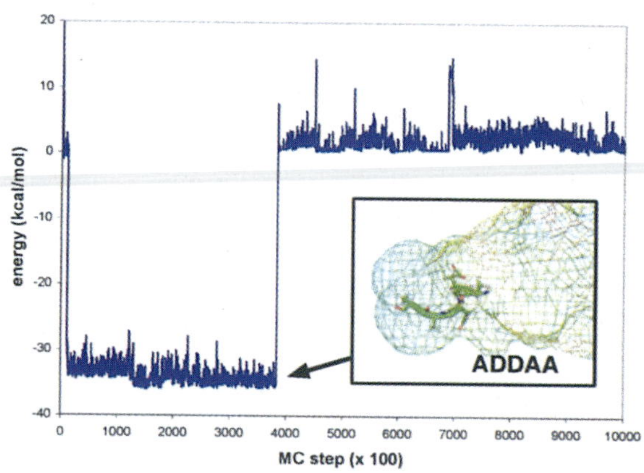

Fig. 4. An energy trace of MSFP sampling of nascent chains in the ribosomal exit tunnel. Generalized ensemble Monte Carlo sampling was performed across all sequences from the alphabet (A,D,I,K,W) and across all chain lengths up to 5-mers, in the presence of MD-derived PMFs for each residue type. A low-energy conformation for the sequence ADDAA was found to be exceptionally stable.

KNOWLEDGE-DRIVEN ANALYSIS AND DATA INTEGRATION FOR HIGH-THROUGHPUT BIOLOGICAL DATA

M. F. OCHS[1], J. QUACKENBUSH[2], R. DAVULURI[3], H. RESSOM[4]

[1]*The Sidney Kimmel Comprehensive Cancer Center, Johns Hopkins University, Baltimore, MD 21205, USA*, E-mail: mfo@jhu.edu,

[2]*Dana Farber Cancer Institute, Harvard Medical School, Boston, MA, USA*, E-mail: johnq@jimmy.harvard.edu

[3]*The Wistar Institute, Philadelphia, PA, USA*, E-mail: rdavuluri@wistar.org,

[4]*Lombardi Cancer Center, Georgetown University, Washington, DC, USA*, E-mail: hwr@georgetown.edu,

Keywords: Genomics, Controlled vocabulary, Database, Bayesian analysis

Introduction

There is a data explosion overtaking medicine and biology due to the development of high-throughput technologies for measuring most molecular constituents of cells. Novel sequencing approaches are now nearing the $1000 genome. SNPchips already measure individual genetic variation for the same cost, and microrarrays provide a global picture of the transcripts, with miRNAs and exon level analyses beginning. Proteomic measurements, both from mass spectrometry and from antibody arrays, are probing increasing amounts of the proteome, and NMR techniques are doing the same for metabolic components. The goal is a deeper view into the complexity of biological systems, with the ultimate prize being personalized medicine and finely targeted therapies.

The deluge of data has created new problems, some held in common with older fields dealing with high-throughput data (e.g., particle physics) and others unique to the field (i.e., context dependent data from cell types, environment, etc.). Already it is impossible for any individual or research group to review the available data, bring it together, integrate it across the various molecular domains either directly, such as DNA exons to specific mRNA variant, or in context, such as by pathway, and analyze it. It is necessary to develop methods to encode data in forms usable by automated systems, to encapsulate our knowledge of the biology and medicine, and to leverage this knowledge during analysis.

Data integration can be done at a number of levels.[1] Encoding of data will ideally use a formal ontology that encodes our knowledge,[2] and at least will utilize standards as provided by controlled vocabularies[3] or mediated schemas.[4] Future data systems are likely to comprise both data warehouses, for subsets of data within a single institution,[5] and data federation, for automated querying of data from multiple sources.[6]

This session focuses on steps along the path to the integration of genomic data, the automated capture of biological information, the development of analyses that leverage biological knowledge, and the creation of integrated and federated data resources.

Papers

The first three papers in this session focus on integrating data from multiple data sources. Carey and Gentleman present a new data structure within the R/Bioconductor system that integrates SNP, expression, annotation, and phenotypic data and which links to the many existing Bioconductor tools for analysis, as well as to the UCSC genome browser for visualization. Hart and Mukhyala describe the UNISON database system and web interface, which integrates proteomics data from many sources, precomputes structures and other values, and provides a rich set of query tools. Shen-Orr and colleagues introduce a knowledgebase of the adaptive immune system and its cytokine mediated cell-cell interactions. This knowledgebase was generated through text mining using an approach focused on a minimal number of terms and their synonyms.

The next two papers integrate data across multiple species to gain insight into cellular processes. Bidaut and Stoeckert integrate expression data from mouse and human tissues using homologene and cell hierarchy. They then utilize an artificial neural network classifier to deduce transcriptional signatures for different points along the differentiation of stem cells. Fox, Taylor and Slonim link protein interaction data across four species and identify novel features of hub proteins, showing that the interaction partners of hub proteins are more conserved across evolution than other proteins.

The final four papers utilize prior information from biological studies to refine biomarker discovery and genome wide association studies. Phan and colleagues refine support vector machine learning of biomarkers by modifying the distance metric comparing expression profiles based on reliable biomarkers reported in the literature. Webb-Robertson and colleagues use a Bayesian framework to integrate proteomic and metabolomic data for improved detection of pathogens. Bush, Dudek, and Ritchie address the combinatoric explosion in GWAS when looking for combinations of SNPs linked to a disease by focusing analysis on SNPs related to pathways or genes linked to the disease under study. Pan and colleagues use local phylogenic alignments in GWAS to guide the choice of regions under study.

These papers demonstrate all aspects of the focus of this PSB session. While it is very early in the era of knowledge-driven analysis in the biomedical sciences, these papers show that difficult problems are beginning to be solved by new approaches that rely on data integration. These approaches include both tools for linking diverse data across molecular domains and species and methods that leverage prior knowledge to provide new insights into biology. As our data resources continue to grow exponentially, such methods will grow in importance and should provide ever greater insight.

References

1. B. Louie, P. Mork, F. Martin-Sanchez, A. Halevy and P. Tarczy-Hornoch, *J Biomed Inform* **40**, 5 (2007 Feb).
2. M. Ashburner, C. A. Ball, J. A. Blake, D. Botstein, H. Butler, J. M. Cherry, A. P. Davis, K. Dolinski, S. S. Dwight, J. T. Eppig, M. A. Harris, D. P. Hill, L. Issel-Tarver, A. Kasarskis, S. Lewis, J. C. Matese, J. E. Richardson, M. Ringwald, G. M. Rubin and G. Sherlock, *Nat Genet* **25**, 25 (2000 May).
3. O. Bodenreider, *Nucleic Acids Res* **32**, D267 (2004 Jan 1).
4. R. Shaker, P. Mork, M. Barclay and P. Tarczy-Hornoch, *Proc AMIA Symp* , 692 (2002).
5. R. Dhaval, J. Buskirk, J. Backer, C. K. Sen, G. Gordillo and J. Kamal, *AMIA Annu Symp Proc* **11**, p. 939 (2007).
6. N. Stolba, T. M. Nguyen and A. M. Tjoa, *Conf Proc IEEE Eng Med Biol Soc* , 4355 (2007).

CHARACTERIZATION OF UNKNOWN ADULT STEM CELL SAMPLES BY LARGE SCALE DATA INTEGRATION AND ARTIFICIAL NEURAL NETWORKS

G.BIDAUT [1,2] AND C.J. STOECKERT JR.[1]

1: Center for Bioinformatics, Department of Genetics,
University of Pennsylvania School of Medicine
423 Guardian Drive, Philadelphia, PA 19104, USA
2: Centre de Recherche en Cancérologie de Marseille
INSERM U891 - Institut Paoli-Calmettes
Université de la Méditerranée
27 Boulevard Leï Roure, 13009 Marseille, France
E-mail: ghislain.bidaut@inserm.fr, stoeckrt@pcbi.upenn.edu

Stem cells represent not only a potential source of treatment for degenerative diseases but can also shed light on developmental biology and cancer. It is believed that stem cells differentiation and fate is triggered by a common genetic program that endows those cells with the ability to differentiate into specialized progenitors and fully differentiated cells. To extract the *stemness* signature of several cells types at the transcription level, we integrated heterogeneous datasets (microarray experiments) performed in different adult and embryonic tissues (liver, blood, bone, prostate and stomach in *Homo sapiens* and *Mus musculus*). Data were integrated by generalization of the hematopoietic stem cell hierarchy and by homology between mouse and human. The variation-filtered and integrated gene expression dataset was fed to a single-layered neural network to create a classifier to (i) extract the *stemness* signature and (ii) characterize unknown stem cell tissue samples by attribution of a stem cell differentiation stage. We were able to characterize mouse stomach progenitor and human prostate progenitor samples and isolate gene signatures playing a fundamental role for every level of the generalized stem cell hierarchy.

1. Introduction

A wide variety of stem cell types have been recently reported to exist in several adult organs and are suspected to be present in most tissues. Well known stem cells types include hematopoietic stem cells (HSCs), neural stem cells (NSCs), myogenic progenitors (muscle), and others having a more restricted potential (such as gut and skin[1]) The consensus among stem cell researchers is that stem cell fate decision and renewal are triggered by several mechanisms that do not completely overlap among different stem cells type in the same organism[2], and in different species[3]. Two key questions are still unresolved: (i) Despite the knowledge of their existence, the location and differentiation capabilities of stem or progenitor cells are unknown for the vast majority adult organs ; methods to confirm the presence of stem/progenitor cells in adult tissues are needed. For instance, there is a large uncertainty on whether pancreatic beta-cell progenitor cells reside within the pancreatic ductal epithelium, the pancreatic small cells, or acinar tissue, or all of the previous[4]. (ii) The list of early markers that drive differentiation and self-renewal properties is still not agreed upon.

In this paper, we present a computational method to decipher common mechanisms of stem cells differentiation at the transcriptomic level using microarray data. We therefore integrated several datasets obtained from heterogeneous stem cells studies and trained an artificial neural network to extract a list of common gene markers triggering stem cell differentiation and fate decisions. Our hypothesis is that pathways triggering differentiation are at least partially conserved among adult stem cells/progenitors, forming a molecular signature reflecting embryonic and adult stem cells plasticity - the *stemness*. This hypothesis has been formulated several times[5], and is supported by measured transcriptome data[6], following a controversy about whether the *stemnesss* property could really be verified at the transcriptional level[7]. Discovering a shared signature should help characterize unknown stem cells populations, i.e. to determine whether a cell population contains stem cell/progenitors, and what is their differentiation potential/stage. In addition, the publication of a catalog of genes involved in differentiation of several cell types would be a valuable resource for stem cell researchers and developmentalists.

To extract molecular signatures for stem/progenitor cells developmental stages by gene expression profiling, we trained a multiclass single-layer linear ANN that subsequently allowed us to characterize unknown samples and position them in a hierarchy ranging from totipotent stem cells to fully differentiated cells. We tested the predictions made by our neural network using two tissues that were only partially characterized: Mouse stomach epithelium and human prostate (Figure 1). The ability of our system to generalize its classification to unknown stem cell types was assessed with a one-leave out cross-validation procedure on the training data.

ANNs represent a class of machine learning algorithm that were successfully applied to a large range of open ended problems. Their basic structure has been inspired from neurobiology and takes the form of a feed forward network of neurons modeled by a transition function. In a typical setup, an initial network topology is chosen depending on the nature of the problem; neurons are then trained for several epochs, leading to an ANN model, which is in turn applied to the classification of unknown samples. In biology, they were applied early for multiclass tumor classification in cancer discovery of cancer molecular subtypes and identification of biomarkers[8].

To properly use a classification system on the selected dataset we consistently labeled the stem and differentiated cells in tissues according to their differentiation stage/properties using a controlled vocabulary. The hematopoietic stem cell hierarchy[5] was used as a model: in this differentiation system, the different cells identified and isolated thanks to functional assays, are hierarchically positioned according to their differentiation potentials. At top of the hierarchy is the HSC, which can give rise to all blood cells and has self-renewal capacity, and finally the various types of mature blood cells. We defined a more general model (see section *Data integration*) applicable to all stem cell types, including adult stem cells and embryonic stem cells. Five stages are recognized, as shown in Table 1

Table 1. Properties of the generalized stem cell hierarchy and the controlled vocabulary used for classification.

Code	Stem Cell Type	Properties
A	Totipotent Stem Cell	Capable of self-renewal and able to generate all cell types
B	Multipotent Stem Cell	Capable of self-renewal and able to generate most cell types
C	Progenitor Cell	Capable of generating several cell types
D	Lineage-Committed Progenitor (LCP) Cell	Capable of generating a single or a restricted number of cell types
E	Differentiated Cell	Cell displaying final phenotype

The whole scheme operates in three steps. First we generated a training dataset by integration of several gene expression datasets generated by the SCGAP (Stem Cell Genome Anatomy Projects) consortium in different tissues, and projected it in a predefined space of basis vectors using the technique of vector projection[9], to group genes having similar dynamics of differentiation. We then used this integrated dataset to train a single-layer artificial multi-class neural network to classify unknown tissues potentially containing stem/progenitors cells in one of the predefined categories. Cross validation over training dataset led to 31 independent ANN models, trained independently by iteratively pulling out a tissue from the training data. The optimal number of genes to be employed in the classification was found by performing the training while reducing the number of genes used for classification. At each iteration, the top genes were retained by sorting the ANN weights and were kept before next training iteration. Classification error rate was minimized for 63 genes, and this set of genes was retained as a minimal core representing the *stemness* property shared by the set of stem/progenitor cells in the data. Finally, we tested the predicting power of the 31 models on two tissues not used for training that are potentially a mixture of adult and progenitor cells, namely mouse stomach progenitors[10] and human prostate progenitors[11].

The datasets analyzed measure gene expression in mouse and human, stem or progenitor cell in various tissues at different development stages. Gene coverage varies with the type of chip (from 6K-to 18K genes, see Table 2). Data were integrated at the gene level using the NCBI *Homologene* database[12] and at the experimental level using a controlled vocabulary of Table 1. The resulting data matrix contains 18720 genes profiled across 82 samples in total.

The 82 arrays are grouped by tissues. A unique tissue sample is characterized by a group of at least two arrays, up to five arrays covering previously stem/progenitor cells categories A-E - this is a necessary condition to be able to perform vector projection. This lead us to a final table of 40 tissues.

Table 2. Summary of tissues, platforms and cover. The two last tissues marked in bold are used for testing. Stem cell categories are marked with code (From A to E), corresponding to the previously defined stem cell stages in Table 1.

Author/Lab	Tissue	Platform	N genes	Categories
Darlington *et al.* (2007)[13]	Mouse Embryonic Liver	Affy. 430A	12798	C,E
Rowe *et al.* (unpublished)	Mouse Bone	Affy. U74 Av2, Bv2, Cv2	15127	C,D,E
Ivanova *et al.* (2002)[5]	Human Fetal Liver (HSCs)	Affy. U95 Av2,B,C,D,E	17024	B,D,E
Ivanova *et al.* (2002)[5]	Mouse Fetal Liver (HSC)	Affy. U74 Av2, Bv2, Cv2	15127	B,D,E
Ivanova *et al.* (2002)[5]	Mouse Adult Bone Marrow	Affy. U74 Av2, Bv2, Cv2	15127	B,C,D,E
Ivanova *et al.* (2002)[5]	Mouse Embryonic Stem Cells (ESCs)	Affy. U74 Av2, Bv2, Cv2	15127	A
Ivanova *et al.* (2002)[5]	Mouse Neural Stem Cells (NSCs)	Affy. U74 Av2, Bv2, Cv2	15127	B
Ivanova *et al.* (unpublished)	Human Coord Blood (HSCs)	Affy. U133 A,B	17275	B,D
Ivanova *et al.* (unpublished)	Human Adult Bone Marrow (HSCs)	Affy. U133 A,B	17275	B,D
Ivanova *et al.* (unpublished)	Mouse Adult Bone Marrow (HSCs)	Affy. 430 A,B	18626	B,C
Oudes *et al.* (2006)[11]	**Human Prostate Progenitors**	Affy. U133plus2.0	18806	X,E
Mills *et al.* (2002)[10]	**Mouse Stomach progenitors**	Affy. Mu11K A,B	6975	X,E
Total: 5 distinct	12 distinct	6 distinct	18720	5 distinct

The inherent architecture of the neural network-based single layer system allows for detailed exploration of two important parameters: (i) ranking weights for each of the five stages allow extraction of genes reported by the classifier to be a marker for this particular stem cell stage. (ii) ranking genes on a vector y resulting from the expression projection multiplied by the corresponding weight for a given cell population ($y_n = w_n.p_n$, w being the highest level neuron weight and p the expression projection) allows to delineate gene profiles critical for proper classification of this tissue. We additionally performed an analysis of highly ranked genes and statistically enriched gene ontologies (GO) for each of the stem/progenitor stage (unpublished data). A hidden layer to the architecture was not added even though it might decrease classification error because of the concern of overtraining and losing the ability to interpret the gene weights associated with each stem/progenitor cell stage. Other classifier types have also been considered but ANNs have been retained to keep the ability to rank genes according to their weights.

2. Material and Methods

2.1. *Dataset*

The dataset is an integration of stem cell transcriptional data generated by the 7 members of the SCGAP (Stem Cell Genome Anatomy Projects) consortium to train and test the system. Every SCGAP members performed large scale gene expression analysis on AffyMetrix arrays in given cell types and in two organisms, *M. musculus* and *H. sapiens* after specific purification steps (see Table 2). Data are available for interactive search on the Consortium's web portal (http://www.scgap.org), for download from every SCGAP member's web site, and in an integrated form the supportive web site (http://scann.sourceforge.net).

2.2. *Data integration*

Data were integrated at two level to take into account the heterogeneous nature of the consortium platforms. Data were first normalized within single datasets using iterative *lowess* normalization to compensate for differences in hybridization among the different platforms. The *mas* summarization method implemented in the bioconductor (R) software package (library *affy*) provided us with the expression values. We labeled the training dataset with a controlled vocabulary describing the five stem cell/progenitor differentiation stages (See Figure 1(b) and Table 1).

2.3. *Vector Projection*

Vector projection is a technique previously used for feature extraction in time course pancreas development gene expression data[9]. This technique allows for quick gene identification from pre-defined expression profiles that model the expected behavior of the genes we wish to isolate as most representative. In our case, vector projection captures genes with expression profiles peaking in different stem/progenitor differentiation stages for a given tissue (See Figure 1). The mathematical definition of vector projection is the inner product of a gene profile within a tissue to each model vector. We designed all model vector for the extraction of gene profile that were primarily expressed at one of the five progenitor/stem cell stage. These projection values were used to filter out genes without enough expression variation across tissues, and after filtering, projection values were presented to the ANN for training. Vector projection could be extended to extract gene profiles peaking in two or mode population - for instance to extract genes expressed during totipotency, turned off during self-renewal, and expressed again lated during lineage committed differentiation.

2.4. *Missing values and Data Organization*

A critical issue of the data we are analyzing are the missing values. This is particularly concerning here since projecting gene expression profiles characterized by missing value points can lead to results dramatically different from those expected. To cope with this issue, we performed vector projection using only the determined values, and re-normalize the model vector to a value of 1.0 using only the non-missing populations. After projection, the obtained dataset is organized both by tissue and projection values and reports the projection value for every stem/progenitor stem cell stage within every tissue. The projection values are then submitted to the ANN.

2.5. *System Architecture and Parameters*

The classifier used here is an extension of the well known single layer associative memory (Figure 1). It is characterized by a single neuronal layer holding the input weights to be learned from the data. Five neurons in total have been associated to the five levels of stem/differentiated cell hierarchy previously defined. The output for each neuron is the scalar product of the projection input by the weights: $y_i = \sum_{n=1}^{N} w_{i,n} . p_n$, with y being the neuron output, i the neuron number, n the gene index, N the total number of genes, $w_{i,n}$ the nth value of the ith neuron weight, and p_n the nth value of the gene expression projection to be classified. At every cross-validation step, one tissue is selected for testing and the others as a training set. A new neural network model is created (the weights are reinitialized) and the training set is presented to the network for 200 epochs. At each epoch, tissues from the training set are presented in a random order to the network, and weights are updated with a classic gradient descent-type training rule: $\Delta w_n = a(k)[y_n - yd_n]$, w_n being the current weight value, $a(k)$ the learning rate (typically decreasing over time, we used $a(k) = 1/(k+1)$, k being the epoch number), y_n the output and yd_n the desired neuron output. Each ANN model trained during cross validation is kept and used at a later stage for final classification of unknown samples. The number of weights to be conserved is updated until a minimal classification error rate is reached during cross-validation.

We then performed the cross-validation and estimated the optimal number of features for tissue classification by iteratively reducing the number of weights (or input genes) used by the ANN during training. The sum of misclassified test samples on the 31 ANN models versus the number of genes is represented figure 2(a) and is minimized for 63 genes (16 genes on individual neurons).

2.6. *Classification algorithm*

For a given tissue that we wish to classify and attribute to one of the five cell types defined in Table 1, classification was performed as follows: The tissue tested must meet several conditions: At least two samples must be available to perform vector projection, a fully differentiated sample known *a priori*, and the second sample containing stem/progenitor cells to be characterized. The system then characterizes every sample by first assigning it to a category as described in Table 1 for each of the 31 ANN models generated during leave one out training. A consensus is given by majority voting on the 31 ANN models. To understand the biological functions involved in the various differentiation stages, we examined the genes used in the classification and the enriched ontologies with ClutrFree[14] (data not shown).

3. Results and Discussion

3.1. *classification results*

Figure 2(a) shows the global misclassification rate versus the number of genes. The quadratic error rate for the training set and tests sets are respectively represented Figures 2(b) and 2(c). The error rate of every individual ANN trained with 63 genes is shown Figures 2(d) for the training set and the corresponding held out tissue. Misclassified tissues are pointed with an arrow. The minimal misclassification rate was obtained for 63 genes (See arrow on Figure 2(a), yielding a set of genes representing the core of *stemness* genes necessary for cell differentiation from ESCs to fully differentiated cells in both *H. sapiens* and *M. musculus* data. Error progression

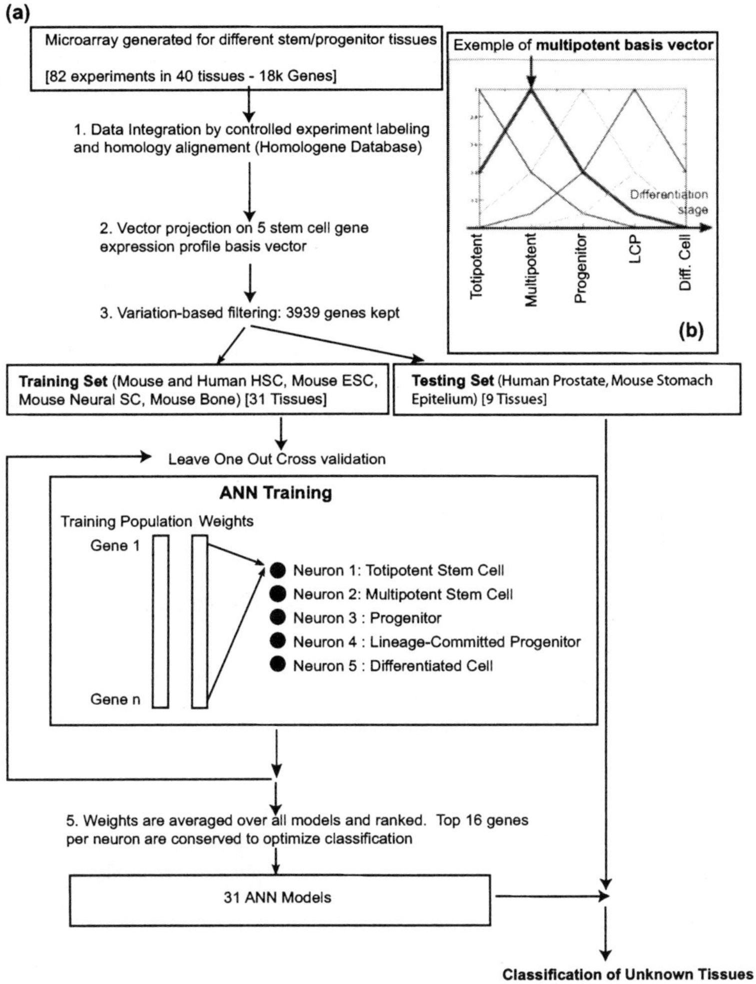

Figure 1. (a): ANN System Architecture used in the analysis: The data is initially projected onto the space defined by the 5 stem/progenitor cell basis vector. Splitting of the input dataset in Training/Testing dataset is shown. (b): Model vectors used for vector projection.

curves decreased rapidly and did not revealed overtraining in the neural network, even though we went up to 200 epochs, except for one neural stem cell tissues (Mouse NSC sample 2, not pictured).

We used the network to characterize two tissues potentially containing progenitors, namely human prostate and mouse stomach. Those two tissues were classified as expected by the system and are thus likely a mixture of fully differentiated cells and stem cells (Table 3).

Table 3. Classification by majority vote on the 31 Neural Network models generated for 63 genes

Tissue to be tested	Majority Vote
Mouse Stomach Progenitor	Progenitor
Human Prostate Progenitor	Multipotent Stem Cells

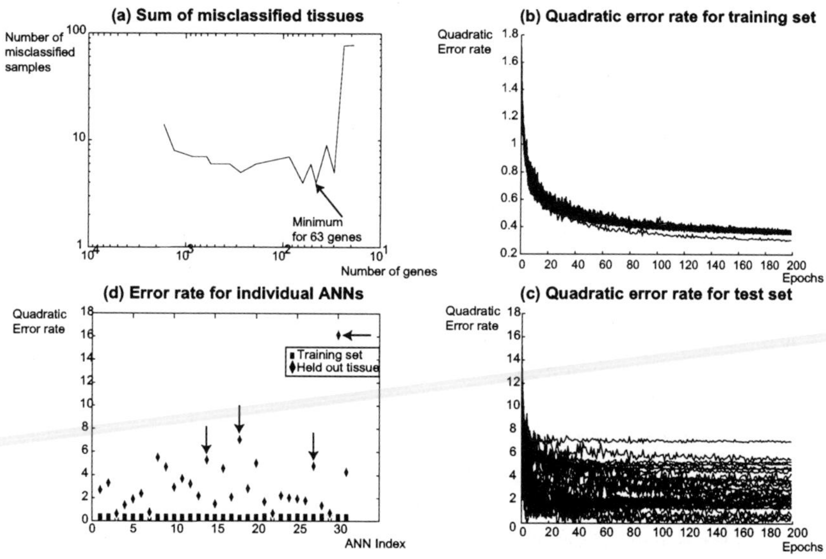

Figure 2. Classification results obtained for the ANN system.

The ANN models allowed us not only to characterize unknown tissues but also to extract genes that back the classification in those categories, by ranking the genes according to their weights - a higher weight meaning the gene plays an essential role in the classification. The training of 5 neurons allowed us to extract a total of 5 gene sets, each of them related to a stem cell developmental stage (totipotent, multipotent, progenitor, lineage-committed progenitors (LCP) and differentiated cell). The analysis of markers was done on the first model given by the leave-one out run for a training on 63 genes. The top 10 genes lists ranked by weights for each population are presented Table 4.

Table 4. Top 10 genes sets for the 5 stem cell stages ranked by weights.

Totipotent	Multipotent	Progenitor	LCP	Diff. Cells
Gene List	Gene List	Gene List	Gene List	Gene List
Dbn1	Procr	Letmd1	Coq3	Aqp1
1110001A23Rik	Gprasp2	Lrp8	Mgst1	Rhced
BC053917	AI661017	Kpna3	PDZK8	Rhbdl4
5830405N20Rik	Ctso	Ptprm	Cybb	MGI:1933403
Iqwd1	Gkap1	Rbp4	Rbbp9	Wnt11
5730420B22Rik	Lrrc16	Ass1	Cst7	Eif2ak3
Rbp4	Nrbp2	Itgb2	4932441K18	Nfe2
Rpp40	Adam8	Med6	Fli1	Fech
Nfe2	Irak1bp1	Cd109	Anxa1	AI661017
Rbbp9	MGI:1916782	5830405N20Rik	Gpr124	Gzma

For the totipotent population, the top genes include *Rbbp9*, a gene that may play a role in cell proliferation and differentiation.

In the multipotent population, we found MGI:1916782 (*Hopx*) - expressed in embryonic myocardium and other mesoderm, but not in endocardium or great vessels. *HOP* is also highly expressed in the developing heart, dependent on the cardiac-restricted homeodomain protein Nkx2.5. Inactivation of Hop in mice by homologous recombination results in a partially penetrant embryonic lethal phenotype with severe developmental cardiac defects involving the myocardium.

In tissues containing progenitor cells, we found several interesting markers: *Letmd1* is an oncoprotein that has a role in the development of human breast cancer. The protein encoded by *ptprm* is a member of the protein tyrosine phosphatase (PTP) family which is known to be signaling molecules regulating a variety of cellular processes including cell growth, differentiation, mitotic cycle, and oncogenic transformation. We also found the cell surface antigen *CD109*. This is a glycosylphosphatidylinositol (GPI)-linked glycoprotein found on a subset of hematopoietic stem and progenitor cells[15].

Other markers of interest were also found in the two other categories: In lineage-committed progenitors, *4932441K18* (*FIAT*) is a transcriptional regulator of osteoblastic functions. *Anxa3* encodes a protein member of the annexin family. Members of this family are involved in cellular growth, and this particular protein may play a role in anti-coagulation. *Sfrp4* is a Wnt pathway inhibitor and plays a central role in cell fate decisions. We also found *CD34*, known to be expressed in hematopoietic stem cells, and a marker for stem cell purification (ranked 13th in the LCP category).

Interestingly, exploration of the gene list for 87 genes shows *Socs2* in the top 10 genes for the Multipotent category. *Socs2* is ubitiquously expressed in most tissues in *M. musculus* and was shown to play an important regulatory role in neural development, neural stem cell differentiation and neuronal growth[16]. It is also a growth-hormone inducible and novel inhibitor of intestinal epithelial cell proliferation and intestinal growth. It was also shown to play a role in mammary gland development and regulate the fate of mesenchymal precursor cells[17], which is the nature of the cells we are studying here. Phenotype of Socs2-/- mice are describe therein and includes enlargement of bone and skeletal muscles.

The full list of markers obtained with the training for 63 genes is available from the supporting web site (scann.sourceforge.net).

3.2. *Discussion*

We present a novel approach that involved large scale integration of heterogeneous microarray datasets and pattern recognition analysis based on a vector projection technique to create a neural network-based classifier for characterization the differentiation properties of unknown stem cell tissues and extraction a molecular signature of *stemness*. Analysis of genes obtained by weight ranking highlighted gene product involved in several steps of cell differentiation. Thus, differentiation recruits a large panel of different genes and pathways and results from subtle expression level changes in a large number of genes. However, there is obviously more work to precisely validate the markers found. We were able to perform correct classification of unknown tissues (mouse stomach progenitors and human prostate progenitors) with a signature of 63 genes representing a core of genes involved in the differentiation process. As a natural extension of this work, we plan to include other stem cell studies publicly available on repositories such as Gene Expression Omnibus and ArrayExpress. On the technical side, we plan to improve the generalization capabilities of the classifier by boosting techniques, and make it accessible through a publicly available web server to perform the classification of unknown stem cell type. Other classifiers types will also be considered and their performance on this dataset studied.

4. Acknowledgments

We thank the SCGAP Consortium members for sharing data and providing insights for the analysis, and Dr Françoise Birg, Dr Patrice Dubreuil and

Wahiba Gherraby for their valuable comments on the manuscript. Dr Ghislain Bidaut is funded by the Fondation pour la Recherche Médicale (FRM) and the Institut National du Cancer (INCa). This work was initially funded by grant U01 DK63481.

References

1. K. D. Bunting and R. G. Hawley, *Biol Cell* **95**, 563(Dec 2003).
2. A. V. Molofsky, R. Pardal and S. J. Morrison, *Curr Opin Cell Biol* **16**, 700(Dec 2004).
3. M. Rao, *Dev Biol* **275**, 269(Nov 2004).
4. K. L. Seeberger, J. M. Dufour, A. M. J. Shapiro, J. R. T. Lakey, R. V. Rajotte and G. S. Korbutt, *Lab Invest* **86**, 141(Feb 2006).
5. N. B. Ivanova, J. T. Dimos, C. Schaniel, J. A. Hackney, K. A. Moore and I. R. Lemischka, *Science* **298**, 601(Oct 2002).
6. A. C. Piscaglia, T. Shupe, A. Gasbarrini and B. E. Petersen, *Curr Pharm Biotechnol* **8**, 167(Jun 2007).
7. N. B. Ivanova, J. T. Dimos, C. Schaniel, J. A. Hackney, K. A. Moore, M. Ramalho-Santos, S. Yoon, Y. Matsuzaki, R. C. Mulligan, D. A. Melton and I. R. Lemischka, *Science* **302**, p. 393d(Oct 2003).
8. B. T. Greer and J. Khan, *Ann N Y Acad Sci* **1020**, 49(May 2004).
9. L. M. Scearce, J. E. Brestelli, S. K. McWeeney, C. S. Lee, J. Mazzarelli, D. F. Pinney, A. Pizarro, C. J. Stoeckert, S. W. Clifton, M. A. Permutt, J. Brown, D. A. Melton and K. H. Kaestner, *Diabetes* **51**, 1997(Jul 2002).
10. J. C. Mills, N. Andersson, C. V. Hong, T. S. Stappenbeck and J. I. Gordon, *Proc Natl Acad Sci U S A* **99**, 14819(Nov 2002).
11. A. J. Oudes, D. S. Campbell, C. M. Sorensen, L. S. Walashek, L. D. True and A. Y. Liu, *BMC Genomics* **7**, p. 92 (2006).
12. D. L. Wheeler, T. Barrett, D. A. Benson, S. H. Bryant, K. Canese, V. Chetvernin, D. M. Church, M. Dicuccio, R. Edgar, S. Federhen, M. Feolo, L. Y. Geer, W. Helmberg, Y. Kapustin, O. Khovayko, D. Landsman, D. J. Lipman, T. L. Madden, D. R. Maglott, V. Miller, J. Ostell, K. D. Pruitt, G. D. Schuler, M. Shumway, E. Sequeira, S. T. Sherry, K. Sirotkin, A. Souvorov, G. Starchenko, R. L. Tatusov, T. A. Tatusova, L. Wagner and E. Yaschenko, *Nucleic Acids Res* **36**, D13(Jan 2008).
13. S. A. Ochsner, H. Strick-Marchand, Q. Qiu, S. Venable, A. Dean, M. Wilde, M. C. Weiss and G. J. Darlington, *Stem Cells* **25**, 2476(Oct 2007).
14. G. Bidaut and M. F. Ochs, *Bioinformatics* **20**, 2869(Nov 2004).
15. L. J. Murray, E. Bruno, N. Uchida, R. Hoffman, R. Nayar, E. L. Yeo, A. C. Schuh and D. R. Sutherland, *Exp Hematol* **27**, 1282(Aug 1999).
16. A. M. Turnley, *Trends Endocrinol Metab* **16**, 53(Mar 2005).
17. X. Ouyang, M. Fujimoto, R. Nakagawa, S. Serada, T. Tanaka, S. Nomura, I. Kawase, T. Kishimoto and T. Naka, *J Cell Physiol* **207**, 428(May 2006).

BIOFILTER: A KNOWLEDGE-INTEGRATION SYSTEM FOR THE MULTI-LOCUS ANALYSIS OF GENOME-WIDE ASSOCIATION STUDIES*

WILLIAM S. BUSH, SCOTT M. DUDEK, AND MARYLYN D. RITCHIE

Center for Human Genetics Research, Vanderbilt University
Nashville, TN 37232, USA

Genome-wide association studies provide an unprecedented opportunity to identify combinations of genetic variants that contribute to disease susceptibility. The combinatorial problem of jointly analyzing the millions of genetic variations accessible by high-throughput genotyping technologies is a difficult challenge. One approach to reducing the search space of this variable selection problem is to assess specific combinations of genetic variations based on prior statistical and biological knowledge. In this work, we provide a systematic approach to integrate multiple public databases of gene groupings and sets of disease-related genes to produce multi-SNP models that have an established biological foundation. This approach yields a collection of models which can be tested statistically in genome-wide data, along with an ordinal quantity describing the number of data sources that support any given model. Using this knowledge-driven approach reduces the computational and statistical burden of large-scale interaction analysis while simultaneously providing a biological foundation for the relevance of any significant statistical result that is found.

1. Introduction

1.1. *Genome-Wide Association Studies (GWAS)*

Over the last five years, genome-wide association studies (GWAS) have become a very popular study design for identifying genetic variants that incur disease risk in human populations. The overall strategy of the GWAS approach is inherently high-throughput, allowing investigators to blanket the genome with hundreds of thousands of single nucleotide polymorphisms (SNPs) in many individuals with the general goal of elucidating genetic causes of common human phenotypes – complex diseases in particular.

Traditional methods of genetic study design and analysis which excelled at identifying the rare mutations that cause Mendelian genetic disease have not performed as well for common complex disease, such as sporadic breast cancer or autism. Numerous candidate gene studies have been conducted for complex diseases, where particular genes of interest are investigated, but in many cases

* This work was supported by National Institutes of Health grants HL65962, and AG20135.

the results of these studies fail to replicate in other samples. While GWA studies are beginning to unravel the genetics of these complex diseases, one possible explanation for the lack of consistent findings from traditional studies is epistasis, or gene-gene interaction, and unless explicitly assessed, it may affect GWA studies also.

1.2. *Epistasis in GWA Studies*

Epistasis was first described by Bateson as the effect of one gene masking (or literally *standing upon*) the effect of another [1]. The Bateson view of epistasis has also been described as *biological epistasis* [2], where variation in the physical interaction of biomolecules affects a phenotype [3]. From a statistical perspective, epistasis was also observed as multi-allelic segregation patterns by Fisher who mathematically described the phenomenon as deviation from additivity in a linear model of genotypes [4]. Statistical epistasis and biological epistasis eventually converge as scientific understanding progresses. For example Bridges discovered statistical epistasis in *Drosophila* eye color, where collections of alleles Mendelize with various eye color phenotypes [5]. These alleles influence a common set of biochemical pathways controlling eye pigmentation that was elucidated many years later [6]. Epistasis can cause non-replication of single-SNP effects. If the effect of one allele is conditional on the presence of a second unknown allele, that second allele may not be present in a new population, and the effect of allele one will not replicate.

As epistasis is believed to play an important role in the genesis of complex disease, analysis strategies for detecting epistasis in large-scale data are increasingly important. A major hurdle in discovering epistasis, however, is the variable selection problem. Exhaustively evaluating all two-marker models in whole-genome data is a computational and statistical challenge, as processing the 5.00e11 possible two-marker models from a set of 1 million SNPs requires extensive computing resources and produces a plethora of statistically significant results with limited biological interpretability.

Two approaches are commonly suggested to address the variable selection problem. One approach is to select SNPs based on the strength of independent main effects, evaluating interactions only between SNPs that meet a certain effect size threshold. Another approach is to evaluate multi-marker combinations based on biological criteria [7]. Each of these strategies imposes a specific bias into the analysis, and neither strategy will be optimal in all cases. If we select or filter variables based on their main effects, we bias the analysis using statistical information, and assume that relevant interactions occur only between markers that independently have some effect on the phenotype alone.

Several studies have proposed complex theoretical penetrance models that influence the trait only through the interaction of two or more genetic variants [8-10], and filtering based on main effects would potentially miss these types of discoveries. If we filter variables using biological information – i.e. only examine interactions between SNPs in a common pathway or with a common structure or function – we bias the analysis in favor of models with an established biological foundation in the literature, and novel interactions between SNPs would be missed. Furthermore, the entire analysis is conditional upon the quality of the biological information used.

Several new tools have recently been developed to incorporate biological information with analytical approaches for GWAS data. Prioritizer is a Bayesian approach to incorporate multiple sources of gene interrelationships in a global "functional gene network". This network is used to prioritize significant single-SNP results by gene function [11]. Others methods use structured knowledge as a way to guide (but not restrict) variable selection for regression-based modeling. Province and Borecki propose a Bayesian re-sampling approach to select collections of SNPs that may have very small independent effects but function in aggregate explain a more substantial portion of trait variance [12]. Conti proposes a hierarchical modeling approach that uses an expert knowledge ontology to search for and test complex multi-SNP models. This Bayesian modeling process is flexible, allowing SNPs outside the knowledge-base to also be used in models (Pharmacogenomics Research Network Presentation 2008).

We propose a strategy that steps beyond the annotation and grouping of independent SNP effects, but does not attempt to jointly model large numbers of SNPs simultaneously. Also, we believe that ultimately data from multiple sources will better facilitate a comprehensive analysis, providing a biological foundation for testing specific multi-SNP association models in GWAS data. In this work, we present the Biofilter, a tool for knowledge-driven multi-SNP analysis of large scale SNP data. The Biofilter fundamentally differs from other methods in the way knowledge is incorporated into the analysis pipeline. The Biofilter uses biological information about gene-gene relationships and gene-disease relationships to construct multi-SNP models before conducting any statistical analysis. Rather than annotating the independent effect of each SNP in a GWAS dataset, the Biofilter allows the explicit detection and modeling of interactions between a set of SNPs. In this manner, the Biofilter process provides a tool to discover significant multi-SNP models with non-significant main effects that have established biological plausibility. This approach has the added benefit of reducing both the computational and statistical burden of exhaustively evaluating all possible multi-SNP models.

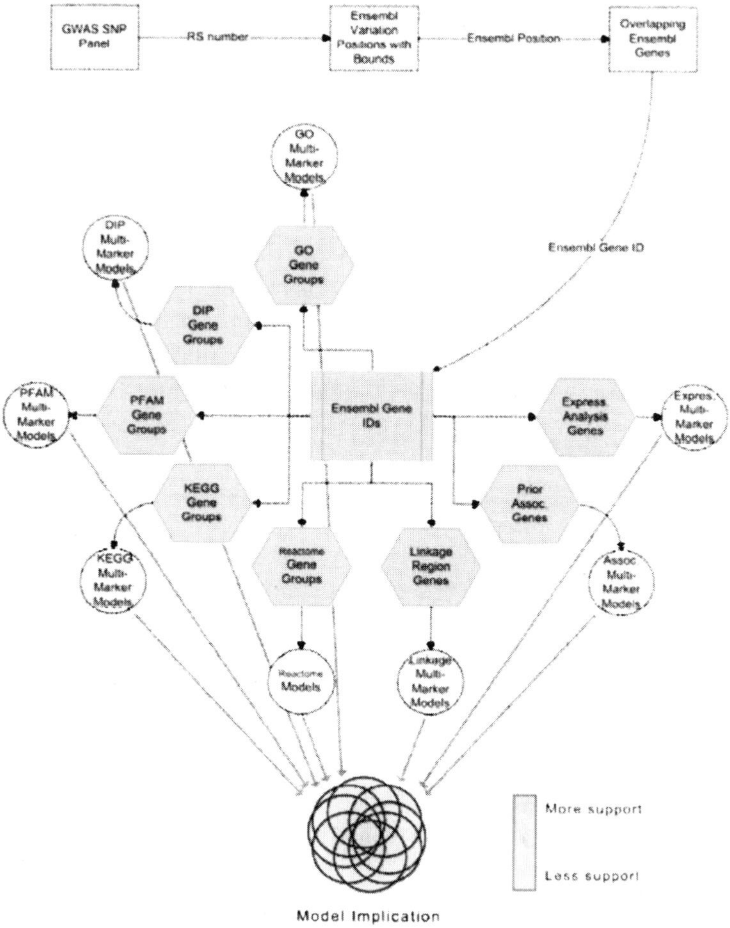

Figure 1. Overview of the Biofilter process. GWAS platform SNPs are mapped to Ensembl gene IDs and related to disease-independent sources (left) and to disease-dependent sources (right). Multi-marker models are generated from SNPs within knowledge-related genes. Derived models are overlaid to assess overall model implication.

2. Methods

2.1. *Overview*

An overview of the Biofilter is show in figure 1. The Biofilter model generation process is gene-centric, and as such, SNPs from GWAS genotyping platforms must first be assigned to genes. Relationships between the genes represented by a platform can then be translated to multi-SNP models. Structured biological knowledge relevant to GWAS interaction analysis can come from various sources. We have partitioned relevant knowledge into two basic types: disease-dependent and disease-independent. *Disease-dependent* knowledge is information that relates a gene to the disease phenotype being studied, such as a previously associated SNP or a gene that is over-expressed in cases. *Disease-independent* knowledge is information that relates genes to one another, or defines collections of genes, such as a metabolic pathway or a common structural motif. These two types of information can be combined to form different classes of multi-SNP models and provide a measure of how strongly implicated a given model is based on the current available knowledge.

2.2. *Database Integration*

We chose Ensembl as our source of gene and SNP positional information due to ease of access and its clearly defined database schema. RS numbers for SNPs used in genotyping are matched to records within the Ensembl variation database (Release 49) to retrieve position information. The probe positions were joined to gene information tables to determine if the SNPs lie within one of the 32,000 known Ensembl genes that physically map to the autosomes or the X and Y chromosomes. This SNP-to-gene mapping is stored as a derived table in the Biofilter database.

Disease-dependent knowledge sources link individual genes to a disease phenotype. The goal of using disease-dependent knowledge is to identify genes that have some prior evidence of putative influence on the phenotype. One systematic source of disease-dependent knowledge is the Genetic Association Database (GAD). GAD is an archive of human genetic association studies of complex diseases and disorders established in 2004 [13]. GAD contains a list of prior associated genes for a variety of specific disease phenotypes and broader phenotype classes. Other types of disease-dependent knowledge may require manual selection from literature. Previous regions of genetic linkage, studies of differential gene expression, and hypothetical disease etiologies are possible sources of disease-gene relationships.

Disease-independent sources link two or more genes together. The goal of using disease-independent knowledge is to identify pairs of genes with some prior evidence of putative epistasis. The Gene Ontology project (GO, accessed on 3/16/08) is a collaborative effort to characterize and describe gene products in a collection of three hierarchical ontologies: cellular component, biological process, and molecular function. Because of its hierarchical structure, some broad ontology categories contain many hundreds of genes. For this analysis, smaller, more precisely defined gene categories (< 30 genes) were used as these presumably contain stronger gene relationships. The Database of Interacting Proteins (DIP, 1/14/08 update) documents experimentally determined protein-protein interactions from more than 80 organisms [14]. We used the pair-wise human protein-protein interaction set contained in DIP to produce gene-gene pairs. The Protein Families Database (PFAM, Release 22) uses multiple sequence alignments and hidden Markov models to identify common protein domains and families based on structural and functional sequence patterns [15]. Generating pairs of genes using this data relies on the hypothesis that members of the same protein family are more likely to jointly influence disease risk. As such, we generated gene-gene pairs within proteins having the same domain, the same protein family, the same structural motif, and the same sequence repeat. The Kyoto Encyclopedia of Genes and Genomes (KEGG, 3/6/08 update) Pathway set is a collection of manually drawn pathway maps for a variety of metabolic and signaling pathways [16]. Reactome (Version 24) is a database of curated core pathways and reactions in human biology [17]. Netpath is a relatively new source of curated immune signaling and cancer pathways provided by the Pandey Lab at Johns Hopkins University and the Institute of Bioinformatics [18]. With these pathway collections, all possible gene-gene pairs were generated within each pathway-based gene group.

Relational data sources were downloaded and reconstituted in their original form within a MySQL database using Perl scripts. Using the schema for each data source, proteins and/or genes were translated to Ensembl gene IDs, and derivative tables containing gene groupings (such as protein families) were generated within the Biofilter database. Non-structured data sources, such as gene lists from publications, were manually imported into the Biofilter database. These gene lists were then translated to Ensembl gene IDs and used to establish gene groupings.

2.3. *Model Types and Generation*

Using both disease-dependent and disease-independent data sources, there are four types of two-SNP models possible: disease-independent, disease-dependent, hybrid with one disease-dependent gene, and hybrid with two disease-dependent genes. Figure 2 illustrates each of these model types. Disease-dependent information is based on a set of genes that are related to disease, visualized in the figure as a collection of dashed boxes, or unconnected nodes of a graph. Disease-dependent models are generated by exhaustively pairing all possible combinations of disease-related genes. Disease-independent information is based on relationships between sets of genes, visualized in the figure as a set of lines, or edges in a graph. To build disease-independent models, we generate pair-wise combinations of SNPs located in genes that are related, illustrated by as edges in figure 2. Hybrid models blend disease-dependent and disease-independent information, and can contain either one or two disease-related genes. In the figure, one-gene hybrids must be connected by an edge (disease-independent connection) and contain at least one dashed square (disease-dependent gene). Two-gene hybrids must contain two dashed squares connected by an edge, meaning that there is evidence for biological interaction of two disease-related genes. For each pair-wise combination of genes, all possible two-SNP models across the two genes are built. For example, if there are two SNPs (A and B) in gene one and there is one SNP (C) in gene two, two models are generated (A,C and B,C).

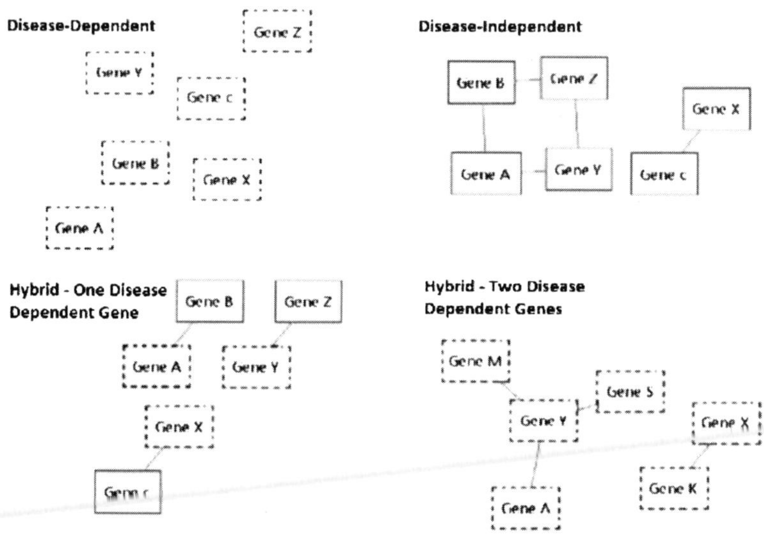

Figure 2. Two-gene model types. Each box represents a gene, and each line a connection between genes. Boxes that are dashed have been previously linked to disease by at least one data source.

2.4. *Model Implication*

Each model constructed has a set of Biofilter data sources that support it. If a combination of genes is supported by multiple data sources, it is likely more accepted by the scientific community and therefore may be more biologically plausible. We quantify the degree of knowledge-based support for a model with an *implication index.* The implication index is simply the number of data sources that provide evidence of gene-gene interaction or gene-disease relationship, and is calculated simply by summing the number of data sources supporting each of the two genes and the connection between them. For example, if one disease-related gene supported by two data sources is connected to another non-disease related gene, and that connection is supported by three data sources, the implication index of the model is five. In this manner, the implication index provides an ordinal value representing the strength of the biological plausibility of a multi-SNP model.

Table 1. General GWAS platform statistics

	Illm1M	Affy60
Total SNPs (with RS IDs)	1,055,373	924,689
SNPs within genes	493,854	353,913
Genes represented	21,024	17,418
Common SNPs	267,900	
Common SNPs within genes	118,355	
Common genes represented	16,908	

3. Results

3.1. *GWAS Platform Representation*

Two large-scale genotyping platforms were assessed in this study: the Illumina Human1M-Duo BeadChip (Ilmn1M) and the Affymetrix Genome-Wide Human SNP Array 6.0 (Affy60). For the purposes of our assessment, only probes with vendor-specified Reference Sequence (RS) numbers were used in order to assure continuity of genomic position. General statistics for these two platforms are shown in Table 1.

As described in the methods, genes can be defined as disease-dependent and disease-independent. Because the disease-independent genes can be used in an analysis of any phenotype, we focused on how those genes are represented in the different data sources. Table 2 shows the number of gene pairs represented from each platform in each disease-independent data source. Table 3 shows the pair-wise overlap across the six different public databases.

3.2. *Generalized Disease Independent Models*

As disease-independent models are universally applicable to any phenotype, we provide files containing all derived gene-gene pairs, and platform specific two-SNP models for the Illm1M and Affy60 (available: http://chgr.mc.vanderbilt.edu/ritchielab/method.php?method=biofilter). Counts of these two-SNP models by implication index are shown in table 4. The Illm1M platform covers over a million more two-SNP models than the Affy60. This is most likely due to a lack of RS numbers for many Affymetrix probes, rather than a lack of gene coverage.

Table 2. Gene pairs represented by platform.

Data Source	Illm1M	Affy60
PFAM	14911	12837
DIP	747	638
GO	6129	5359
KEGG	4058	3543
Reactome	1799	1610
Netpath	3704	3246

Table 3. Pair-wise overlap of all known genes in disease-independent Biofilter data sources. The data source listed on each row contains genes that overlap the data source listed on each column. Cell values indicate the proportion of the genes in the column source that are represented in the row source.

	PFAM	DIP	GO	KEGG	Reactome	Netpath
PFAM	1	0.95	0.92	0.95	0.95	0.92
DIP	0.05	1	0.09	0.08	0.15	0.12
GO	0.37	0.73	1	0.63	0.64	0.56
KEGG	0.01	0.02	0.01	1	0.02	0.02
Reactome	0.12	0.36	0.19	0.25	1	0.21
Netpath	0.22	0.6	0.34	0.41	0.41	1

Performing an exhaustive analysis of all possible two-SNP models within genes represented by these two platforms would result in 1.22e11 models for the Illumina 1M and 6.26e10 models for the Affymetrix 1M. By reducing the interaction search space to only models with established biological plausibility via the disease-independent data sources, only 2.23e9 (Illm1M) and 1.2e9 (Affy60) model evaluations are required. Applying a Bonferroni correction to the exhaustive approach would require a model fit p-value of 4.10e-13 (Illm1M) and 7.98e-13 (Affy60) to be statistically significant. In contrast, using the knowledge-based approach, a Bonferroni correction of 2.25e-11 (Illm1M) or 4.16e-11 (Affy60) is required. In this manner, reducing the search space not only improves computation time, but also reduces the statistical burden of conducting biologically non-relevant statistical tests. Further model restriction (such as using models with an implication index > 1) would further reduce the Bonferroni adjusted significance threshold.

4. Discussion

When examining epistasis in genome-wide association studies, there are several variable selection strategies. Exhaustive evaluation of all multi-SNP models is generally computationally impractical. Exploring epistasis within a set of SNPs with detectable main effects may prevent the discovery of complex genetic models where trait variance is explained largely by the interaction of SNPs. Using biological knowledge to perform SNP selection provides two key benefits simultaneously: it reduces the multi-SNP model search space, and it provides a biologically plausible foundation for the models to be evaluated. We developed the Biofilter to systematically reduce model search space based on multiple sources of structured biological knowledge.

We mapped the disease-independent models generated by the Biofilter to two GWAS genotyping platforms, the Affymetrix 1M and the Illumina 1M. The final evaluated model search space was 0.241% of the exhaustive model space for Affy60 and 0.40% of the exhaustive model space for Illm1M when requiring at least one source of structured biological knowledge connecting the two genes in a two-SNP model, with further reductions possible by adjusting the number of required knowledge sources implicating the model.

Table 4. Disease-independent gene pairs and model counts by implication index.

Implication Index	Illm1M Gene Pairs	Illm1M Two-SNP Models	Affy60 Gene Pairs	Affy60 Two-SNP Models
1	4,679,363	2,174,328,700	3,505,773	1,162,090,222
2	87,163	44,960,600	67,341	36,825,703
3	8,065	6,425,788	6,094	4,102,173
4	715	397,966	546	171,075
5	45	11,033	40	4,122
6	1	569	1	757
Total	4,775,352	2,226,124,656	3,579,795	1,203,194,052

The Biofilter method of variable selection can be implemented with a variety of analysis techniques, including logistic regression, classification and regression trees, and basic categorical statistics, among many others. To this end, the Biofilter is being developed as a knowledge-based filter in part of a larger analysis framework. The collection of multi-SNP models generated by the Biofilter can be passed seamlessly to several analytical methods. Statistical properties of each multi-SNP model are then stored, allowing retrieval of results with complete annotation of the SNPs, genes, gene grouping information, and in some cases, PubMed references to the original articles implicating the model. The end result of this analysis pipeline is a set of biologically plausible, statistically relevant multi-SNP genetic models.

Some approaches may be adapted to incorporate the implication index into the analysis plan. Prioritized subset analysis, for example, partitions statistical results based on prior biological knowledge. The false discovery rate (FDR) for the "prioritized subset" is estimated separately, improving power when the prior knowledge is accurate [19]. Applying this strategy to subsets defined by the implication index could improve statistical power via p-value correction.

Ranking models based on the number of supporting data sources may introduce unknown literature-based biases. Some data sources may have inter-dependencies, where one source was referenced in the creation of another. The breakdown of gene overlap for the 6 disease-independent data sources shows the diversity of gene-pairs represented, though notably PFAM contains nearly all of the gene-pairs established by the other sources. This is likely because PFAM contains the largest number of genes. When using disease-dependent data sources, there are certainly many factors that influence the inclusion and promotion of specific genes in relation to a phenotype, such as reporting bias.

Overall, the Biofilter provides a systematic way to assess the level of knowledge-based support for a given genetic model, provide a ranked list of all possible knowledge-based models, and to statistically test each of these hypotheses in genome-wide association data.

Acknowledgments

This work was supported by National Institutes of Health grants HL65962, and AG20135.

References

1. Bateson W: *Mendel's Principles of Heredity.* Cambridge: Cambridge University Press; 1909.
2. Moore JH, Williams SM: **Traversing the conceptual divide between biological and statistical epistasis: systems biology and a more modern synthesis.** *Bioessays* 2005, **27:**637-646.
3. Moore JH: **The ubiquitous nature of epistasis in determining susceptibility to common human diseases.** *Hum Hered* 2003, **56:**73-82.
4. Fisher RA: **The Correlation Between Relatives on the Supposition of Mendelian Inheritance.** *Transactions of the Royal Society of Edinburgh* 1918, **52:**399-433.
5. Bridges CB: **Specific modifiers of eosin eye color in Drosophila melanogaster.** *J Experimental Zoology* 1919, **28:**337-384.
6. Lloyd V, Ramaswami M, Kramer H: **Not just pretty eyes: Drosophila eye-colour mutations and lysosomal delivery.** *Trends Cell Biol* 1998, **8:**257-259.
7. Carlson CS, Eberle MA, Kruglyak L, Nickerson DA: **Mapping complex disease loci in whole-genome association studies.** *Nature* 2004, **429:**446-452.

8. Culverhouse R, Suarez BK, Lin J, Reich T: **A perspective on epistasis: limits of models displaying no main effect.** *Am J Hum Genet* 2002, **70:**461-471.
9. Frankel WN, Schork NJ: **Who's afraid of epistasis?** *Nat Genet* 1996, **14:**371-373.
10. Moore J, Hahn L, Ritchie M, Thornton T, White B: **Routine discovery of complex genetic models using genetic algorithms.** *Applied Soft Computing* 2004, **4:**79-86.
11. Franke L, van BH, Fokkens L, de Jong ED, Egmont-Petersen M, Wijmenga C: **Reconstruction of a functional human gene network, with an application for prioritizing positional candidate genes.** *Am J Hum Genet* 2006, **78:**1011-1025.
12. Province MA, Borecki IB: **Gathering the gold dust: methods for assessing the aggregate impact of small effect genes in genomic scans.** *Pac Symp Biocomput* 2008,190-200.
13. Becker KG, Barnes KC, Bright TJ, Wang SA: **The genetic association database.** *Nat Genet* 2004, **36:**431-432.
14. Xenarios I, Salwinski L, Duan XJ, Higney P, Kim SM, Eisenberg D: **DIP, the Database of Interacting Proteins: a research tool for studying cellular networks of protein interactions.** *Nucleic Acids Res* 2002, **30:**303-305.
15. Finn RD, Tate J, Mistry J, Coggill PC, Sammut SJ, Hotz HR, Ceric G, Forslund K, Eddy SR, Sonnhammer EL et al.: **The Pfam protein families database.** *Nucleic Acids Res* 2008, **36:**D281-D288.
16. Kanehisa M, Araki M, Goto S, Hattori M, Hirakawa M, Itoh M, Katayama T, Kawashima S, Okuda S, Tokimatsu T et al.: **KEGG for linking genomes to life and the environment.** *Nucleic Acids Res* 2008, **36:**D480-D484.
17. Vastrik I, D'Eustachio P, Schmidt E, Joshi-Tope G, Gopinath G, Croft D, de BB, Gillespie M, Jassal B, Lewis S et al.: **Reactome: a knowledge base of biologic pathways and processes.** *Genome Biol* 2007, **8:**R39.
18. **NetPath** [http://www.netpath.org]
19. Li C, Li M, Lange EM, Watanabe RM: **Prioritized subset analysis: improving power in genome-wide association studies.** *Hum Hered* 2008, **65:**129-141.

INTERPRETING GENETICS OF GENE EXPRESSION: INTEGRATIVE ARCHITECTURE IN BIOCONDUCTOR *

V. J. CAREY

Channing Laboratory
Brigham and Women's Hospital
Harvard Medical School
181 Longwood Ave.
Boston MA 02115 USA
E-mail: stvjc@channing.harvard.edu

R. GENTLEMAN

Program in Computational Biology
Division of Public Health Sciences
Fred Hutchinson Cancer Research Center
1100 Fairview Ave. N, M2-B876
Seattle WA 98109 USA

Several influential studies of genotypic determinants of gene expression in humans have now been published based on various populations including HapMap cohorts. The magnitude of the analytic task (transcriptome vs. SNP-genome) is a hindrance to dissemination of efficient, thorough, and auditable inference methods for this project. We describe the structure and use of Bioconductor facilities for inference in genetics of gene expression, with simultaneous application to multiple HapMap cohorts. Tools distributed for this purpose are readily adapted for the structure and analysis of privately-generated data in expression genetics.

1. Introduction

Figure 1 depicts findings in a general population study of genetics of gene expression. On the left we plot distributions of expression of gene HLA-DRB1 (measured on an Illumina WG-1 platform, as distributed by Sanger Institute in the Genevar project[1]) against genotypes for SNP rs9271367, assayed in 60 individuals from the CEU (central european ancestry) CEPH

*This work was supported in part by NIH P41 HG004059-01 (R Gentleman, PI), NIH R01 HL086601-01 (B Raby, PI), NIH R01 HG003646-01 (R Lazarus, PI).

HapMap cohort. On the right we plot a gender-adjusted measure of association of HLA-DRB1 expression with SNP rare allele counts for all phase II HapMap SNP (3.9 million loci). Interest in HLA-DRB1 stems from the report of Schadt et al.[2] who used a cohort of human liver samples to identify an eQTL at rs9272723. We display the third most-associated SNP (gender-adjusted nominal $p = 1.8 \times 10^{-9}$) in the CEU cohort as it decomposes the expression distribution among three genotype groups most effectively. This association was not reported in the multipopulation eQTL tables published by Stranger et al.[3], presumably for lack of formal genome-wide significance. It is noteworthy that the immortalized B-cell samples assayed by Stranger et al. carry signal similar in nature to that reported for the liver samples by Schadt et al.[2] Using tools described in this paper, it is straightforward to show that among 60 founders in the Yoruba CEPH cohort, the rare allele copy number for rs9271367 is also associated with HLA-DRB1 at a nominal p of 1.5×10^{-7}, but that the primary distinction in expression distributions falls between homozygous common and all others.

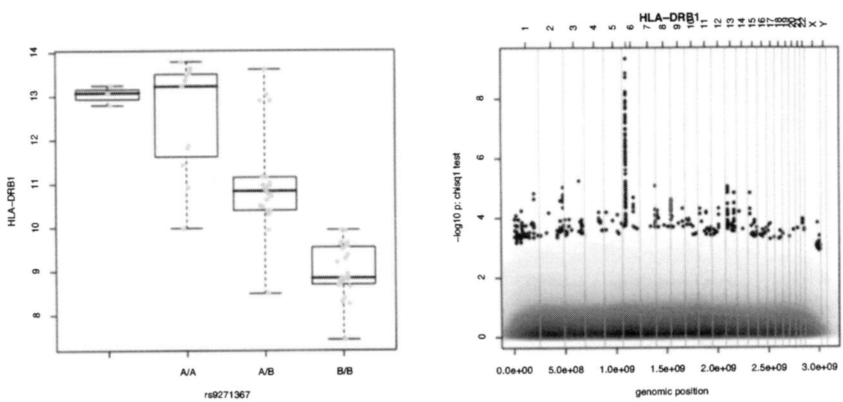

Figure 1. Left: y axis: Expression of HLA-DRB1 in 60 CEU HapMap individuals; x axis: genotype groups formed using SNP rs9271367. Right: y axis: measure of association between HLA-DRB1 expression and rare allele count ($-\log_{10} p$-value for Cochran-Armitage test for trend); x axis: locations of HapMap Phase II SNP according to build 36 release 23a.

In the work of Stranger and colleagues "a detailed association analysis identified at least 1,348 genes with association signals in *cis* and at least 180 in *trans*". The analyses involved filtering of expression probes on the basis of overall "variance and population differentiation", filtering of SNPs on the basis of minor allele frequency (confined to 5% and above in all CEPH cohorts) and considered both linear regression and rank correlation measures of association between allele counts and expression, applied only to unrelated individuals. The *trans* investigation was limited by confining attention to nonsynonymous SNP known to have *cis*-associtions, SNP thought to be involved in splicing (via Ensembl v41 annotation), and SNP found in sequences defining microRNA.

The combination of Genevar and HapMap resources provides potent tools for increasing our knowledge of genomic structures contributing to expression variation. A number of the steps taken in the pioneering analyses of Stranger *et al.* constitute concessions to computational, inferential and annotational barriers that will be lowered as research and computational prowess mature. We have undertaken in Bioconductor (www.bioconductor.org) to design and disseminate data structures, algorithms, and concrete software packages that simplify research into genetics of gene expression, both with public Genevar/HapMap data, and with privately generated data.

2. Data structures

2.1. *Abstract data types and methods*

An instance `x` of the `smlSet` class (`sml` denoting SNP-matrix list) defined in Bioconductor package *GGBase* satisfies the following basic constraints. If `x` holds information on N individuals assayed for expression through G features of a transcriptome-wide array, then `exprs(x)` returns a $G \times N$ matrix of expression measures, and `featureNames(x)` is a G-vector of strings encoding expression probe identifiers. If the individuals were genotyped on S_c SNP loci on chromosome c, $c = 1, \ldots, C$ (or, more generally, $c \in C$ a set of tokens enumerating chromosomes in an organism), then `snps(x, c)` returns an $S_c \times N$ matrix of SNP genotype assignments of the form A/A, A/B, B/B, NA, where the latter token denotes unavailable genotype. Identifiers for SNPs are retrieved using `snpNames(x, c)`. Genome-wide coordinates of SNP are retrieved using `getSnpLocs(x)`.

A primary method for conducting genome-wide association tests for phenotypes defined by gene expression (mRNA abundance) measures is `gwSnpScreen`, defined in the Bioconductor *GGtools* package. The call `res = gwSnpScreen(formula, smlSet [, chrnum])` returns an object `res` with S_c inference results if `chrnum` identifies chromosome `c`; if `chrnum` is omitted, then the object has $S = \sum_c S_c$ results. The type of analysis conducted depends upon the `formula` passed. In general this will have the form `gs ~ x1 [+ x2 + ...]` where `gs` is an instance of the `GeneSet` class defined by Bioconductor package *GSEABase*, and the right-hand side is a linear predictor expressed in Wilkinson-Rogers notation, using N-dimensional variables typically found in the `phenoData` component of the `smlSet`. These variables can be employed in addition to the rare allele counts for all SNP in $K \times S$ individual tests for association of SNP genotype with expression values for each of the K elements of the gene set `gs`. Finally, metadata about expression reporters can be housed in a `featureData` component, metadata about sample-level variable can be housed in a `varMetadata` component, and the MIAME schema for the experiment can reside in an `experimentData` component; designs for all these structures are defined in package *Biobase*.

The entity `res` returned by `gwSnpScreen` is an instance of a formal class that can be interrogated to identify most strongly associated SNP (adjusting for covariates in the linear predictor), and that can be easily visualized as in the left panel of Figure 1. Inferential summaries are those generated by the *snpMatrix* package genome-wide association test functions `single.snp.tests` (no covariate adjustment) or `snp.rhs.tests` (generalized linear model with covariates)[4]. The return object also includes `call` information to indicate how it was created.

2.2. *Statistical internals*

Of primary interest are collections of p-values summarizing the statistical strength of the genotype-expression associations. These are derived from percentiles of the Chi-squared distribution with one d.f. evaluated at the score statistic for the gene- and SNP-specific parameter β_{gs} in the linear model for log expression of gene g

$$Y_{gi} = \alpha_g + \beta_{gs} c_{si} + \gamma_{gs}^t Z_i + e_{gsi},$$

where c_{si} is the copy number of the rare allele for SNP s on subject i, Z_i is a q-vector of confounders of the association betwen genotype and expression, and e_{gsi} is a Gaussian disturbance with zero mean and constant (gene- and SNP-specific) variance over all subjects. These p-values are provided in nominal form; of them may be transformed to false discovery rates, or to other corrected versions accounting for multiple comparisons, using the `multtest` package of Bioconductor.

2.3. *Concrete illustrations*

The *GGtools* package provides an exemplar `smlSet` instance called `hmceuB36.2021`:

```
> library(GGtools); data(hmceuB36.2021)

> hmceuB36.2021

snp.matrix-based genotype set:
number of samples:  90
number of snp.matrix:  2
annotation:
 exprs: illuminaHumanv1.db
 snps: snp locs package: GGBase ; SQLite ref: hmceuAmbB36_23a_dbconn
Expression data: 47293 x 90
Phenodata: An object of class "AnnotatedDataFrame"
  sampleNames: NA06985, NA06991, ..., NA12892  (90 total)
  varLabels and varMetadata description:
    famid: hapmap family id
    persid: hapmap person id
    ...: ...
    isAdad: logical TRUE if person is a father
    (9 total)
```

This structure is easy to work with interactively as it maintains information on only two chromosomes, 20 and 21. The genome-wide data are available separately in the *GGdata* package as `hmceuB36`.

Decoding and translation of expression reporter nomenclature is carried out using standard platform-specific Bioconductor SQLite annotation data packages. Maintenance of metadata on millions of SNP reporters is more challenging. We investigated several approaches to storing and responding to queries about SNP identifiers, locations, and allele assignments, including web services and netCDF. At present we use SQLite tables; when an `smlSet` instance is brought into scope, a SQLite connection is created and linked to the instance for query resolution. The primary use cases of interest concern visualization and computation of SNP-gene distances, and so only retrieval of large blocks of locations (genome-wide or chromosome-wide) are provided at present. Fine-grained queries on SNP metadata can be carried out using special packages devoted to dbSNP data serialization or the Bioconductor *biomaRt* package.

To conduct chromosome-wide inference on a small gene set, we create `hmFou`, a restriction of the samples in `hmceuB36.2021` to the cohort founders (parents), and then:

```
> library(GSEABase)
> s1 = GeneSet(c("CPNE1", "RPS26"),
     geneIdType = SymbolIdentifier())

> f1 = gwSnpScreen(s1 ~ male, hmFou, chrnum(20))

> f1

multi genome-wide snp screen result:
gene set used as response:
setName: NA
geneIds: CPNE1, RPS26 (total: 2)
geneIdType: Symbol
collectionType: Null
details: use 'details(object)'
there are 2 results.
the call was:
gwSnpScreen(sym = s1 ~ male, sms = hmFou, cnum = chrnum(20))
```

The formula dictates that each gene will be analyzed correcting for gender. The resulting object is an instance of class multiGwSnpScreenResult, a list of elements of class cwSnpScreenResult. Each of these elements can be plotted, or interrogated for significant SNP.

Genome-wide inference on a gene set of even modest size can be mechanically challenging. In the phase II HapMap context, each gene yields 4 million inference measures, many of which are of no importance. The `gwSnpScreen` method for gene sets can receive a `snpdepth` parameter. If this parameter has value D, results on all SNPs but those yielding the D smallest p-values, per chromosome, are discarded at the earliest possible moment.

3. Knowledge-driven applications

3.1. *In silico appraisal of putative eQTL*

In Stranger's published list of multi-population *cis*-eQTL, gene UTS2 on chromosome 1 is distinguished as having SNP determinants of expression only in the Asian populations CHB-JPT. Figure 2 depicts the findings, chromosome-wide, in the CEU founders. The right-hand panel shows that the tests statistics in customary use are sensitive to outlying values for expression in small subsets of the data.

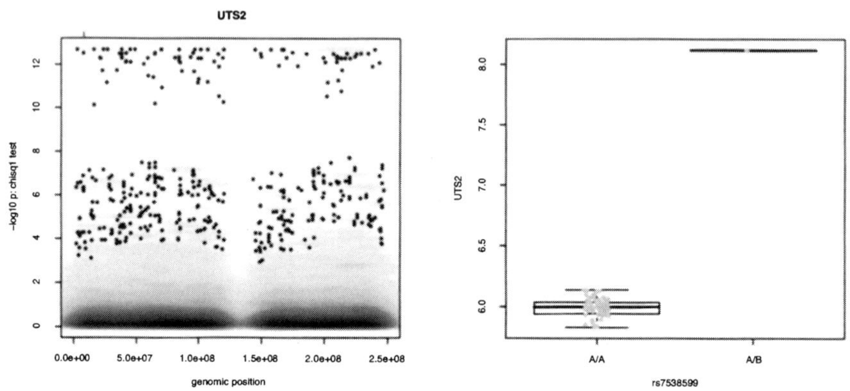

Figure 2. Left: whole-chromosome association analysis for gene UTS in the CEU founders. Right: the expression-genotype configuration that gives rise to the "many eQTL" appearance in the whole-chromosome analysis.

3.2. *Surveying a gene set for eQTL*

The *GSEABase* package provides convenient facilities for defining and translating gene sets between diverse nomenclatures. There are also facilities for importing reference gene set collections such as the Broad Institute's msigDB. We chose to study the motif-based set `V$FREAC2_01`, containing genes with promoter regions including a motif related to FOXF2 (forkhead box F2), because this set includes a gene (CPNE1) with a well-documented eQTL, and because FOXF2 is involved in activation of lung-specific proteins. Probes on the Illumina WG-1 expression array for the CEU founders were filtered to satisfy 1) membership in this gene set, 2) existence of unique Entrez identifier, and 3) in the case of multiple probes sharing an Entrez identifier, the probe with greatest IQR over all samples was retained. This yielded 201 probes; at time of writing, 140 have been analyzed as described here. Using genome-wide testing with snpdepth (as described above) set to 500 per chromosome, each gene is analyzed for eQTL in about three minutes on a Sun Blade with 8GB RAM.

Figure 3 gives lightweight visualizations of per-chromosome distributions of assocation statistics for four genes from the FOXF2 gene set. Some association statistic had to satisfy $-\log_{10} p > 6$ to be included; HABP4 seems to possess a straight *cis*-eQTL; PIK3C2A has a complex appearance; MCM7 appears to have a *trans*-eQTL; AKT2 may possess several.

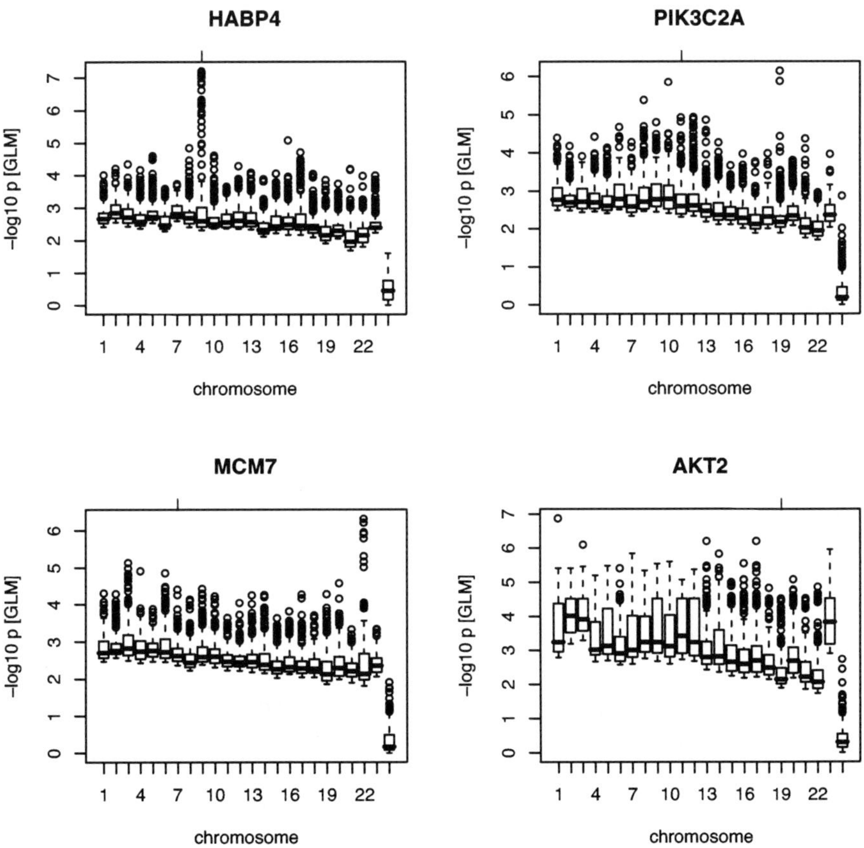

Figure 3. Left: whole-genome association analyses for four members of the FOXF2 motif-based gene set. Tick marks on upper bounding box are approximate location of coding region for each gene for which eQTL were assessed.

3.3. *Combining SNP-expression association scores with reference information on regulatory elements*

Results of **gwSnpScreen** can be transformed to UCSC browser track inputs (WIG format) using the **toTrackSet** method in conjunction with the *rtracklayer* package. Figure 4 shows a fairly coarse view of SNP and putative regulatory regions in the vicinity of CPNE1. There are many nonsynonymous coding SNP lying under the CNPE1-associated hump, along with various locations where there is evidence of regulatory elements. Much more information on functional impacts and correlates of polymorphic DNA must be brought to bear to further our understanding of diversity in gene expression.

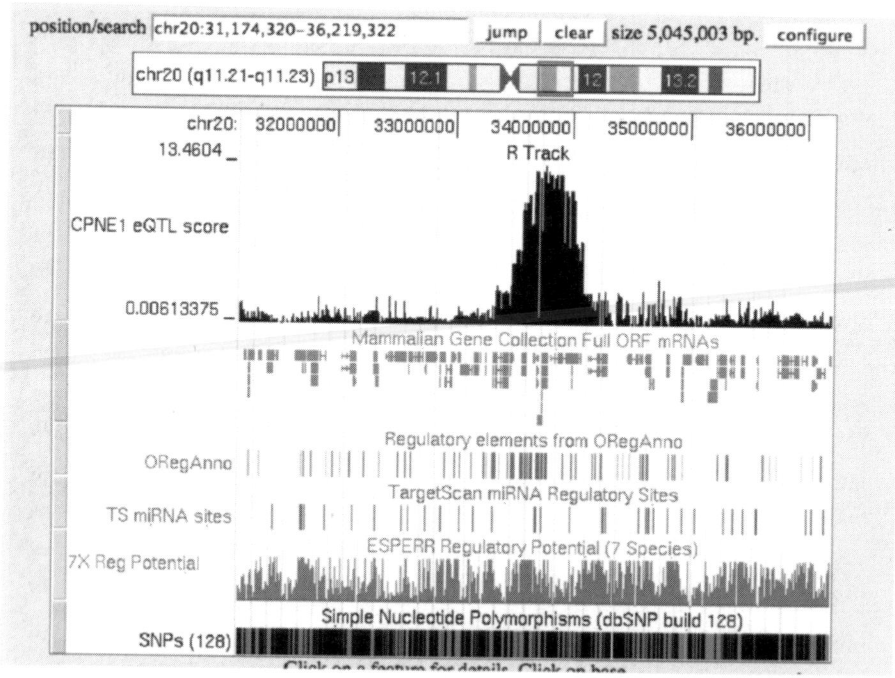

Figure 4. UCSC browser with custom track based on tests for eQTL for CPNE1. The score is $-\log_{10} p$ for the linear regression of log expression on copy number of SNP rare allele.

4. Discussion

In a recent survey of expression genetics, Williams and colleagues[5] suggest that the search for genetic explanation of expression variation is "somewhat simplistic", citing the many non-genetic determinants along with the wide variety of mechanisms by which genetic variation could affect gene expression. These authors also identify a number of technical problems of interpretation of eQTL statistics, including the effects of polymorphisms in hybridization probes, expression array batch effects, and effects due to expression array normalization. They complain that "a more disappointing general observation is that the ability to combine independent studies, even those carried out upon the same organism, is severely compromised by the multiplicity of mapping panels, genetic markers, statistical methodology, genes on arrays, and array platforms".

The approach described in this paper to investigating the relationships between expression variation and genotypic variation represents a step towards facilitating broader integration of multiple experiments and multiple forms of biologic metadata in studies of expression genetics.

- First, multiassay surveys of cohorts are represented in unified and coordinated objects with relatively simple but rich query resolution support. These objects can contain hundreds of samples with millions of SNPs and be manipulated interactively on commodity hardware.
- Second, genome-wide statistical analyses of expression-genotype associations are conducted using high-level facilities (including general covariate adjustments, and formulas involving gene sets as dependent variables) with good performance thanks to detailed programming with byte-level representations of SNP genotypes due to D. Clayton (package *snpMatrix*). These analyses also occupy coordinated computational objects that may be programatically transformed, queried, visualized as needed to identify biologically important interpretations.
- Third, a specific mechanism for integrating expression-genotype analysis results with biologic metadata available in the UCSC genome browser has been created on the basis of the *rtracklayer* package. The importation and visualization shown in this paper are complemented by the bidirectional aspect of the browser interface. Information on regulatory structures can be imported back into R for numerical and statistical analysis, to permit detailed

interpretation of observed *cis*- and *trans*- relationships.

It is well-acknowledged that much work remains to be done to create knowledge from the results of expression genetics experiments. Transparent and extensible computational architectures for representing and interpreting these experiments will play a fundamental role in these efforts.

References

1. URL: `www.sanger.ac.uk/humgen/genevar/`
2. E. Schadt, C. Molony, E. Chudin, et al., *PLoS Biology* 6(5):e107 (2008).
3. B. Stranger, A. Nica, M. Forrest, et al., *Nat. Genet.*, 39(10):1217 (2007).
4. D. Clayton, H. Leung, *Hum. Hered.*, 64:41 (2007).
5. R. Williams, E. Chan, M. Cowley et al., *Genome Res.*, 17:1707 (2007).

HIGH THROUGHPUT INTERACTION DATA REVEALS DEGREE CONSERVATION OF HUB PROTEINS

A. FOX[*,1], D. TAYLOR[1], and D. K. SLONIM[1,2]

[1]*Department of Computer Science, Tufts University,*
161 College Avenue, Medford, MA 02155, USA
[2]*Department of Pathology, Tufts University School of Medicine,*
145 Harrison Ave., Boston, MA 02111, USA
[*]*E-mail: afox01@cs.tufts.edu*

Research in model organisms relies on unspoken assumptions about the conservation of protein-protein interactions across species, yet several analyses suggest such conservation is limited. Fortunately, for many purposes the crucial issue is not global conservation of interactions, but preferential conservation of functionally important ones. An observed bias towards essentiality in highly-connected proteins implies the functional importance of such "hubs". We therefore define the notion of *degree-conservation* and demonstrate that hubs are preferentially degree-conserved. We show that a protein is more likely to be a hub if it has a high-degree ortholog, and that once a protein becomes a hub, it tends to remain so. We also identify a positive correlation between the average degree of a protein and the conservation of its interaction partners, and we find that the conservation of individual hub interactions is surprisingly high. Our work has important implications for prediction of protein function, computational inference of PPIs, and interpretation of data from model organisms.

1. Introduction

The power of comparative genomic research has grown steadily with the availability of genomic sequence and annotation for increasing numbers of organisms. A variety of techniques for solving such diverse problems as motif discovery,[1] gene expression analysis,[2] regulatory network inference[3] and interactome discovery[4] rely on the assumption that functionally important protein-DNA and protein-protein interactions (PPIs) are preferentially conserved across species. Yet much remains to be done to fully understand the conservation of protein interaction modules and functions.[5] Although studies of 'interologs' (orthologous pairs of proteins whose interactions are conserved)[6] report preferential conservation of interactions between highly similar orthologous protein pairs,[7] Mika and Rost[8] point out that inference of PPIs based on sequence homology across species continues to be quite inaccurate, even at very high levels of sequence identity.

For the interpretation of data from model organisms, such low interolog conservation might still be acceptable, provided that the most *functionally important* interactions for which orthologs exist in both species are conserved at an acceptably high rate. Yet even this remains to be shown. Recent work[9] shows that functional modules are no better conserved between yeast and fly than would be expected by chance. Furthermore, previous work in studying transcriptional regulation shows that despite the preferential sequence conservation of functionally important proteins, functional regulatory modules are not especially well conserved.[10]

Here, we focus not on the conservation of interactions among all proteins, but on the most highly-connected proteins, the "hubs" of a species' interaction network. We select these hub proteins precisely because their high connectivity may serve as an indication of functional importance. Indeed, proteins with many interactions are more likely to be essential,[11] and among these, hubs connected to other hubs may be more essential still.[12] If hub proteins are more likely functionally important, perhaps their roles as hubs are preferentially conserved, even when many of their individual interactions are not. We therefore investigate whether the property of being a hub protein is conserved across organisms, and we determine the level of PPI conservation between hubs and their neighbors. We show that interactions with high degree proteins are preferentially conserved and that even when specific interactions with hubs are lost or gained, the *high degree* of a hub protein is nonetheless conserved. Thus, there is greater hope that the the functional importance of these hubs is conserved as well.

We point out that there is ample reason to expect *a priori* that the high degree of hub proteins would be preferentially conserved during evolution.[13] This is based on the argument that once a protein has evolved functional interactions with many other proteins, any dramatic change would affect all of its many partners, and thus is likely to be evolutionarily disadvantageous. However, given that recent work[9,10] casts doubt upon modular functional conservation, this hypothesis requires proof.

There are also reasons to be skeptical of the quality and quantity of the available PPI data. Estimates of the false-positive rates for high throughput interaction data sets range from 45% to 90%,[14,15] and coverage estimates for the human and yeast interactomes are around 10% and 50%, respectively.[14,16]

Thus, any attempts to identify high degree proteins will be affected by noise, coverage, and testing bias.[16] Methods for estimating the true degree of interactions[17] are available, but are not directly applicable to the data sets studied here, because in many cases it is not clear which interactions were tested. Nonetheless, with only modest data filtering, we obtain clear evidence here of several important trends.

Our results provide additional evidence that identifying biologically relevant high-degree proteins in model organisms should shed light on human response. Our work also has important implications for protein function prediction, comparative genomics, and network inference and analysis.

2. Background and Definitions

2.1. *Data Sources*

All PPIs in *Homo sapiens*, *Drosophila melanogaster*, *Caenorhabditis elegans*, and *Saccharomyces cerevisiae* were downloaded from the BioGRID,[18] IntAct[19] and MINT[20] online databases in July, 2008 and combined to form a single large data set.

To map orthologous proteins between species, we used the InParanoid database[21] because it discriminates true orthologs and 'in' paralogs that differentiated after the relevant speciation event from 'out' paralogs that arose before speciation.

2.2. *Filtering Out PPI Noise*

To improve the reliability of the data,[17] we built a PPI graph for each assay method, with directed edges (b, p) indicating the roles of each protein as (b)ait or (p)rey where appropriate. We define the *baitrank* for a protein p in species s for assay type a as the fraction of proteins in species s having (non-zero) out-degree less than p's out-degree for assay a. The *preyrank* is defined analogously on the in-degree. If a given out-degree or in-degree is zero then the corresponding baitrank or preyrank is undefined.

We concluded that assay type a was *inconsistent* for protein p if both p's baitrank and preyrank were defined and $|baitrank(p) - preyrank(p)| > 0.1$. Under these conditions we removed data from that assay type for p. After this filtering, degrees were estimated using the naive method[17] from the remaining data. Overall, this approach removed 36% of the data, and led to stronger results. Full datasets and descriptive statistics are available at *http://bcb.cs.tufts.edu/hubsPSB09.*

2.3. *Ranking Protein Degrees*

Degrees of proteins in different species cannot be directly compared because each species has a different degree distribution. To address this issue, we ranked the proteins within each species by their degree in the filtered dataset for that species, and worked only with these ranks instead of the raw degree. We define r, the rank of protein p with degree d in species s, as the fraction of proteins in species s having non-zero degree less than d.

Individual proteins are labeled as hubs using a straightforward threshold on the rank. The hub threshold t is defined on the interval [0.5,1) since intuitively, it makes little sense to talk about a protein with rank less than 0.5 as a hub.

2.4. *Degree Conservation*

A protein is considered to have been *evolutionarily conserved* if it has one or more orthologous proteins in one or more of the other species under consideration. Most of the literature on conservation of hub proteins has focused on this type of sequence conservation.[22,23] At the other end of the spectrum, one can consider whether specific PPIs have been conserved, given pairs of proteins known to have orthologs in both species. (Such conserved interactions are frequently referred to as *interologs*.[6])

In this paper, we focus on the middle ground - looking at whether the interaction degree of a protein is conserved. This conservation can occur whether or not the specific interologs are maintained. We define a pair of orthologous proteins as *degree-conserved* at hub threshold t if both proteins are hubs at hub threshold t in their respective species. This definition can be naturally extended to more than two species.

3. Having an Orthologous Hub Increases Hub Likelihood

3.1. *Degree Conservation in Ortholog Pairs*

We first analyzed all pairs of orthologous proteins among the four species. We created a dataset D containing all proteins in the filtered data such that there are at least two species in which the proteins are orthologous. We were interested in determining whether proteins that have a hub ortholog have a significantly higher probability of being a hub protein.

To do this, we compare the observed posterior probability that a protein is a hub, given that it has an ortholog in another species that is a hub, to the prior probability of a protein being a hub in our data set. A simple way to define the prior probability of being a hub is just $(1-t)$, where t is the hub threshold. We call this the *uninformed prior*, $U(t)$. However, since we are considering only proteins with orthologs in 1 of the 3 other species under consideration, and since it has been shown that proteins with distant orthologs have higher average degree,[24] the true prior may be higher. Thus, we compute the *informed prior*, $I(t)$. Let $hub_t(p)$ indicate that p is a hub at threshold t and let $orth(x, y)$ indicate that x and y are orthologs. Then $I(t) = Pr[hub_t(p) = 1]$, which represents the observed fraction of proteins in the data set that are hubs. We compute $I(t)$ by dividing the number of proteins in D with rank $\geq t$ by the total size of D.

We then define the posterior probability $P(t) = Pr[hub_t(p_i) = 1 \mid \exists p_j$ in another species, s.t. $orth(p_i, p_j) \wedge hub_t(p_j) = 1]$. To compute $P(t)$, we first create the set S_t containing all proteins that have an ortholog in another species that is a hub at threshold t. Then $P(t) = |\{p_i \in S_t \mid hub_t(p_i) = 1\}|/|S_t|$. Intuitively, $P(t)$ is the fraction of proteins in S_t that are hubs at threshold t. Figure 1a shows $P(t)$, $I(t)$ and $U(t)$.

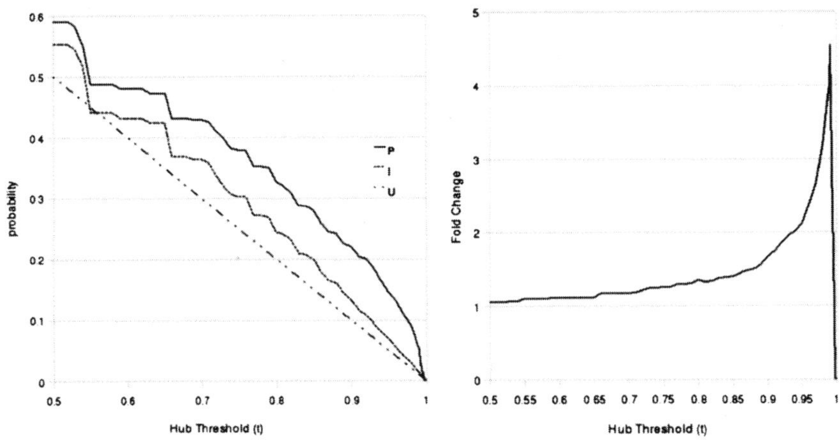

Fig. 1. Hub likelihood plots. a) Conditional probability of being a hub given that an ortholog is a hub, compared to the prior probability of being a hub. The solid line represents P(t), the posterior (conditional) hub probability; the dot-dash line U(t) is the uninformed prior probability, and dashed line I(t) is an informed prior based on the observed data. b) Fold Increase F in hub likelihood, given an orthologous hub protein, as a function of hub threshold t.

We compare the observed distribution P to the informed prior I using a Kolmogorov-Smirnov (KS) Test.[25] We used a 1-sided KS Test to determine whether $[P > I]$ on the intervals $t \in [x, 1)$ for each $x \in \{0.5, 0.55, \ldots, 0.95\}$. Tests were significant at $p < 10^{-6}$ on all intervals. These results support our hypothesis that the probability of a protein being a hub is higher if its ortholog is a hub, implying that hub proteins are degree-conserved more often than expected by chance. In the same way, we determined that I is significantly greater than U, as expected for the reason described above.

To determine the increase in hub probability given the existence of an orthologous hub, we plot the fold-change in probabilities $F(t) = P(t)/I(t)$ in Figure 1b. From the graph of $F(t)$ we can see that the probability a given protein is a hub is greater if it has a hub ortholog and the magnitude of this effect increases with higher hub threshold. Indeed, $F(t)$ peaks at approximately $t = \gamma = 0.99$. When $t > \gamma$ the expected number of conserved hubs quickly approaches zero, due to the diminishing expectation of the number of *degree-conserved* pairs in the data set (the numerator of $P(t)$) for $t > \gamma$. For hub thresholds above $t \approx 0.93$ we see a two-fold increase in hub likelihood.

3.2. *Degree Conservation Across Four Species*

We next extend the pairwise ortholog analysis to look at ortholog groups across all four species. We create 4-component rank vectors $\mathbf{x_i} = [rank(y_i), rank(w_i), rank(f_i), rank(h_i)]$ where y_i, w_i, f_i and h_i are proteins that form an ortholog group across yeast, worm, fly and human. Let n_T be the total number of rank vectors, $n(t)$ be the number of rank vectors that exhibit degree-conservation at hub threshold t, and $O(t) = n(t)/n_T$ be the proportion of ortholog groups that exhibit hub conservation at hub threshold t. We also compute the expected rate of conservation if degree is not conserved by orthology and call this $E(t)$. Then $E(t) = \prod_{c \in M} Pr[r > t | r \in c]$, where M is the full matrix of rank 4-vectors generated from the orthology groups and c represents the column of ranks specific to a particular species. If $O(t)$ is significantly greater than $E(t)$ then we can conclude that degree is in fact conserved by orthology.

We compare the observed rate of 4-species degree-conservation $O(t)$ to the expected rate $E(t)$ using a series of KS tests. We tested each interval $t \in [a, b]$ where $a \in \{0.5, 0.51, \ldots, 0.95\}$ and $b \in \{0.55, 0.56, \ldots, 1.0\}$ and $a \leq (b - 0.05)$. The KS test results showed that $[O > E]$ for all t-intervals fully subsumed by the interval $t \in [0.66, 0.95]$ ($p < 10^{-6}$) and we conclude that high degree is conserved across the four considered species.

The lack of 4-species degree-conservation for $t > 0.95$ may be attributed to incomplete interaction data, low overlap of interactome coverage between species,[26] or a combination of these effects.

To gain an understanding of the functional classes of proteins that are degree-conserved, we performed functional enrichment analysis (DAVID[27]) of all yeast proteins that are degree-conserved ($t = 0.8$) in at least 1 of the other 3 species. Enriched functions included primary metabolism, protein synthesis, splicing, DNA repair and regulation of the cell-cycle (FDR < 0.05). Enrichment of basic cellular processes is expected given that these hubs are required to be degree-conserved between yeast and higher eukaryotes. In contrast, degree-conserved ($t = 0.8$) proteins between human and fly were enriched for processes such as cell signalling (the p53, Wnt, ErbB, Notch and TGF-β pathways), cell differentiation, cellular developmental processes, post-translational protein modification and the regulation of protein kinase activity (FDR < 0.05). Full details of functional enrichment results are available online at *http://bcb.cs.tufts.edu/hubsPSB09*.

4. Once a Hub, Always a Hub

Our hypothesis that hubs exhibit degree-conservation relies on the intuition that once a protein has evolved many connections, it is difficult for that protein to lose its importance because so many neighbors would be affected by the change. However, our analyses so far fail to account for orthologs that may have diverged *before* one of the proteins evolved its observed "hub" role. A simple evolutionary model that expects hubs to remain highly connected once they become so might account for a larger fraction of the data. We call this the "once a hub, always a hub" hypothesis.

To test this hypothesis, we describe a simple phylogenetic tree (Figure 2a) relating the four species in our data set. We then seek to determine how closely the data matches this phylogenetic model. For each ortholog group

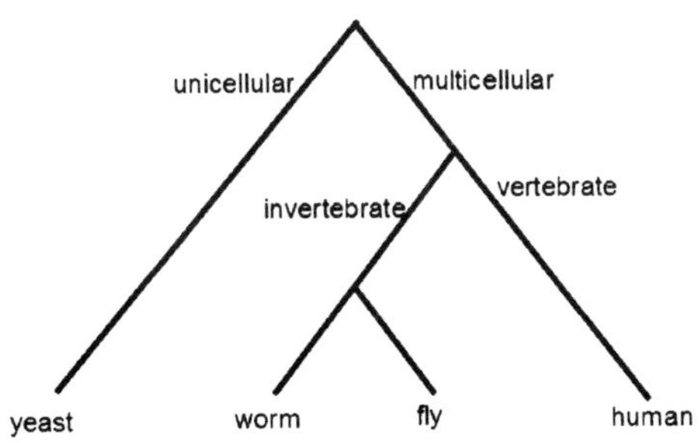

Consistent YWFH	Inconsistent YWFH
0000	0011
0001	0101
0010	1001
0100	1010
1000	1011
0110	1100
0111	1101
1111	1110

Fig. 2. a) Simple phylogenetic model of the four species in our data set, based on data from the Tree of Life database (www.tol.org). b) Consistency of hub-status bitstrings (yeast,worm,fly,human) with the 'Once a hub, always a hub' hypothesis.

spanning all four species, we form a rank vector as in Section 3.2. We then create a 4-bit bitstring describing the hub status of its component protein in each species. (If a given protein's rank is greater than the hub threshold t, then that species' component of the bitstring is 1; otherwise it is 0.) An ortholog group is considered *consistent* with the model (with assumption of parsimony) if its bitstring can be explained by the evolution of the hub property at at most one point in the tree (Figure 2b).

We then consider all possible ortholog groups over four species. Figure 3 shows how the fraction of ortholog groups consistent with the model grows as the hub threshold increases. For $t > 0.73$, the majority of the data can be explained by this straightforward phylogenetic model in which the hub property is under selective pressure to be retained once it has evolved.

One caveat is that for sufficiently high t, all the data will have an all-zero bitstring which is consistent with the model. Thus, potentially all the growth that we see in the fraction of consistent ortholog groups is due to that one category. However, we show that this is not the case by repeating the analysis but excluding the all-zero data (Figure 3). Full details of the consistency analysis are provided at *http://bcb.cs.tufts.edu/hubsPSB09.*

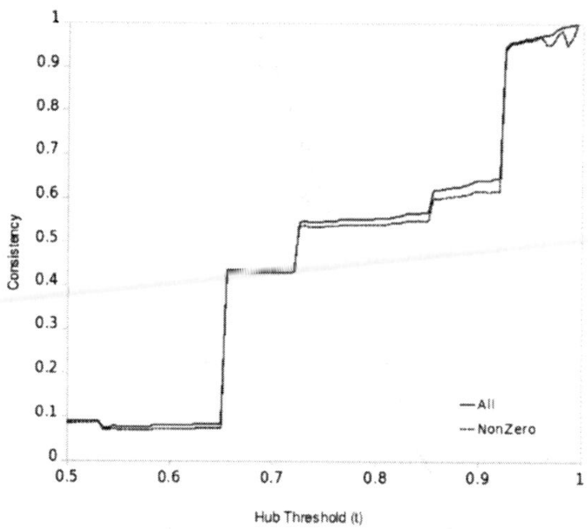

Fig. 3. Consistency with the "once a hub, always a hub" hypothesis, as a function of hub threshold. The y-axis shows the percentage of ortholog groups that are consistent with this hypothesis, given the phylogenetic model from Figure 2. The solid line represents all consistent ortholog groups; the dashed line excludes the all-zero vectors (proteins that are not hubs in any species) from the computation.

5. Hub Interactions Are Preferentially Conserved

Finally, we examine whether the retention of the hub property is due to the retention of individual PPIs, or simply to the general importance of high degree proteins. We consider two models. In one, interologs are conserved sufficiently well - at least in high degree proteins - that the "module" structure of a hub and its neighbors is retained. This suggests that the function of the protein and its partners is likely to be conserved as well. The alternative hypothesis assumes that once a protein has evolved a large number of partners, it is under pressure to evolve more slowly.[24] Although many of the protein's interaction partners may be lost over time, the high sequence stability of hubs may allow new interactions to arise and be retained more easily. Note that the former hypothesis is somewhat at odds with a growing body of work suggesting that interologs are conserved at only a moderate level,[7,26] and functional modules are generally *not* conserved intact.[9]

To address this question, we analyzed the conservation of individual interactions as a function of protein interaction degree. To measure the conservation of interactions for a given protein p in species s_1, in a designated second species s_2, we rely on the following definitions. First, let G_1 be the protein-protein interaction graph for species s_1: $G_1 = \langle V_1, E_1 \rangle$, and let G_2 be similarly defined for species s_2. Furthermore, let the relation $orth$, defined on $V_1 \times V_2$, denote orthologous pairs of proteins between the two species. Let the potential number of conserved interactions of p, $Poten(p)$ be the number of protein interaction partners q of p in s_1 such that orthologs of both p and q exist in s_2. Also let the actual number of conserved interactions of p, $Conserv(p)$ be the number of protein interactions of p in s_1 that are conserved in s_2. Formally, for protein $p \in V_1$, we define $Poten(p) = |\{q \in V_1 | \langle p, q \rangle \in E_1 \wedge (\exists p', q' \in V_2 | orth(p, p') \wedge orth(q, q'))\}|$ and $Conserv(p) = |\{q \in V_1 | \langle p, q \rangle \in E_1 \wedge (\exists p', q' \in V_2 | orth(p, p') \wedge orth(q, q') \wedge \langle p', q' \rangle \in E_2)\}|$. Then we can compute the conservation rate $R(p) = Conserv(p)/Poten(p)$ which is the proportion of interactions conserved out of all interactions that could possibly have been conserved.

We measured the conservation rate $R(p)$ of each protein in each pair of species compared these to $rank(p)$. Results for the fly proteins are shown in Figure 4. The plot reveals that the proportion of a protein's interactions that are conserved increases as protein rank (or degree) increases. In other words, hubs show preferential conservation of individual protein interactions over non-hubs.

For other species pairs, the graphs are similar in shape, though the overall conservation of the highest degree proteins is often lower (65%-85%). The especially high conservation between fly and human interologs may in part reflect a bias in how protein interactions were chosen as test assays for these particular species. Even for the less similar species pairs, however, we find that at high hub thresholds, more than half of interologs are conserved. This contrasts sharply with the more general analysis of Mika and Rost,[8] which shows interolog conservation hovering around 25% even at the highest levels of sequence similarity, using data that overlaps significantly with ours. We conclude that the interolog conservation we see with hub proteins may not simply be due to the fact that highly-connected proteins have higher sequence conservation, but that the interactions are truly preferentially conserved.

We also note that for a number of the species pairs, a significant fraction (up to 40%) of the interactions are *not* conserved even at high hub thresholds. This suggests that the degree-conservation of hubs may not rely solely on the conservation of individual pairwise interactions, but on the functional roles of the hub proteins themselves.

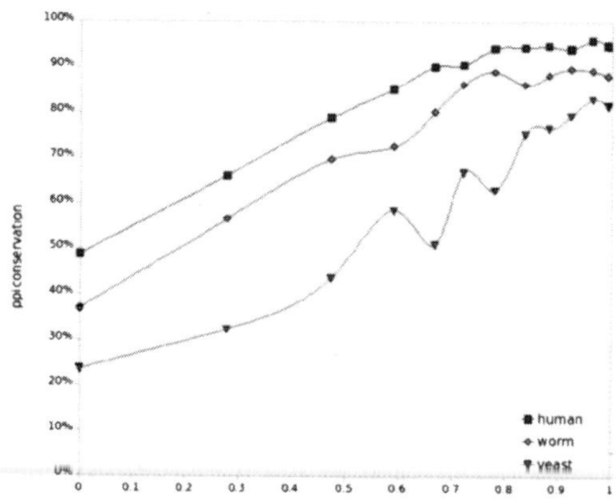

Fig. 4. Interolog conservation as a function of protein interaction degree rank. The data shown are for all fly proteins with orthologs in at least one of the other three species. Similar-rank proteins are binned and their conservation rates averaged so the trend can be seen clearly. Conservation of interologs is low for low-degree proteins, but increases as a function of degree-rank, and is high for proteins above a typical hub threshold of 0.8 or greater.

6. Discussion

We have demonstrated three important and novel facts about conservation of protein interactions and function. First, the likelihood of a protein being a hub increases if it has an ortholog that is a hub. Second, once a protein evolves to become a hub, it tends to remain so, even though specific interaction partners may be gained or lost. Finally, the rate of interolog conservation for hub proteins is much higher than for low-degree proteins.

These results imply that, whatever can be said about functional conservation of modules defined by various clustering methods (e.g., Wang and Zhang[9]), functional modules around network hubs are likely to persist across species. This result has crucial implications for the unspoken assumption underlying most research in model organisms: that the functional relationships between key proteins are conserved between humans and the model organisms being studied. Furthermore, the "once a hub, always a hub" property implies that it might be possible to infer the identity of hub proteins even in organisms where sufficient protein interaction data sets are not yet available. For example, if two orthologous proteins in worm and fly are both PPI hubs, it may be reasonable to assume that the evolution of the hub property occurred in some ancestor of both worm and fly; therefore, an orthologous protein in bees is likely to be a hub as well.

In addition, the stronger-than-expected conservation of individual interactions with hub proteins has important implications for the computational prediction of protein-protein interactions. Although the growth of experimentally-generated protein interaction data is substantial, the interactomes of many organisms are likely to remain largely incomplete for some time, so computational inference of interactions will continue to play an important role. Furthermore, as suggested by Matthews *et al.*,[4] computational predictions can be used to guide the selection of experimental assays, so higher predictive accuracy can reduce the cost of such screening, allowing the generation of more data.

Despite these results, there are many issues surrounding the data that require some caution in interpreting any results based on existing protein interaction networks. Most importantly, the data are both noisy and extraordinarily incomplete. Since different subsets of the interactomes of the four species have been investigated, it is possible that bias in species coverage may affect any conclusions based on the data currently available. In addition, some methods of identifying protein interactions (such as coimmunoprecipitation and affinity precipitation[9]) identify protein complexes rather than pairwise interactions. Many pairwise interactions are sometimes inferred for the participants in these complexes,[28] potentially creating high-degree nodes erroneously. Thus, it will be informative to revisit these conclusions as the quality and quantity of interactome data grow.

Acknowledgments

This research was supported by grant number 004911 to DKS from the National Library of Medicine. We thank the members of the Tufts BCB research group for helpful discussions and comments.

References

1. M. Blanchette and M. Tompa, *Genome Res.* **12:5**, p. 739 (2002).
2. M. Eisen, P. Spellman, P. Brown and D. Botstein, *PNAS* **95**, p. 14863 (1998).
3. J. Yu, V. Smith, P. Wang, A. Hartemink and E. Jarvis, *Bioinformatics* **20**, p. 3594 (2001).
4. L. Matthews, P. Vaglio, J. Reboul, H. Ge, B. Davis, J. Garrels, S. Vincent and M. Vidal, *Genome Res.* **11**, 2120 (2001).
5. J. Han, N. Bertin, T. Hao, D. Goldberg, G. Berriz, D. Dupuy, A. Walhout, M. Cusick, F. Roth and M. Vidal, *Nature* **88**, p. 6995 (2004).
6. A. Walhout, R. Sordella, X. Lu, J. Hartley, G. Temple, M. Brasch, N. Thierry-Mieg and M. Vidal, *Science* **287**, 116 (2000).
7. H. Yu, N. Luscombe, H. Lu, X. Zhu, Y. Xia and J. Han, *Genome Res* **14** (2004).
8. S. Mika and B. Rost, *PLoS Comput. Biol.* **2**, p. e79 (2006).
9. Z. Wang and J. Zhang, *PLoS Comp Biol* **3** (2007).
10. M. Babu, N. Luscombe, L. Aravind and M. Gerstein, *Curr Opin Cell Biol* **14**, 283 (2004).
11. H. Jeong, S. Mason, A. Barabasi and Z. Oltvai, *Nature* **411**, 41 (2001).
12. N. Batada, T. Reguly, A. Breitkreutz, L. Boucher, B. Breitkreutz, L. Hurst and M. Tyers, *PLoS Biol* **4**, p. e317 (2006).
13. S. Wuchty, A. Barabasi and M. Ferdig, *BMC Evol Biol* **6**, p. 1471 (2006).
14. G. Hart, A. Ramani and E. Marcotte, *Genome Biol.* **7**, p. 120 (2006).
15. P. D'haeseleer and G. Church, 216(Aug 2004).
16. T. Chiang, D. Scholtens, D. Sarkar *et al.*, *Genome Biol* **8**, p. 186 (2007).
17. D. Scholtens, T. Chiang, W. Huber and R. Gentleman, *Bioinformatics* **24**, 218 (2008).
18. C. Stark, B. Breitkreutz, T. Reguly, L. Boucher, A. Breitkreutz and M. Tyers, *Nucleic Acids Res* **34**, D535(Jan 2006).
19. H. Hermjakob, L. Montecci-Palazzi, C. Lewington, S. Mudali, S. Kerrien *et al.*, *Nucleic Acids Res* **32**, D452 (2004).
20. A. Zanzoni, L. Montecci-Palazzi, M. Quondam, G. Ausiello, M. Helmer-Citterich *et al.*, *FEBS Letters* **513**, 135 (2002).
21. M. Remm, C. Storm and E. Sonnhammer, *JMB* **314**, p. 1041 (2001).
22. H. Fraser, A. Hirsh, L. Steinmetz, C. Scharfe and M. Felman, *Science* **296**, 750 (2002).
23. I. Jordan, I. Wolf and E. Koonin, *BMC Evol Biol* **3**, p. 1 (2003).
24. K. Brown and I. Jurisica, *Genome Biol* **8** (2007).
25. W. Press, S. Teukolsky, W. Vetterling and B. Flannery, *Numerical Recipes in C, the Art of Scientific Computing* (Cambridge Univeristy Press, 1992).
26. T. Gandhi, J. Zhong, S. Mathivanan *et al.*, *Nature Gen* **38** (2006).
27. G. D. Jr., B. Sherman, D. Hosack *et al.*, *Genome Biol* **4**, p. 3 (2003).
28. G. Bader and C. Hogue, *Nature Biotechnol* **20**, 991 (2002).

UNISON: AN INTEGRATED PLATFORM FOR COMPUTATIONAL BIOLOGY DISCOVERY

REECE K. HART, KIRAN MUKHYALA
Genentech, Inc.
1 DNA Way
South San Francisco, CA, 94080, USA
E-mail: rkh @ gene.com
Revision 187 (2008-10-07 23:10)

This paper describes the design and applications of Unison, a comprehensive and integrated warehouse of protein sequences, diverse precomputed predictions, and other biological data. Unison provides a practical solution to the burden of preparing data for computational discovery projects, enables holistic feature-based mining queries regarding protein composition and functions, and provides a foundation for the development of new tools. Unison is available for immediate use online via direct database connections and a web interface. In addition, the database schema, command line tools, web interface, and non-proprietary precomputed predictions are released under the Academic Free License and available for download at http://unison-db.org/. This project has resulted in a system that significantly reduces several practical impediments to the initiation of computational biology discovery projects.

1. Introduction

Computational biology projects frequently begin with the time-consuming process of downloading, formatting, characterizing, and integrating data of disparate types and sources [1]. These data preparation tasks require significant effort while contributing little to the distinguishing intellectual value of a project. Furthermore, these efforts are often duplicated for other projects, by other scientists, and even by the same scientist for the same project when new source data are released. The data preparation burden is a barrier to efficient and reproducible computational discovery efforts.

From a computational biologist's point of view, there are several important functional criteria for integrative databases: completeness and currency of the data, the breadth of source data and data types, query speed, technical accessibility, legal accessibility (*i.e.*, licensing restrictions), and the extent of semantic integration. Semantic integration means that data of the same type are modeled similarly and that appropriate relationships are established among data, regardless of source. Importantly, users experience semantically integrated databases as a representation of familiar concepts rather than a collection of proprietary data models (even if freely available). Reliable reasoning with data necessitates consistent, well-defined and well-understood definitions of the modeled data.

There are many techniques for data integration (for review, see [2][3]). Link integration, such as that provided by web pages, enables users to follow prescribed links between data sources. Although link integration is useful for web browsing, it is insufficient for reliable and semantically precise querying. Similarly, full text indexing greatly facilitates searching, but it is insufficient for the reliable integration of concepts.

Database federation semantically integrates data that are stored remotely. The principle advantage of federation is that queries are always based on current data, but the drawbacks are the run-time dependencies on external resources, poor query performance compared to that for locally stored data, and the difficulty of devising interfaces that translate the semantics of the remote data models to those of the integrated schema.

Data warehouses, in which source data are aggregated within one database environment, eliminate external database dependencies and generally provide better query performance than other integration methods. The simplest data warehouses replicate source data locally but provide little or no semantic integration of the data. Because data are local, performance is improved relative to that of remote databases. More frequently, data warehouses provide a semantically integrated schema by creating database views to local replicas of source data ("view integration" [3]) or by materializing source data within an integrated schema during loading.

Despite the existence of several high-quality and well-known integrative databases (*e.g.*, ATLAS [4], BioMart [5], InterPro [6], RefSeq [7], STRING [8], UniProt [9], and others), data preparation for new computational discovery projects remains burdensome. There are technical, practical, and legal reasons that current databases do not meet the needs of computational biologists [2][3]. A few of these reasons are: limited content/project specificity, inaccurate or out-of-date data, limited access methods, necessity for local deployment, and licensing restrictions. Although no system will meet all needs of all users, there is a significant and unmet need for a standardized integration platform that lessens the data preparation burden of a broad audience.

Unison is a comprehensive data warehouse of a superset of nearly all available protein sequences (currently, 12M from 20 sources), extensive precomputed proteomic predictions (200M of 18 distinct types), and diverse auxiliary data. Unison includes predictions of protein domains and motifs, signal and transmembrane domains, secondary and tertiary structure, disorder, cellular localization, phosphorylation, and genomic alignment and clustering. The motivation for Unison is to lessen the data preparation burden of computational biologists by providing a standardized integration platform of commonly-used source databases and computationally expensive proteomic predictions. The integration enables the same resource to be used for traditional per-sequence domain analysis and for complex, holistic data mining queries regarding protein domain composition, structure, and function. Unison's schema is designed for incremental updating with respect to source sequences, models, and computational methods. The entire update process is fully automated. The complete Unison package – schema, tools, web site, underlying database, and precomputed data – are freely available online for immediate use and for download.

2. Methods

2.1. *Schema overview*

Unison is principally a data warehouse of sequences, annotations, and precomputed predictions in an integrated schema. Familiar concepts that transcend multiple data sources, such as protein sequences and Hidden Markov Model (HMM) alignments, are modeled as abstract types in the semantically integrated core schema. Specialized data, such as NCBI GeneRIF and SCOP, are incorporated as auxiliary data without remodeling.

One of the tenets of Unison's core schema is to represent only the salient features of an entity or concept rather than to fully represent the character of each source database. As a result, most tables in the Unison schema are concise representations of the essential features of a biological entity or piece of information. This design decision is consistent with the goal of including data only when it is likely to inform queries or analysis. Links to source data enable users to pursue source-specific content. Unison makes extensive use of table inheritance to model abstract data types and their concrete descendants.

Normalization is essential to enabling efficient incremental updates of source data, which is one of Unison's primary goals. Nearly all tables in Unison are third normal form [10]. In rare cases when normalization leads to unacceptable performance loss, materialized views are provided. In practice, this level of normalization is readily understood by Unison's users and does not impose a barrier to use. Nonetheless, Unison provides a functional layer of views that provide a simplified, pragmatic, and stable query interface for users.

2.2. *Essential schema objects*

The essential objects modeled within the Unison schema are shown in Figure 1 and described below. The full schema and schema documentation are available online.

The `origin` table stores the provenance of all data, the version of the source database, a Uniform Resource Identifier (URI) to information about the source (*e.g.*, a project's "home page"), a URI of the file or directory from which data were loaded, and a URI template that is used to construct web page links to source records. A flag also indicates whether the data are publicly distributable.

Protein sequences are stored non-redundantly in the `pseq` table and referred to by a the primary key, `pseq_id`. A database trigger on this table computes the MD5 checksum on the protein sequence during loading, and a uniqueness constraint on the checksum ensures that sequences are distinct. A second trigger prohibits updating or deleting any sequence so that predictions are

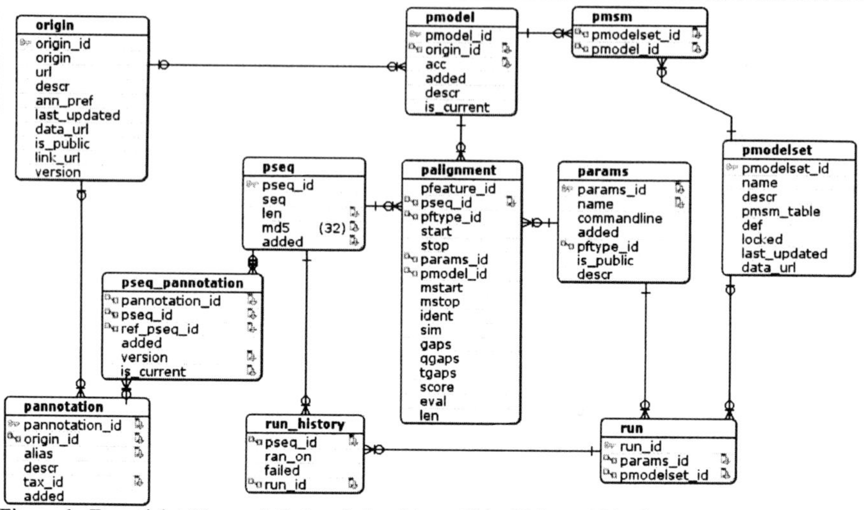

Figure 1: Essential tables and their relationships within Unison. This figure represents a small subset of the schema and a majority of the content of Unison.

guaranteed to reflect the stored sequence; sequence changes require inserting a new sequence.

The `pannotation` table represents the “alias” (accession or identifier from a source database), description, and taxonomic annotation of a sequence. A uniqueness constraint prohibits duplicate aliases within the same origin. The `pseq_pannotation` table maps and versions annotations in the `pannotation` table to sequences in the `pseq` table. Annotation versioning enables Unison to properly track obsolete aliases and changes to a sequence changes that is associated with a single alias. The current annotations for sequences are available through a view.

The `params` table represents prediction methods and their invocation. Each row includes a primary key, `params_id`, the command line that specified how a program is invoked, and a flag that indicates whether the results of the method are publicly distributable. The command line provides important traceability of all predictions in Unison.

Some prediction methods depend on additional external input, generically called models and represented using an abstract base table, `pmodel`. For example, Position Specific Scoring Matrices (PSSMs), HMMs, and Prospect Pro (http://bioinformaticssolutions.com/) protein structure templates are represented by `pmpssm`, `pmhmm`, and `pmprospect` respectively, each of which inherits from the abstract table, `pmodel`. The `pmodel` table and its derived tables contain a primary key, `pmodel_id`. Sets of models are represented by `pmodelset`.

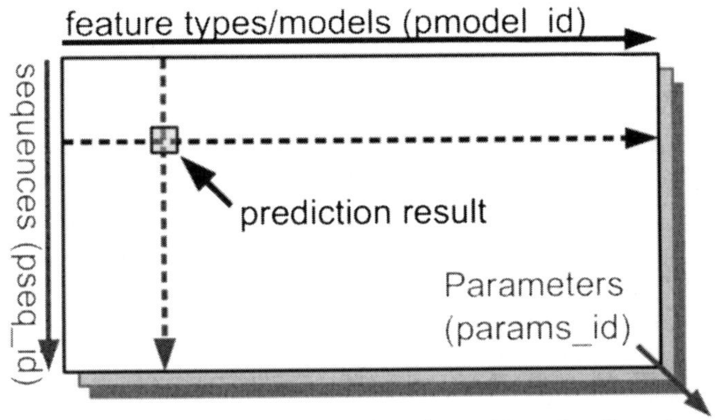

Figure 2: Visualization of the prediction results volume. The cube dimensions are unique sequences, feature types/models, and parameters. An analysis of protein features corresponds to horizontal slices of data. Feature-based mining corresponds to a vertical slices of data.

The `run` table is used to associate parameters and appropriate model sets, referenced by `params_id` and `pmodelset_id` respectively. The `params`, `pmodelset`, and `run` tables together permit Unison to distinguish how a method is invoked on a sequence and which models a sequence is run against.

Prediction results in Unison are represented using an abstract base table, `pfeature`, that represents a localizable feature on a protein sequence, specified by the foreign key `pseq_id`, using a specific prediction method, specified by the foreign key `params_id`. By including `params_id` in this table, Unison explicitly models alternative invocations of a predictive method and this, in turn, enables specialized or exploratory work using multiple prediction parameters with a single method. The `palignment` table is a subclass of `pfeature` that represents an alignment of sequence to a model, as specified by the foreign key `pmodel_id`. Results for each prediction method are modeled as distinct types by subclassing `pfeature` or `palignment` as appropriate and adding columns that are specific to the prediction method. Readers may benefit from visualizing the arrangement of prediction results as a volume of distinct predictions, each of which depends on sequence (`pseq_id`), method (`params_id`), and an optional model (`pmodel_id`), as shown in Figure 2.

Unison's schema includes a `run_history` table to track which sequence analyses – that is, runs from the `run` table – were performed on which sequences and the date of the execution. Run histories enable Unison to provide incremental updates with respect to new sequence releases, new versions of predictive methods, and new models/modelsets.

Hidden Markov Models and sequence alignments to them permit a concrete and representative discussion of the schema in practice. HMMs for HMMer [11] are available from many sources, such as PANTHER [12], Pfam [13], Superfamily [14], and in-house efforts. Such sources are stored in the `origin` table. HMMs from those sources may be loaded into `pmhmm` with an obligatory reference to the model origin. Details specific to HMMer-built HMMs, such as score cutoffs, are included in the model table to enable advanced queries using these criteria. HMM alignments are represented using `pahmm` (Protein Alignment HMM), which is a subclass of `palignment` and therefore contains foreign keys to the query sequence (`pseq_id`), the parameters (`params_id`), and the HMM (`pmodel_id`), in addition to the positions of the alignment on the sequence and the model, the score, the E-value for the individual alignment, and other alignment details. When the alignments are loaded, the `run_history` table is updated to reflect this. The `pahmm_v` view provides a denormalized relation that simplifies the querying of HMM data.

2.3. *Loading and Updating Unison*

Most of the Unison's initial build process and all of the update process, including data downloads and submission of prediction jobs to a compute cluster, are automated by a series of makefiles and Perl scripts. The loading process consists of several phases that must proceed in order and without error before continuing to the next phase.

In phase 1, auxiliary data are loaded. These data include cytogenetic bands from UCSC [15], Gene Ontology [16], Human Genome Organization official gene names [17], NCBI Entrez Gene, GeneRIF, HomoloGene and taxonomy files [7], PDB [18], and SCOP [19]. The PDB schema includes structure summary data, ligands with canonical names and various descriptors, and an explicit sequence-to-residue mapping to facilitate the localization of primary sequence features on structures.

In phase 2, sequences and annotations are loaded from approximately 30 public sequence sources; the list of origins and currency is available at http://unison-db.org/contents.pl. Sequence loading from the commercial GENESEQ patent database [20] is also supported, but these data are not in the public release. Within Genentech, in-house sequences and annotations are also loaded.

In phase 3, three nested sets of sequences – runA, runB, and runC – prioritize sequences by origin, species, size and other criteria. Because predictions take widely varying time, expensive computations such as protein structure predictions are maintained for reliable human sequences of moderate size (runA); methods of intermediate computational cost are maintained for certain eukaryotic sequences from reliable sources (runB); and inexpensive computations are maintained for sequences from a wider selection of sources and species (runC). Ad hoc analyses of any type may be arranged for any sequence.

In phase 4, sequences are submitted to a compute cluster queue for analysis and the resulting data are loaded. The sequences to be run are determined by subtracting the set of sequences for which data are already available, as determined by the `run_history` table, from the run set that is appropriate for the analysis type. The 18 methods and command lines are listed at http://unison-db.org/contents.pl. Unison itself does not infer any features by sequence similarity, although many methods use sequence similarity internally.

In phase 5, materialized views that depend on phase 4 data are built. In phase 6, database summary statistics, maintenance, and cleanup are performed. The public version of Unison is built by specialized scripts that extract only the publicly distributable sequences, annotations, and predictions.

2.4. *Hardware and software requirements*

Unison is implemented in the PostgreSQL relational database running in a Linux/GNU environment. Deploying the database server and Unison on other operating systems should be straightforward. Installation requires approximately 200 GB of disk space. Although the extensive use of table inheritance and server-side functions in C and Perl would make porting to other database systems difficult, this is under consideration.

The Unison Perl API facilitates access to the Unison database and provides many convenience functions. The API makes extensive use of BioPerl [21] and other modules readily available from the Comprehensive Perl Archive Network.

The Unison web interface runs on Apache in a Linux/GNU environment. All web pages are currently implemented as Perl CGI scripts to enable Kerberos authentication. This configuration causes user authentication and authorization to be delegated to the database and security policies to be enforced there rather than in the middle tier. Neither registration nor authentication is required for the public version.

2.5. *Accessing Unison*

Direct database access to Unison is available at host `unison-db.org`, port `5432`, database `unison`, and username `PUBLIC` via the PostgreSQL native protocol, ODBC, JDBC (for Java), and SDBC (for OpenOffice). ODBC access within R has been tested. These connection methods are not specific to the client or server operating system. The number of concurrent connections and statement query times are limited to promote equitable access.

Web access is available at http://unison-db.org/. Unison web pages use URLs with simple “GET”-method query arguments, thereby facilitating programmatic linking to content. For example, all sequence analysis pages allow specification of the sequence by any recognized alias, protein sequence, MD5 checksum, or Unison's internal `pseq_id`. MD5 checksums provide an intrinsic key that obviates inter-database coordination of protein sequence accessions.

```
CREATE OR REPLACE VIEW itim_search_v AS
SELECT IG.pseq_id,
       IG.start as ig_start,IG.stop as ig_stop,
       TM.start as tm_start,TM.stop as tm_stop,
       ITIM.start as itim_start,ITIM.stop as itim_stop
  FROM pahmm_current_pfam_v IG
  JOIN pftmhmm_tms_v T ON IG.pseq_id=TM.pseq_id   AND IG.stop<TM.start
  JOIN pfregexp_v ITIM ON TM.pseq_id=ITIM.pseq_id AND TM.stop<ITIM.start
 WHERE IG.name='ig'
       AND IG.eval<1e-2
       AND IG.params_id=params_id('HMMer')
       AND ITIM.acc='MOD_TYR_ITIM';
```

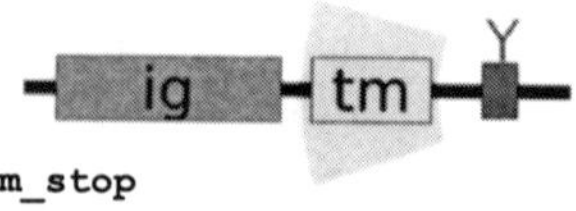

Figure 3. A canonical ITIM-containing protein and a query which mines Unison for such proteins. The query identifies sequences that contain an extracellular ig domain, a TMHMM transmembrane domain, and an intracellular Immunoreceptor Tyrosine Inhibitory Motif using the regular expression from ELM.

The Unison schema, tools, makefiles, web pages and documentation are released under the Academic Free License and available for local deployment. Non-proprietary data (sequences, annotations, and methods) are available as a PostgreSQL database dump. The source code repository is also publicly accessible. All users are advised to join the mailing list. Links to all of these resources are provided at the Unison web site.

3. Application of Unison to Immunoreceptor Tyrosine Inhibitor Motif (ITIM) proteins

A demonstration of the utility of a comprehensive set of sequences, precomputed predictions, and auxiliary data is best served by example. Immunoreceptor Tyrosine Inhibitory Motif-containing proteins are an important class of immune system regulators (see [22] for review). ITIMs are short (6-8 amino acid) sequences in the intracellular domain of immune receptors. ITIM-containing receptors were initially identified on the surface of NK cells and macrophages where they are believed to prevent self reactivity. Phosphorylation of the tyrosine within an ITIM leads to the binding of, for example, SHP1 or SHP2 phosphatases and the attenuation of a corresponding activating receptor.

Canonical ITIM-containing proteins possess an extracellular immunoglobulin domain and a transmembrane domain in addition to the intracellular ITIM. This simple biological model may be translated into a Structured Query Language [10] command that searches for such proteins within Unison, as shown in Figure 3. This query may be modified to use other HMMs, structure prediction or PSSMs to identify immunoglobulin domains, or to identify related Immunoreceptor Tyrosine Activating or Switch Motifs. This query takes approximately 20 seconds on an otherwise idle system.

Unison's extensive precomputed data enables queries that develop new hypotheses. The first known ITIM-containing proteins possessed immunoglobulin extracellular domains. Suspecting that diverged ITIM protein homologs might contain alternative extracellular interaction domains, Unison was used to identify and count all extracellular Pfam domains in the context of a transmembrane domain and intracellular ITIM. Models of immunoglobulin subfamilies, fibronectin III, and cadherin domains, all members of the same beta-sandwich structural superfamily, stood out as obvious alternative extracellular domains and informed subsequent searches that resulted in several candidates, at least one of which was later demonstrated to be an ITIM protein.

The HMM, TM, and ITIM criteria used in Figure 3 are individually non-specific. In such cases, evolutionary conservation provides compelling evidence for the functional significance of domains or motifs. The query in Figure 3 may be extended using the HomoloGene data within Unison to identify Human ITIM-containing proteins that have an ortholog of similar domain composition. The queries in this section are available online in the Unison tutorial.

4. The Unison web interface

The Unison web interface presents an easy-to-use portal to much of Unison's content. In addition to the sequence analysis pages, several special-purpose tools are likely to be useful to a broad audience. BabelFish is Unison's sequence accession translator. For a list of protein sequence accessions, identifiers, and/or MD5 checksums, BabelFish returns a table of identifiers and links to other databases. This tool is aimed at users who have protein accessions from gene expression analysis, mass spectrometry or other high-throughput method and wish to identify identical sequences in other databases.

Unison's AliAn tool provides tabulated protein annotations. For a set of protein accessions, identifiers, and/or MD5 checksums, AliAn will return a table of "best" (*i.e.,* most reliable) annotations, NCBI Gene information, human locus, Gene Ontology terms, Affymetrix probes, and other data. As with BabelFish, this tool is intended for users who have protein identifiers from high-throughput experimental methods.

The Jmol-based structure viewer, shown in Figure 4, enables users to explore the structural context of sequence features. Clicking on features or SNPs in the causes those features to be highlighted on the structure. The URL syntax enables users and programmers to define and display features not currently in Unison. The same interface can also be used to display protein structure prediction results, but these data are not included in the public release.

5. Discussion

Unison is a data warehouse of 12M protein sequences from 20 sequence sources, 200M precomputed predictions of 18 distinct types, and an extensive collection of auxiliary data. The integration of these data facilitates computational discovery projects that require mining based on protein domain composition. Upon this foundation, Unison also provides a variety of web and command-line tools that address "real world" problems such as the annotation of large sets of protein accessions from high-throughput data. A Firefox browser "search plugin"

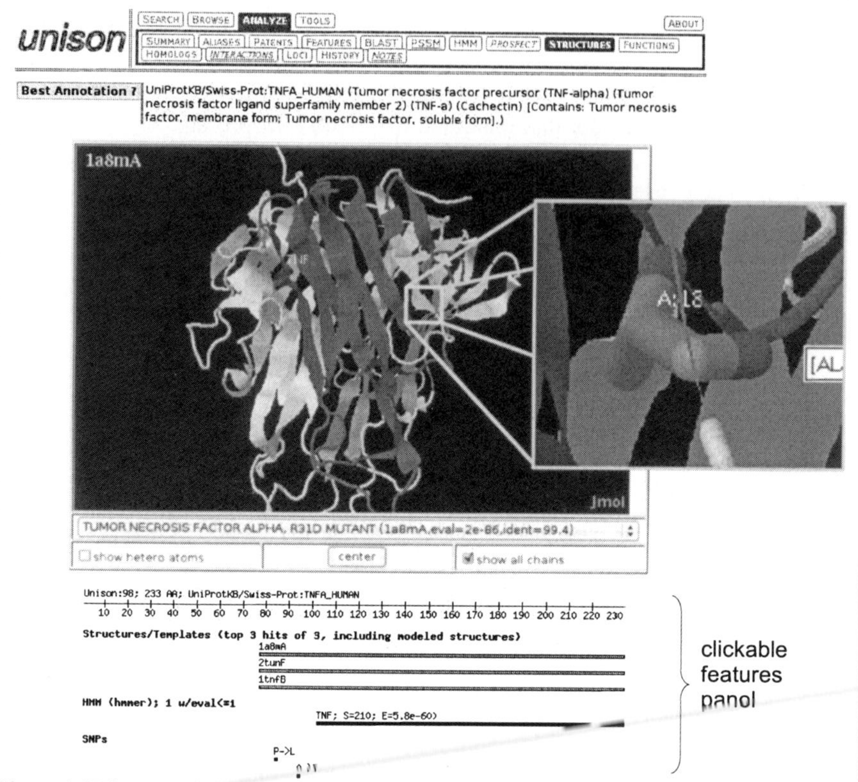

Figure 4: Unison structure tab for Tumor Necrosis Factor alpha (TNFA). Clicking features in the lower panel highlights the corresponding structural region, as shown for the TNF Pfam domain (beta sheets in main figure) and for a SNP (inset).

is available at the Unison web site. Within Genentech, Unison has improved the consistency and currency of data used in discovery and sequence analysis projects for five years. It has become a foundation upon which many in-house tools and projects are based, including a recent discovery of four novel Bcl-2 family members in zebrafish [23], a systematic study of the evolutionary constraints of single nucleotide polymorphisms in terms of protein structure and function [24], and a comprehensive search for novel immune system regulators (in preparation).

Unison uses two approaches for data integration. Familiar concepts are abstracted into well-defined types within the semantically integrated core schema. An important advantage of Unison's integrated schema is that the table definitions are transparent to users, which facilitates semantically correct queries. Unison incorporates other useful data into auxiliary schema. The entire Unison database is updated incrementally and automatically by build scripts; at Genentech, Unison is updated biweekly by a Unix cron job.

Unison excels at identifying sequences that match combinations of features. Feature-based mining necessitates a comprehensive set of sequences, including speculative sequences beyond those in curated databases, and precomputed features for those sequences. The depth of Unison's content enables expert-level queries that are not possible in other systems, such as filtering HMM alignments based on the "trusted cutoff" score.

Unison is a practical solution to a significant problem and it addresses many common shortcomings of integrative databases [2][3]. Unison data models are concise and readily understood. The content and currency of source data, and the status of sequence analyses, are stored in the database and easily discerned by users. Unison is automatically updated. Documentation, a bug report system, and an announcement list are available to facilitate corrections and communication. Unison is available for immediate use online through a direct database connection and a web interface. Unison is released under an approved Open Source™ license and is available for local installation. Sites may load additional sequences and predictions into a local deployment using the tools included with the source package.

Several new features are being considered for forthcoming releases of Unison, including free text searching, pathways and gene expression data and new prediction types from InterPro. Several web interface improvements and a distributed annotation system [25] server are in progress. Protein family classification is being investigated using machine learning and the diverse predictions that are available in Unison.

Acknowledgments

Unison was made possible by numerous Open Source projects and by the freely available contributions of data and methods by scientific colleagues. RKH thanks Matthew Brauer and Josh Kaminker for comments on the manuscript, the Bioinformatics and Protein Engineering department for feedback on Unison, and Simran Hansrai and Dave Windgassen for exceptional computing support.

References

1. L. D. Burgoon and T. R. Zacharewski, *Methods Mol. Biol.* **460**, 145 (2008).
2. S. Philippi and J. Köhler, *Nat. Rev. Genet.* **7**, 482 (2006).
3. L. D. Stein, *Nat. Rev. Genet.* **4**, 337 (2003).
4. S. P. Shah, Y. Huang, T. Xu, M. M. S. Yuen, J. Ling and B. F. F. Ouellette, *BMC Bioinformatics* **6**, 34 (2005).
5. S. Durinck, Y. Moreau, A. Kasprzyk, S. Davis, B. De Moor, A. Brazma and W. Huber, *Bioinformatics* **21**, 3439 (2005).
6. N. J. Mulder and R. Apweiler, *Curr Protoc Bioinformatics* **Chapter 2**, Unit 2.7 (2008).
7. D. L. Wheeler, T. Barrett, D. A. Benson, S. H. Bryant, K. Canese, V. Chetvernin, D. M. Church, M. DiCuccio, R. Edgar, S. Federhen, L. Y. Geer,

Y. Kapustin, O. Khovayko, D. Landsman, D. J. Lipman, T. L. Madden, D. R. Maglott, J. Ostell, V. Miller, K. D. Pruitt, G. D. Schuler, E. Sequeira, S. T. Sherry, K. Sirotkin, A. Souvorov, G. Starchenko, R. L. Tatusov, T. A. Tatusova, L. Wagner and E. Yaschenko, *Nucleic Acids Res.* **35**, D5 (2007).

8. C. von Mering, L. J. Jensen, M. Kuhn, S. Chaffron, T. Doerks, B. Krüger, B. Snel and P. Bork, *Nucleic Acids Res.* **35**, D358 (2007).
9. The UniProt Consortium, *Nucleic Acids Res.* **35**, D193 (2007).
10. C. J. Date. An introduction to database systems. Addison-Wesley (1999).
11. R. Durbin, S. Eddy, A. Krogh and G. Mitchison. Biological sequence analysis: probabilistic models of proteins and nucleic acids. Cambridge University Press (1998).
12. H. Mi, N. Guo, A. Kejariwal and P. D. Thomas, *Nucleic Acids Res.* **35**, D247 (2007).
13. R. D. Finn, J. Mistry, B. Schuster-Böckler, S. Griffiths-Jones, V. Hollich, T. Lassmann, S. Moxon, M. Marshall, A. Khanna, R. Durbin, S. R. Eddy, E. L. L. Sonnhammer and A. Bateman, *Nucleic Acids Res.* **34**, D247 (2006).
14. D. Wilson, M. Madera, C. Vogel, C. Chothia and J. Gough, *Nucleic Acids Res.* **35**, D308 (2007).
15. http://hgdownload.cse.ucsc.edu/goldenPath/.
16. M. Ashburner, C. A. Ball, J. A. Blake, D. Botstein, H. Butler, J. M. Cherry, A. P. Davis, K. Dolinski, S. S. Dwight, J. T. Eppig, M. A. Harris, D. P. Hill, L. Issel-Tarver, A. Kasarskis, S. Lewis, J. C. Matese, J. E. Richardson, M. Ringwald, G. M. Rubin and G. Sherlock, *Nat. Genet.* **25**, 25 (2000).
17. E. A. Bruford, M. J. Lush, M. W. Wright, T. P. Sneddon, S. Povey and E. Birney, *Nucleic Acids Res.* **36**, D445 (2008).
18. H. M. Berman, J. Westbrook, Z. Feng, G. Gilliland, T. N. Bhat, H. Weissig, I. N. Shindyalov and P. E. Bourne, *Nucleic Acids Res.* **28**, 235 (2000).
19. A. Andreeva, D. Howorth, J. Chandonia, S. E. Brenner, T. J. P. Hubbard, C. Chothia and A. G. Murzin, *Nucleic Acids Res.* **36**, D419 (2008).
20. http://scientific.thomsonreuters.com/pharma/geneseq/.
21. J. E. Stajich, *Methods Mol. Biol.* **406**, 535 (2007).
22. E. Vivier, E. Tomasello, M. Baratin, T. Walzer and S. Ugolini, *Nat. Immunol.* **9**, 503 (2008).
23. E. Kratz, P. M. Eimon, K. Mukhyala, H. Stern, J. Zha, A. Strasser, R. Hart and A. Ashkenazi, *Cell Death Differ.* **13**, 1631 (2006).
24. J. Liu, Y. Zhang, X. Lei and Z. Zhang, *Genome Biol.* **9**, R69 (2008).
25. R. D. Dowell, R. M. Jokerst, A. Day, S. R. Eddy and L. Stein, *BMC Bioinformatics* **2**, 7 (2001).

TREEQA: QUANTITATIVE GENOME WIDE ASSOCIATION MAPPING USING LOCAL PERFECT PHYLOGENY TREES

FENG PAN[1], LEONARD MCMILLAN[1], FERNANDO PARDO-MANUEL DE VILLENA[2], DAVID THREADGILL[2] AND WEI WANG[1]

[1]*Department of Computer Science,* [2]*Department of Genetics*
University of North Carolina at Chapel Hill
Chapel Hill, NC, USA
E-mail: [1]*{panfeng,mcmillan,weiwang}@cs.unc.edu,* [2]*{fernando,dwt}@med.unc.edu*

The goal of genome wide association (GWA) mapping in modern genetics is to identify genes or narrow regions in the genome that contribute to genetically complex phenotypes such as morphology or disease. Among the existing methods, tree-based association mapping methods show obvious advantages over single marker-based and haplotype-based methods because they incorporate information about the evolutionary history of the genome into the analysis. However, existing tree-based methods are designed primarily for binary phenotypes derived from case/control studies or fail to scale genome-wide.

In this paper, we introduce TreeQA, a quantitative GWA mapping algorithm. TreeQA utilizes local perfect phylogenies constructed in genomic regions exhibiting no evidence of historical recombination. By efficient algorithm design and implementation, TreeQA can efficiently conduct quantitative genom-wide association analysis and is more effective than the previous methods. We conducted extensive experiments on both simulated datasets and mouse inbred lines to demonstrate the efficiency and effectiveness of TreeQA.

1. Introduction

Genome wide association (GWA) mapping locates genes or narrows regions in the genome that have significant statistical connections to phenotypes of interest. The discovery of these genes and regions offers the potential to increase understanding of biological processes controlling manifestation of phenotypes.

The most frequent genetic variants are single nucleotide polymorphisms (SNPs), in which a single nucleotide in the genome differs between individuals within a species. With the development of low-cost genotyping technologies, extensive SNP data can be cheaply and efficiently produced, which further increases the computational complexity of GWA mapping. Thus, there is an evident need for fast and effective GWA mapping methods.

Existing methods of association mapping look for similarities among samples (chromosomes, haplotypes, etc.) that are correlated with the phenotypes. If strong associations are present, the variance of the phenotype within groups of similar samples is substantially smaller than the variance over all samples.

For example, in single marker-based [17,5] and haplotype-based association mapping [10,4,12], samples are grouped according to their genetic variation at a single marker or a set of markers. For case/control phenotypes, markers that can divide samples into (almost) pure classes are reported. Though these methods employ different strategies for grouping samples, the derived groups are evaluated without further consideration of the intergroup similarities or alternate groupings.

In observation of this, tree-based association methods [14,18,13] utilize phylogenies constructed over the samples. The phylogeny tree is a rich yet compact representation of genetic similarities of the samples. It provides sensible groupings of samples at multiple resolutions. However, the existing methods either handle only case/control phenotypes [14,18] or do not scale to GWA mapping [13].

In this paper, we introduce TreeQA, a tree-based quantitative GWA mapping algorithm. TreeQA utilizes local perfect phylogeny trees constructed in genomic regions exhibiting no evidence of historical recombination by the 4-gamete test [2]. Given a perfect phylogeny, TreeQA evaluates all implied groupings and finds the strongest associations to the phenotype. Furthermore, TreeQA can identify and remove outliers during association analysis.

A brute-force implementation consists of a double loop: for every phylogeny tree, and for every grouping represented by the tree, we conduct a separate ANOVA test to measure its association to the phenotype, and keep track of the best groupings and trees. This approach is inefficient and prone to multiple test errors[1]. Both the number of trees and number of groupings per tree can be very large[a]. This large number of possible groupings requires many ANOVA tests, which is not only expensive computationally, but also gives rise to spurious associations[b]. Thus, permutation tests are necessary to ensure the statistical significance of the discovered associations, which will further increase the computational burden.

TreeQA exploits the following properties: (1) Groupings generated from the same tree obey a partial order, thus allowing reuse of intermediate computations; (2) A grouping may be derived from different trees, but only need to be evaluated once; and (3) Different phenotype permutations may share a substantial number of common computations that need to be computed only once. Thus, TreeQA employs two prefix-tree structures [21] to organize all observed sample subsets and groupings to facilitate the caching and retrieval of reusable computations and guide the enumeration and evaluation of groupings. As a result, TreeQA is able to handle quantitative GWA mapping very efficiently and is more effective and robust in association mapping than previous methods.

2. Related Work

Single-feature association mapping [17,5] considers the sample groupings induced independently by each single marker. Statistical tests such as χ^2 and F-tests are used to measure the association between the phenotype and each grouping. These methods are computationally efficient, however, they do not utilize the additional

[a]For example, the number of trees can exceed tens of thousands in a chromosome-wide association study. And there are up to 2^{2n-2} groupings that can be generated from a tree of n samples.

[b]With ε error rate, the risk of reporting at least one spurious association from x tests is $1-(1-\varepsilon)^x$.

information content carried by haplotypes over single markers.

To address this shortcoming, haplotype-based methods have been developed. HAM [15] considers combinations of three consecutive SNPs along the genome. QHPM [8] uses frequent pattern mining methods to find haplotype patterns in the data, upon which sample groupings are created and evaluated. HapMiner [10] clusters samples using consecutive subsets of markers, and then assess the phenotype's association strength.

The utility of local phylogenies in association mapping has been recently explored in TreeLD [13], Blossoc [14], and TreeDT [18]. These methods use trees to represent sample similarities. Their approach is to exhaustively examine all possible groupings implied by the given phylogenies without explicitly excluding any outliers. Both Blossoc and TreeDT assume simple categorical (binary) phenotypes. TreeLD handles quantitative phenotypes but is not scalable to GWA analysis.

Some other work [6,7,16] uses a global phylogeny structure, e.g., ancestral recombination graph, over all markers in association mapping. However, because of the high computational cost of global phylogeny construction, these methods are not scalable to genome-wide analysis.

3. Preliminaries

We use a binary matrix $H = S \times M$ to represent a SNP dataset, where $S = \{s_1, s_2, ..., s_n\}$ is the set of samples, and $M = \{m_1, m_2, ..., m_z\}$ is the SNP marker set. Each sample is represented by a binary vector, in which '0' represents the majority alleles and '1' represents the minority alleles. We use $f(s_i)$ to denote the phenotype value of a sample s_i and $F(S')$ to denote the phenotype values of samples in a subset S'. An example matrix H containing 10 samples and 10 SNP markers with phenotype is shown in Fig. 1(a).

	m1	m2	m3	m4	m5	m6	m7	m8	m9	m10	f(s)
s1	1	0	1	0	0	0	0	0	0	0	109
s2	1	0	0	0	1	0	0	0	1	0	97
s3	1	0	0	0	1	0	0	0	0	0	86
s4	0	0	0	0	0	0	0	0	0	0	108
s5	1	0	1	0	0	0	0	0	0	0	85
s6	0	0	0	0	0	0	0	0	0	0	56
s7	0	0	0	0	0	1	0	1	0	0	78
s8	0	1	0	1	0	1	1	0	1	1	79
s9	0	1	0	1	0	1	1	0	1	0	61
s10	0	0	0	0	0	1	0	1	0	0	54

(a) SNP data & phenotype

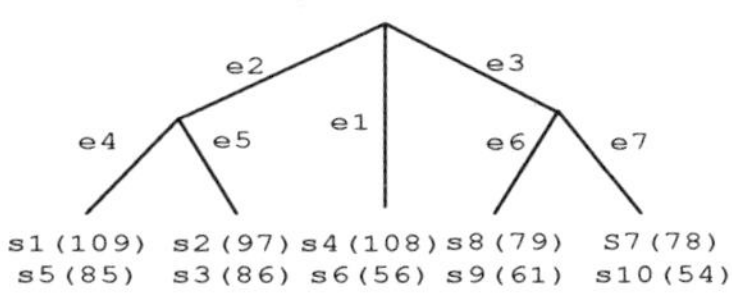

(b) Tree $T_{1,8}$

Figure 1. Example: a SNP dataset and a perfect phylogeny tree

Compatible region: A consecutive region of the genome is called a compatible region *iff* any pair of markers in that region are compatible by the 4-gamete test [2]. That is, among the 4 possible haplotypes formed by the two markers, at most three of them occur.

A compatible region is a genomic region exhibiting no evidence of historical recombination. In Fig. 1(a), the region from markers m_1 to m_8 is a compatible region. We use $C_{u,v}$ to denote a compatible region from markers m_u to m_v.

Maximal Compatible region: A compatible region is a maximal compatible region *iff* it can not be extended on either side to include more SNPs and remains compatible.

Perfect Phylogeny Tree: A phylogeny tree for a set of samples is *perfect* if the phylogeny avoids homoplasy. Every SNP is introduced by a mutation and is represented by an edge of the tree. Given a genomic region, a perfect phylogeny exists *iff* the region is a compatible region.

We use $T_{u,v}$ to denote the perfect phylogeny tree of compatible region $C_{u,v}$. Given $C_{1,8}$ in Fig. 1(a), its tree $T_{1,8}$ is shown in Fig. 1(b). All samples are at the leaf nodes. Samples having identical haplotypes in the region share the same leaf node in the tree, e.g., s_1 and s_5. Each internal node represents a hypothetical common ancestor of a subset of samples. Each edge uniquely corresponds to a SNP (or a historical mutation). Interested readers may refer to paper [3] for inferring perfect phylogenies from a set of SNPs.

Let $E(T_{u,v}) = \{e_1, e_2, ..., e_p\}$ denote the set of edges in $T_{u,v}$. The removal of each edge partitions the samples into two subsets denoted by $S^{(0)}(e_i)$ and $S^{(1)}(e_i)$. Given a tree $T_{u,v}$, we can generate $2|E(T_{u,v})|$ sample subsets by removing each edge separately. We denote this set of sample subsets by $S^{(E)}(T_{u,v})$, $S^{(E)}(T_{u,v}) = \{S^{(j)}(e_i)|j = \{0,1\}, e_i \in E(T_{u,v})\}$.

Definition 3.1. A grouping of a sample subset S', $G(S')$, is formed by a set of disjoint subsets of S', $G(S') = \{S'_1, S'_2, ..., S'_k\}$, $S'_i \subset S'$, $S'_i \cap S'_j = \emptyset$, $\bigcup_{i=1}^{k} S'_i = S'$. Given a tree $T_{u,v}$, we say a grouping $G(S')$ *follows* $T_{u,v}$ *iff* $\forall S'_i \in G(S')$, $S'_i \in S^{(E)}(T_{u,v})$.

For example, grouping $G(S') = \{\{s_1, s_5, s_2, s_3\}, \{s_8, s_9, s_7, s_{10}\}\}$ follows the tree in Fig. 1(b), while grouping $G(S') = \{\{s_1, s_2\}, \{s_8, s_4\}\}$ does not.

Definition 3.2. Given a sample subset S', $G_1(S')$ is called a *parent-grouping* of $G_2(S')$ ($G_2(S')$ called a *child-grouping* of $G_1(S')$) *iff* $\forall S'_i \in G_1(S')$

$$\exists S'_j \in G_2(S'), s.t. S'_i = S'_j.\ OR\ \exists\{S'_{j_q}|S'_{j_q} \in G_2(S'), q = 1, ..., u\}, s.t. S'_i = \bigcup_{q=1}^{u} S'_{j_q}$$

A child-grouping represents a finer partition of its parent-grouping on the same set of samples. For example, grouping $\{\{s_1, s_5, s_2, s_3\}, \{s_4, s_6\}\}$ is the parent-grouping of $\{\{s_1, s_5\}, \{s_2, s_3\}, \{s_4, s_6\}\}$. We summarize the notations in Table 1.

Association between a Compatible Region and a Phenotype

We use the one-way ANOVA test with permutations to measure the association between a grouping of samples and a quantitative phenotype. To accelerate the execution, we re-derive the formula of the ANOVA test.

Given a grouping $G(S') = \{S'_1, ..., S'_k\}$, for every $S'_i \in G(S')$, we calculate

$$SQ(S'_i) = \sum_{s_j \in S'_i} f(s_j)^2, \quad SM(S'_i) = \sum_{s_j \in S'_i} f(s_j) \tag{1}$$

Table 1. Summary of Notations

S, s_i, S_i'	the sample set, a sample, a subset of samples
M, m_i	the marker set, a marker
H	a binary matrix representing the data
$C_{u,v}$	a compatible interval of H
$f(s_i)$	phenotype value of sample s_i
$F(S_i')$	the set of phenotype values of the samples in S_i'
$G_i(S')$	a grouping of a sample subsets S'
$T_{u,v}$	the perfect phylogeny tree of $C_{u,v}$
$E(T_{u,v})$	the edge set of $T_{u,v}$
$S^{(E)}(T_{u,v})$	the set of sample subsets implied in tree $T_{u,v}$ (leaf-sets)

$$SSE_i = SQ(S_i') - SM(S_i')^2/|S_i'|, \quad SSB_i = SM(S_i')^2/|S_i'| \tag{2}$$

Combining all subsets together, we have $MM = \frac{1}{|S'|}\sum_{i=1}^{k} SM(S_i')$ and

$$MSE = \frac{1}{|S'|-k}\sum_{i=1}^{k} SSE_i, \quad MSB = \frac{1}{k-1}(\sum_{i=1}^{k} SSB_i - |S'|\cdot MM^2) \tag{3}$$

We obtain a base score for grouping $G(S')$

$$\mathcal{F}_0(G(S')) = \frac{MSB}{MSE} \tag{4}$$

A higher score indicates a stronger association between the grouping and the phenotype. Given the tree and the data in Fig. 1 and the following two groupings: $G(S_1') = \{\{s_2, s_3\}, \{s_4, s_6\}, \{s_8, s_9\}\}, G(S_2') = \{\{s_2, s_3\}, \{s_8, s_9\}\}$, the scores are $\mathcal{F}_0(G(S_1')) = 0.44, \mathcal{F}_0(G(S_2')) = 4.16$. Thus, grouping $G(S_2')$ has a stronger association with the phenotype than grouping $G(S_1')$.

To correct the multiple test errors, we apply a permutation test on $G(S')$ to calculate a significance score. To permute the phenotype, the phenotype values in $F(S')$ are randomly re-assigned to samples in S'. Then we calculate an $\mathcal{F}$-score using the permuted phenotype following Eqs. 1 to 4.

Assume that we conduct *nPerm* random permutations in total, for each permutation, we get score $\mathcal{F}_j(j = 1...nPerm)$. Among the $nPerm$ $\mathcal{F}$-scores, let p be the number of scores which are greater than or equal to the base score $\mathcal{F}_0(G(S'))$, i.e., $p = |\{\mathcal{F}_j | \mathcal{F}_j \geq \mathcal{F}_0(G(S')), j \in 1...nPerm\}|$. Then the significant score (P score) of $G(S')$ is

$$P(G(S')) = \log_{10}\left(\frac{nPerm}{p}\right) \tag{5}$$

A higher P score indicates that the association between grouping $G(S')$ and the phenotype is more significant.

Definition 3.3. The association between a compatible region and a phenotype: For a compatible region $C_{u,v}$, the highest P score achieved by any grouping following $T_{u,v}$ is regarded as the P score of $C_{u,v}$. The P score represents the association between the compatible region and the phenotype,

$$P(C_{u,v}) = max\{P(G_j(S'))|\forall G_j(S')\ follows\ T_{u,v}, S' \subseteq S\}. \tag{6}$$

Problem Definition: Given a SNP data and a quantitative phenotype, calculate the P-score of every maximal compatible region and report the most significant ones.

4. TreeQA Algorithm

TreeQA takes two major steps: 1) identify maximal compatible regions in the genome and construct the perfect phylogenies of the regions; 2) compute the association between each compatible region and the phenotype.

4.1. *Maximal Compatible Region and Phylogeny Construction*

TreeQA scans the markers in a left to right order. In order to find the maximal compatible regions, it continuously extends the current region by adding the next marker until the new marker is incompatible with some markers in the region. And it maximizes the overlap between two consecutive regions. Assume that the current compatible region is $C_{u,v}$, and marker m_{v+1} is incompatible with markers $m_{i_1}, ..., m_{i_k}$, $u \le i_1 < ... < i_k \le v$, then TreeQA starts the next compatible region at marker m_{i_k+1}. For each maximal compatible region, TreeQA utilizes the inferring algorithm [3] to construct the local perfect phylogeny.

4.2. *Association Computing*

In the second step, TreeQA takes as input a quantitative phenotype and a set of local perfect phylogenies. It considers all possible groupings following the phylogenies and systematically explores the search space of these groupings in a carefully designed order such that intermediate computations can be maximally reused.

According to Definition 3.1, any grouping of a sample subset[c] that follows a tree $T_{u,v}$ can be created from non-overlapping subsets in $S^{(E)}(T_{u,v})$. By utilizing the lexicographical order[d] of subsets in $S^{(E)}(T_{u,v})$, TreeQA can enumerate and evaluate all combinations of non-overlapping subsets systematically.

TreeQA enumerates all groupings via a depth-first recursive procedure. TreeQA extends the current grouping by including a new sample subset which does not overlap with any subsets in the current grouping. The association of each new grouping to the phenotype via a permutation test is computed. The P score of the corresponding maximal compatible region is updated accordingly. The enumeration continues recursively for each newly extended grouping.

Consider the tree in Figure 1. There are 14 sample subsets in $S^{(E)}(T_{1,8})$. Assume that the subsets have the following order,

---

[c]Considering groupings of a sample subset allows TreeQA to exclude potential outliers from the ANOVA test.

[d]Any other ways of defining a total order of the subsets would also work.

$$se_1 = \{s_1, s_5\}, se_2 = S - se_1, se_3 = \{s_2, s_3\}, se_4 = S - se_3, se_5 = \{s_4, s_6\}$$
$$se_6 = S - se_5, se_7 = \{s_8, s_9\}, se_8 = S - se_7, se_9 = \{s_7, s_{10}\}, se_{10} = S - se_9$$
$$se_{11} = \{s_1, s_5, s_2, s_3\}, se_{12} = S - se_{11}, se_{13} = \{s_8, s_9, s_7, s_{10}\}, se_{14} = S - se_{13}$$

TreeQA first generates a grouping containing se_1 only. Among the remaining sample subsets, $\{se_2, se_3, se_5, se_7, se_9, se_{12}, se_{13}\}$ do not overlap with se_1. In the next step, a grouping $\{se_1, se_2\}$ is formed by adding se_2 into the current grouping and its P score is calculated. $P(C_{1,8})$ is updated accordingly. Since all other sample subsets overlap with se_1 or se_2. Thus, no new grouping can be extended from $\{se_1, se_2\}$. Then, TreeQA examines the next grouping extended from $\{se_1\}$, $\{se_1, se_3\}$, and all groupings extended from it. After examining all groupings containing se_1, TreeQA will start from the grouping $\{se_2\}$ and extend it recursively to generate all groupings containing se_2 but not se_1. This process continues until all distinct groupings are enumerated.

4.3. *Effective Permutation*

We found that more than 90% of the execution time of TreeQA is spent in permutation tests. Given a grouping $G(S')$, a permutation test is conducted in two steps: 1) randomly re-assigning the phenotype values in $F(S')$ to samples in S'; 2) calculating the corresponding $\mathcal{F}$ score by Eq. 4.

Given a subset S', both steps take $O(|S'|)$ time. TreeQA exploits maximal reusability of intermediate computation shared by permutation through the following two optimizations:

1) ***inTree***: Common computation units shared by permutation tests of parent/child-groupings in a tree.

2) ***amgTree***: Common computation units shared by permutation tests on groupings following multiple trees.

We use two global prefix-tree structures [21], $Tree_{grouping}$ and $Tree_{subset}$ to organize groupings and sample subsets examined thus far respectively to enable effective permutation tests.

4.3.1. *inTree: Effective permutation tests within a tree*

A pair of parent/child-groupings always involve the same set of samples. Let S' denote a set of samples. For the permutation tests of the parent/child groupings of S', instead of re-assigning the phenotype values in $F(S')$ independently for each grouping, they can share the same set of random permutations of $F(S')$.

For example, given the example in Fig. 1 and a pair of parent/child-groupings, $G_1(S') = \{\{s_1, s_5, s_2, s_3\}, \{s_8, s_9, s_7, s_{10}\}\}$ and $G_2(S') = \{\{s_1, s_5\}, \{s_2, s_3\}, \{s_8, s_9, s_7, s_{10}\}\}$, their $\mathcal{F}_0$ scores are: $\mathcal{F}_0(G_1(S')) = 9.79$ and $\mathcal{F}_0(G_2(S')) = 4.32$. Assume that after a random permutation, the new phenotype values for the samples are: $f(s_1) = 85$, $f(s_2) = 79$, $f(s_3) = 109$, $f(s_5) = 61$, $f(s_7) = 86$, $f(s_8) = 97$, $f(s_9) = 78$, $f(s_{10}) = 54$. Using this new assignment, we can calculate the new $\mathcal{F}$ scores for both groupings: $\mathcal{F}(G_1(S')) = 0.12$ and $\mathcal{F}(G_2(S')) = 0.7$. By reusing the phenotype permutation between $G_1(S')$ and $G_2(S')$, we save $O(|S'|)$ runtime in each permutation.

A child-grouping represents a finer partition of sample subsets in its parent-grouping. We say a grouping is at the finest level if it does not have any child-groupings. We use a global prefix-tree $Tree_{grouping}$ to index all groupings and maintain the parent/child relationship through auxiliary links (from a child-grouping to its parent-groupings). For each permutation of the phenotype, the F scores of a finest grouping and all of its parent-groupings are calculated together. We examine the finest grouping immediately followed by the examination of its parent groupings for maximum computation reuse. If a finest child-grouping has n parent-groupings, we save $O(n|S'|)$ time in each permutation.

4.3.2. *amgTree: Effective permutation among trees*

The same grouping occurs repeatedly in different trees. We only need to compute its P score at its first occurrence. We use $Tree_{grouping}$ to store and retrieve the P score of all examined groupings. If the grouping formed by TreeQA can be found in $Tree_{grouping}$, its P score is directly used. Otherwise, its P score is calculated and stored in $Tree_{grouping}$.

4.4. *Reuse of Intermediate Computation of Statistical Tests*

For any sample subset S', $SQ(S')$ and $SM(S')$ calculated using the original phenotype values (with no permutation) may be reused in any grouping containing S' and all its parent-groupings. We denote them by $SQ_0(S')$ and $SM_0(S')$ respectively in the following discussion.

We employ a global prefix-tree $Tree_{subset}$ to keep track of all sample subsets in any groupings examined thus far. Three values are stored at the leaf node corresponding to the subset S': (subset ID, $SQ_0(S')$, $SM_0(S')$).

For example, given the 10 samples and their phenotype values in Fig. 1(a), we calculate the base score F_0 of grouping $G_1(S') = \{\{s_1, s_5\}, \{s_2, s_3\}, \{s_7, s_{10}\}\}$.

$SQ_0(S'_{1_1}) = 19106, SQ_0(S'_{1_2}) = 16805, SQ_0(S'_{1_3}) = 9000.$
$SM_0(S'_{1_1}) = 194, SM_0(S'_{1_2}) = 183, SM_0(S'_{1_3}) = 132.$
$F_0(G_1(S')) = 547.17/212.17 = 2.58.$

The SQ_0 and SM_0 values of the three subsets are then stored in $Tree_{subset}$. Given a parent-grouping of $G_1(S')$, $G_2(S') = \{\{s_1, s_5, s_2, s_3\}, \{s_7, s_{10}\}\}$, we can retrieve the values of SQ_0 and SM_0 and use them to calculate $F_0(G_2(S'))$,

$SQ_0(S'_{2_1}) = SQ_0(S'_{1_1}) + SQ_0(S'_{1_2}) = 35911, SQ_0(S'_{2_2}) = SQ_0(S'_{1_3}).$
$SM_0(S'_{2_1}) = SM_0(S'_{1_1}) + SM_0(S'_{1_2}) = 377, SM_0(S'_{2_2}) = SM_0(S'_{1_3}).$
$F_0(G_2(S')) = 1064.08/166.69 = 6.38.$

The reuse of $SQ_0(S')$ and $SM_0(S')$ between parent/child groupings may work in conjunction with the *inTree* effective permutation. Besides, $SQ_0(S')$ and $SM_0(S')$ can also be reused by any groupings that contain the subset S'.

5. Results

We compare TreeQA with the following algorithms: 1) **SMA**, our implementation of the Single Marker Association algorithm [17,5]; 2) **HAM**, our implementation of the Haplotype Association Mapping algorithm [15] that slides a 3-SNP window through the genome; 3) **HapMiner** [10], downloaded from the website[e]; and 4) **TreeLD** [13], downloaded from the website[f]. Both **SMA** and **HAM** use the one-way ANOVA test for fair comparison.

QHPM [8] is not used for comparison because it is not scalable to large data sets. Blossoc[14] and TreeDT [18] are not used because they require categorical phenotypes.

5.1. *Experiments on Simulated Data*

We use Coasim [11] to simulate 1000 sequences with scaled recombination rate $\rho = 400$ that corresponds roughly to 10 cM. 10,000 SNP markers are placed uniformly at random over the sequences.

SNP markers on the sequences are randomly selected as causative loci with one, two and three causative mutations. The first SNP is always selected randomly from all SNPs. In the cases of two and three mutations, the second and third causative SNPs are selected from compatible SNPs that are located less than 10 SNPs away from the first SNP. Phenotype values are sampled from four Gaussian distributions: $N_1(140, 35)$, $N_2(90, 35)$, $N_3(50, 40)$, and $N_4(10, 35)$. The one-mutation case uses N_1 and N_3. The two-mutation case uses N_1, N_2 and N_3. The three-mutation case uses all four Gaussian distributions. After assigning the phenotype values, all causative SNPs are removed from the data and we randomly select 100 sequences for our experiments.

SMA, HAM and HapMiner output the top one scoring locus as a point estimation of the causative locus, while TreeQA outputs the top one compatible region. We compare the effectiveness of the algorithms by measuring the distance (in cM) from the top one scoring locus or the center of the top one region to the causative SNP (or the average distance to every causative SNP). We call the distance the **Prediction error**.

Since HapMiner can not finish processing 10,000 SNP markers in a reasonable time, we only use the first 1,000 markers of each sequence when applying HapMiner on the simulated data.

The comparison of SMA, HAM, HapMiner and TreeQA is shown in Figure 2. The x-axis represents the prediction error (distance) to the causative locus and the y-axis represents the percentage of causative loci which are found in distance less than x. In all three cases, the estimated loci by TreeQA are closer to the causative loci than those by SMA, HAM and HapMiner.

The TreeLD algorithm uses local phylogenies and analyzes quantitative phenotypes. However, TreeLD can only process a very small amount of data in rea-

---

[e] http://vorlon.case.edu/ jxl175/HapMiner.html
[f] http://pritch.bsd.uchicago.edu/treeld.html

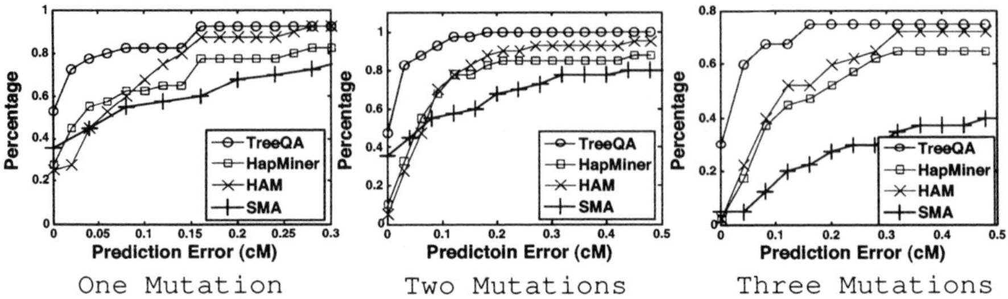

Figure 2. Comparison of SMA, HAM, HapMiner and TreeQA on the simulated data

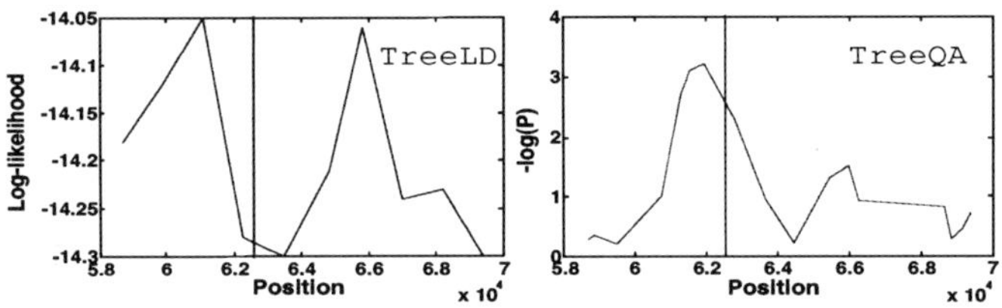

Figure 3. Comparison of TreeLD and TreeQA on the simulated data

sonable time. Therefore, we select 36 samples and 20 SNP markers from the simulated data for performance comparison. A one-mutation causative locus is selected from the 20 SNPs. For TreeQA, instead of generating maximal compatible regions as discussed in Sec. 4, a compatible region is generated around each SNP and contains up to five SNPs. TreeLD takes about two hours to analyze this small data while TreeQA finishes in seconds. Figure 3 plots the results from TreeLD and TreeQA. The x-axis represents the simulated positions in the genome and the y-axis represents the scores of the SNPs. The vertical line demonstrates the causative locus. TreeQA detects a peak near the causative locus while TreeLD identifies two spurious peaks.

5.2. *Experiments on Mouse Genotype Data*

We used a set of mouse genotypes that combines experimental and imputed data[g] [20] from the Jackson Laboratory, consisting of 74 samples. The dataset contains over 7 million SNP markers distributed over all 20 chromosomes. We removed wild derived mouse inbred strains since they are quantitatively and qualitatively different than other laboratory inbred strains and we only used in our experiments the remaining 55 samples that have a share set of common ancestral relationships[19].

We used high density lipoprotein cholesterol (HDL-C) levels in blood as the test phenotype, downloaded from the Mouse Phenome Database[h]. Several HDL-C datasets are available, each of which was collected under different conditions,

[g]http://cgd.jax.org/ImputedSNPData/imputedSNPs.htm

[h]http://phenome.jax.org/pub-cgi/phenome/mpdcgi?rtn=meas/catlister/req=Cblood+lipids

and are thus treated as separate phenotypes. Some candidate genes that may play a role in regulating HDL-C levels are reported in [9].

We apply SMA, HAM and TreeQA on the data and examine how close they can identify the top peak near the locus of those candidate genes.

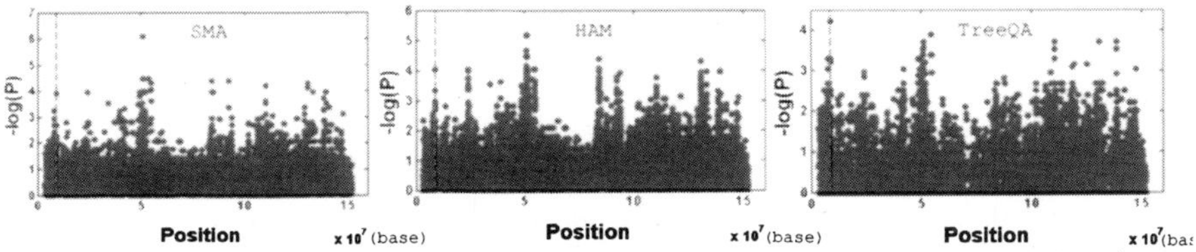

Gene: *Abcb4* on chromosome 5. Genome Coordinates: 8893717-8959226 (base)
Phenotype: Paigen 2, baseline of HDL cholesterol (female)

Figure 4. Compare SMA, HAM and TreeQA on the mouse genotype data

TreeQA detects top peaks near the locations for over 10 of the candidate genes[9], including *Ppara*, *Abcb4* and *Rxrb*. The top peaks reported by SMA and HAM are often far from the locations of these genes. Due to space limitation, we only show the results for one of them, *Abcb4*, in Figure 4.

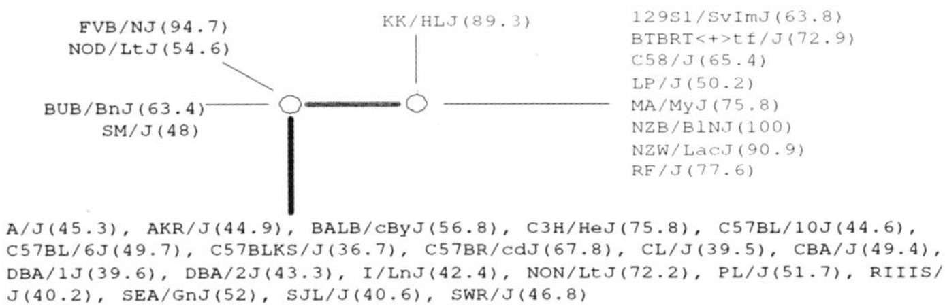

Figure 5. The perfect phylogeny at the peak point found by TreeQA in Figure 4

The perfect phylogeny corresponding to the peak point (compatible region from 8799298 to 8801558 (base)) found by TreeQA is plotted in Fig. 5. The phenotype values of the samples are in parentheses. Samples with unknown phenotype values are omitted from the tree. The subtree on the right contains samples having high phenotype values while the subtree at the bottom contains samples having low values. Other subtrees are considered as outliers and are excluded from the grouping. SMA and HAM fail to identify the locus because they only examine sample groupings that can be generated from single SNPs or 3-SNP windows, which are a small subset of the groupings examined by TreeQA.

TreeQA takes about 10 minutes to analyze each chromosome which contains around 40000 SNPs on average. SMA and HAM take slightly less time than TreeQA. Both HapMiner and TreeLD are unable to finish in reasonable time.

6. Conclusion

In this paper, we present a tree-based quantitative GWA mapping algorithm, TreeQA. TreeQA utilizes local perfect phylogenies in detecting associations. Perfect phylogenies provide sensible groupings of samples at multiple resolutions. TreeQA explores the space of all possible groupings implied by the perfect phylogenies in a carefully designed order so that intermediate computations can be maximally reused. Our experimental results on both simulated and real data show that TreeQA can efficiently conduct quantitative GWA analysis and is more effective than the previous methods.

References

1. R. G. Miller. *Simultaneous Statistical Inference*. Springer Verlag New York, 1981.
2. R. R. Hudson and N. L. Kaplan. Statistical properties of the number of recombination events in the history of a sample of dna sequences. *Genetics*, 111(1):147C164, 1985.
3. R. Agarwala, D. Fernandez-Baca, and G. Slutzki. Fast algorithms for inferring evolutionary trees. *Journal of Computational Biology*, 2(3):397–408, 1995.
4. H. Toivonen, P. Onkamo, K. Vasko, and et al. Data mining applied to linkage disequilibrium mapping. *Am J Hum Genet*, 67:133–145, 2000.
5. J. Akey, L. Jin, and M. Xiong. Haplotypes vs single marker linkage disequilibrium tests: what do we gain? *Eur J. Hum Genet.*, 9(4):291–300, 2001.
6. F. Larribea, S. Lessarda, and N. J. Schork. Gene mapping via the ancestral recombination graph. *Theoretical Population Biology*, 62(2):215–229, 2002.
7. A. P. Morris, J. C. Whittaker, and D. J. Balding. Fine-scale mapping of disease loci via shattered coalescent modeling of genealogies. *Am J Hum Genet*, 70(3), 2002.
8. P. Onkamo, V. Ollikainen, P. Sevon, and et al. Association analysis for quantitative traits by data mining: Qhpm. *Ann. Hum. Genet.*, 66:419–429, 2002.
9. X. Wang and B. Paigen. Quantitative trait loci and candidate genes regulating hdl cholesterol. *Arteriosclerosis, Thrombosis, and Vascular Biology*, 22:1390, 2002.
10. J. Li and T. Jiang. Haplotype-based linkage disequilibrium mapping via direct data mining. *Bioinformatics*, 21(24):4384–4393, 2005.
11. T. Mailund, M. H. Schierup, and et al. Coasim: A flexible environment for simulating genetic data under coalescent model. *BMC Bioinformatics*, 6(252), 2005.
12. E. Waldron, J. Whittaker, and D. Balding. Fine mapping of disease genes via haplotype clustering. *Genetic Epidemiology*, 30:170–179, 2005.
13. S. Zöllner and J. K. Pritchard. Coalescent-based association mapping and fine mapping of complex trait loci. *Genetics*, 169(2):1071–92, 2005.
14. T. Mailund, S. Besenbacher, and M. H. Schierup. Whole genome association mapping by incompatibilities and local perfect phylogenies. *BMC Bioinformatics*, 7:454, 2006.
15. P. McClurg, M. T. Pletcher, T. Wiltshire, and A. I. Su. Comparative analysis of haplotype association mapping algorithms. *BMC Bioinformatics*, 7:61, 2006.
16. M. J. Minichiello, and R. Durbin. Mapping Trait Loci by Use of Inferred Ancestral Recombination Graphs. *Am J Hum Genet*, 79(5):910–922, 2006.
17. I. Pe'er and et al. Evaluating and improving power in whole-genome association studies using fixed marker sets. *Nature Genetics*, 38:663–667, 2006.
18. P. Sevon, H. Toivonen, and V. Ollikainen. Treedt: Tree pattern mining for gene mapping. *IEEE Transactions on Computational Biology and Bioinformatics*, 3(2), 2006.
19. H. Yang, T. A. Bell, G. A. Churchill, and F. P.-M. de Villena. On the subspecific origin of the laboratory mouse. *Nature Genetics*, 39, 2007.
20. J. P. Szatkiewicz, G. L. Beane, Y. Ding, L. Hutchins, F. P.-M. de Villena, and G. A. Churchill. An imputed genotype resource for the laboratory mouse. *Mammalian Genome*, 19(3), 2008.
21. T. H. Cormen, C. E. Leiserson, R. L. Rivest, and C. Stein. Introduction to algorithms.

IMPROVING THE EFFICIENCY OF BIOMARKER IDENTIFICATION USING BIOLOGICAL KNOWLEDGE

JOHN H. PHAN

The Wallace H. Coulter Department of Biomedical Engineering, Georgia Institute of Technology, 313 Ferst Drive
Atlanta, GA 30332, USA

QIQIN YIN-GOEN
ANDREW N. YOUNG

Department of Pathology and Laboratory Medicine, Emory University
Atlanta, GA 30322, USA

MAY D. WANG

The Wallace H. Coulter Department of Biomedical Engineering, Georgia Institute of Technology, 313 Ferst Drive
Atlanta, GA 30332, USA

Identifying and validating biomarkers from high-throughput gene expression data is important for understanding and treating cancer. Typically, we identify candidate biomarkers as features that are differentially expressed between two or more classes of samples. Many feature selection metrics rely on ranking by some measure of differential expression. However, interpreting these results is difficult due to the large variety of existing algorithms and metrics, each of which may produce different results. Consequently, a feature ranking metric may work well on some datasets but perform considerably worse on others. We propose a method to choose an optimal feature ranking metric on an individual dataset basis. A metric is optimal if, for a particular dataset, it favorably ranks features that are known to be relevant biomarkers. Extensive knowledge of biomarker candidates is available in public databases and literature. Using this knowledge, we can choose a ranking metric that produces the most biologically meaningful results. In this paper, we first describe a framework for assessing the ability of a ranking metric to detect known relevant biomarkers. We then apply this method to clinical renal cancer microarray data to choose an optimal metric and identify several candidate biomarkers.

1. Introduction

The subjective nature of traditional medical techniques limits the accuracy of cancer subtype classification and, subsequently, the effectiveness of therapy. Clinicians visually examine cancer specimens to determine their subtypes before proposing treatment regimens. However, cancers with similar characteristics may behave very differently despite similar treatment conditions [1]. Because cancer is the result of genetic anomalies, emerging diagnostic research has primarily focused on genetic and proteomic expression. This research generally involves the use of high throughput technology (e.g. microarrays and mass spectrometry) to generate large amounts of genetic and proteomic expression data. We typically reduce this data using one of many analysis algorithms with the goal of identifying a subset of features (corresponding to genes or proteins) with high predictive accuracy [2-4]. We hope that these feature subsets will both enhance our understanding of the biological mechanisms as well as provide us with an accurate diagnostic system. When validated, we call these differentially expressed features biomarkers. Unfortunately, even the selection of a ranking metric is subjective, as different metrics may identify different subsets of features [5]. Feature ranking affects both the efficiency of identifying relevant genes and the accuracy of subsequent predictive models. We address this issue by presenting a method that uses existing biological knowledge to identify the best feature ranking metric for a particular gene expression dataset. The optimal metric maximizes the probability of correctly ranking differentially expressed and previously validated genes.

Despite numerous feature selection studies, there is still a lack of clinically validated and proven biomarkers for most cancers. Thus, the use of "correct" genes as knowledge for algorithm selection is subjective and we should choose these genes carefully. Sources of biological knowledge are abundant, but vary in terms of reliability. We consider a knowledge source to be reliable if genes (or the corresponding expressed proteins) from that source have been clinically validated as differentially expressed. The majority of knowledge is contained in the literature and roughly falls into four levels of reliability, adapted from a review of post-analysis validation methods by Chuaqui et al. [6]:

1. **No biological validation.** As the lowest level of reliability, this includes studies that develop feature selection algorithms and present the selected list of genes without a stringent interpretation of the biological results.
2. ***In silico* validation.** Also known as computational validation, these studies compare their feature selection results to the results of other studies. They may also identify Gene Ontology (GO) categories that are statistically overrepresented as a result of feature selection.
3. **Same-sample validation.** These studies validate their microarray experiments by performing additional assays on the same samples from which their microarrays were derived. These assays typically include quantitative real-time PCR (qRT-PCR) or northern analysis and serve to validate the technical reliability of the microarrays.
4. **Independent or clinical validation.** As the highest level of reliability, these studies validate the results of their microarray experiments using independent biological samples, usually from a clinical source. Independent

validation ensures that the selected features are not a result of over-fitting. These validations often take the form of qRT-PCR and *in situ* hybridization (ISH) for RNA products, or immunohistochemistry (IHC) and western analysis for protein products.

Despite frequent disagreement between qRT-PCR and microarray results, qRT-PCR is the most common method for validation of differentially expressed genes. Genes with large fold-change in microarray data are consistently correlated with qRT-PCR while those with smaller fold change are more susceptible to technical variability [7]. The detection of differentially expressed genes is generally reproducible across several microarray platforms [8]. However, in light of a recent study illustrating the pervasiveness of technical artifacts in microarray data [9], we only consider a knowledge source reliable if it falls into category three or four.

Investigators have attempted to improve feature selection by using biological knowledge. Their knowledge sources often fall into category two of reliability, *in silico* validation, and include Gene Ontology and pathway databases, published literature, microarray repositories, and sequence information. Generally, these studies identify genes that cluster or correlate with genes from the knowledge sources [10-12]. Another study developed a theoretical framework to compare feature ranking metrics in the presence of control features [13]. However, this study also neglected to focus on the reliability of the control features. Indeed, the wealth of available information in the form of gene and protein interactions, functional annotation, and genetic and pathways can improve the results of data analysis [14]. Furthermore, microarray data analysis has shifted from purely data driven methods to methods that use additional knowledge, even in the feature selection process [14].

We develop a method to quantify the efficiency of detecting biomarkers by feature ranking. This method maximizes the biological relevance of feature ranking by choosing the best metric from a population of metrics. The chosen ranking metric is optimal with respect to knowledge obtained from reliable sources. We test the effectiveness of our method using clinical gene expression data. Results indicate that the choice of ranking metric significantly affects feature ranking, which, in turn, affects the efficiency of discovering and validating novel biomarkers.

2. Methods

2.1. *Modeling Knowledge in Feature Selection*

Throughout this paper, the term 'feature set' denotes a group of one or more features or genes that act in concert. A 'sample' refers to measurements of a feature set from a single microarray or molecular profile. The entire microarray sample contains ℓ features while a feature set may contain p features (where $p \ll \ell$). We represent samples for feature set i as jointly distributed random vectors, $\vec{X}^i \in \Re^p$, and labels, $Y^i \in \{0,1\}$. The class label, Y, indicates the clinical source of the microarray sample. In most cancer problems, $Y = 1$ indicates, for example, samples measured from patients with cancer and $Y = 0$ indicates samples from patients with no cancer. For a microarray dataset with N samples, feature set i for a particular dataset is the vector $\vec{d}_i = ((y_1^i, \vec{x}_1^i), (y_2^i, \vec{x}_2^i), \ldots, (y_N^i, \vec{x}_N^i))$ from the random variable $\vec{D}$, which represents all feature sets in a dataset. Each feature set is associated with a relevance variable, r_i, from the random variable $R \in \{0,1\}$. r_i represents the biological relevance of the feature set and the reliability of the knowledge source. $\vec{D}$ and R are jointly distributed.

For each feature set, we assign a score that represents the predictive ability of that feature set:

$$A = h(\vec{D}, \theta) \tag{1}$$

where $A \in \Re$ is a random variable and θ is a meta-parameter that characterizes the scoring function, or ranking metric. Although θ may represent the space of all ranking methods, we use a reduced set of wrapper-based methods in our simulations. Specifically, we use a support vector machine (SVM) classifier with the linear and radial basis kernels and estimate the classification accuracy of biomarkers using the 0.632 bootstrap [5, 15]. The SVM classifier depends on a cost parameter, C, which determines the penalty of misclassification. The radial basis kernel depends on γ, which is proportional to the complexity of the classifier. For the radial basis kernel, the pair of parameters, (C, γ), represents θ. We discretely vary C and γ over the log scale range of 0.1 to 10^3 and 0.01 to 10^5, respectively. For the linear kernel, only the single parameter, C, represents θ. We vary this parameter over the log scale range of 0.01 to 10^2.

In practice, a gene expression dataset will have N samples, each with ℓ features. We separately examine m (m can be different from ℓ and include, for example, all pairs, triplets, or a subset of feature combinations) feature sets, corresponding to $\{\vec{d}_1, \vec{d}_2, \ldots, \vec{d}_m\}$ and $\{r_1, r_2, \ldots, r_m\}$. From the mapping defined in eq. 1, we compute the set of values $\{\alpha_1, \alpha_2, \ldots, \alpha_m\}$ where each α is an observation from A. Using a simple selection method, we can then conclude that the best feature sets and potential biomarkers are in the set

$$G = \{i : \alpha_i < \tau\} \tag{2}$$

where τ is a threshold.

We want to choose a θ that produces the most biologically relevant ranking of the m feature sets, $\{\vec{d}_1, \vec{d}_2, \ldots, \vec{d}_m\}$, with respect to a given set of knowledge. Assuming that lower scores are better, the best θ assigns scores such that $\alpha_i < \alpha_j$ for $r_i = 1$ and $r_j = 0$, i.e., feature set i is known to be more relevant than feature set j for this particular dataset. Although we may never know the relevance of all features in a dataset, we may infer from literature that the k feature sets, $G_k = \{g_1, g_2, \ldots, g_k\}$, are relevant, where $k << m$. This implies that the elements of the set $\{\alpha_i : i \in G_k\}$ should generally be smaller than those of $\{\alpha_j : j \notin G_k\}$. If the knowledge is reliable, we want to choose a θ that maximizes the probability that the score of a feature set from G_k is less than that of a feature set that is not from G_k. Explicitly, this probability is

$$P(\alpha_i < \alpha_j \mid \theta) \tag{3}$$

for $i \in G_k$ and $j \notin G_k$. The estimated optimal ranking method is

$$\hat{\theta} = \arg\max_\theta P(\alpha_i < \alpha_j \mid \theta), \tag{4}$$

keeping in mind that $\hat{\theta}$ is only optimal, or maximizes the probability, with respect to the given knowledge set. For m feature sets, k of which are in our knowledge set, G_k, we can empirically approximate the probability of eq. 3 with

$$P(\alpha_i < \alpha_j \mid \theta) = \frac{1}{k(m-k)} \sum_{i \in G_k} \sum_{j \notin G} I(\alpha_i < \alpha_j) \tag{5}$$

where $I(x)$ evaluates to one when x is true and zero when x is false. Eq. 5 is equivalent to computing the area under an ROC curve (AUC) for classifying feature sets as either relevant or irrelevant [13].

2.2. *Iteratively Updating Knowledge*

It may be difficult to compile a comprehensive list of knowledge from literature and independent validation. Consequently, we can expect that some feature sets that are not in our knowledge set, $j \notin G_k$, are, in fact, relevant biomarkers. If V is the set of all relevant biomarkers, regardless of whether their relevance is known, we define the knowledge update function, $S_{\hat{\theta}}$, as

$$G_{k+1} = S_{\hat{\theta}}(G_k) = \{\{G_k, \arg\min_i \alpha_i\} : i \in V, i \notin G_k\}. \quad (6)$$

This function adds to G_k a relevant biomarker with the best rank according to the estimated optimal metric, $\hat{\theta}$. Of course, a feature set is known to be in the set V only after performing a validation procedure such as qRT-PCR.

If we know all feature sets in V, we can quantify any improvement in efficiency due to optimization of the ranking metric. Using bootstrap resampling, we randomly and repeatedly partition the feature sets in V into a group of known relevant feature sets (training) and a group of unknown relevant feature sets (testing). If there are K elements in V, we randomly select K elements with replacement, resulting in K^* $(K^* < K)$ unique elements for the testing set. We use the group of $K - K^*$ known relevant feature sets to optimize the ranking metric, then iteratively detect feature sets from the unknown set of K^* features and update our knowledge using eq. 6. Every validation test requires a finite amount of time and resources. Plotting the fraction of correctly validated biomarkers (y-axis) vs. total validation time (x-axis), reveals that higher detection efficiency corresponds to a larger area under this curve. This curve is similar to a ROC curve, so we also call the area under this curve the AUC. We repeat this bootstrap sampling of feature sets 100 times in order to compute the significance of the differences among three conditions: optimal metric selection, sub-optimal metric selection, and sub-optimal initial knowledge. For the sub-optimal metric selection condition, we use correct initial knowledge selected from V via bootstrap, but use a modified equation to choose $\hat{\theta}$ with median AUC:

$$\hat{\theta} = \arg median_{\theta} P(\alpha_i < \alpha_j \mid \theta). \quad (7)$$

Selection of a ranking metric with median AUC represents the common practice of arbitrarily selecting a metric with no regard for biological relevance and efficiency. This median AUC algorithm also serves as a reference point for assessing the potential improvement of efficiency when using the optimal algorithm.

For the sub-optimal initial knowledge condition, we begin the simulation with incorrect knowledge selected via bootstrap and use eq. 4 to optimize the ranking algorithm before updating the current knowledge set. We expect the average AUC of the optimal selection condition to be higher than that of both of the sub-optimal conditions. Figure 1 illustrates this process.

To determine whether the optimization procedure is over-fitting to the knowledge set, we conduct additional tests using randomly selected knowledge sets. If over-fitting is occurring, results of the optimal, suboptimal, and suboptimal knowledge tests for randomly selected knowledge should be similar to those of the true knowledge set.

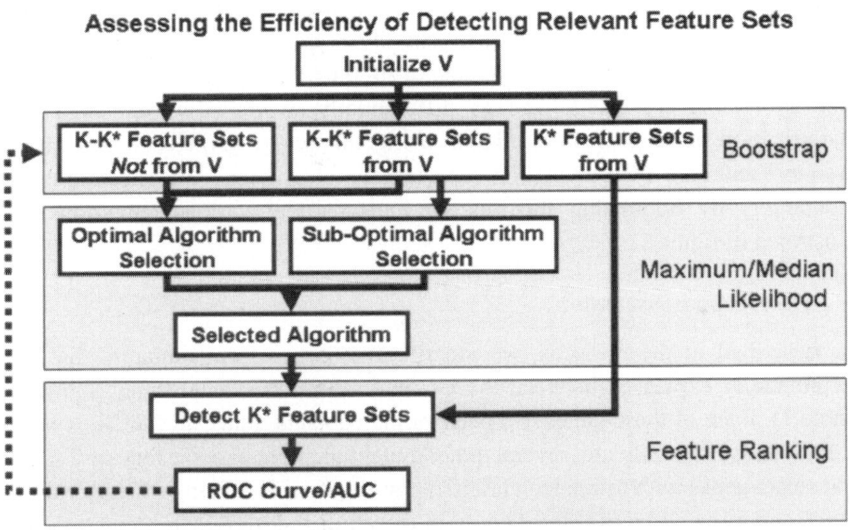

Figure 1. Quantifying the efficiency of detecting relevant feature sets. For clinical data, we define V as the set of K known differentially expressed feature sets. Using bootstrap cross validation, we partition V into K^* and K-K^* samples. K^* is the number of unique samples after sampling from V K times with replacement. We optimize the ranking algorithm using K-K^* feature sets and assess the algorithm's efficiency in detecting the remaining K^* feature sets. For each of the three conditions—optimal metric selection, sub-optimal metric selection, and sub-optimal initial knowledge—we perform this bootstrap sampling 100 times in order to compute the significance of any differences between mean AUC values.

2.3. *Microarray Data Analysis and qRT-PCR Validation*

We examine two clinical case studies using renal tumor microarray datasets. The first dataset, from a study by Schuetz *et al.*, uses Affymetrix microarrays (HG-Focus, 8793 probesets) to profile samples from three subtypes of renal tumors: 13 clear cell (CC) renal cell carcinoma (RCC), 4 chromophobe (CHR) RCC, and 3 oncocytoma (ONC, benign) [2]. The second dataset, from a study by Jones *et al.*, uses a different model of Affymetrix microarrays (HG-U133A, 22283 probesets reduced to 8793 that are common to HG-Focus) to examine similar renal tumor subtypes with 32 CC, 6 CHR, and 12 ONC samples [16]. We are interested in biomarkers that differentiate the CC class from the combined group of ONC and CHR.

Using literature, we identify genes that have been validated (via qRT-PCR or IHC) as differentially expressed between the CC and ONC/CHR subtypes. We then validate an additional 94 genes using qRT-PCR (using RNA from 34 CC and 18 CHR tissue samples). These 94 genes were selected by a renal cancer pathologist based on his knowledge and previous research. Only some of the 94 genes assayed with qRT-PCR are differentially expressed as assessed by a linear SVM with classification error estimated using 0.632 bootstrap. Genes measured with qRT-PCR are categorized as differentially expressed if the estimated classification error is less than 10%. Using the set of knowledge from both literature and qRT-PCR validation, we examine the efficiency of detecting these biomarkers by optimizing the ranking metric under various conditions, as illustrated in figure 1.

3. Results and Discussion

As described in the methods, we identify five genes from literature that are differentially expressed between the CC and ONC/CHR renal tumor subtypes (table 1). Each of these genes had been validated using either qRT-PCR or IHC. Additionally, we validate several other potential biomarkers using qRT-PCR and select genes with estimated classification errors of less than 10% (table 2).

Combining all knowledge from both literature and qRT-PCR validation, we examine the effect of optimizing the feature ranking metric using the method illustrated in figure 1. Box plots of the 100 iterations for each of the three tests indicate that optimal selection outperforms sub-optimal selection (figure 2, left column). The comparison of optimal to suboptimal metrics may seem to always favor the optimal metric. However, the optimal metric is not always a simple linear classifier. In fact, during the iterative gene detection process, θ changes frequently as V is updated. Moreover, suboptimal selection may represent the common practice of arbitrarily selecting ranking metrics with no regard to their potential disadvantages for particular datasets. The box plots represent the median and quartiles of the AUC values for each of the 100 iterations. Correspondingly, the ROC curves also indicate that the optimal selection method improves the efficiency of biomarker detection (figure 2, right column).

For the Schuetz data (figure 2, top row), the performance difference between the optimal and suboptimal ranking metrics seems small according to the box plots. However, the ROC curve of the optimal metric initially rises much more quickly compared to that of the suboptimal. The region of low specificity boosts the performance of the suboptimal metric. However, this region should be neglected when assessing performance since the number of false positives at this point is very high. Validation procedures would likely consider only the biomarkers detected in the high specificity region. Results are similar for the Jones data (figure 2, bottom row).

The high variance of the suboptimal initial knowledge condition indicates that optimization of the ranking metric is sensitive to the initial conditions. Some of the randomly selected initial knowledge may, in fact, be differentially expressed, resulting in good performance. However, these random initial knowledge sets are more likely to be irrelevant. Thus, box plots for this condition illustrate this mixture of knowledge quality. These results stress the importance of the quality of biomarker knowledge.

The control tests using random knowledge sets for V show that our method does not over-fit to the knowledge (figure 2, box plots CO, CSO, and CSK). None of the algorithms considered in our space of θ are able to favorably rank these randomly selected genes. AUCs of these control tests are close to 0.5 as expected for random classification.

Using all knowledge from literature and the first round of qRT-PCR, we optimize the ranking metric and select the top genes that have not been previously validated and that have estimated classification errors of less than 5% (table 3). We can link a few of these genes directly to previous literature pertaining to renal cancer. For example, CXCR4 has been linked to kidney cancer. Using qRT-PCR, Schrader *et al.* shows that this gene is over-expressed in kidney cancer tissue compared to normal kidney tissue [17]. IGFBP3 and KLF10 has also been linked to renal cell carcinoma [18, 19]. Validation of these genes using qRT-PCR may yield additional knowledge to iteratively refine the biomarker selection process. However, since we want to primarily focus on the methodology here, we reserve the actual validation of these results for a future study.

Table 1. Genes validated as differentially expressed between CC and ONC/CHR renal tumor subtypes from various knowledge sources.

Gene Symbol	Knowledge Source	Validation Method
CA9	Chen *et al.*, Clin Cancer Res, 2005	qRT-PCR
CLCNKB	Chen *et al.*, Clin Cancer Res, 2005	qRT-PCR
DEFB1	Schuetz *et al.*, J Mol Diagn, 2004	qRT-PCR, IHC
LRP2	Schuetz *et al.*, J Mol Diagn, 2004	qRT-PCR, IHC
PVALB	Chen *et al.*, Clin Cancer Res, 2005	qRT-PCR

Table 2. Genes that we validated with qRT-PCR. These genes have estimated classification errors of less than 10% as assessed by a linear SVM classifier using 0.632 bootstrap estimation.

Gene Symbol	Error	Gene Symbol	Error
STC1	2.43E-05	COX5A	0.0394058
SLC25A4	0.00186696	BAG1	0.0548365
CFTR	0.00279081	LY6E	0.0596081
PDHA1	0.0133316	CD99	0.0600892
PFKM	0.0279739	AKAP12	0.0624445
NNMT	0.0289622	ACAT1	0.0687972
CP	0.0300157	SPTBN2	0.077287
CFB	0.0387219	GOT1	0.0784855

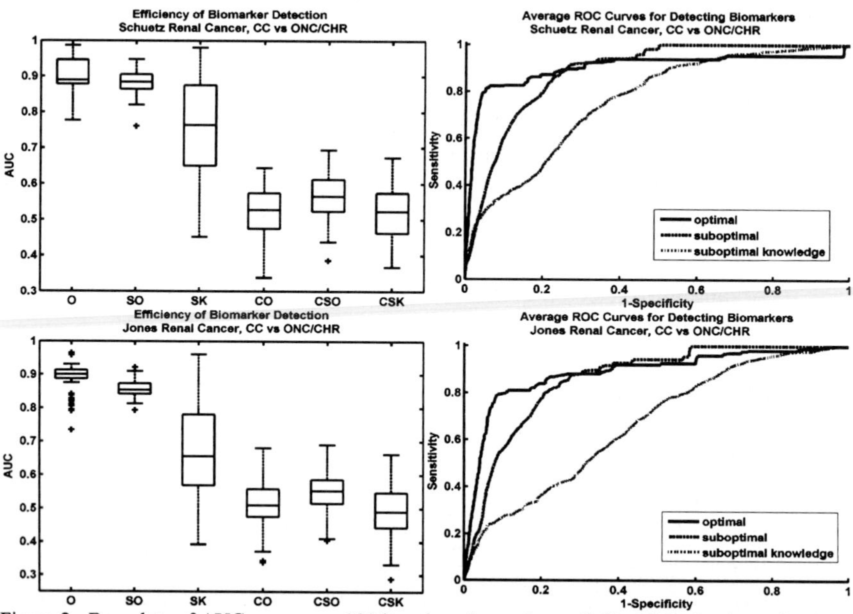

Figure 2. Box plots of AUC areas over 100 iterations for each test (left). AUCs for the optimal test (O) are higher than both the sub-optimal (SO) and sub-optimal knowledge (SK) tests (differences are statistically significant with p-values very close to 0). The control tests, using randomly selected knowledge indicate that optimizing the ranking metric does not over-fit (CO=control optimal, CSO=control suboptimal, CSK=control suboptimal knowledge). Average ROC curves for each test, illustrate the differences in biomarker detection efficiency (right). The ROC for the optimal metric test (solid line) indicates more accurate biomarker detection for both the Schuetz (top row) and Jones (bottom row) renal cancer datasets.

Table 3. Proposed list of genes for further qRT-PCR validation.

Gene Symbol	Error	Gene Symbol	Error
ACLY	0	PCCB	0.03274
CXCR4	0.013907	TMSB10	0.034201
C4A /// C4B	0.0187	HCLS1	0.034415
FLNA	0.019903	ACTA2	0.039398
PMP22	0.023798	IGFBP3	0.040989
PFKFB3	0.026506	NFKBIA	0.042332
KLF10	0.027801	CD44	0.049095
PRG1	0.03003	IER3	0.049571
LGALS1	0.030617		

4. Conclusion

We have shown that biomarker identification by feature ranking benefits from knowledge integration at key points. Using this knowledge—whether from clinical observations, laboratory experiments, or existing literature—we can intelligently choose an optimal ranking metric for a specific gene expression dataset. The use of an optimal metric for ranking and identifying novel biomarkers reduces the number of false discoveries, increases the number of true discoveries, reduces the required time for validation, and increases the overall efficiency of the process.

The results of our simulations indicate that knowledge integration improves biomarker selection for clinical microarray data. Although this study assumes independent gene expression, the method is general and we can use it to rank combinatorial gene expression data as well. Furthermore, we test this method using only a limited set of wrapper-based feature ranking metrics. However, it is easily expandable to encompass a variety of metrics, including the commonly used filter methods such as t-tests and fold change. We hope that the proposed method will impact biomarker identification practices and improve the effectiveness of resulting clinical applications.

Acknowledgments

This research has been supported by grants from National Institutes of Health (R01CA108468, P20GM072069, U54CA119338), Microsoft Research Funding, and Georgia Cancer Coalition (Distinguished Cancer Scholar Award to MDW).

References

1. Golub, T., et al., *Molecular Classification of Cancer: Class Discovery and Class Prediction by Gene Expression Monitoring.* Science, 1999. **286**: p. 531-537.
2. Schuetz, A., et al., *Molecular classification of renal tumors by gene expression profiling.* J Mol Diagn, 2004.

3. Singh, D., et al., *Gene expression correlates of clinical prostate cancer behavior.* Cancer Cell, 2002. **1**: p. 203-209.
4. van't Veer, L., et al., *Gene expression profiling predicts clinical outcome of breast cancer.* Nature, 2002. **415**: p. 530-536.
5. Braga-Neto, U. and E. Dougherty, *Is cross-validation valid for small-sample microarray classification?* Bioinformatics, 2004. **20**: p. 374-380.
6. Chuaqui, R., et al., *Post-analysis follow-up and validation of microarray experiments.* Nature Genetics, 2002. **32**: p. 509-514.
7. Morey, J., J. Ryna, and F. Van Dolah, *Microarray validation: factors influencing correlation between oligonucleotide microarrays and real-time PCR.* Biol. Proced. Online, 2006. **8**(1): p. 175-193.
8. Shi, L., et al., *The MicroArray Quality Control (MAQC) project shows inter- and intraplatform reproducibility of gene expression measurements.* Nat Biotechnol, 2006. **24**(9): p. 1151-61.
9. Stokes, T., et al., *chip artifact CORRECTion (caCORRECT): A Bioinformatics System for Quality Assurance of Genomics and Proteomics Array Data.* Annals of Biomedical Engineering, 2007. **35**: p. 1068-1080.
10. Aerts, S., et al., *Gene prioritization through genomic data fusion.* Nature Biotechnology, 2006. **24**(5): p. 537-544.
11. Kuffner, R., K. Fundel, and R. Zimmer, *Expert knowledge without the expert: integrated analysis of gene expression and literature to derive active functional contexts.* Bioinformatics, 2005. **21**: p. ii259-ii267.
12. Kong, S., W. Pu, and P. Park, *A multivariate approach for integrating genome-wide expression data and biological knowledge.* Bioinformatics, 2006. **22**(19): p. 2373-2380.
13. Mukherjee, S. and S. Roberts, *A theoretical analysis of the selection of differentially expressed genes.* J Bioinformatics Comput Biol, 2005. **3**: p. 627-643.
14. Bellazzi, R. and B. Zupan, *Towards knowledge-based gene expression data mining.* Journal of Biomedical Informatics, 2007. **40**: p. 787-802.
15. Efron, B. and R. Tibshirani, *Improvements on Cross-Validation: The .632+ Bootstrap Method.* Journal of the American Statistical Association, 1997. **92**(438): p. 548-560.
16. Jones, J., et al., *Gene signatures of progression and metastasis in renal cell cancer.* Clin Cancer Res, 2005. **11**(16): p. 5730-9.
17. Schrader, A., et al., *CXCR4/CXCL12 expression and signalling in kidney cancer.* British Journal of Cancer, 2002. **86**: p. 1250-1256.
18. Rosendahl, A. and G. Forseberg, *IGF-I and IGFBP-3 augment transforming growth factor-beta actions in human renal carcinoma cells.* Kidney International, 2006. **70**: p. 1584-1590.
19. Ivanov, S., et al., *Two novel VHL targets, TGFBI (BIGH3) and its transactivator KLF10, are up-regulated in renal clear cell carcinoma and other tumors.* Biochem Biophys Res Commun, 2008.

TOWARDS A CYTOKINE-CELL INTERACTION KNOWLEDGEBASE OF THE ADAPTIVE IMMUNE SYSTEM

SHAI S. SHEN-ORR[1,2], OFIR GOLDBERGER[2], YAEL GARTEN[3], YAEL ROSENBERG-HASSON[4], PATRICIA A. LOVELACE[4,5], DAVID L. HIRSCHBERG[4], RUSS B. ALTMAN[6], MARK M. DAVIS[2,7], ATUL J. BUTTE[1,8]

[1] Stanford Biomedical Informatics Research, Department of Medicine, [2] Department of Microbiology & Immunology, [3] Stanford Biomedical Informatics Training Program, [4] Stanford Human Immune Monitoring Center, [5] Stem Cell Biology and Regenerative Medicine, [6] Departments of Bioengineering and Genetics, [7] The Howard Hughes Medical Institute, [8] Department of Pediatrics
Stanford University, Stanford, CA 94305-5479, USA

The immune system of higher organisms is, by any standard, complex. To date, using reductionist techniques, immunologists have elucidated many of the basic principles of how the immune system functions, yet our understanding is still far from complete. In an era of high throughput measurements, it is already clear that the scientific knowledge we have accumulated has itself grown larger than our ability to cope with it, and thus it is increasingly important to develop bioinformatics tools with which to navigate the complexity of the information that is available to us. Here, we describe ImmuneXpresso, an information extraction system, tailored for parsing the primary literature of immunology and relating it to experimental data. The immune system is very much dependent on the interactions of various white blood cells with each other, either in synaptic contacts, at a distance using cytokines or chemokines, or both. Therefore, as a first approximation, we used ImmuneXpresso to create a literature derived network of interactions between cells and cytokines. Integration of cell-specific gene expression data facilitates cross-validation of cytokine mediated cell-cell interactions and suggests novel interactions. We evaluate the performance of our automatically generated multi-scale model against existing manually curated data, and show how this system can be used to guide experimentalists in interpreting multi-scale, experimental data. Our methodology is scalable and can be generalized to other systems.

1. Introduction

Modern biology has benefited greatly from reductionism, perhaps most notably in the last century by the decision by Delbruck and colleagues to focus on the lowly bacteriophage in their efforts to discern the relationship between genes and DNA on a chromosome, thus escaping the complexities of higher organisms. But to understand the complexities of systems that are only present in higher organisms, such as the immune system, reductionism can only be taken so far before it veers into oversimplification. This is especially true in human immunology, where it is much more difficult to perform experiments as tightly controlled with respect to the many possible variables as in the mouse model. For these reasons we believe that high throughput assays and their accompanying large datasets will be increasingly the norm in immunology, and thus it is critical to develop bioinformatics tools that can assimilate and interpret these datasets as efficiently and broadly as possible, ideally not just the output of one laboratory, but of thousands, and over the several decades that could be considered the modern era.

With respect to how this might be accomplished, the complex and concise nature of the scientific literature means that the use of information extraction tools developed for generic tests is often insufficient. Driven by the utility of high-throughput technologies such as microarrays to measure whole genome gene expression data and yeast two hybrid screens to measure protein-protein interactions, many of the information extraction systems were developed with the aim of extracting genetic[1] or protein-protein interactions[2] from the literature or for annotating groups of differentially expressed genes. Results from these have been used either for construction of searchable knowledgebases[1,3] or for validation of high-throughput experimental results[4].

Here, we construct an information extraction system, partly based on the earlier Textpresso[1] and later Pharmspresso[3] knowledgebases, which we have named ImmuneXpresso, to search abstracts of primary immunology literature for interactions between adaptive immune cells and the cytokines which they secrete and/or affect them. Using the identified interactions, we assemble an inter-cellular network of cells and cytokines to which we integrate with cell type specific gene expression data. Identification of expressed cytokines and receptors in specific cells provides support for ImmuneXpresso identified interactions and allows, in many instances, to assign them directionality. We evaluate the performance of our automatically generated model against existing manually curated data, and use it to guide the identification of novel findings.

2. Methods

2.1 *Immune related corpus and lexicon for information extraction*

To define a comprehensive list of relationships between cells and cytokines, we queried the NCBI journals database to identify journals reporting findings in immunology. 326 journals are annotated under the subject terms: "Immunology & Allergy", or "General Science". Our corpus thus consists of all abstracts published in these journals circa-1960 (PubMed limitation) and onwards, downloaded using NCBI's *eFetch* utility. We established a lexicon of immunological terms, and a comprehensive list of their synonyms (e.g., RANTES = CCL5). Our lexicon is currently limited to six adaptive immune system cell types: B-cells, Cytotoxic T-cell lymphocytes (CTLs), T helper cells, T-regulatory cells, γδ T-cells, and dendritic cells) and 38 cytokines and growth factors. We also constructed a list of verb stems which describe interactions between terms (e.g., induc, stimulat, repres).

2.2 *Automatic construction of immune-centric knowledgebase of cells and cytokines*

To extract information from the primary literature on interactions of immune system components, we built on the Textpresso extracting and processing package[1]. Given a corpus of literature and a lexicon of concepts, ImmuneXpresso identifies occurrences and co-occurrences of concepts within those sentences as well as relations between them. Importantly, owing to its rules, lexicon and corpus, ImmuneXpresso is tailored for the immunology knowledge domain with its unique and developing jargon and expressions. Our corpus of immune related text was tokenized into single sentences. For each sentence we marked all appearances of terms from our lexicon of cells, cytokines and verbs. We then identified all sentences in which two or more terms of different categories co-occurred (e.g. IL-2 stimulates T-helper cells). Next, a filtering step was applied to remove sentences not containing interaction terms or sentences from which it was difficult to infer a regulatory interaction, such as those containing negation. We categorized the verbs in our lexicon as describing positive or negative interactions. For example, sentences containing interactions described using terms such as 'stimulate', 'increase' and 'induce' were all considered descriptive of a positive interaction, whereas sentences containing terms such as 'decrease', 'interferes with production of' or 'deactivates' were categorized as describing negative interactions. A third category of 'undetermined' interactions were those for which we could not assign either a positive or negative role. These included sentences in which terms such as 'alter', 'mediates' and 'cooperatively' appeared. The resulting output is a list of all interactions ImmuneXpresso detected in the corpus. Each interaction described a relationship of a certain type (positive, negative or undetermined) between a cell and a cytokine. For each we also kept track of the number of sentences an interaction was detected in, for use as measure of confidence in an interaction. The resultant interactions were visualized as a bipartite network of cells and cytokines using Cytoscape[5].

2.3 *Data for cell specific gene expression*

Ultimately, cellular communication through cytokines is mediated by the binding of a cytokine secreted from one cell to a receptor expressed on the surface of another. The signature of cytokine and receptor expression should be detectable in cell specific gene expression patterns. We identified publically available microarray studies in GEO in which specific cell subsets were isolated from human blood and their gene expression measured. Data for stimulated and unstimulated dendritic cells was obtained from Jeffery et al.[6] (GSM IDs: 90666, 90836, 90837, 90664). Data for T-regulatory cells was obtained from Ocklenburg et al.[7] (GSM IDs: 101521, 101519). In both of these studies, hybridizations were done on an Affymetrix GeneChip Human Genome U133 Array Set HG-U133A. Palmer et al. [8] (GSE4889) isolated B-cells, T-helper cells and CTLs from healthy individuals and hybridized them on an array. These are two-color printed microarrays in which a single individual's cells (Cy5 labeled) were hybridized against a standard reference – a mixture of RNA from 11 human cell lines. Blood samples from 3 males and 3 females were drawn, and cells selected for using magnetic beads. B-cell selection was positive, whereas for T-helper and CTLs selection was negative. To the best of our knowledge, no high quality $\gamma\delta$ T-cell study for humans is publicly available.

For each cell type, we performed quantile normalization[9] across all replicate and/or multiple measurements. We matched microarray platform probes to Entrez GeneIDs[10]. Those probes not mapped to any Entrez GeneID were discarded, whereas the intensity values of those mapping to the same gene were averaged. We converted the expression values into binary form by considering, for each array, only the genes in the top 40% of the intensity values, as expressed. We picked the 40th percentile as this is where the log ratio turns positive on most two color arrays in this data. For each cell type, we generated a cell type specific gene expression signature by considering a gene as expressed if it was detected as expressed in arbitrarily 50% or more of the replicate or multiple array measurements of that cell type. This resulted in 4338, 4468, 4931, 5803 and 6122 genes considered as expressed in B-cells, T-helper cells, CTLs, dendritic cells and T-regulatory cells, respectively.

2.4 *Linking inter-cellular interactions with intra-cellular content*

To link our inter-cellular communication network to the intra-cellular networks which drive them, for each cytokine listed in the ImmuneXpresso lexicon, we identified in the Entrez Gene database one or more receptors to which it binds.

This complete list included a total of 79 genes, 38 of which were cytokines or growth factors and 41 receptors, with 68 binary cytokine-receptor interactions between them. For example, the binding of IL-13 to the IL13RA1 receptor as well as IL13RA2 and IL4RA was considered as three separate binding events. All 79 genes had probes on at least one of the array platforms (see 2.3), but only 57 had probes on both array platforms. We use these to compile a cellular gene expression driven cell-cytokine network and contrast it with a literature-derived network generated by ImmuneXpresso.

2.5 Validation and Experimental data

It is critical to assess the quality of the output ImmuneXpresso extracts, namely how well do we automatically extract cell-cytokine interactions from literature. The Cytokine Online Pathfinder Encyclopedia (COPE) database [11] and the Cytokine Reference – Online Database [12] are the two largest publicly available references for cytokine functionality. Both are manually curated by experts. We used these databases to search for each of the cell-cytokine interactions identified by ImmuneXpresso and check its validity.

Unlike the free text which it mines, ImmuneXpresso output is in machine interpretable form that can easily be utilized for downstream analysis. We contrasted ImmuneXpresso output to two experimentally derived datasets: in the first, blood was drawn from 29 human individuals, males and females of varying ages, and analyzed for cell frequency and serum cytokines. The frequency of B-cells, γδ T-cells, CTLs and T-helper cells was assessed using antibody staining and flow cytometry. Serum cytokines were analyzed using the Luminex 200 assay system with MilliPore 37-plex kits. Employing our network algorithm (relevance networks)[13] and using a Pearson's correlation cutoff of ±0.8, we identified 668 associations in the data, of which 41 were between cells and cytokines.

In the second dataset, a cytokine response assay was performed on CD4+ T cells from spleens of 10 week old MRL.MpJ mice. These were stimulated with various dilutions of 5 cytokines (IL-2, IL-3, IL-12, IFN-γ and TNF-α) by incubation for 16 hours at 10^6 cells per well. Secretion blockers were then added for an additional 2 hours in order to accumulate intracellular cytokines and then the cells were lysed and lysates used for Luminex cytokine multiplex assays with Panomics Procarta 21-plex kits. An increase, appearance or decrease in the abundance of the 21 measured cytokines compared with their abundance with no stimulation, was indicative of an increased production above background, *de novo* production and inhibition of production of cytokines respectively.

3. Results

A total of 3695 sentences reporting interactions between the six cell types used here and 38 cytokines were identified by ImmuneXpresso (Fig. 1). These were unified into 217 unique interactions, with an average number of 17 sentences per interaction. The resultant network is dense, with 54% of cell-cytokine pairs connected to one another directly. In comparison with other biological networks, this is a very dense network[14], in agreement with the high density value observed for the cellular immune network (nodes are cells, edges are cytokines) that Frankenstein et al.[14] derived from curated databases.

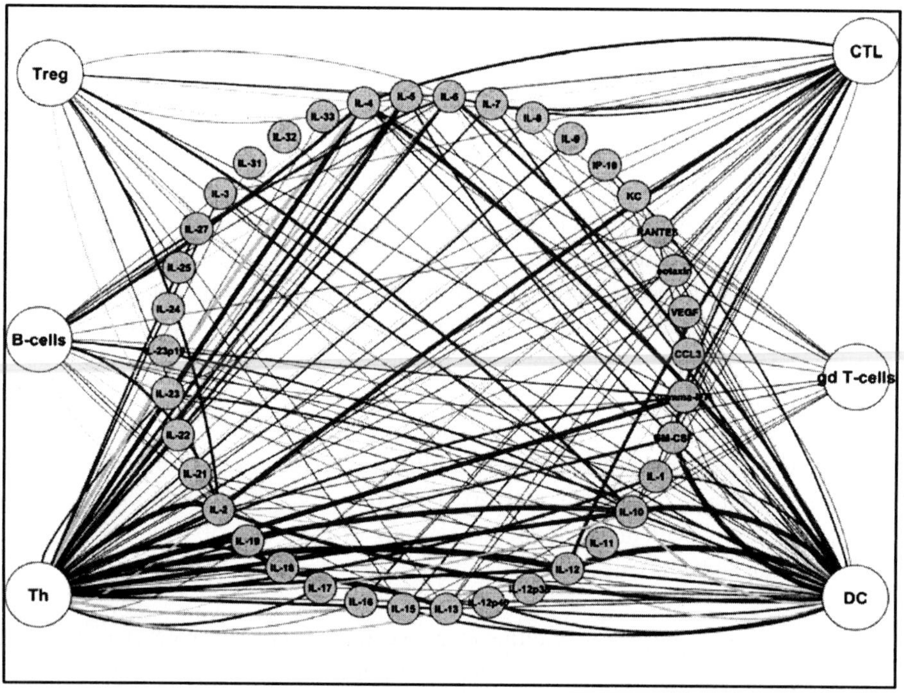

Figure 1. An automatically derived network of cells and cytokines from the immunological literature. Six immune system cell subsets (white nodes) and their relationship, through 217 edges, with 38 cytokines (grey nodes). Edge color, from light to dark denotes the type of interaction, positive, undetermined, and negative. Edge width denotes the confidence of the relationship, represented as the log of the number of sentences in which the interactions was identified. Treg – T-regulatory cell, Th – T-helper cell, γδ – gamma delta, DC- dendritic cell.

Despite the high density we observe, the theoretical density which could be reached with 217 edges in a bipartite network of this size is 95%. However, many of the cell-cytokine pairs have more than one edge type spanning between them which likely reflects either incorrect semantic parsing of the text or the multiple functionality of cytokines under different conditions or in specific cell subsets. From the 217 edges, 113 were positive, 50 were negative and 54 were of undetermined type. Examination of the number of sentences shows that the evidence for most cell-cytokine relationships is predominantly of one type (Fig. 2). Furthermore, there is a strong correlation (Pearson's 0.91 for cell types, 0.59 for cytokines) between the degree of a node and the number of sentences which support its interactions. This is to be expected in a literature based network as the more research is done in a field, the more separate components become associated with one another.

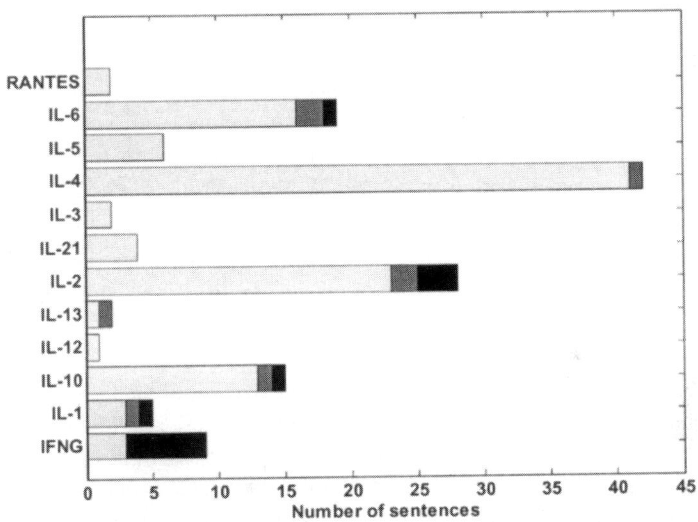

Figure 2: Interactions between B-cells and twelve cytokines extracted from abstracts by ImmuneXpresso. X-axis shows the number of sentences the interaction was observed in. Color denotes interaction type. White, gray and black denote positive, negative and undetermined interaction respectively. All twelve cytokines are true effectors or products of B-cells.

To evaluate our ability to correctly capture cell-cytokine interactions automatically from abstracts, we compared our results to the relevant subset of a manually curated set of cytokine-cell interactions from the COPE database and the Online Cytokine Reference[12]. Checking specifically by cell type, we verified each of our detected B-cell interactions with those available on the two websites. Overall, we had a 40% false-negative rate, but no false positives, better results than are usually observed for co-occurrence systems which search abstracts only[1]. Two of the cytokines interacting with B-cells, IL-21 and RANTES, did not appear in the Online Cytokine Reference, but did appear in COPE. Similar false negative rates were observed for dendritic cells. Interestingly, more recent discoveries in cytokine-cell regulation such as the regulatory interactions of γδ-T-cells, or IL-33 did not appear in either database, showing the power of automatic annotation for keeping up to date with published literature.

Edges in the ImmuneXpresso network link cells and cytokines to one another (Fig. 1). In reality, these cells produce and secrete cytokines which bind to membrane receptors expressed on the surface of another cell or even the same one. We hypothesized that inter-cellular interactions between cytokines and their receptors would be evident, in cell specific gene expression data. If we could detect cytokines and receptors gene expression, than we may be able to assemble gene-expression-based inter-cellular communication networks. Further, if cytokines and their binding partner receptors were exclusively expressed between cells, we may be able to assign directionality to currently undirected ImmuneXpresso reported interactions. To test this, we assembled a compendium of cell specific gene expression signatures from publically available microarray studies (see 2.3) matching the cell types in the present version of the ImmuneXpresso lexicon, as well as identified the one or more receptors each cytokine binds (see 2.4)

We asked how many cytokines in a given cell type are exclusively expressed from their receptor and how many are expressed in the same cell. The majority of cytokines and receptors were expressed exclusively of their binding partner. For example, the 8 cytokines and 21 receptors we detect expressed in T-regulatory cells may participate in 35 different binary interactions, both within our model framework (5 immune cell subsets, no data for γδ-T-cells) and outside, but for only two (IL-13 binds IL4RA and IL-16 binds CD4) are both the receptor and the cytokine expressed by T-regulatory cells. We note that the cytokine IL-2, known to both be expressed by T-regulatory cells, and regulate them via IL2Ra (CD25), is not detected as expressed in T-regulatory cells under our criteria. Similarly, for only 7 out of 40, 10 out of 30, 6 out of 40 and 5 out of 34 for B-cells, T-helper, CTL and dendritic cells respectively, both cytokine and receptor are expressed on the same cell. Expression of both cytokine and receptor in the same cell may be indicative of auto-regulation, or of the expression of either the cytokine or its' receptor under different conditions or subsets.

The ImmuneXpresso network is bipartite, with each edge representing an interaction between cell and cytokine. Each edge in this network has been extracted from the literature and thus is supported by experimental evidence. On the other hand, an interaction between two cells must always consist of at least two edges. Unlike single edges, interpretation of a path of two or more edges as a cytokine mediated interaction between two cells, is not necessarily warranted, as it requires one cell to be a producer of the cytokine and the other to be affected by it. Here, gene expression data may come to the aid, as it allows one to identify cells expressing or producing cytokines, and those with the potential to be affected by them by expressing the corresponding cytokine receptors.

Ignoring edge types, the ImmuneXpresso network has 119 edges. In theory, these could encode for 309 possible undirected regulatory paths between the 6 cells in our system, 119 of which are auto-regulatory. Using the information obtained from the gene expression data, we can now estimate how many of the pathways theoretically derived from the ImmuneXpresso output are supported by the gene expression data, such that a cytokine and its receptor are expressed in the two communicating cells. To do so, we consolidated the gene specific cytokine-receptor information we obtained from Entrez to match the details of the lexicon ImmuneXpresso uses. For example, the information 'B-cells express both IL-10 and the IL-10 receptor IL-10Ra and IL-10Rb' is simplified in the consolidated form to 'B-cells express the IL-10 cytokine and a receptor to which it can bind'.

Filtering the 309 ImmuneXpresso paths by requiring one of the cells to express a cytokine while the other expresses its receptor drops the number of interactions to 158 of which 30 are auto-regulatory (same cell expresses both the cytokine and the receptor). The remaining 151 non-functional paths are either a byproduct of the network representation, or appear as such due to the threshold we set as to which genes should be considered expressed (see 2.2). Furthermore, as we observed, many of the cytokines and receptors are exclusively expressed on one of the two cells. Therefore, unlike the directionless interactions ImmuneXpresso currently reports, many of the cytokine mediated cell-cell interactions we infer from the gene expression data are directional. Of the 158 ImmuneXpresso paths supported by the gene expression data, we can assign directionality to 76. Last, we can ask the reverse question, namely how many of the possible cytokine mediated, cell-cell interactions appear in the gene expression data, and cannot be traced to any path in ImmuneXpresso. We find 27 such paths, 21 of which we could not find in the manually curated cytokine databases[11,12]. Each represents a hypothetical cytokine mediated cell-cell interaction that can be tested experimentally.

The utility of a knowledgebase in machine computable format stands out when conducting high throughput discovery driven experiments. In such experiments, researches rarely have expert knowledge in all of the variables being assayed and the number of results from experiments is often very high. Thus, it may be difficult to prioritize findings and link them to one another or to previous discoveries, to establish a comprehensive perspective. As a proof of principle, we analyzed serum cytokine and cell subset frequency data measured at the Stanford Human Immune Monitoring Core for 29 individuals, males and females of varying ages. Of the 41 detected interactions (see 2.5), 18 were between a cell and a cytokine covered in the present ImmuneXpresso lexicon version (see Data sources 2.1). Remarkably, ImmuneXpresso could verify each of those 18 and match it with a reported interaction in the literature. For example, an interaction between IL-15 and CTLs, which was deduced by our algorithm by the positive correlation observed between IL-15 and CTLs, was detected 9 times in abstract sentences by ImmuneXpresso, from such sentences as 'These findings identify a novel CTL costimulatory pathway regulated by IL-15 and suggest that tissues can fine-tune the activation of effector T cells based on the presence or absence of stress and inflammation'[15]. Comparison of a second dataset, in which interactions of 13 cytokines were experimentally identified with spleen derived CD4+ T-helper cells, showed that ImmuneXpresso could validate 10 out of the 13 observed interactions. Manual searches for the other three interactions was able to confirm one additional interaction not captured by ImmuneXpresso. Each such interaction suggests a testable hypothesis which requires considerable time and resources to test. As datasets grow larger, machine identification and prioritization of novel findings to follow up on is key.

4. Discussion

The multi-scale nature of the immune system and its high complexity challenge us to find new ways to analyze it. Discoveries in immunology and cellular biology are occurring at an unprecedented rate. Despite this, we are far away from understanding how the immune system of higher organisms, and humans in particular, are able to mount an immune response. Here, we take the first steps towards building bioinformatics tools that are specifically geared towards extracting information from primary immunology literature and comparing it to high throughput data. For the primary literature in this field, a particular value of ImmuneXpresso is that it facilitates 'connecting the dots' between very different areas of immunology. Its low false positive rate, unlike the 35% false positive rate[1] observed in extraction systems specializing in genetic interactions, may be due to the types of associations we are extracting (i.e. cells with cytokines) which may be easier to extract from natural language text than other types of associations. The false negative rate matches that of other systems using abstracts to extract information. Expansion to full text analysis is likely to drop the false negative rate[1], but will likely result in an increase in false positives.

However, solely relying on information extraction systems to establish interaction networks always run the risk of misleading. As the accuracy of each individual reported interaction may not to be perfect, confidence in correctness of any given path decreases as a function of path length. To aid in this, as well as validate our findings, we use cell specific gene expression data to establish a biological context to our network. These provide both supporting evidence for extracted information and utilization of the extracted information for the identification of novel associations. We note that at present, our approach discards low abundance genes, and is sensitive to the conditions under which the gene expression data was assayed. These limit our ability to infer on cytokine mediated cell-cell interaction. Furthermore, extension of this approach to rare cell types may be complicated by the smaller body of literature and the lack of expression data. Thus we advocate systems integrating data from multiple data sources. For example, the requirements for high accuracy in natural language text processing, may be lifted to some extent by the integration of machine formatted data from other electronic sources. Future work should address this by integration of protein interaction data, enriching the lexicon by the use of predefined ontologies, and expanding the system to integrate not only cytokines and receptors, but also activation pathways and downstream targets.

Acknowledgments

We thank Alex Skrenchuk and Rong Chen from Stanford University for computer support and database support, respectively. The Ellison Foundation and the NIH specifically U19 AI057229-06, as well as the Hedco and the Russell Family Foundation for funding HIMC research. YG is supported by training grant NIH LM007033 from the National Library Medicine. RBA is supported by NIH/NIGMS Pharmacogenetics Research Network and Database and the PharmGKB resource (NIH GM61374). AJB is supported by the Lucile Packard Foundation for Children's Health.

References

1. Muller, H.M., Kenny, E.E. & Sternberg, P.W. Textpresso: an ontology-based information retrieval and extraction system for biological literature. *PLoS Biol* **2**, e309 (2004).
2. Rzhetsky, A. et al. GeneWays: a system for extracting, analyzing, visualizing, and integrating molecular pathway data. *J Biomed Inform* **37**, 43-53 (2004).
3. Garten, Y. & Altman, R.B. Pharmspresso: a text mining tool for extraction of pharmacogenomic concepts and relationships from full text. *BMC Bioinformatics* (Accepted).
4. Rual, J.F. et al. Towards a proteome-scale map of the human protein-protein interaction network. *Nature* **437**, 1173-8 (2005).
5. Shannon, P. et al. Cytoscape: a software environment for integrated models of biomolecular interaction networks. *Genome Res* **13**, 2498-504 (2003).
6. Jeffrey, K.L. et al. Positive regulation of immune cell function and inflammatory responses by phosphatase PAC-1. *Nat Immunol* **7**, 274-83 (2006).
7. Ocklenburg, F. et al. UBD, a downstream element of FOXP3, allows the identification of LGALS3, a new marker of human regulatory T cells. *Lab Invest* **86**, 724-37 (2006).
8. Palmer, C., Diehn, M., Alizadeh, A.A. & Brown, P.O. Cell-type specific gene expression profiles of leukocytes in human peripheral blood. *BMC Genomics* **7**, 115 (2006).
9. Irizarry, R.A. et al. Exploration, normalization, and summaries of high density oligonucleotide array probe level data. *Biostatistics* **4**, 249-64 (2003).
10. Chen, R. & Butte, A.J. AILUN: reannotating gene expression data automatically. *Nature Methods* **4**, 879 (2007).
11. Ibelgaufts, H. COPE: Cytokines Online Pathfinder Encyclopaedia. . (1997).
12. Oppenheim JJ et al. The Online Cytokine Reference Database. (2000).
13. Butte, A.J., Tamayo, P., Slonim, D., Golub, T.R. & Kohane, I.S. Discovering functional relationships between RNA expression and chemotherapeutic susceptibility using relevance networks. *Proc Natl Acad Sci U S A* **97**, 12182-6 (2000).
14. Frankenstein, Z., Alon, U. & Cohen, I.R. The immune-body cytokine network defines a social architecture of cell interactions. *Biol Direct* **1**, 32 (2006).
15. Roberts, A.I. et al. NKG2D receptors induced by IL-15 costimulate CD28-negative effector CTL in the tissue microenvironment. *J Immunol* **167**, 5527-30 (2001).

A BAYESIAN INTEGRATION MODEL OF HIGH-THROUGHPUT PROTEOMICS AND METABOLOMICS DATA FOR IMPROVED EARLY DETECTION OF MICROBIAL INFECTIONS

BOBBIE-JO M. WEBB-ROBERTSON[†], LEE ANN MCCUE, NATHANIAL BEAGLEY, JASON E. MCDERMOTT, DAVID S. WUNSCHEL, SUSAN M. VARNUM, JIAN ZHI HU, NANCY G. ISERN, GARRY W. BUCHKO, KATHLEEN MCATEER, JOEL G. POUNDS

Pacific Northwest National Laboratory
Richland, WA 99352, USA

SHAWN J. SKERRETT, DENNY LIGGITT, CHARLES W. FREVERT

Pulmonary Research Laboratory, University of Washington
Seattle, WA 98108, USA

High-throughput (HTP) technologies offer the capability to evaluate the genome, proteome, and metabolome of an organism at a global scale. This opens up new opportunities to define complex signatures of disease that involve signals from multiple types of biomolecules. However, integrating these data types is difficult due to the heterogeneity of the data. We present a Bayesian approach to integration that uses posterior probabilities to assign class memberships to samples using individual and multiple data sources; these probabilities are based on lower-level likelihood functions derived from standard statistical learning algorithms. We demonstrate this approach on microbial infections of mice, where the bronchial alveolar lavage fluid was analyzed by three HTP technologies, two proteomic and one metabolomic. We demonstrate that integration of the three datasets improves classification accuracy to ~89% from the best individual dataset at ~83%. In addition, we present a new visualization tool called Visual Integration for Bayesian Evaluation (VIBE) that allows the user to observe classification accuracies at the class level and evaluate classification accuracies on any subset of available data types based on the posterior probability models defined for the individual and integrated data.

1. Introduction

Developing molecular markers of disease is a subject of intense interest, however knowing *a priori* the appropriate analytical methods to target the correct biomolecules is challenging. With recent advances in high-throughput (HTP) experimental methods, the need for statistical methods to integrate data of different types to provide biological models that can be used to make predictions about the underlying systems has become paramount. Integrating the

[†] Corresponding Author.

data from multiple technologies into an interpretable form, to identify either a diagnostic pattern or the underlying molecular response, requires developing statistical methods that can mange the differences in formats, resolutions, and data sizes from the different instruments. The promise of data fusion or data integration is that different types of data about a system can be integrated to give biological models that are more complete and accurate than those obtained by using any of the individual data sources, i.e. the whole is greater than the sum of its parts. Data integration requires a number of choices to be made: e.g., the biological question to be asked, the input data that will enable that question to be answered, and the learning model.

Despite the obvious importance of fusing HTP biological data, the heterogeneity of the data (widely varying size, scale, specificity and format) presents many challenges. To date, most methods for biological data fusion use a combination of experimental data (e.g., gene expression measurements) and predictive information (e.g., sequence homology). The most common methods use kernel fusion, such as a support vector machine (SVM), or Bayesian integration approaches [1-11]. The success of these methods depends on the data from different streams being independent enough to improve elucidation of the underlying system. For example, Lu *et al.* [10] showed that the predictive potential of their integrative model was reached after integrating the top few data types; many other kinds of data examined did not add any value to the model. Therefore it is important to have a way to evaluate the contribution of each type of input data to the final model.

We present a Bayesian integration strategy that uses statistical learning algorithms, such as partial least squares discriminant analysis (PLS-DA) [12], to build initial likelihood probability models which are then transformed into the posterior probability models of interest, in particular, the probability of class membership given the sample and data stream [13]. In addition, the Bayesian model can be used directly to integrate disparate data types using posterior probabilities. We demonstrate that the integration of datasets can increase the classification accuracy of the model, although integration of pairs of datasets does not always improve classification accuracy. We also present a visualization tool, Visual Integration for Bayesian Evaluation (VIBE), which allows the user to easily determine the class for which each dataset holds the most power, and allows the user to investigate different combinations of the datasets in an automated fashion.

One area of particular interest for data integration is early detection of exposure to pathogens. We present a mouse aerosol exposure experiment for which both proteomic and metabolomic biosignatures were collected. In this experiment mice were exposed to one of three organisms, *Pseudomonas aeruginosa*, *Francisella novicida*, and *Francisella tularensis subsp. novicida*, an avirulent mutant of *F. novicida*. The mice were evaluated at 4 and 24 hour time points to determine if markers of exposure were present at a pre-symptomatic state. *F. novicida* infection in mice serves as a model of human infection with the category A pathogen, *F. tularensis subsp. tularensis* [14]. In mice the disease course with aerosol exposure to *F. novicida* is fatal, with necrosis in the lungs as a typical pathology, whereas *P. aeruginosa* infection results in a distinct pneumonia. The goal of this study was to determine if integration of the HTP proteomic and metabolomic data can more accurately predict the infection class than any data source individually.

2. Sample Preparation and Analysis Methods

2.1. *Pathogen Exposure*

Young male mice (C57/BL) were subjected to aerosol exposure to initiate infection with one of three organisms (Table 1): virulent strains of *F. novicida* (FTN) and *P. aeruginosa* (PA), and an avirulent *F. novicida* (MGLA) containing a mutation to the transcriptional regulator gene *mglA*. The C57/BL mice were obtained from Jackson Labs (Bar Harbor, ME), exposed to one of the above pathogens using In-tox snout-only restraining tubs (In-Tox Products LLC, Moriarty, NM), and sacrificed at one of three time points. This resulted in seven classes of interest shown in Table 1: (1) Control at times 0, 4 and 24 Hrs, (2) PA at 4 Hrs, (3) PA at 24 Hrs, (4) FTN at 4 Hrs, (5) FTN at 24 Hrs, (6) MGLA at 4 Hrs, and (7) MGLA at 24 Hrs. The study was conducted using an experimental design to limit effects of exposure time or order prior to sample collection.

Table 1. Pathogen Exposure Experiment.

	0 Hrs	4 Hrs	24 Hrs
Control[a]	$n = 4$	$n = 4$	$n = 4$
PA[b]		$n = 4$	$n = 4$
FTN[b]		$n = 4$	$n = 4$
MGLA[b]		$n = 4$	$n = 4$

[a]Control animals were exposed to phosphate buffered saline.
[b]PA = *P. aeruginosa*, FTN = *F. novicida* (wild-type) and MGLA = *F. novicida* mglA^{-}.

Bronchial alveolar lavage fluid (BALF) was collected from each animal and subjected to three HTP analytical methods: Matrix Assisted Laser Desorption/Ionization (MALDI) mass spectroscopy (MS) to evaluate large proteins, accurate mass and time tag (AMT) proteomics on an Orbitrap mass spectrometer to evaluate protein fragments, and nuclear magnetic resonance (NMR) spectroscopy to evaluate metabolites.

2.2. *MALDI Mass Spectrometry*

The MS analysis of purified proteins was performed using an Autoflex II MALDI tandem time-of-flight mass spectrometer equipped with a HIMAS™ detector (Bruker Daltonics, Billerica MA). Raw data was processed using available functions in the vendor software, which allows for spectral recalibration and alignment. Peaks are extracted using an automated peak detection algorithm that creates a list of peak locations (m/z) and associated intensity values from the mass spectrum of each sample [15]. The full sets of spectra (all technical replicates from all experimental replicates) are aligned for each peak, identifying its location (or non existence) over all replicates [16]. The code to perform this processing runs in MatLab® Version R2008a. Replicate peak identifications within a sample class were averaged and subjected to an occurrence filter of 60% [17], which resulted in a final dataset of peak intensities for 51 locations. Due to the variability of peak intensities and missing data, MALDI data are commonly converted to binary values. Thus, the final dataset consisted of 51 locations marked with presence/absence over the 36 samples.

2.3. *AMT-based Proteomics*

MS analysis was also performed on protein fragments (peptides) using an LTQ-Orbitrap™ mass spectrometer (Thermo Electron Corp., Waltham, MA) with nanoelectrospray ionization. Orbitrap™ spectra were collected from 400-2000 *m/z* at a resolution of 100k and analyzed using the accurate mass and elution time (AMT) tag approach [18]. Briefly, the theoretical mass and the observed normalized elution time (NET) of peptides identified by LC−MS/MS have been used previously to construct a reference database of murine AMT tags [19]. Features from the LC−MS analyses (i.e., *m/z* peaks deconvolved of isotopic and charge state effects and then correlated by mass and NET) were matched to AMT tags to identify peptides, using a tolerance of +/-6 ppm for mass and 0.025% for the LC NET. The mass deisotoping and alignment process was performed using Decon2LS, and the matching process was performed using VIPER [20].

Peptide abundance data was further processed to remove peptides identified with low confidence. First, a uniqueness filter of a SLiC score of 0.5 with DELSLiC of 0.2 was applied to the data [21]. All peptides were then filtered using an occurrence filter that required peptides to have been observed in at least one exposure class for at least 75% of the samples. After filtering, the missing observations for the remaining 2023 non-redundant peptides were imputed using a tiered strategy, where if a peptide was observed in more than 50% of the samples of a particular class the missing values were imputed as the mean of the observed values, otherwise it was imputed as ½ of the minimum observed abundance for that peptide. Thus, the final reduced dataset consisted of the 36 samples with measured or imputed abundance values for 2023 peptides.

2.4. *Proton (^1H) NMR*

NMR analysis of metabolites was performed by adding 200µL of D_20 containing 0.4 mM trimethylsilylpropionic-2,2,3,3,-d_4 acid (internal reference) to 400 mL of each BALF sample and recording a one-dimensional ^{1}H NMR spectrum on a Varian Unity-600 NMR spectrometer (Varian Inc., Palo Alto, CA) at 10°C. The data were collected with a sweep width of 8000 Hz, 32k of data points, a delay of 3.0 s, and 4096 transients, and processed with Felix 97 software (Accelrys, San Diego, CA). After converting the Varian free-induction decay data into Felix format, the data were adopidized with a square sinebell function prior to Fourier transformation into the frequency domain. The spectral files were then imported to Chenomx NMR Suite 5.0 software (Chenomx Inc, Alberta, Canada) for baseline correction, normalization, and binning using 0.01 ppm bins. This resulted in a final dataset that consisted of 324 bins. A large fraction of these bins had considerable variability and were further processed by a Kruskal-Wallis test across the seven exposure classes defined in Table 1 [20]. A bin inclusion p-value of 0.01 was selected, which resulted in a final reduced dataset of measured intensity values for 27 bins for the 36 samples.

3. Statistical Methods

3.1. *Statistical Learning Algorithms*

Bayesian integration is based on the ability to define probability models for independent sources of data. We used partial least squares discriminant analysis (PLS-DA) [12] to evaluate the Orbitrap and NMR datasets. The MALDI data is in a binary format, not well suited to PLS-DA, thus, a spectral fingerprinting approach called degree of association was used [17]. Additionally, leave-one-out cross-validation (LOOCV) was used to assure that the probabilities obtained for a specific sample were independent from the training data [23].

3.1.1. *Partial Least Squares– Discriminant Analysis (PLS-DA)*

The primary goal of PLS [24] is to build a linear model between a set of independent variables and a predictor variable (e.g., exposure and time). In general, PLS produces factor scores that are linear combinations of the original variables in such a manner that the factor score variables are uncorrelated. PLS-DA is used for categorical variables, such as the binned NMR and proteomics values in this analysis. PLS-DA was run in MatLab® Version R2008a using Version 4.2 of the PLS_Toolbox© from Eigenvector Research. PLS-DA gives a number nominally between zero and one that is associated with the likelihood that the sample i (s_i) from dataset Q belongs to event k, $P(s_i, D_Q|E_k)$, where $Q \in [O,N]$ for the Orbitrap and NMR datasets, respectively and k=1,…,7 (classes associated with Table 1). The LOOCV allows these likelihood values to deviate outside of the zero-to-one boundary of probability values, therefore a piecewise function was used to assure that appropriate probability values were obtained.

$$P(s_i, D_Q \mid E_k) = \begin{cases} 0 \text{ if } PLS_{LOOCV}(s_i, D_Q \mid E_k) < 0 \\ 1 \text{ if } PLS_{LOOCV}(s_i, D_Q \mid E_k) > 1 \\ PLS_{LOOCV}(s_i, D_Q \mid E_k) \text{ otherwise} \end{cases} \quad (1)$$

3.1.2. *Degree of Association*

Degree of Association is a statistical algorithm that returns a probability associated with the similarity of a MALDI reference library and the spectra of interest [17]. In particular, the null hypothesis (H_O) that a specific sample is from class k (e.g., FTN at 4 hours) is considered versus the alternative that the specific sample is not from class k. Assuming H_O, the sample under consideration can be described by the probability of observing peak i (p_i). These probabilities are compared to the reference fingerprint of class k based on the set of peaks that differ and those that are in common between the sample fingerprint and the reference fingerprint:

$$da(k) = P(s_i, D_M \mid E_k) = 1 - \prod_{i \in A^C} p_i \left[\prod_{i \in A} (1 - p_i) \right] \quad (2)$$

where A is the fingerprint peaks that are not observed in the sample and A^C is the complement, i.e., observed peaks. The code to compute the Degree of Association runs in MatLab® Version R2008a. The result of the model is the likelihood that that the sample is in the modeled class based on LOOCV.

3.2. *Bayesian Statistics, Integration, and Classification*

Bayesian statistics is an attractive approach for making probabilistic inferences from biological data because all data are defined as random variables, allowing the removal of nuisance parameters via integration or summation [13]. Let ϑ denote the set of unknown parameters and let y_{obs} denote the observed data, e.g., resulting from an experimental technique like NMR. The likelihood function is defined as the probability of the observed data given the unknown parameters:

$$L(\vartheta; y_{obs}) = P(y_{obs} \mid \vartheta).$$

Thus, the joint probability distribution of ϑ and y_{obs} is defined as:

$$Joint = Likelihood * Priors$$

$$P(y_{obs}, \vartheta) = L(\vartheta; y_{obs})P(\vartheta) = P(y_{obs} \mid \vartheta)P(\vartheta).$$

The Bayesian inference is made by obtaining and inspecting the posterior distributions of the unknown quantities of interest. These posterior distributions are obtained from Bayes theorem:

$$P(\vartheta \mid y_{obs}) = \frac{P(y_{obs}, \vartheta)}{P(y_{obs})},$$

where $P(y_{obs})$ is computed by integrating over ϑ in the joint distribution. Suppose the unknown parameter is of n dimension, $\vartheta = (\vartheta_1,\ldots,\vartheta_n)$. Those parameter components that are not of immediate interest, but necessary to the model, must be integrated out from the joint distribution to provide a proper inference on the unknown variable of interest.

For our experiment, the probabilities obtained in Eqs. (1) and (2) are the likelihood of observing sample i (s_i) associated with dataset Q given a specific exposure/time class (Table 1). However, in predicting an exposure/time class, the posterior probability is the one of primary interest:

$$P(E_k \mid s_i, D_Q) = \frac{P(s_i, D_Q \mid E_k)P(E_k)}{\sum_{k=1}^{NG} P(s_i, D_Q \mid E_k)P(E_k)}, \quad Q \in [0,N,M] \tag{3}$$

where NG is the number of classes, seven for our mouse infection experiment. Thus, for each sample a set of NG probabilities are obtained which sum to one. Class membership is assigned to the class that has maximum probability:

$$s_i \in E_k \ \text{ where } k = \max_{k=1}^{NG}\left(P(E_k \mid s_i, D_Q\right). \tag{4}$$

Bayes formula can also be used to integrate and extract posterior probabilities for multiple data streams. Under the assumption that each data stream is independent but shares a common sample source, then the integrated likelihood is the product of the individual likelihoods:

$$P(s_i, D_1, \ldots, D_{NQ} \mid E_k) = P(s_i, D_1 \mid E_k) * P(s_i, D_2 \mid E_k) * \cdots P(s_i, D_{NQ} \mid E_k),$$

where *NQ* is the number of data streams. Given that the datasets share common samples, the datasets likely have some correlation; however, since the end goal is classification this assumption simplifies the integration step and the visualization tool can be used to assess the accuracy of the model and the probability of a particular sample is

$$P(E_k \mid s_i, D_1, \cdots, D_{NQ}) = \frac{\left[\prod_{Q=1}^{NQ} P(s_i, D_Q \mid E_k)\right] P(E_k)}{\sum_{k=1}^{NG} \left[\prod_{Q=1}^{NQ} P(s_i, D_Q \mid E_k)\right] P(E_k)}. \quad (5)$$

For our experiment, the maximum of *NQ* was three, however any subset of data streams can be analyzed. For our experiment, the prior probabilities $P(E_k)$ were set to be equal (1/NG). Classification of the integrated data is performed in an analogous manner to Eq. (4).

4. Results and Discussion

In our pathogen exposure experiment, three data sources representing the proteomic and metabolomic profiles were collected, each of varying size and content. For example, the Orbitrap data has multivariate form of $n = 36$ samples by 2023 abundance values, while the MALDI data is in the form of 36 samples by 51 binary values. To transform the three datasets into a common form, we first used statistical learning algorithms, PLS-DA (Section 3.1.1) and Degree of Association (Section 3.1.2), to transform these datasets into probability metrics associated with class membership using Eqs. (1) and (2). The second step consisted of using Bayesian statistics to modify these probability models into the posterior form associated with the probability of observing a specific class given the observed data (Eq. (3)). This posterior probability model was then used directly to perform classification on individual datasets (Eq. (4)) and the product of sets of probability models used to compute integrated posterior probability models (Eq. (5)). This process is described in Figure 1 with respect to the pathogen exposure experiment and is easily generalizable.

We describe the results for the individual and integrated datasets with respect to classification accuracy (CA). CA is simply measured as the number of correctly classified observations (*#TP*) divided by the total number of samples: (CA = *#TP*/*n*). Samples were classified into whichever class they fell with maximum probability (Eq. (4)). We observed that the individual data source with the highest LOOCV CA was the AMT-based Orbitrap data, with on overall CA of ~83.3%. Second was the MALDI data with an overall CA of

~75%, followed by the NMR data at ~61.1%. Integration of the three datasets increased the overall CA of the model to ~89%. The results broken down by class are given in Table 2.

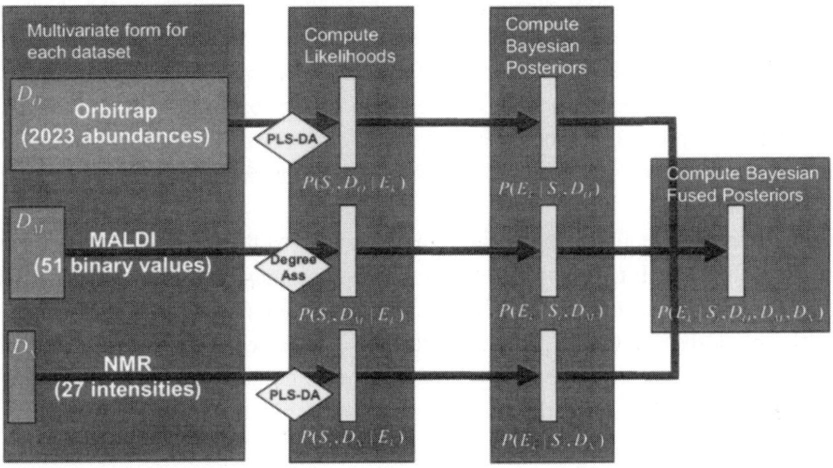

Figure 1. Standard analysis pipeline to obtain integrated posterior probabilities across the heterogeneous data types.

Table 2. Overall class accuracy results for each class using LOOCV.

	NMR	MALDI	Orbitrap	Integrated
Control	75%	75%	92%	100%
PA – 4	75%	100%	100%	100%
PA – 24	50%	75%	50%	50%
FTN – 4	50%	75%	75%	100%
FTN – 24	75%	25%	75%	100%
MGLA – 4	75%	100%	75%	75%
MGLA – 24	0%	75%	100%	75%
Total	61.1%	61.1%	83.3%	89.0%

We were interested not simply in the CA values for each dataset, but also in which samples were misclassified into which classes, as well as which dataset offered the most advantage to the integration. To explore these questions we developed a visualization tool in MatLab® Version R2008a called Visual Integration for Bayesian Evaluation (VIBE) that ingests the posterior probability models from Eq. (3), computes the CA and performs the integration. VIBE plots the CA matrices, where the x-axis is the true class and the y-axis is the predicted class, and the color indicates the fraction of samples that were classified into specific classes (see Figure 2). For example, we observed that the

NMR data misclassified a few samples into the *P. aeruginosa* class at 24 hours and the *F. novicida* class at 4 hours, and the MALDI data misclassified control samples into the *P. aeruginosa* and the *F. novicida* mutant (MGLA) 4 hour classes. The visualization allows the user to observe these apparent differences, which presumably allow the integrated probabilities to attain the correct classification and point out both correlated and orthogonal structures between the datasets at the classification level. Thus, in this manner the complementary or overlapping nature of specific datasets can be explored. In addition, VIBE computes the posterior probability of each dataset in the integrated model (below the integrated CA plot), which gives a rough measure of the contribution of each dataset.

It is not always the case that integration improves the CA of a system. For example, the overall CA drops to ~64% when using only the NMR and MALDI data to perform the integration (Figure 3), and to ~81% when integrating only NMR and Orbitrap data. These values are lower than individual CAs for MALDI and Orbitrap, respectively. However, we do see an increase in CA to ~86% when the MALDI and Orbitrap data are combined. Thus, it appears that the NMR data does not complement the MALDI and Orbitrap datasets as we might have expected, but instead validates misclassifications and, in some cases, changes correct classifications to incorrect ones.

Observations from the CA plots, such as the lack of correct classifications of the MGLA (24h) by NMR could be used to derive prior probabilities that would account for these biases. The small sample size of the data used here is not adequate to derive completely disjoint training and testing data, but with an additional independent experiment such priors could be validated. Nonetheless, the observations in the CA plots give significant insight into the potential biomarkers of infection; future work would include methods to identify which biomolecules are working cooperatively across datasets to improve the classification accuracy of the model. They also yield insight into which datasets may not be adequately independent for the model in Eq. (5).

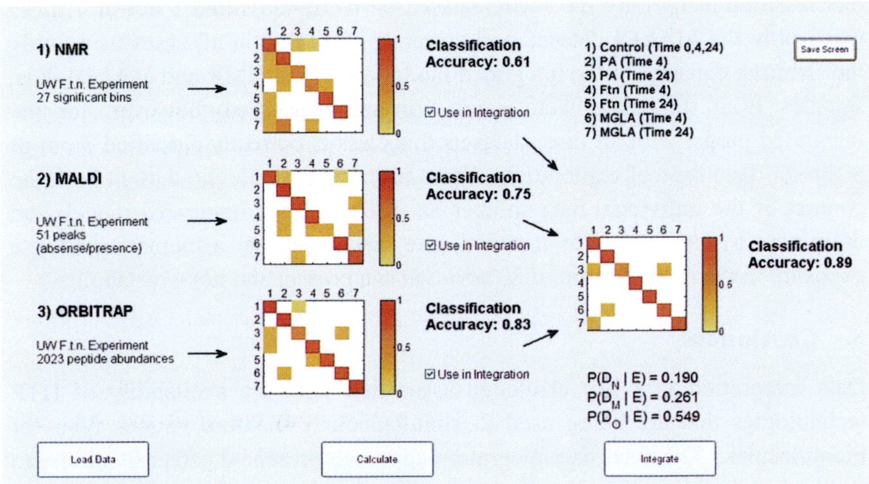

Figure 2. VIBE screenshot of the integrated classification and accuracy calculation over the Orbitrap, MALDI, and NMR datasets.

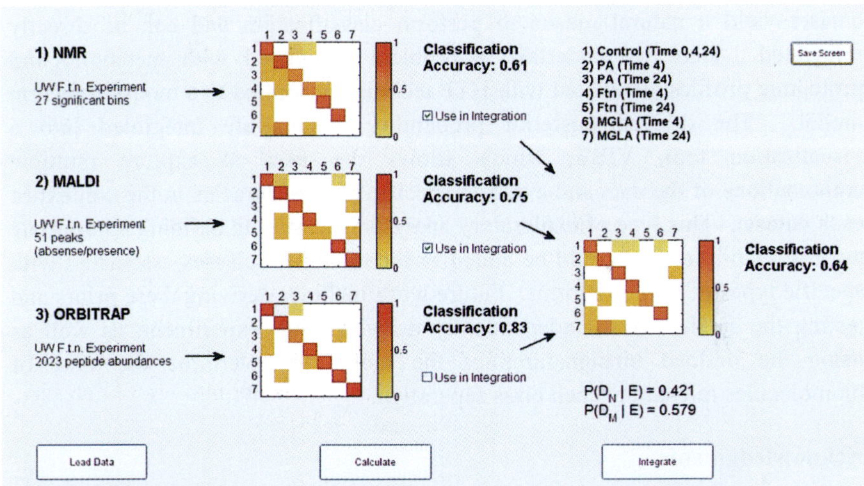

Figure 3. VIBE visualization of the integrated classification and accuracy calculation over two datasets: NMR and MALDI.

In evaluating the CA plots in VIBE, we also observed that many of the misclassifications in each dataset remain misclassified in the integrated model (see Figure 3). This suggests that the probabilities associated with these misclassifications are relatively strong and only with all three datasets can many of the probabilities be pushed toward a maximum value for the correct class. The MGLA at 24 hours is a good example. This class is completely misclassified using only the NMR dataset, correctly classified 3 out of 4 times using only the MALDI dataset, and correctly classified in all cases using only the Orbitrap dataset. In the integrated model using just NMR and MALDI data, samples from this class were never correctly classified, however, for the integrated model with all three datasets this class is correctly classified 3 out of 4 times. This type of exploration allows users to evaluate the data in both the context of the individual data sources as well as in the integrated model, and draw new hypotheses. For instance, one may infer that a biomarker of late exposure to the mutant strain of *F. novicida* is a protein and not a metabolite.

5. Conclusions

Data integration is a key challenge associated with the availability of HTP technologies that are being used to simultaneously measure various types of biomolecules. The Bayesian integration approach presented offers flexibility to tackle this challenge, by allowing any appropriate statistical learning algorithm to be used to derive the likelihood values associated with samples originating from common classes. The posterior probability models associated with each dataset yield a natural means to perform classification and can be directly integrated. Here we undertake a problem associated with metabolic and proteomic profiles associated with HTP techniques applied to a mouse infection model. The derived posterior probability models are integrated into a visualization tool, VIBE, which allows the user to explore multiple combinations of the data and evaluate specific class accuracies in the context of each dataset. This type of exploratory analysis is helpful in defining appropriate prior probabilities that could be added to the model for classes associated with specific types of HTP platforms. Future work includes deriving these priors and testing the model on an independently derived sister experiment, as well as using the defined biosignatures of the model to determine the suite of biomolecules relevant to each class separation.

Acknowledgments

This work was supported by the U.S. Department of Energy (DOE) through the Environmental Biomarkers Initiative of the Laboratory Directed Research and Development program at Pacific Northwest National Laboratory (PNNL) and the National Institutes of Health through grant U54 AI057141 awarded to the University of Washington. PNNL is a multiprogram national laboratory operated by Battelle for the U.S. DOE under Contract DE-AC06-76RL01830.

References

1. N. Friedman, *Bioinformatics.* **19 Suppl 2**, II57 (2003).
2. O.G. Troyanskaya et al., *PNAS* **100**, 14 (2003).
3. C. Agaton, M. Uhlen and S. Hober, *Electrophoresis* **25**, 9 (2004).
4. M. Deng, T. Chen and F. Sun, *J Comput Biol* **11**, 2-3 (2004).
5. G.R. Lanckriet, et al., *Bioinformatics* **20**, 16 (2004).
6. G.R. Lanckriet, et al., *Pac Symp Biocomput* (2004).
7. D.M. Reif, B.C White and J.H. Moore, *Expert Rev Proteomics* **1**, 1 (2004).
8. A. Bernard and A.J. Hartemink. *Pac Symp Biocomput* (2005).
9. D. Hwang et al., *PNAS* **102**, 48 (2005).
10. L.J. Lu et al., *Genome Res* **15**, 7 (2005).
11. D.P. Lewis, T. Jebara and W.S. Noble, *Bioinformatics* **22**, 22 (2006).
12. H. Martens and T. Naes, Wiley (1989).
13. J.M. Bernardo and A.F. Smith, Wiley (2000).
14. A.M. Hajjar et al., *Infect Immun* **74**, (2006).
15. K.H. Jarman et al., *Chemom Intell Lab Sys* **69** (2003).
16. K.H. Jarman et al., *Rapid Commun Mass Spectrom* **13** (1999).
17. K.H. Jarman et al., *Anal Chem* **72**, 6 (2000)
18. E.F. Strittmatter et al., *J Am Soc Mass Spectrom* **14**, 9 (2003).
19 J.G. Pounds et al., *J Chrom B Analyt Tech Biomed Life Sci* **684**, 1-2 (2008)
20. M.E. Monroe et al., *Bioinformatics* **23**, 15 (2007).
21. K.K. Anderson, M.E. Monroe and D.S. Daly, *Proteome Sci* **4**, 1 (2006).
22. R.L. Ott and M. Longnecker, Duxbury (2001).
23. P.A. Devijver and J. Kittler, Prentice-Hall (1982).
24. F. Ildiko and J. Friedman, *Technometrics* **35**, 2 (1993).

MOLECULAR BIOINFORMATICS FOR DISEASE

YVES A. LUSSIER*, YOUNGHEE LEE

Center for Biomedical Informatics and Section of Genetic Medicine, Dept. of Medicine, Computational Institute,[4]Institute for Genomics and Systems Biology, and UC Cancer Research Center; The University of Chicago, IL, 60637, U.S.A.

PREDRAG RADIVOJAC*

School of Informatics, Indiana University Bloomington, IN 47408, U.S.A.

YANAY OFRAN*

Faculty of Life Sciences, Bar Ilan University, 52900 Israel

ATUL BUTTE*

Depts. of Pediatrics and of Medicine, Stanford University, Sanford, CA 94305, U.S.A.

MARICEL KANN*

Dept. of Biological Sciences, University of Maryland, Baltimore County, MA, USA

This session focuses on the emerging field of molecular bioinformatics techniques applied to understand disease, and establish better prognosis or treatments. With the advent of whole genome association studies, there is a clear need to amass and think about and relate phenotypes in a computational way. For example, Brunner and van Driel showed how syndromes sharing clinical features could be related to the same protein complex (Nature Reviews Genetics 2004 5:545). While the application of methods from the domain of systems biology to the phenotypes of model organisms has enabled the prediction of molecular pathways and gene functions, it is not yet obvious how such methods will scale to phenotypes and diseases of higher organisms. Investigating methods to improve either genetics or phenomics methods independently carries enormous scientific and societal merit in this quest for disease genes. Yet taken together, powerful joint methods are likely to emerge such as phenome-anchored whole genome association studies, where prior probabilities stemming from independent studies can be used to increase, *ab initio*, the sensitivity of association studies or interpret the associations a posteriori through meta-analyses.

* Session co-chairs

The accepted papers for the session on Molecular Bioinformatics for Diseases comprises original research projects that span human genome variation, host-pathogen interactions, phenotypic relationship, and proteomics.

The first two papers address methodological improvements in human genome variation studies and provide insight in the molecular mechanisms human diseases. **Daemen et al.** proposed a new methodology in which copy number variations (CNV) resulting from array Comparative Genomic Hybridization (CGH) are transformed into features for classification purpose using a special variant of hidden Markov Models (HMM) able to capture recurrent copy number alterations by coupling the HMMs of individual samples. They demonstrate that the combination of HMMs for the detection of copy number alterations with Weighted Least Squares Support Vector Machines (LS-SVM) classifiers offers a novel methodological approach for classification based on copy number alterations and also suggest that copy number variations are not sufficient in survival-related studies and that other high-throughput data sources should be investigated. Conversely, **Zhang et al.** were interested in the effects of genetic variation across multiple loci. They developed an algorithm, FastChi, that utilizes an upper bound of the two-locus chi-square test in that an upper bound enables the algorithm to perform the two-locus chi-square test on a smaller prioritized number of candidate SNP pairs. Moreover, they contribute to solve the computational challenge of controlling for multiplicity of tests in genome-wide association study. The computational aspect of the chi-square test was examined on a combined dataset consisting of SNPs from the Broad/MIT mouse database and of the GNF mouse dataset. They demonstrate their method is orders of magnitudes faster than the classical - brute force - alternative.

The second group consists of three papers that explore associations between host-pathogen with particular focus on intracellular host proteins that relied on extensive data integration. The pathogen species ranged from viruses (e.g. HIV, Influenza) to prokaryotes (e.g. *Vibrio cholera*). As protein-protein interactions (PPI) between human pathogens and their host are vital at every step of the pathogen life cycle, pathogens are and particularly viruses are dependent on human host proteins. However, the novelty of some studies is to introduce a systems biology approach to the problem by focusing on the human PPIs or pathways. Two research groups, **Tastan et al.** and **Kanabar et al.,** both focused on uncovering host-pathogen PPIs and associated protein networks. **Tastan et al.** propose a supervised learning framework by utilizing multiple information data sources such as co-occurrence of functional motif, interaction domains, ontology, posttranslational modifications and class, gene expression profiles, topological properties, and the similarity of HIB-1 proteins to human proteins' known binding partners. They trained and tested a Random Forest (RF) classifier with this extensive feature set. This approach achieved 23% of average Mean Average Precision (MAP) score. This work is among the first to predict the global set of interactions between HIV-1 and human host cellular proteins. In contrast, **Kanabar et al** apply a probabilistic model, the Joint Intervention Network (JIN), to the relative gene expression changes of a set of indicator genes in regulatory network of *V. cholerae* to discover mechanisms critical to its environmental persistence. This method implements an epistasis analysis for quantitative, multivariate phenotypes. They analyzed the expression response of several indicator genes under various single and multiple knock-outs of three known biofilm regulators (e.g. hapR, upsR, and upsT). Another other described by **Kasson and Pande**, proposes the use of mutual information to examine the functional regulation of influenza hemagglutinin. The approach consist of employing mutual information over structure-based protein sequences from crystallographic data which allows to extract clusters of concerted mutation sites and comparison of influenza isolates from two distinct Influenza strains, 5N1 and human H3N2, which yield substantial differences in spatial localization of the clustered residues of proteins sequences abstracted from crystallographic data. They suggest that the residues identified on common between H5N1 and H3N2 may be important to the basic functional roles of hemagglutinin and also be useful targets for future mutagenesis experiments. Therefore, the combined sequence-information and structural analysis help generate novel functional hypotheses that would not occur via either method alone.

Finally, the last paper describes the regulatory network of transcriptional factors (TF) related to breast cancer metastasis classifier. **Lim and et al.** aim to develop a computational systems biology approach that infers robust prognostic markers by identifying upstream "Master Regulator Genes", causally related to the presentation of the phenotype of interest. Recently the research group of **Dr. Califano** introduced ARACNe (Nature Genetics 2005 37:382), an algorithm for the dissection of transcriptional networks that can infer the targets of TF from microarray expression profile. ARACNe is interrogated for the unbiased inference of TFs that induce or suppress specific gene signatures associated with poor prognosis in breast cancer.

Acknowledgements

The session co-chairs would like to thank numerous reviewers for their help in select the best papers for the session Molecular Bioinformatics for Disease and Dr. Andrea Califano for his participation as a Session Keynote.

MOLECULAR INTERACTION NETWORK BASED DISCOVERY OF MASTER REGULATOR GENES

ANDREA CALIFANO
KEYNOTE SPEAKER FOR
"MOLECULAR BIOINFORMATICS FOR DISEASE"

Center for Computational Biology and Bioinformatics, Dept. of Biomedical Informatics, Columbia University, 1130 Saint Nicholas Avenue, New York, NY 10032, USA,

The identification of genes acting synergistically as master regulators of physiologic and pathologic cellular phenotypes is a key open problem in systems biology. Here we use a molecular interaction based approach to identify the repertoire of transcription factors (TFs) of a master regulatory module responsible for synergistic activation of a tumor-specific signature. Specifically, we used the ARACNe algorithm and other computational tools to infer regulatory interactions responsible for initiating and maintaining the mesenchymal phenotype of Glioblastoma Multiforme (GBM), previously associated with the poorest disease prognosis. Expression of mesenchymal genes is a hallmark of aggressiveness but the upstream regulators of the signature are unknown. Starting from the unbiased analysis of all TFs, we identify a highly interconnected module of six TFs jointly regulating >75% of the genes in the signature. Two TFs (Stat3 and C/EBPb), in partic ular, display features of initiators and master regulators of module activity. Biochemical validation confirms that the TFs in the module bind to the inferred promoters in vivo and ectopic expression of the master TFs activates expression of the mesenchymal signature. These effects are sufficient to trigger mesenchymal transformation of neural stem cells, which become highly tumorigenic in vivo, and promote migration and invasion. Conversely, silencing of Stat3 and C/EBPb in human glioma cells leads to collapse of the mesenchymal signature and reduction of tumor aggressiveness. Our results reveal that activation of a small transcriptional module is necessary and sufficient to induce a mesenchymal phenotype in malignant brain tumors.

SUPERVISED CLASSIFICATION OF ARRAY CGH DATA WITH HMM-BASED FEATURE SELECTION

ANNELEEN DAEMEN[1]*, OLIVIER GEVAERT[1], KARIN LEUNEN[2], ERIC LEGIUS[3], IGNACE VERGOTE[2] AND BART DE MOOR[1]

[1] *Department of Electrical Engineering (ESAT), Katholieke Universiteit Leuven SCD-SISTA (BIOI), Kasteelpark Arenberg 10 - bus 2446, B-3001 Leuven (Heverlee), Belgium*

[2] *Department of Obstetrics and Gynaecology, Division of Gynaecologic Oncology Multidisciplinary Breast Centre, University Hospital Leuven, Herestraat 49 - bus 7003 , B-3000 Leuven, Belgium*

[3] *Department of Human Genetics, O&N I, University Hospital Leuven, Herestraat 49 - bus 603, B-3000 Leuven, Belgium*

Motivation: For different tumour types, extended knowledge about the molecular mechanisms involved in tumorigenesis is lacking. Looking for copy number variations (CNV) by Comparative Genomic Hybridization (CGH) can help however to determine key elements in this tumorigenesis. As genome-wide array CGH gives the opportunity to evaluate CNV at high resolution, this leads to huge amount of data, necessitating adequate mathematical methods to carefully select and interpret these data.

Results: Two groups of patients differing in cancer subtype were defined in two publicly available array CGH data sets as well as in our own data set on ovarian cancer. Chromosomal regions characterizing each group of patients were gathered using recurrent hidden Markov Models (HMM). The differential regions were reduced to a subset of features for classification by integrating different univariate feature selection methods. Weighted Least Squares Support Vector Machines (LS-SVM), a supervised classification method which takes unbalancedness of data sets into account, resulted in leave-one-out or 10-fold cross-validation accuracies ranging from 88 to 95.5%.

Conclusion: The combination of recurrent HMMs for the detection of copy number alterations with LS-SVM classifiers offers a novel methodological approach for classification based on copy number alterations. Additionally, this approach limits the chromosomal regions that are necessary to classify patients according to cancer subtype.

*To whom correspondence should be addressed: anneleen.daemen@esat.kuleuven.be

1. Introduction

In cancers, many gains and losses of chromosomes and chromosomal segments have been described. These aberrations defined as regions of increased or decreased DNA copy number can be detected at high resolution using an array comparative genomic hybridization (CGH) technology. This technique measures variations in DNA copy number within the entire genome of a disease sample compared to a normal sample[1]. This makes array CGH ideally suitable for a genome-wide identification and localization of genetic alterations involved in human diseases. An overview of algorithms for array CGH data analysis is given by Lai *et al.*[2] Segmentation approaches divide each single-sample signal into regions of constant copy number, called segments. Subsequent classification labels the segments as gain or loss. A popular method combining both tasks is the hidden Markov Model (HMM) with states defined as loss, neutral, one-gain, and multiple-gain. Recently, this traditional procedure has been extended to a recurrent HMM in which a class of samples instead of an individual sample is modeled by sharing information on copy number variations across multiple samples[3]. Here, we present a method to identify copy number alterations with the recurrent HMM which goes beyond the exploratory phase by using these alterations as features in a supervised classification setting and by validating these features biologically.

For classification, we started from the class of kernel methods which is powerful for pattern analysis[4]. Their rapid uptake in bioinformatics is due to their reliability, accuracy, and computational efficiency[5]. More specifically, we made use of the weighted Least Squares Support Vector Machine (LS-SVM), taking the unbalancedness of data sets into account[6–7].

We applied our method on two publicly available data sets of lung cancer and oral squamous cell carcinoma, and subsequently on our data set of ovarian cancer. The knowledge of different copy number variations between specific groups of patients may help to better understand tumorigenesis of these cancers. When applied to larger study groups, this method could result in a better comprehension of the different clinical behaviour of both groups, probably necessitating different treatment strategies.

The outline of this article is as follows. In section 2, we describe the data sets, the applied methods for segmentation, feature selection, and classification, the workflow of our proposed methodology, as well as the functional annotation analysis for biological validation. We describe our results on the different data sets in Section 3 and conclude in Section 4.

2. Materials and Methods

2.1. *Data Sets*

In the study of Snijders *et al.*[8] a genome-wide analysis of copy number aberrations was carried out in 89 patients with oral squamous cell carcinoma, using the HumArray2.0 genome scanning array with 2464 clones from the University of California San Francisco. Missing copy number values were imputed unsupervisedly using the k-nearest neighbours method with k set to 15[9]. The final data set contained 2056 unique clones. Patients were subdivided according to *TP53* mutation with 59 of them being wildtype and 16 having a mutation for the *TP53* gene. The mutation status was unknown for the remaining 14 patients.

Garnis and colleagues[10] measured genome copy number profiles for 28 commonly used non-small cell lung carcinoma (NSCLC) cell lines across the whole genome using the submegabase resolution tiling array, consisting of 32433 BAC clones. 29781 unique clones remained after preprocessing. Twenty-two of the cell lines belonged to one of the 2 main subgroups of NSCLC: adenocarcinoma (13) and squamous cell carcinoma (9).

Our data were collected from patients treated for ovarian cancer at the University Hospital of Leuven, Belgium. Eight patients had a sporadic tumour without a positive family history for breast and/or ovarian cancer, while five patients were carrier of a mutation in the tumour suppressor gene *BRCA1*, involved in DNA damage repair and transcriptional regulation[11]. Array comparative genomic hybridization was performed using a 1Mb array CGH platform, version CGH-SANGER 3K 7, containing 3593 clones and developed by the Flanders Institute for Biotechnology (VIB), Department of Microarray Facility, Leuven, Belgium.

2.2. *Array Comparative Genomic Hybridization*

Array comparative genomic hybridization (array CGH) is a high-throughput technique for measuring DNA copy number variations (CNV) within the entire genome of a disease sample relative to a normal sample[1]. In an array CGH experiment, total genomic DNA from tumour and normal reference cell populations are isolated and subsequently labeled with different fluorescent dyes before being hybridized to several thousands of probes on a glass slide. This allows calculating the log ratios of the fluorescence intensities of the tumour to that of the normal reference DNA. Because the reference cell population is normal, an increase or decrease in the log intensity ratio indicates a DNA copy number variation in the genome of the tumour cells such that negative log ratios correspond to deletions (losses), positive log ratios to gains or amplifications and zero log ratios to neutral regions in which no change occurred.

2.3. *Recurrent HMM*

As was stated in the introduction, we will use a recurrent hidden Markov Model (HMM) proposed by Shah *et al.*[3] for the identification of extended chromosomal regions of altered copy numbers, labeled as gain or loss. The goal of this model is to construct features that distinguish two groups of patients and subsequently to use them in a classifier (see Section 2.4). Due to sensitivity of traditional HMMs to outliers being measurement noise, mislabeling, and copy number polymorphisms within the normal human population, a robust HMM was proposed by Shah *et al.*[12] and extended to a multiple sample version in which array CGH experiments from a cohort of individuals are used to borrow statistical strength across samples instead of modeling each sample individually[3]. This reduces the influence of various sources of noise on the detection of recurrent copy number alterations and makes even copy number alterations in a small number of adjacent clones reliable when shared across many samples.

In this study, a recurrent HMM is constructed on a chromosomal basis for each group of patients separately, resulting in regions with the probabilities of genetic alterations across these patients. A clone was labeled as gain or loss when its probability of occurring was more than 70%. An important parameter of the recurrent HMM that needs to be set is the recurrent variable ϵ. This variable represents the probability with which samples will get reflected in the recurrent CNV profile - accounting for sample-specific random effects - and is inversely proportional to the sparseness of the profile. As a trade-off between sparseness and false positive rate, ϵ was set to 0.8, although ϵ=0.7 is already sufficient for small data sets with less heterogeneity within each group of samples (i.e. our data set with 8 and 5 samples, respectively). A differential region was defined as a chromosomal region which is gained/lost in one group while not being gained/lost in the other group.

2.4. *Kernel Methods and Weighted Least Squares Support Vector Machines*

The differential regions that result from the recurrent HMM are used as features in a classifier for which we chose kernel methods. These methods are a group of algorithms that do not depend on the nature of the data by representing data entities through a set of pairwise comparisons called the kernel matrix[13]. This matrix can be geometrically expressed as a transformation of each data point x to a high dimensional feature space with the mapping function $\Phi(x)$. An explicit representation of $\Phi(x)$ is not needed when defining a kernel function $k(x_k, x_l)$ as the inner product $\langle\Phi(x_k), \Phi(x_l)\rangle$ of two data points x_k and x_l.

A kernel algorithm for supervised classification which can tackle high dimensional data and contains regularization, is the Support Vector Machine (SVM) developed by Vapnik[14] and others. Given a training set $\{x_k, y_k\}_{k=1}^{N}$ of N samples with feature vectors $x_k \in \mathbb{R}^n$ and output labels $y_k \in \{-1, +1\}$, the SVM forms a linear discriminant boundary $y(x) = \text{sign}[w^T\Phi(x) + b]$ in the feature space with maximum distance between samples of the two considered classes. This corresponds to a non-linear discriminant function in the original input space. A modified version of SVM, the Least Squares Support Vector Machine (LS-SVM), was developed by Suykens *et al.*[6]. On high dimensional data sets, this modified version is much faster for classification because a linear system instead of a quadratic programming problem needs to be solved.

In many two-class problems, data sets are skewed in favour of one class such that the contribution of false negative and false positive errors to the performance assessment criterion are not balanced. We therefore used a weighted LS-SVM in which a different weight ζ_k is given to positive and negative samples in order to account for the unbalancedness in the data set[7]. The constrained optimization problem for the weighted LS-SVM has the following form:

$$\min_{w,b,e} J_P(w, e) = \frac{1}{2}w^T w + \gamma\frac{1}{2}\sum_{k=1}^{N} \zeta_k e_k^2$$

with

$$\zeta_k = \begin{cases} \frac{N}{2N_P} & \text{if } y_k = +1 \\ \frac{N}{2N_N} & \text{if } y_k = -1 \end{cases}$$

and N_P and N_N representing the number of positive and negative samples, respectively. Variables e_k represent the error variables, tolerating misclassifications in case of overlapping distributions, and γ the regularization parameter which allows tackling the problem of overfitting.

2.5. *Feature Selection*

Because it has been shown by Lai and colleagues[15] that univariate gene selection methods lead to good and stable performances across many cancer types and yield in many cases consistently better results than multivariate approaches, we used the method DEDS (Differential Expression via Distance Synthesis)[16]. This technique is based on the integration of different test statistics via a distance synthesis scheme because features highly ranked simultaneously by multiple measures are more likely to be differential expressed than features highly ranked by a single measure. The statistical tests combined are ordinary fold changes, ordinary t-statistics, SAM-statistics and moderated t-statistics (i.e. B-statistics[17]). DEDS is available as a BioConductor package in R.

2.6. *Proposed Methodology*

For the data set of Snijders, a 10-fold cross-validation (CV) strategy was applied, while a leave-one-out (LOO) cross-validation strategy was chosen for the other, rather small data sets. The 4 different steps that have to be accomplished in each CV iteration are shown in Figure 1. After leaving out m samples (i.e. N/10 for 10-fold CV; 1 for LOO), a recurrent HMM (see Sect. 2.3) is constructed for both groups in step 1 to determine the chromosomal regions with genetic alterations that characterize each group. Combining these regions results in the chromosomal regions that are differential between the remaining N-m samples from both groups. Because multiple clones can be located within each differential region, the clones need to be combined. This is done per sample in the second step by taking the median of the log ratios of the clones in each region. Afterwards, a standardization is performed per sample (i.e. meanshifting to 0 and autoscaling to 1) because the raw log ratios cannot be compared in absolute values between the samples. In step 3, DEDS determines which preprocessed log ratios, called features, best discriminate the N-m samples (see Sect. 2.5). The number of included features is iteratively increased according to the obtained feature ranking without including more features than the number of samples on which the optimal number of features is determined[18]. This subset of features forms the input for classification in the last step (see Sect. 2.4). For all possible combinations of γ and number of features, an LS-SVM is built on the training set and validated on the m left out samples. This is repeated until each sample has been left out once. The parameter combination is chosen corresponding to the highest AUC (area under the

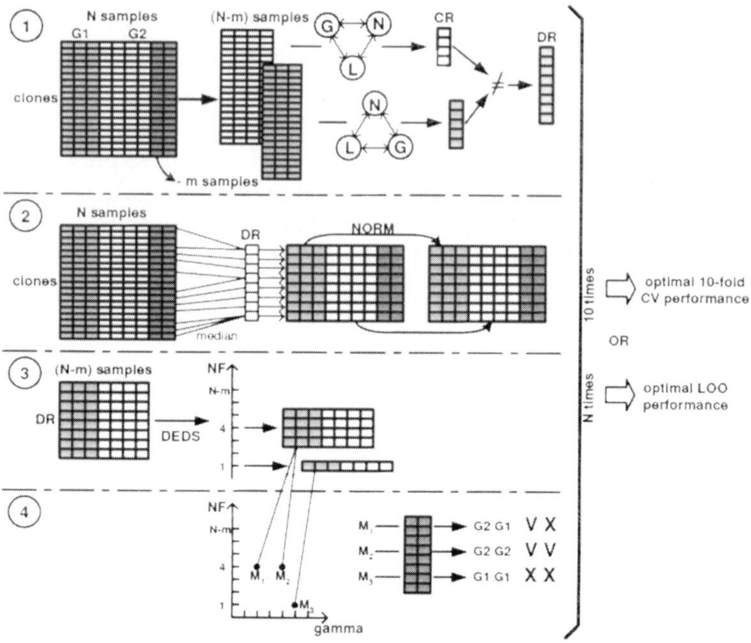

Figure 1. The methodology, applied in a 10-fold or LOO CV setting, consists of 4 steps: step 1 - recurrent HMMs applied on both groups of samples, resulting in differential regions; step 2 - conversion of clones to differential regions, and normalization (performed for each sample separately); step 3 - feature selection using DEDS; step 4 - LS-SVM training on 2-dimensional grid for regularization parameter γ and number of features, and validation on m left out samples (G1 = group 1; G2 = group 2; CR = Chromosomal Region; DR = Differential Region; NORM = Normalization; DEDS = Differential Expression via Distance Synthesis; NF = Number of Features)

ROC curve). When multiple such combinations, the combination with the lowest balanced error rate and an as high as possible sum of sensitivity and specificity is considered as optimal. For the LS-SVM, a linear kernel function $k(x_k, x_l) = x_k^T x_l$ was chosen. A kernel function that accounts for the correlation between neighbouring clones[19] did not lead to a better performance because the HMM already takes the properties of CGH data into account.

2.7. *Functional Annotation Analysis*

To validate the selected chromosomal regions, gene set enrichment was performed as an indication for agreement with "known" biology. Curated gene sets as defined in the Molecular Signatures Database (MSigDB) (i.e. sets of co-regulated genes from online pathway databases, publications in PubMed and knowledge of domain experts) were used[20]. Using the HUGO gene nomenclature[21], genes within the differential chromosomal regions were divided into 9 gene signatures, depending on the group (G1 versus G2 versus both) and CNV type (gain versus loss versus both). For each signature, the overlap was calculated between all gene sets and the signature as well as 5000 equally-sized signatures containing genes randomly selected from the genome. The corrected method of North *et al.*[22] was used to calculate the empirical p-value for each gene set as $(r+1)/(n+1)$ with n the number of random signatures (i.e. 5000) and r the number of them with an equal or higher overlap with the gene set than obtained with the actual signature. Only gene sets with r smaller than 10 (p-value < 0.002) were further investigated.

3. Results

We applied our method on two publicly available array CGH data sets, as well as on our own data set of ovarian cancer for the prediction of cancer subtype.

The data set of Snijders was used to distinguish patients being wildtype for the *TP53* gene from those having a mutation for *TP53*. 55 of the 59 wildtype patients and 11 of the 16 *TP53*-mutated patients could be classified correctly based on 10 differential regions, none of them containing *TP53* located on chromosome 17 (see Table 1). The wildtype group was characterized by 3 gained regions on chromosome arms 2p, 11q, and 18q, and 1 lost region on 12p. The mutated group was distinguishable with 1 gained region on 10p and 5 lost regions on chromosome arms 8p, 11q, and 18q. Five of these regions were selected in 8 to 10 of the 10 CV iterations while the remaining 5 regions appeared in 1 to 5 CV iterations. Loss of chromosome arms 8p, 11q, and 18q in *TP53*-mutated samples was mentioned in the original manuscript as well.

When applying our methodology on the data set of Garnis to distinguish adenocarcinomas from squamous cell carcinomas, 21 out of the 22 cell lines could be classified correctly based on 8 regions on chromosome arms 1q, 2p, 3q, 5p, 7q, 11p, 13q, and 20p, all lost in adenocarcinoma (see Table 1). Five regions were selected in all LOO iterations, the regions on 7q and 11p in 19 and 15 LOO iterations, respectively while the lowest ranked region according to DEDS appeared in 9 of the 22 LOO iterations. The

Table 1. Performances of the three considered data sets

Data set	Nb regions	Accuracy$^{\mu}$	Sensitivity	Specificity	AUC$^{\mu}$
Snijders[8]	10	88 (66/75)	93.2 (55/59)	68.8 (11/16)	0.840
Garnis[10]	8	95.5 (21/22)	92.3 (12/13)	100 (9/9)	0.983
own data	11	92.3 (12/13)	100 (5/5)	87.5 (7/8)	0.875

$^{\mu}$Represents 10-fold CV performance (Snijders) or LOO performance (own data, Garnis)

chromosome arms 2p and 13q were also found by Garnis and colleagues to be frequently deleted in adenocarcinoma while being amplified in the cell lines of the squamous cell carcinomas.

Eight sporadic and five *BRCA1* mutated ovarian cancer patients were included and profiled using array CGH technology. When applying the proposed methodology on this data set, CNVs in 11 chromosomal regions were sufficient to correctly classify 12 out of 13 samples (see Table 1). Five regions on chromosome arms 3p, 4p, 6p, 12q, and Xp were gained and 3 regions on 10p, 13q, and 19q were lost in *BRCA1* mutated samples. The sporadic ovarian cancer patients were characterized by loss of 3 regions on 4p and 16. The top 5 of features with the lowest p-value according to DEDS appeared in 8 to 11 of the 13 LOO iterations while the lower ranked regions were selected in 4 to 7 LOO iterations.

Because we hypothesize that genes in the chromosomal regions participate in processes that distinguish subtypes of cancer, a gene set enrichment-based approach was followed (see Sect. 2.7). This analysis highlighted enriched gene sets containing genes that may act as an oncogene or tumour suppressor gene or for which is shown that they repress tumour suppressor genes. The gene *LAMA3* (alpha-3 subunit of laminin 5) was proposed by Snijders and colleagues as one of the candidate driver genes in the development of oral squamous cell carcinoma and appears to be important for classification of *TP53* status as well. It may play an active role in cell adhesion, migration, and proliferation of the carcinoma cells[23]. Furthermore, the protein cortactin, contributing to tumour progression and metastasis formation[24], was gained in the wildtype samples, while the *TP53*-mutated samples were characterized by loss of the Smad transcription factors *Smad2*, *Smad4*, and *Smad7*. Members of the *TGFβ/Smad* signaling pathway are potent tumour suppressor genes[25] and the inactivation of Smad factors is correlated with loss of responsiveness to *TGFβ*-mediated signal transduction[26].

Important genes lost in the adenocarcinoma cell lines were *CUTL1* which activates a transcriptional program regulating genes involved in cell motility, invasion, and extracellular matrix composition[27], and *FGF-10*, a key regulator of lung branching morphogenesis[28]. In our data set the genes *BAF57* (important in transcriptional repression of tumour suppressor genes among which *BRCA1*[29]) and *HOXA5* (suggested to act as a tumour suppressor gene in breast cells[30]) seemed to be correlated with hereditary ovarian cancer, whereas loss of the *v-myb* oncogene seemed more characteristic for the sporadic group.

4. Conclusion

In this manuscript, a new methodology is proposed in which copy number variations resulting from array CGH are transformed into features for classification purpose. This general method which is independent of cancer site allows identifying a small set of class-specific aberrations that can distinguish patients, and biologically validating them. It can also result in clinically relevant models based on a limited set of features. As increasing amounts of array CGH data become available, there is a need for algorithms to identify recurrent gains and losses based on statistically sound methods and to use them for classification. A large number of approaches for the analysis of array CGH data have already been proposed recently, ranging from mixture models and HMMs to wavelets and genetic algorithms[2]. However, most cancer studies that gather array CGH data only apply methods for exploratory analysis. Often a fixed threshold is used for defining gains and losses, making these studies less robust against systematic changes in the baseline copy number measurements between samples[31]. A HMM on the contrary is a probabilistic method that can handle the uncertainty in the data in a formal way compared to deterministic algorithms. This makes the HMM more robust against outliers such as measurement noise and wrong recordings of locations of clones. Moreover, we used a special variant of HMM able to capture recurrent copy number alterations by coupling the HMMs of individual samples. This makes weak copy number alterations but shared across many samples reliable features. In our setup we used this property by first modeling the copy number variations in each group of patients separately. Subsequently, the alterations that were different between these two groups were used as features in an LS-SVM for classification. In our opinion this is one step further compared to many other studies that only perform an exploratory analysis.

A comparison of the regions found in each of the CV or LOO iterations showed a limited variability in the selected regions. This strengthens our confidence that the chromosomal regions found with our methodology are robust. Regions lacking genes with an annotated HUGO symbol seem uninteresting at first sight. However, recent research findings on 1% of the genome indicated that 93% of the bases are transcribed, increasing the importance of non-protein-coding RNA[32]. For each data set the remaining regions were validated biologically using a gene set enrichment-based approach. Keep in mind that, because the number of features is minimized, one can expect that biological validation using pathways may fail because not all genes belonging to a certain pathway may be needed in a classification setting.

It was not self-evident to find appropriate data sets because current studies utilize array CGH, not for classification purpose but for characterizing a specific set of patients. Clinical or histopathological variables that can be used to define two classifiable subgroups of patients are lacking in most of these studies. We applied our method on two extra data sets for the prediction of disease specific survival and distant recurrence, even though the main purpose of the original manuscripts was not survival[33–34] (results not shown). Although previous studies on CGH have shown evidence for an association of amplification with poor prognosis[35–36], we were not able to classify the patients in these two data sets properly according to survival using array CGH data. Further research is required on the appropriateness of array CGH data in survival-related studies.

The proposed method performed well in distinguishing adenocarcinoma from squamous cell carcinoma. This can partly be caused by the use of cell lines instead of clinical specimens, although a number of studies have characterized clinical cases of NSCLC with comparable CNV profiles as found by Garnis *et al.* in cell lines[10]. This good performance may also be due to the use of a high resolution tiling array. The use of such arrays yields potential for utilizing CNV profiles in classification for which a method has been presented here.

Acknowledgements

AD is research assistant of the Fund for Scientific Research - Flanders (FWO-Vlaanderen). BDM is a full professor at the Katholieke Universiteit Leuven, Belgium. This work is partially supported by: **1.** Research Council KUL: GOA AMBioRICS, CoE SymBioSys, PROMETA. **2.** Flemish Government: **a.** FWO projects G.0241.04, G.0499.04, G.0318.05, G.0302.07; **b.** IWT: GBOU-McKnow-E, GBOU-ANA, TAD-BioScope-IT,

Silicos, SBO-BioFrame, SBO-MoKa, TBM-Endometriosis, TBM IOTA3. **3.** Belgian Federal Science Policy Office: IUAP P6/25. **4.** EU-RTD: ERNSI, FP6-NoE Biopattern, FP6-IP e-Tumours, FP6-MC-EST Bioptrain, FP6-STREP Strokemap.

References

1. D Pinkel, D G Albertson, *Nature Genetics*, **37**(Suppl.), 11 (2005).
2. W R Lai, M D Johnson *et al.*, *Bioinformatics*, **21**(19), 3763 (2005).
3. S Shah, W L Lam *et al.*, *Bioinformatics*, **23**, i450 (2007).
4. J Shawe-Taylor, N Cristianini, *Cambridge University Press*, (2004).
5. H Bhaskar, D C Hoyle *et al.*, *Comput Biol Med*, **36**(10), 1104 (2006).
6. J A K Suykens, T Van Gestel *et al.*, *World Scientific*, (2002).
7. G C Cawley, *Proceedings of the International Joint Conference on Neural Networks*, 1661 (2006).
8. A M Snijders, B L Schmidt *et al.*, *Oncogene*, **24**, 4232 (2005).
9. O Troyanskaya, M Cantor *et al.*, *Bioinformatics*, **17**(6), 520 (2001).
10. C Garnis, W W Lockwood *et al.*, *Int J Cancer*, **118**, 1556 (2006).
11. L M Starita, J D Parvin, *Current Opinion in Cell Biology*, **15**(3), 345 (2003).
12. S Shah, X Xuan *et al.*, *Bioinformatics*, **22**(14), e431 (2006).
13. B Schölkopf, K Tsuda *et al*, *MIT Press*, (2004).
14. V Vapnik, *Wiley*, (1998).
15. C Lai, M J T Reinders *et al.*, *BMC Bioinformatics*, **7**, 235 (2006).
16. Y H Yang, Y Xiao *et al.*, *Bioinformatics*, **21**(7), 1084 (2005).
17. I Lönnstedt and T P Speed, *Statist Sinica*, **12**, 31 (2001).
18. W Li, Y Yang, *Kluwer Academic*, 137 (2002).
19. J Liu, S Ranka *et al.*, *Bioinformatics*, **24**, i86 (2008).
20. A Subramanian, P Tamayo *et al.*, *Proc Natl Acad Scie*, **102**(43), 15545 (2005).
21. H M Wain, E A Bruford *et al.*, *Genomics*, **79**(4), 464 (2002).
22. B V North, D Curtis *et al.*, *Am J Hum Genet*, **71**, 439 (2002).
23. J Lohi, *Int J Cancer*, **94**, 763 (2001).
24. L Buday, J Downward, *Biochim Biophys Acta*, **1775**(2), 263 (2007).
25. M M Reinholz, M W An *et al.*, *Breast Cancer Res Treat*, **86**(1), 75 (2004).
26. A Kretschmer, K Moepert *et al.*, *Oncogene*, **22**(43), 6748 (2003).
27. P Michl, A R Ramjaun *et al.*, *Cancer Cell*, **7**(6), 521 (2005).
28. W Y Park, B Miranda *et al.*, *Dev Biol*, **201**(2), 125 (1998).
29. L Wang, R A Baiocchi *et al.*, *Mol Cell Biol*, **25**(18), 7953 (2005).
30. H Chen, E Rubin *et al.*, *J Biol Chem*, **280**(19), 19373 (2005).
31. C Klijn, H Holstege *et al.*, *Nucleic Acids Research*, **36**(2), e13 (2008).
32. The ENCODE Project Consortium, *Nature*, **447**, 799 (2007).
33. J Fridlyand, A M Snijders *et al.*, *BMC Cancer*, **6**, 96 (2006).
34. K Chin, S De Vries *et al.*, *Cancer Cell*, **10**, 529 (2006).
35. K Al-Kuraya, P Schraml *et al.*, *Cancer Research*, **64**, 8534 (2004).
36. H Blegen, J S Will *et al.*, *Anal Cell Pathol*, **25**, 103 (2003).

INFERRING DISEASE-RELATED PATHWAYS USING A PROBABILISTIC EPISTASIS MODEL

P. N. KANABAR*[1], C. J. VASKE *[1], C. H. YEANG[2], F. H. YILDIZ[3], AND J. M. STUART[1†]

[1]*Department of Biomolecular Engineering, University of California Santa Cruz 1156 High Street, Santa Cruz, CA 95062, USA*

[2]*Institute for Advanced Study, Princeton, New Jersey 08540, USA*

[3]*Department of Microbiology and Environmental Toxicology, University of California Santa Cruz*

Motivation: We present a probabilistic model called a Joint Intervention Network (JIN) for inferring interactions among a chosen set of regulator genes. The input to the method are expression changes of downstream indicator genes observed under the knock-out of the regulators. JIN can use any number of perturbation combinations for model inference (e.g. single, double, and triple knock-outs). **Results/Conclusions**: We applied JIN to a *Vibrio cholerae* regulatory network to uncover mechanisms critical to its environmental persistence. *V. cholerae* is a facultative human pathogen that causes cholera in humans and responsible for seven pandemics. We analyzed the expression response of 17 *V. cholerae* biofilm indicator genes under various single and multiple knock-outs of three known biofilm regulators. Using the inferred network, we were able to identify new genes involved in biofilm formation more accurately than clustering expression profiles.

1. Introduction and Previous Work

DNA microarray expression data provides a rich source of phenotypes to infer gene regulatory networks. Recently, several methods have been developed to learn a cellular network from quantitative observations[9,2]. Notably, Bayesian Networks (BNs) have been used to infer networks from both observational[7,16] and interventional[11] gene expression microarray data. Interventional data can be used to identify causal networks[10]. In addition to gene expression data, BNs have been used to learn signaling networks from phosphorylation data collected from single-cells under perturbations[12]. Al-

*equal contributions.

†correspondence to jstuart@soe.ucsc.edu

gorithms for exact inference of BN structure with uncertain perturbations have been applied to similar data[5].

In contrast to the standard BN approach, the Nested Effects Model (NEM)[9] infers gene networks from expression changes of "secondary genes" observed under the knock-out of regulatory genes. NEMs search for a network of regulators that are consistent with a nested hierarchy among the expression changes. Like BNs, NEMs provide a generative model that can score a candidate network in proportion to the data likelihood.

Geneticists have used epistasis analysis of complex phenotypes to order genes into pathways for nearly a century[3]. The models considered here assume phenotypes (or expression states in our case) result from the loss of a signal propagating along a genetic network. Epistasis analysis uses the phenotypes of single- and multiple-gene deletions to place genes into a pathway order. In this work, we assume the genes of interest signal to one another according to a "switch regulatory," rather than a "substrate dependent," model as defined by Huang and Sternberg (2006)[8]. In switch regulatory epistasis, if the knock-out of gene A produces phenotype P, the knock-out of B produces phenotype Q, and the double knock-out of AB produces phenotype Q, B is said to be epistatic to, and is placed downstream of, A. In the context of using microarray expression data, the up- or down-regulation of a gene across a panel of mutant regulators does not directly convey the information about the network of regulators. Instead, a secondary gene's relative expression differences in single- versus double-mutants may be informative. Recently, Van Driessche *et al.*[14] used an *ad hoc* approach based on Euclidean distance to compare double and single knock-out expression profiles.

In this paper, we describe the Joint Intervention Network (JIN) model that implements a probabilistic epistasis analysis for quantitative, multivariate phenotypes. JIN differs from previous methods in the following aspects. First, it employs relative expression changes formed from the comparison of all possible pairs of perturbation datasets. Second, it extracts epistatic information from data on multiple perturbation combinations. Third, it is able to determine the order of genes as well as their functional dependencies (e.g., inhibition, activation, multiplicative, or additive) from the downstream effects, and can therefore find functional dependencies between regulators that may not be observed in direct transcript levels. Fourth, it identifies new targets of the regulators that have expression profiles consistent with the predicted regulatory network.

The survival of *Vibrio cholerae* both outside and inside a host, has

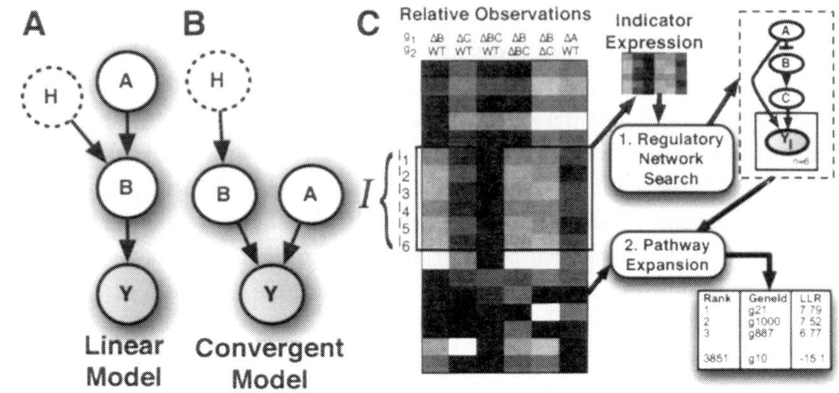

Figure 1. **A–B**. Signaling networks used in the toy example discussed in the text. A and B are known regulators, while H is a hidden regulatory influence, Y. **A**. Linear network. B directly controls Y's expression; A indirectly controls Y via B. **B**. Convergent network. A and B both directly control Y's expression. **C**. The input to the Joint Intervention Network model are relative gene expression changes of a set of indicator genes, I, measured under different comparisons of knock-out combinations of a set of regulator genes (A, B, and C), and the genotypes of the comparisons (g_1 and g_2). First, a high scoring network is found that connects regulators (nodes) to each other by predicted interactions (edges) and regulators to indicators (shown in plate notation using a solid box to indicate replicated indicator nodes and edges). Next, this regulatory model is used to find other genes in the genome that have consistent expression changes.

been linked to its ability to form biofilm. We applied JIN to the perturbation expression data of three *vps* biofilm regulators and automatically reconstructed the model consistent with the network proposed by the experimentalists that collected the data[4]. Furthermore, the model identified new target genes involved in biofilm formation.

2. Methods

We assume n signaling genes participate in the regulation of a pathway of interest and that the effect of their regulatory input is observable through indicator expression states, Y. Under different combinations of knock-outs to the signaling genes, indicator expression is used as quantitative phenotypes to order the signaling genes into a network.

2.1. *Theoretical Motivation*

We motivate our approach with an example to illustrate the use of using (1) knock-out combinations and (2) relative expression changes observed under knock-outs. Multiple knock-outs provide information that cannot be revealed by single knock-outs. For example, the networks in Figures 1A and 1B are indistinguishable using single knock-outs since both networks will exhibit down-regulation of indicator gene Y in both single knock-outs ΔA and ΔB. However, the results of the double knock-out ΔAB can help distinguish between the two models. If the model is linear, ΔB and ΔAB have the same effects on Y, but ΔA and ΔAB have distinct effects. ΔA has weaker effect on Y because (hidden) regulators of B, in addition to A, may be present. If the model is convergent and additive, both ΔA and ΔB have weaker effects on Y compared to ΔAB. If the model is convergent and multiplicative, both ΔA and ΔB have the same effect on Y compared to ΔAB.

Adopting relative changes alleviates the quantization problem inherent in discrete models that use direct expression levels. Let the relative expression change (REC) for an indicator gene i, $x_i(g_1, g_2)$, be the $\log_2$-ratio of i's expression level in genotype g_1 over its level in genotype g_2. Consider the linear model in Figure 1A. H represents a hidden factor such as an unknown regulator or an environmental variable not controlled in the experiment. Both single knock-outs ΔA and ΔB and double knock-out ΔAB down-regulate Y relative to wild-type. In addition, Y is lower in ΔB or ΔAB compared to ΔA, as H may provide a constituent activation of Y in ΔA. It is problematic to apply a binary quantization on the direct expression level of Y under ΔA, since Y is between the level of the double knock-out and wild type. More refined quantization also suffers from the arbitrary quantization levels.

In contrast, it is straightforward to determine whether Y under one condition is higher, lower or equal to another condition. In this example, suppose $B = H + A + \epsilon$ and $Y = B + \epsilon$. Each regulator contributes a mean influence of μ to Y's activity and $\epsilon \ll \mu$ is the background noise level. Then $\frac{y(\Delta A)}{y(WT)} \approx \frac{1}{2}$, and $\frac{y(\Delta AB)}{y(WT)} = \frac{y(\Delta B)}{y(WT)} \approx \frac{\epsilon}{2\mu}$, where $y(g)$ represents the continuous expression of Y under genotype g. Clearly, $x_Y(\Delta AB, \Delta A) \approx \log(\frac{\epsilon}{\mu}) \ll 0$ and $x_Y(\Delta AB, \Delta B) \approx 0$. Epistatic reasoning could now place A above B since B's knock-out phenotype matches that of the double mutant.

2.2. *The Joint Intervention Network Model*

The input to the method consists of two parts: an $n \times p$ matrix x_{ij} of RECs and a $2 \times p$ matrix g of the genotypes of each comparison (Fig. 1C, "Indicator Expression"). A row in X corresponds to one indicator gene. Columns in

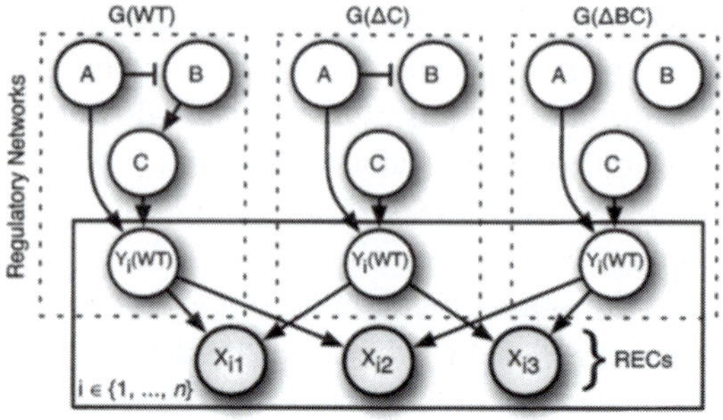

Figure 2. A JIN structure, $J(G)$, for the candidate network G from Fig. 1A, consists of a set of regulatory network instances $G(g)$ connected to the expression state of the indicator genes, Y, and connections from hidden genotype-specific expression $Y(g)$ to X observations representing RECs. A, B, and C are the perturbed signaling genes and the set of $Y_{n=17}$ represents the expression of the indicator genes. The JIN represents three different genotypes, WT (wild-type strain), ΔC (single knock-out strain), and ΔBC (double knock-out strain), and models observations of all possible comparisons. The solid box indicates plate notation to indicate replication of variables and dependencies.

both X and g correspond to a comparison of two of the p phenotypes. Thus, for each indicator i, $x_{ij} = \log(y_i(g_{1j})/y_i(g_{2j}))$, where $y_i(g)$ represents the expression of indicator gene i under genotype g. To form as many RECs as possible, X is augmented with virtual ratios. If a pair of genotypes does not have an observed REC, but there are comparisons to a common reference genotype (i.e. columns j and k where $g_{2j} = g_{2k}$), a new column h was constructed so that $x_{ih} = x_{ij} - x_{ik}$ for all such j and k. The new column(s) represent the pairwise comparison between genotypes g_{1j} and g_{1k} that may be important for detecting epistasis.

We first search for a high-scoring regulatory model by scoring network candidates and then use the highest-scoring candidate to find other genes that are under the same regulation model (Fig. 1C, "Regulatory Network Search" and "Pathway Expansion" respectively). We use a score that reflects how well G explains the RECs recorded in X (either observed or virtual). To score G, a Bayesian Network (BN), $J(G)$, is first constructed, which we refer to as the Joint Intervention Network of G (Figure 2). The log-posterior of the JIN is used as the score for G (see Supp. Methods "Regulatory Network Search"[17]). $J(G)$ is constructed from a candidate network by creating instances of G for each genotype g, referred to as a component network, $G(g)$. For each genotype g, the component network $G(g)$ encodes how the underlying network G is modified under perturbation, i.e. disconnecting perturbed nodes from their parents. Nodes associated with each

$G(g)$ are shown in dashed boxes in Figure 2. Conditional probability tables are shared between component networks. G contains a hidden binary variable for every regulator as well as every indicator gene, representing either *active* or *inactive* gene activity, and connections representing regulatory influences from regulator to regulator states and from regulator to indicator gene states (Y nodes).

In our setting, the state of the indicator genes, $Y = Y_1, Y_2, \ldots Y_n$, are forced to be descendants of the signaling genes since these genes serve as the phenotypic output of signaling regulation. A node, X_{ij}, represents the *discretized* REC of the ith indicator gene under the jth genotype comparison of g_{1j} and g_{2j}. The node X_{ij} is connected to two parents in the BN: $Y_i(g_{1j})$ and $Y_i(g_{2j})$, where $Y_i(g_{1j})$ and $Y_i(g_{2j})$ are binary random variables representing the unobserved expression state of gene i (either 0 "inactive" or 1 "active") under perturbations g_{1j} and g_{2j} respectively. A REC node X_{ij} is a ternary random variable and was set to +1 ("higher") if $x_{ij} > \tau$, -1 ("lower") if $x_{ij} < -\tau$, or 0 ("equal") otherwise. We used $\tau = 0.3$ to provide a conservative estimate of equal expression between two genotypes[4]. With the discrete values from the nodes in the JIN, we use a deterministic conditional probability table for X_{ij}: $\mathrm{P}(X_{ij}|Y_i(g_{1j}), Y_i(g_{2j})) = 1$ iff $X_{ij} = Y_i(g_{1j}) - Y_i(g_{2j})$.

The parameters of $J(G)$, θ, that define how regulators influence gene activity are restricted to encode biological intuitions about influences on gene expression. Each edge in the regulatory network can be either repressing or activating, with activation broken into two classes: *necessary* activators, and a pool of *alternative* activators. (see Supp. Methods "Regulatory Model"[17]).

3. Results

The biofilm regulatory pathway involves a complex interplay of transcription factors HapR, VpsT, and VpsR (see Fig. 3A). These regulators control the expression of *Vibrio* polysaccharide (VPS) genes (*vps*-I and *vps*-II gene clusters) that allow a community of *Vibrio* cells to change its surface properties by modulating cell-matrix and cell-surface contacts. Dual channel microarray gene expression data were obtained from Beyhan *et al.* 2007[4]. Briefly, the dataset consisted of a combination of single, double and triple knock-outs of three biofilm regulatory genes: *hapR*, *vpsR* and *vpsT*, representing eight different genotypes including wild type. A competitive microarray hybridization was performed using RNA isolated from a dele-

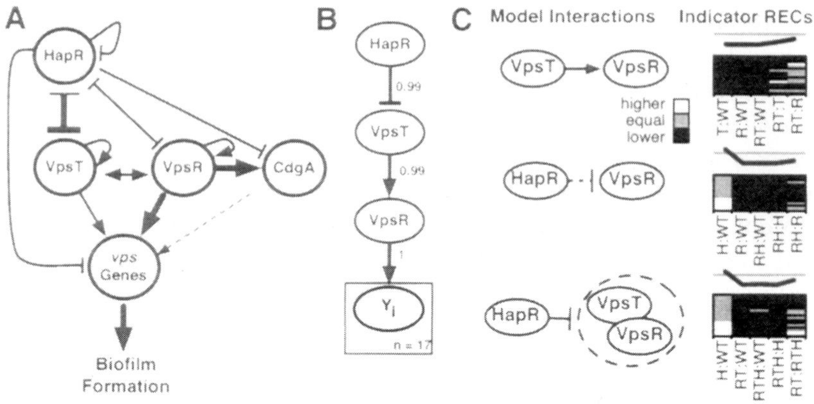

Figure 3. **A. Known biofilm regulation of *V. cholerae*.** HapR, VpsR, and VpsT influence *vps* gene expression. VpsT and VpsR activate the *vps* genes, while HapR represses the expression of *vps* genes. Line thickness indicates magnitude or strength of regulation. **B. Biofilm pathway predicted by JIN.** The regulatory structure learned by JIN. Links are annotated by their feature score (see Results). The solid box indicates plate notation. **C. Predicted interactions and supporting RECs** For three interactions in the predicted regulatory model, we show related discretized RECs. The dashed line from HapR to VpsR indicates an indirect interaction. The dashed circle around VpsT and VpsR indicates treatment as a single genetic unit. The deletions for each REC are coded with H=Δ*hapR*, R=Δ*vpsR*, and T=Δ *vpsT*.

tion mutant and that of wild-type. Six replicates were performed for each of seven deletion strains. Virtual hybridizations (see Methods) were added between compatible deletion strains, resulting in all 28 possible comparisons of strains.

Using known *vps* indicator genes, we used JIN scoring to infer a signaling network among the three regulators. The likelihood of the *vps* data under the learned regulatory network was found to be significant compared to the likelihood of arbitrary, non-*vps* genes. Bayesian feature scoring[11] was used to rank predicted interactions by a confidence measure (see Supp. Methods "Bayesian Feature Scoring"[17]). In addition, the inferred regulatory network was used to implicate new members of the *V. cholerae* biofilm pathway. Several of these genes have been tested and validated for involvement in biofilm formation.

3.1. *Inference of the V. cholerae Biofilm Regulatory Network*

A high-scoring biofilm regulatory network of VpsR, VpsT and HapR was found using known downstream gene clusters *vps*-I (VC0917–VC0927) and *vps*-II (VC0934–VC0939). Their relative expression under each deletion strain compared to wild-type is shown in Figure 3C. Each gene's expression was treated as an independent observation of the biofilm phenotype. A greedy hill-climb with random restarts was used to sample biofilm regulatory networks. We ran 1000 independent hill climbs, each starting from a different G and/or with different parameter values of $J(G)$ (see Supp. Methods "Regulatory Network Search"[17]).

Figure 3B shows the inferred *vps* regulatory network with edges annotated with feature scores. We calculated a feature score[11] for each edge as the proportion of final networks found by hill climbing that contained the edge, weighted by their likelihoods. Three edges in the network were associated with high feature scores.

Consistent with JIN's predictions, two of the three edges are well documented. HapR was found to inhibit VpsT, and VpsR was found to have the most direct influence on the *vps* genes. Several of the relations predicted by the JIN method are consistent with the observed RECs (Figure 3C). The RECs contrasting Δ*vpsT,vpsR* to Δ*vpsR* reveal more "equal" observations than the RECs contrasting Δ*vpsT,vpsR* to Δ*vpsT*, indicating that the regulatory influence from VpsT to VpsR is stronger than the direction from VpsR to VpsT. The model predicts an indirect influence of HapR on VpsR (via VpsT). This prediction is supported by the presence of more "equal" Δ*hapR,vpsR* to Δ*vpsR* RECs compared to the Δ*hapR,vpsR* to Δ*hapR* RECs. If VpsT and VpsR are considered as a single genetic unit, their epistatic relationship to HapR is evident from the triple knock-outs. The Δ*vpsR,vpsT,hapR* to Δ*vpsR,vpsT* RECs have more "equal" observations than the Δ*vpsR,vpsT,hapR* to Δ*hapR* RECs.

3.2. *Significance of the learned network*

To assess the significance of the *vps* network inferred by JIN, we asked how well the expression of *vps* genes fit the model compared to "unrelated" genes. A log-likelihood ratio (LLR) for each gene i was calculated from its vector of data, X_i as $log(P(X_i|J(G),\theta)/P(X_i|Null))$, where the null model was formed by disconnecting all Y_i's from the regulator genes. An LLR for all of the *vps* genes and for genes whose expression was not correlated with

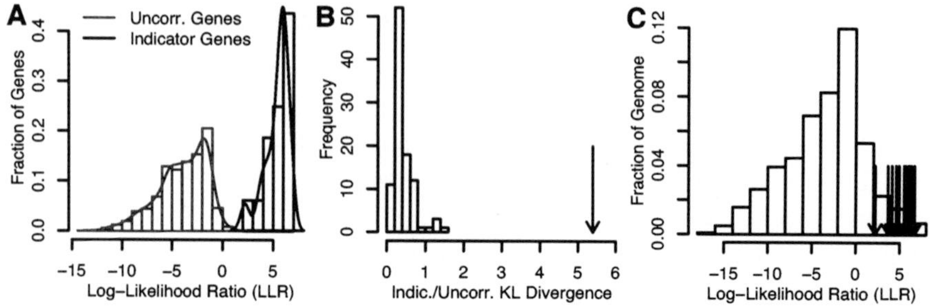

Figure 4. **A. Separation of LLR.** Histograms of the LLRs of indicator genes and of genes uncorrelated to the indicator genes. **B. Significance of learned regulatory network.** Histogram of KL-divergences from sets of randomly drawn indicator genes. The arrow indicates the separation shown in **A**. **C. LLR of entire genome.** Histogram of LLRs from entire genome. Arrows indicated the LLR of the indicator genes.

the *vps* genes was calculated. Uncorrelated genes were selected as those with absolute Pearson correlation to the *vps* genes' mean centroid less than 0.2. The LLRs of the *vps* genes were all above zero, while nearly all of the uncorrelated genes had an LLR less than zero, showing nearly no overlap with the LLRs of uncorrelated genes. The distribution of the *vps* LLRs and the uncorrelated LLRs is shown in Figure 4A.

To quantify the difference between these distributions, we calculated the Kullback-Leibler divergence. We found that the distributions had a KL divergence of 5.4 bits and next asked if this divergence was significant. KL divergences of randomly selected indicators were much lower than the *vps* indicators (see Figure 4B) yielding an empirical P value of 3×10^{-12}. [a]

3.3. *Expansion of the vps Pathway*

We ranked all of the genes by the LLR scores assigned by the JIN *vps* model and chose the top 15 for further analysis. The majority of these top-scoring genes have evidence supporting their association with the biofilm pathway (gray rows in Table 1).

Three genes, *exeA* (VC2445), VC0483, and VC0930, contain a VpsR binding site in their promoter[15], indicating that they are under direct regulation of the furthest downstream biofilm regulator. Additionally, VC0930, along with chromosomal neighbors VC0931 and VC0933 which lie between the *vps*-I and *vps*-II clusters, were recently shown to modulate biofilm

[a] P value computed from a Gamma fit to the empirical distribution.

Table 1. Top predictions for new *vps* pathway members.

LLR	Corr. Rank	Locus	Name	Description
7.14	8	VC2445	*exeA*	general secretion pathway protein A
7.03	34	VC1888	*bap1*	biofilm-associated protein
6.83	116	VC2732	*epsE*	general secretion pathway protein E
6.77	99	VC2730	*epsG*	general secretion pathway protein G
6.77	270	VCA0570		Sui1 family protein
6.74	69	VC0483		hypothetical protein
6.73	287	VC1064		lipoprotein-related protein
6.69	114	VCA0612	*mscL*	large-conductance mechanosensitive channel
6.67	86	VC0930	*rbmC*	rugosity-biofilm modulator
6.67	30	VC0931	*rbmD*	rugosity-biofilm modulator
6.67	12	VC1701		hypothetical protein
6.62	67	VC1320		DNA-binding response regulator
6.51	62	VC1935		CDP-diacylglycerol–glycerol-3-phosphate 3-phosphatidyltransferase-related protein
6.48	133	VC1195		lipoprotein, putative
6.48	9	VC0933	*rbmF*	rugosity-biofilm modulator

formation[6]. VC0930, along with its paralog VC1888, also a top JIN biofilm candidate, are secreted proteins critical to pellicle and biofilm formation.

Several other genes are involved in secretion and are predicted to be *vps* pathway members. VC2730–VC2733, of which JIN predicts *epsE* (VC2732) and *epsG* (VC2730) as top biofilm candidates, are involved in secretion of VPS[1]. *prtV* (rank 18, not shown in Table 1) is a protease that has recently been found in the extra-cellular matrix (Fong and Yildiz, in preparation).

Several regulators were also found. VC1320 negatively regulates biofilm formation in the smooth variant of *V. cholerae* (submitted, Bilecen and Yildiz). *rpoN*, ranked at 25 (not shown in Table 1), has been shown to positively regulate biofilm formation[15]. Other genes predicted are candidates for cell surface sensing and/or modification. *mscL*, predicted to be a high-conductance mechanosensitive channel, may sense surface contact, which is one of the first steps in biofilm formation. Finally, VC1701 and VC1935 have been shown to be differentially regulated between the $\Delta vpsR$ and $\Delta vpsT$ genotypes, indicating that they may play a role in modulating biofilm formation.

We compared the learned regulatory network to that predicted by ModuleNetworks[13] from the same data (see Supplemental File 1[17]). ModuleNetworks only found HapR and VpsR to be significant regulators, with P values of 4.10E-03 and 1.12E-10, respectively. Additionally, robustness analysis revealed that ModuleNetworks used HapR and VpsT as regulators in less than 25% of the trials.

JIN had a higher precision for identifying new candidates than ranking by correlation, which simulates a cluster-based approach. The top 25 candidates were selected from both methods, and genes common to both lists were removed, resulting in 17 genes in each list. Seven genes unique to JIN's list were known to be related to biofilm formation while only three genes unique to the correlation-based list were related.

4. Conclusion and Discussion

The method presented uses a novel probabilistic structure, the Joint Intervention Network, for explicitly linking different genotypes to their phenotypic consequences. It models epistasis analysis by using relative expression changes, rather than direct expression levels, as the quantitative phenotype. The pathway reconstruction method differs from previous approaches, such as NEMs defined by Markowetz *et al.*, in that it uses the downstream effects of knock-out combinations of two or more genes and expression changes of indicator genes already known to be relevant to the pathway rather than the expression of genes selected through data preprocessing.

JIN successfully predicted several genes involved in *V. cholerae* biofilm formation, one of the pathogen's survival mechanisms. With respect to VpsT and VpsR, our model predicts a linear, rather than convergent regulation of the *vps* genes, and the validity of this needs to be tested. Prior studies have found a VpsR binding site in the promoter of *vpsT*, which would suggest VpsR acts upstream of VpsT and not downstream as predicted by our model. The JIN model therefore predicts VpsT regulates VpsR by a different, currently unknown mechanism, and that this regulation has greater strength than VpsR's regulation on VpsT. No transcription factor binding sites have been identified for VpsT. Therefore, it would be interesting to identify targets of VpsT (e.g. through chromatin-immunoprecipitation) and search for the presence of sequence motifs. Binding sites found in the promoter region of *vpsR* but not in *vps* genes would support JIN's prediction that VpsT directly regulates VpsR.

In this work, we assumed the observed epistatic interactions result from the propagation of a signal according to the state of activity of the gene products. However, different signal propagation mechanisms, such as interactions that depend on substrates or other intermediates, are known to reveal different epistatic relationships[8]. Also, feedback present in the network may significantly influence our ability to recover its structure, as may be the case for the predicted interaction between VpsT and VpsR. Thus, further development of computational approaches for ordering genes into pathways using high-dimensional phenotypes, which can incorporate more general models of epistasis than presented here, may hold promise for application to a wide variety of genotype-phenotype investigations to accelerate our understanding of various disease-related processes.

5. Acknowledgments

JMS was supported by NSF grant DBI-0543197 and a fellowship from the Alfred P. Sloan foundation. CHY was supported by the Leon Levy Foundation and the Simons Foundation at the IAS, Princeton, NJ.

References

1. A. Ali, J. Johnson, A. Franco, D. Metzger, T. Connell, J. Morris. JR., S. Sozhamannan, *Infect Immun.* **68:6** 3792 (2000).
2. M. Bansal, V. Belcastro, A. Ambesi-Impiombato, and D. di Bernard, *Molecular Systems Biology* **3:78** (2007).
3. W. Bateson, *Mendel's Principles of Heredity*, Cambridge University Press, Cambridge (1909).
4. S. Beyhan, K. Bilecen, S. Salama, C. Casper-Lindley, and F. Yildiz, *J Bacteriol.*, **189** 388 (2007).
5. D. Eaton, K. Murphy, *Proceedings of the Eleventh International Conference on Artificial Intelligence and Statistics*, (2007).
6. J. Fong, F. Yildiz, *J Bacteriol.*, **189:6** 2319 (2007).
7. N. Friedman, M. Linial, I. Nachman, and D. Pe'er, *Journal of Computational Biology.* **7**, 601 (2000).
8. L.S. Huang and P.W. Sternberg, *WormBook*, **14:1-19** (2006).
9. F. Markowetz and R. Spang, *BMC Bioinformatics* **8** (2007).
10. J. Pearl, *Causality: Models, Reasoning, and Inference*, Cambridge University Press, Cambridge (2000).
11. D. Pe'er, A. Regev, G. Elidan, and N. Friedman, *Bioinformatics* **17S1** S215 (2001).
12. K. Sachs, O. Perez, D. Pe'er, D. Lauffenburger, and G. Nolan, *Science* **308** 523 (2005).
13. E. Segal, R. Yelensky, and D. Koller, *Bioinformatics* **19** i273–i282 (2003).
14. N. Van Driessche, J. Demsar, E. O. Booth, P. Hill, P. Juvan, B. Zupan, A. Kuspa, and G. Shaulsky, *Nature Genetics* **37:5** (2005).
15. F. Yildiz, X. Liu, A. Heydorn, G. Schoolnik, *Molecular Microbiology*, **53** 497 (2004).
16. J. Yu, V.A. Smith, P.P. Wang, A.J. Hartemink, E.D. Jarvis, *Bioinformatics* **20:18** 3594 (2004).
17. http://sysbio.soe.ucsc.edu/projects/knockoutnets

COMBINING MUTUAL INFORMATION WITH STRUCTURAL ANALYSIS TO SCREEN FOR FUNCTIONALLY IMPORTANT RESIDUES IN INFLUENZA HEMAGGLUTININ

PETER M. KASSON

Departments of Structural Biology and Chemistry, Stanford University, Stanford, CA 94305 USA

VIJAY S. PANDE

Department of Chemistry, Stanford University, Stanford, CA 94305 USA

Influenza hemagglutinin mediates both cell-surface binding and cell entry by the virus. Mutations to hemagglutinin are thus critical in determining host species specificity and viral infectivity. Previous approaches have primarily considered point mutations and sequence conservation; here we develop a complementary approach using mutual information to examine concerted mutations. For hemagglutinin, several overlapping selective pressures can cause such concerted mutations, including the host immune response, ligand recognition and host specificity, and functional requirements for pH-induced activation and membrane fusion. Using sequence mutual information as a metric, we extracted clusters of concerted mutation sites and analyzed them in the context of crystallographic data. Comparison of influenza isolates from two subtypes—human H3N2 strains and human and avian H5N1 strains—yielded substantial differences in spatial localization of the clustered residues. We hypothesize that the clusters on the globular head of H3N2 hemagglutinin may relate to antibody recognition (as many protective antibodies are known to bind in that region), while the clusters in common to H3N2 and H5N1 hemagglutinin may indicate shared functional roles. We propose that these shared sites may be particularly fruitful for mutagenesis studies in understanding the infectivity of this common human pathogen. The combination of sequence mutual information and structural analysis thus helps generate novel functional hypotheses that would not be apparent via either method alone.

1. Introduction

Influenza virus is a major cause of both seasonal epidemic respiratory disease and periodic high-mortality pandemics. The most significant of these latter events within recent history, the pandemic of 1918-1919, caused approximately 50 million deaths worldwide.[1] Influenza viruses circulate extensively in birds as well as humans and other mammals, and the three major pandemics of the 20th century (1918, 1957, 1968) were all likely due to epizootic transfer from viruses infecting other species into the human population. This has likely occurred both via adaptation of avian viruses to human hosts and via genetic reassortment between avian or mammalian and human-specific viruses.

More recently, the spread of a highly pathogenic avian influenza virus (HPAI H5N1) and a number of epizootic infections of humans (with a case-fatality rate of approximately 60% [2]) has raised concern of another imminent pandemic. Fortunately, the H5N1 virus has thus far not displayed efficient human-to-human transmission. It has been postulated that the poor human-to-human transmissibility of H5N1 may be due to inefficient viral replication in the upper respiratory epithelium of humans. Since the viral hemagglutinin protein is the primary determinant of both cell entry and antibody-mediated immunity, mutations to the hemagglutinin molecule that increase the efficiency by which human respiratory epithelial cells are infected would be an important permissive factor in human-to-human spread of either an adapted avian H5N1 virus or an avian-human reassortant.

We would therefore like to understand the functional control of influenza hemagglutinin and the means by which the molecule might mutate to alter host range or to evade new therapeutic agents. Informatics-based methods allow computationally efficient screening of the large number of potential hemagglutinin mutations. Such efficiency is required because hemagglutinin is over 500 amino acids in length, yielding a mutation space of $\sim 20^{500}$, and mutations both near and distant from the ligand binding site have been shown experimentally to alter ligand selectivity. We therefore propose a stepwise approach in which informatics-based methods are used to generate an initial set of predictions that can be further refined by a combination of physics-based computational methods and targeted experimental mutagenesis.

Influenza functional regulation differs fundamentally from the canonical systems for which many function-prediction methods were developed. Computational methods that have been used in other systems to predict ligand-binding specificity include shape-based analysis of the ligand-binding pocket [3], analysis of conserved residues [4], evolutionary trace methods [4], and methods that combine phylogenetic and information-theoretic characterizations [5]. For control of hemagglutinin function, particularly ligand specificity switches, experimental characterization of isolates displaying partial specificity switches identified both single point mutations and concerted mutations among several residues[6]. The evolutionary pressure of the host immune response and the frequent recombination events undergone by influenza may also complicate the mutational assumptions of phylogenetic methods, and the lack of crystal structures of a hypothetical human-adapted H5N1 hemagglutinin challenge shape-based methods. While all of these methods may be helpful in studying influenza function, there clearly exists the opportunity for novel methodology to yield additional insight.

Computational prediction methods are particularly helpful for influenza because systematic experimental screening for functionally important mutants is challenging. Hemagglutinin is heavily glycosylated, and the glycan residues affect ligand binding[7-9]. Furthermore, glycosylation patterns vary according to the cell culture system used to express the hemagglutinin protein. Because of these challenges and biosafety issues associated with live H5N1 virus, experimental studies to date have focused on retrospective analysis of mutations observed in clinical isolates rather than prospective screening[6, 10].

Given the complexity of influenza evolution and the absence of phylogenetically distinct lineages of human H5N1 influenza, a statistical approach that does not utilize a model of evolutionary correlation provides a reasonable alternative. Close co-variation between a pair of residues is suggestive of selective pressure and likely functional linkage. This is distinct from conserved residues, which may be required for conformational stability, and hypervariable residues, which may mutate to achieve immune evasion with no functional consequence. In many classic examples of small-molecule recognition by proteins, key residues involved in ligand recognition can be mutated but only in concert. The classic example of this is the catalytic triad in serine proteases; a point mutation to a key residue can destroy activity, but activity can be rescued by a complementary mutation to another residue. In this spirit, co-evolving residues have previously been used as a means of identifying functional significance in other systems [11], including HIV protease drug resistance [12, 13]. We employ a slightly different approach here; instead of introducing a distance matrix for protein mutation and computing co-variance, we compute pair-wise mutual information between residues in a discrete fashion similar to [14]. We then employ crystallographic data to evaluate the sequence-based predictions.

We have used sequence mutual information to analyze closely-linked mutation sites in hemagglutinin from two influenza subtypes: H3N2 isolates that have circulated in humans since 1968 and H5N1 isolates that circulate primarily in birds and have been responsible for millions of poultry deaths and ~400 human deaths since 1997 [2]. To interpret the mutual information data, we employ a combination of hierarchical clustering and visualization using available crystallographic data. Comparison of the two influenza subtypes in this regard is particularly informative: major differences in the residues identified may reflect the different selective pressures on each, while the common set of residues identified may reflect common functional roles. Taken together, these results suggest hypotheses regarding functional regulation of hemagglutinin that are testable via future mutagenesis experiments.

2. Methods

2.1. *Sequences and structures*

Sequences were obtained from the NCBI Influenza Virus Resource [15]; all 2103 full-length human H3N2 hemagglutinin sequences and all 1516 full-length human or avian H5N1 hemagglutinin sequences in the database as of July 2008 were used. Multiple-sequence alignment was performed for each influenza subtype separately and for pooled sequences from both subtypes using MUSCLE.[16] Crystal structures of A/Aichi/2/1968 (H3N2) hemagglutinin (1HGG) [17] and A/Viet Nam/1194/2004 (H5N1; 2IBX)[6] were used for visualization and distance calculations. Since the human H5N1 structure does not include a ligand, ligand coordinates from a related avian H5 hemagglutinin[18] were used for distance calculations after structural alignment of the two hemagglunins.

2.2. *Identification of closely-linked residues*

Pairwise mutual information was calculated in a discrete fashion between each residue position in the H3N2 and H5N1 multiple sequence alignments respectively as follows:

$$I(i,j) = H(i) + H(j) - H(i,j) \tag{1}$$

where

$$H(i) = \sum_{a \in A} -p(x_i = a)\log p(x_i = a) \tag{2}$$

and

$$H(i,j) = \sum_{a \in A}\sum_{b \in A} -p(x_i = a, x_j = b)\log p(x_i = a, x_j = b), \tag{3}$$

for sequence positions *i* and *j*, where the variable x_i represents the values of the multiple sequence alignment at position *i*. Mutual information captures nonlinear relationships more efficiently than covariance-based methods; use of a substitution matrix to give varying mismatch probabilities according to the chemical similarity of the amino acids involved may add further sensitivity.

Symmetric uncertainty[19] was used to normalize the pairwise mutual information matrix as follows: $U(i,j)=2*I(i,j) / (I(i,i) + I(j,j))$ for all positions i and j. Single-linkage hierarchical clustering was performed in MATLAB using U as the distance metric and guaranteeing $U(i,i) = 1$, all i. The 99.9th percentile of all non-self symmetric uncertainty values was calculated, and the corresponding distance metric was used as a threshold for cluster identification.

2.3. *Cluster visualization*

Each residue in a cluster was mapped to the corresponding residue in the crystal structure for H3N2 or H5N1 hemagglutinin as appropriate. These residues were then visualized using Pymol (DeLano Scientific, Palo Alto CA).

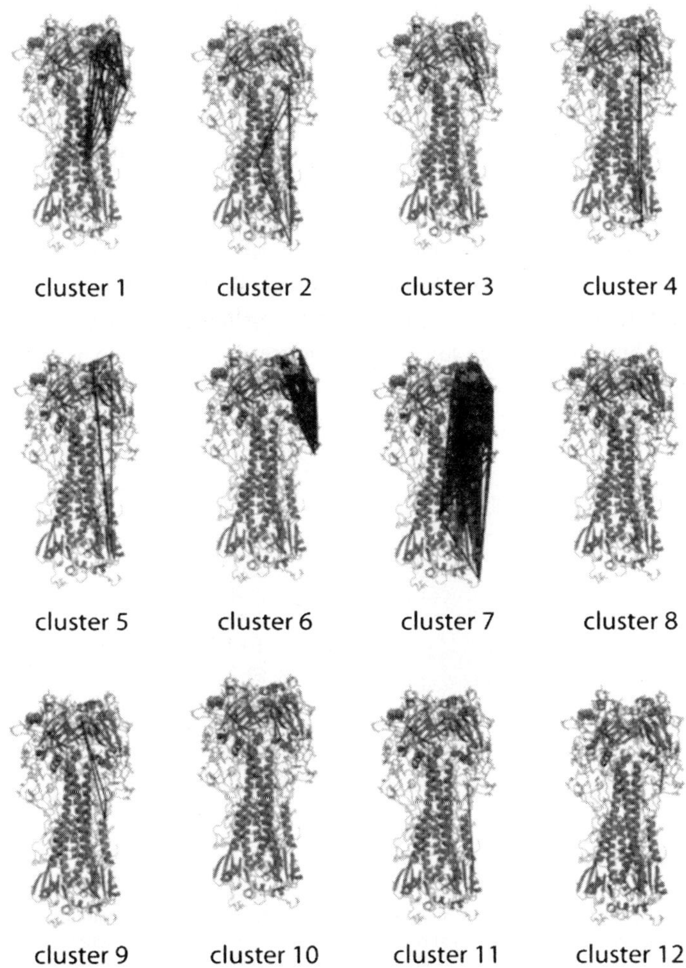

Figure 1. Clusters of closely-linked residues for H3N2 hemagglutinin.
Clusters identified by hierarchical clustering on sequence mutual information are visualized using crystallographic data. Red lines connect α-carbons within each cluster; sialic acid residues of the ligand are visualized via orange spheres.

3. Results

3.1. *Closely-linked residues in H3N2 hemagglutinin*

Closely-linked residues were identified based on hierarchical clustering on normalized sequence mutual information; clusters were selected using a threshold corresponding to the 99.9th percentile of non-self mutual information; for H3N2 hemagglutinin 75 residues in the mature hemagglutinin molecule were identified within 12 clusters. These clusters are visualized on the crystal structure of H3N2 A/Aichi/68 hemagglutinin in Fig. 1. Clusters fall into three general categories: few residues, short-range interactions (clusters 3, 8, 10, 12); few residues, long-range interactions (clusters 2, 4, 5, 9, 11); many residues primarily localized to the globular head (clusters 1, 6, 7). This last category is intriguing because both the ligand-binding sites and the major antibody-recognition epitopes are located here. In isolation, these clusters are interesting but less informative; we have compared the clustering patterns between two influenza subtypes in order to correlate clustering difference with biological differences between the subtypes.

a.   b.

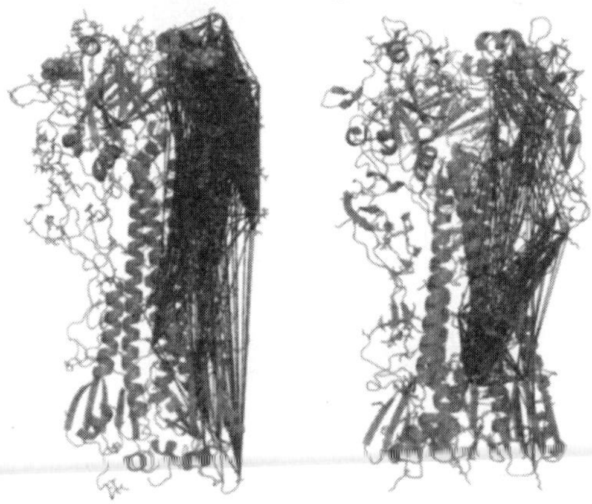

Figure 2. Comparison of closely-linked residues between influenza subtypes.
H3N2 hemagglutinin is rendered in panel (a), H5N1 in panel (b). Clusters are denoted by red lines connecting C-alpha atoms in the A/B monomer. Clusters are obtained via hierarchical clustering on a normalized pairwise mutual information matrix (symmetric uncertainty), with a distance cutoff corresponding to the 99.5th percentile of all non-self interactions in the matrix. The sialic acid of the ligand is shown as orange spheres.

3.2. *Differences between hemagglutinin subtypes*

We have analyzed pooled human and avian H5N1 isolates in the same fashion as the human H3N2 isolates to enable a more informative analysis. Fig. 2 shows all identified clusters for H3N2 and H5N1 hemagglutinin superimposed onto their respective crystal structures. One immediately apparent difference between the two subtypes is the frequency of linked residues on the globular head. This frequency is much greater in human H3N2 isolates than in the predominantly avian H5N1 isolates. We have quantified this relationship by measuring the distribution of radial distances from the sialic-acid-binding pocket to the residues identified as closely-linked for each subtype and comparing that to the distances for all residues of that subtype. This combined structural and sequence-informatic analysis clearly shows an enrichment of residues in the globular head for H3N2 compared to H5N1. While a large number of H3N2 HA residues have been implicated in antibody recognition, crystal structures of neutralizing antibodies show binding primarily to the globular head[20, 21]. It is also believed that the human immune response plays a major role in driving the evolution of the H3N2 virus [22, 23]. Since clusters of closely linked residues involving the globular head are identified in the human H3N2 isolates but not in the primarily avian H5N1 isolates (where the human immune response is not a major factor), we hypothesize that these clusters may be largely involved in immune escape.

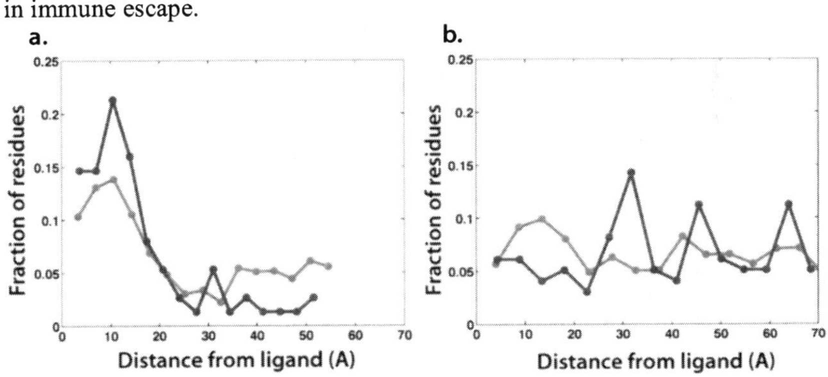

Figure 3. Distance from ligand for closely-linked residues in H3N2 and H5N1 hemagglutinin. Plotted are radial distribution functions for H3N2 hemagglutnin (**a**) and H5N1 hemagglutinin (**b**); closely-linked residues are plotted in red, while all protein residues are plotted in blue.

3.3. *Commonalities between hemagglutinin subtypes*

While human H3N2 isolates differ from H5N1 isolates in the role of the human immune response in driving viral evolution, they share many basic functional characteristics: binding of sialic-acid-terminated glycans, pH-induced conformational activation, and catalysis of membrane fusion. We therefore hypothesize that closely linked residues identified via independent analysis of each of these subtypes may be involved in some of the common functional roles. We have identified 12 residues meeting our detection threshold for both H3N2 and H5N1 isolates; these are listed in Table 1 and mapped to the H3N2 structure and visualized in Figure 4. These residues are an attractive target for future mutagenesis experiments; we hypothesize that single mutants in these positions may have altered activity in one or more of hemagglutinin's functional roles.

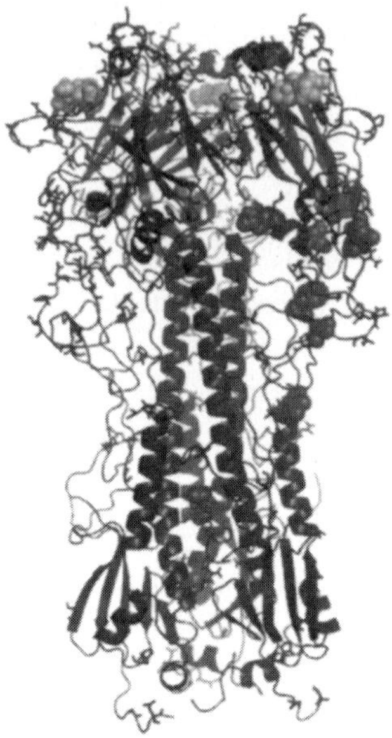

Figure 4. Residues identified in common between influenza subtypes.
Closely-linked residues identified in common between H3N2 and H5N1 subtypes are rendered in red on a structure of H3N2 hemagglutinin. The sialic acid of the ligand is shown as orange spheres.

	Position	A/Aichi/2/68 sequence
HA1	82	Q
HA1	83	T
HA1	106	A
HA1	107	S
HA1	115	S
HA1	164	L
HA1	174	F
HA1	197	Q
HA1	287	S
HA1	302	Y
HA2	2	L
HA2	59	T

Table 1. Closely-linked residues identified via independent analysis of both H3N2 and H5N1 subtypes.

4. Conclusions

We wish better to understand the functional regulation of influenza hemagglutinin, particularly with regard to ligand-binding specificity and catalysis of membrane fusion. Such an understanding will enable better surveillance for host-range changes in influenza and potentially assist in the development of novel small-molecule inhibitors. Using sequence mutual information as a robust metric of co-evolution, we have identified clusters of closely linked residues in both H5N1 and human H3N2 influenza isolates. Structure-based visualization facilitates the interpretation of these clusters in the context of existing functional data. By comparing the clusters for these two subtypes, we hypothesize that the extensive association network on the globular head of H3N2 may be linked to the human immune response, while the residues identified in common between H5N1 and H3N2 may be important to the basic functional roles of hemagglutinin. We suggest that these residues may be useful targets for future mutagenesis experiments. Our combined sequence-informatic and structural analysis thus enables generation of novel functional hypotheses that are not readily apparent via either method alone.

Acknowledgments

The authors thank R. Brandman, K. Branson, and O. Troyanskaya for helpful discussions. This work was supported by a fellowship from the Berry Foundation to P.K.

References

1. N. P. Johnson and J. Mueller, "Updating the accounts: global mortality of the 1918-1920 "Spanish" influenza pandemic," *Bull Hist Med*, vol. 76, pp. 105-15, 2002.
2. "World Health Organization Cumulative number of Confirmed Human Cases of Avian Influenza A(H5N1)," 2008.
3. S. Jones and J. M. Thornton, "Searching for functional sites in protein structures," *Curr Opin Chem Biol*, vol. 8, pp. 3-7, 2004.
4. O. Lichtarge, H. R. Bourne, and F. E. Cohen, "An evolutionary trace method defines binding surfaces common to protein families," *J Mol Biol*, vol. 257, pp. 342-58, 1996.
5. G. Casari, C. Sander, and A. Valencia, "A method to predict functional residues in proteins," *Nat Struct Biol*, vol. 2, pp. 171-8, 1995.

6. S. Yamada, Y. Suzuki, T. Suzuki, M. Q. Le, C. A. Nidom, Y. Sakai-Tagawa, Y. Muramoto, M. Ito, M. Kiso, T. Horimoto, K. Shinya, T. Sawada, M. Kiso, T. Usui, T. Murata, Y. Lin, A. Hay, L. F. Haire, D. J. Stevens, R. J. Russell, S. J. Gamblin, J. J. Skehel, and Y. Kawaoka, "Haemagglutinin mutations responsible for the binding of H5N1 influenza A viruses to human-type receptors," *Nature*, vol. 444, pp. 378-82, 2006.
7. V. P. Marinina, A. S. Gambarian, N. V. Bovin, A. B. Tuzikov, A. A. Shilov, B. V. Sinitsyn, and M. N. Matrosovich, "[The effect of losing glycosylation sites near the receptor-binding region on the receptor phenotype of the human influenza virus H1N1]," in *Mol Biol (Mosk)*, vol. 37, 2003, pp. 550-5.
8. A. S. Gambaryan, V. P. Marinina, A. B. Tuzikov, N. V. Bovin, I. A. Rudneva, B. V. Sinitsyn, A. A. Shilov, and M. N. Matrosovich, "Effects of host-dependent glycosylation of hemagglutinin on receptor-binding properties on H1N1 human influenza A virus grown in MDCK cells and in embryonated eggs," in *Virology*, vol. 247, 1998, pp. 170-7.
9. M. Ohuchi, R. Ohuchi, A. Feldmann, and H. D. Klenk, "Regulation of receptor binding affinity of influenza virus hemagglutinin by its carbohydrate moiety," in *J Virol*, vol. 71, 1997, pp. 8377-84.
10. P. Auewarakul, O. Suptawiwat, A. Kongchanagul, C. Sangma, Y. Suzuki, K. Ungchusak, S. Louisirirotchanakul, H. Lerdsamran, P. Pooruk, A. Thitithanyanont, C. Pittayawonganon, C. T. Guo, H. Hiramatsu, W. Jampangern, S. Chunsutthiwat, and P. Puthavathana, "An avian influenza H5N1 virus that binds to a human-type receptor," *J Virol*, vol. 81, pp. 9950-5, 2007.
11. W. R. Atchley, K. R. Wollenberg, W. M. Fitch, W. Terhalle, and A. W. Dress, "Correlations among amino acid sites in bHLH protein domains: an information theoretic analysis," *Mol Biol Evol*, vol. 17, pp. 164-78, 2000.
12. N. G. Hoffman, C. A. Schiffer, and R. Swanstrom, "Covariation of amino acid positions in HIV-1 protease," *Virology*, vol. 314, pp. 536-48, 2003.
13. Y. Liu, E. Eyal, and I. Bahar, "Analysis of correlated mutations in HIV-1 protease using spectral clustering," *Bioinformatics*, vol. 24, pp. 1243-50, 2008.
14. L. C. Martin, G. B. Gloor, S. D. Dunn, and L. M. Wahl, "Using information theory to search for co-evolving residues in proteins," *Bioinformatics*, vol. 21, pp. 4116-24, 2005.
15. Y. Bao, P. Bolotov, D. Dernovoy, B. Kiryutin, L. Zaslavsky, T. Tatusova, J. Ostell, and D. Lipman, "The influenza virus resource at the National Center for Biotechnology Information," *J Virol*, vol. 82, pp. 596-601, 2008.

16. R. C. Edgar, "MUSCLE: multiple sequence alignment with high accuracy and high throughput," *Nucleic Acids Res*, vol. 32, pp. 1792-7, 2004.
17. N. K. Sauter, J. E. Hanson, G. D. Glick, J. H. Brown, R. L. Crowther, S. Park, J. J. Skehel, and D. C. Wiley, "Binding of influenza virus hemagglutinin to analogs of its cell-surface receptor, sialic acid: analysis by proton nuclear magnetic resonance spectroscopy and X-ray crystallography," in *Biochemistry*, vol. 31, 1992, pp. 9609-21.
18. Y. Ha, D. J. Stevens, J. J. Skehel, and D. C. Wiley, "X-ray structures of H5 avian and H9 swine influenza virus hemagglutinins bound to avian and human receptor analogs," *Proc Natl Acad Sci U S A*, vol. 98, pp. 11181-6, 2001.
19. I. H. Witten and E. Frank, *Data mining : practical machine learning tools and techniques*, 2nd ed. Amsterdam ; Boston, MA: Morgan Kaufman, 2005.
20. T. Bizebard, B. Gigant, P. Rigolet, B. Rasmussen, O. Diat, P. Bosecke, S. A. Wharton, J. J. Skehel, and M. Knossow, "Structure of influenza virus haemagglutinin complexed with a neutralizing antibody," *Nature*, vol. 376, pp. 92-4, 1995.
21. D. Fleury, B. Barrere, T. Bizebard, R. S. Daniels, J. J. Skehel, and M. Knossow, "A complex of influenza hemagglutinin with a neutralizing antibody that binds outside the virus receptor binding site," *Nat Struct Biol*, vol. 6, pp. 530-4, 1999.
22. D. J. Smith, A. S. Lapedes, J. C. de Jong, T. M. Bestebroer, G. F. Rimmelzwaan, A. D. Osterhaus, and R. A. Fouchier, "Mapping the antigenic and genetic evolution of influenza virus," *Science*, vol. 305, pp. 371-6, 2004.
23. B. P. Blackburne, A. J. Hay, and R. A. Goldstein, "Changing selective pressure during antigenic changes in human influenza H3," *PLoS Pathog*, vol. 4, pp. e1000058, 2008.

MASTER REGULATORS USED AS BREAST CANCER METASTASIS CLASSIFIER*

WEI KEAT LIM, EUGENIA LYASHENKO, ANDREA CALIFANO†

Center for Computational Biology and Bioinformatics, Department of Biomedical Informatics, Columbia University, 1130 Saint Nicholas Avenue, New York, NY 10032, USA

Computational identification of prognostic biomarkers capable of withstanding follow-up validation efforts is still an open challenge in cancer research. For instance, several gene expression profiles analysis methods have been developed to identify gene signatures that can classify cancer sub-phenotypes associated with poor prognosis. However, signatures originating from independent studies show only minimal overlap and perform poorly when classifying datasets other than the ones they were generated from. In this paper, we propose a computational systems biology approach that can infer robust prognostic markers by identifying upstream Master Regulators, causally related to the presentation of the phenotype of interest. Such a strategy effectively extends and complements other existing methods and may help further elucidate the molecular mechanisms of the observed pathophysiological phenotype. Results show that inferred regulators substantially outperform canonical gene signatures both on the original dataset and across distinct datasets.

1 Introduction

A key application of genome-wide expression profile analysis is the identification of small gene signatures that effectively classify cancer sub-phenotypes with differential prognosis [1, 2]. For instance, both supervised and unsupervised methods have been extensively used to identify genes whose expression is predictive of patient outcome. In breast cancer, two recent large-scale Gene Expression Profile (GEP) studies identified independent ~70 gene panels that were 60-70% accurate in predicting progression to metastatic cancer in less than five years in their respective patient cohorts [3, 4]. However, paradoxically,

---

* This work is supported by the National Cancer Institute (R01CA109755), the National Institute of Allergy and Infectious Diseases (R01AI066116), and the National Centers for Biomedical Computing NIH Roadmap Initiative (U54CA121852).

† To whom correspondence should be addressed. Email: califano@c2b2.columbia.edu

there was virtually no statistical significance in the overlap between the two signatures (only one gene in common between them) and the signatures performed poorly in classifying samples from the other study. Such apparent paradox cannot be explained based only on technical reasons such as sample collection, microarray technology, and differences in data analysis [5]. Furthermore, such cases are quite the norm rather than the exception [6]. Thus the identification of robust prognostic biomarkers is still an open challenge in cancer research.

We argue that the fundamental reason why genes in prognostic signatures are so unstable and study-dependent is related to their role as "passengers" rather than "drivers" of the phenotypic differences (e.g., poor outcome). Since regulatory networks often act as amplification cascades, genes that are most differentially expressed tend to be further downstream from the somatic or inherited determinants of the prognostic differences. Due to the complex combinatorial interplay of regulatory proteins in the cell, these downstream genes are also the most unstable, as many co-factors and potential noise sources are involved in the transcriptional cascade that leads to their differential expression. Indeed, oncogenes and tumor suppressors are not generally the most differentially expressed genes although they may show an outlier behavior in some samples [7]. Thus, rather than looking for differentially expressed genes between two phenotypes of interest, we argue that one should look instead for regulators that are causally responsible for the implementation of the observed differential expression patterns. This was previously suggested [5, 8] but never implemented because accurate, genome-wide maps of regulatory processes in tumor related human cells have been lacking.

Recently, we introduced and extensively validated ARACNe, an algorithm for the dissection of transcriptional networks that can infer the targets of transcription factors (TF) from microarray expression profile data. ARACNe was shown to scale up to the complexity of mammalian transcriptional networks [9], producing accurate networks that are cell-phenotype specific. The algorithm has been successfully biochemical validated in B and T cells [9, 10] and more recently in glial cells (manuscript submitted), showing an accuracy greater than 80% in identifying targets validated by Chromatin Immunoprecipitation (ChIP) assays. Additional enhancements allow the algorithm to produce transcriptional maps that are fully directed by utilizing explicit knowledge both of which genes encode specific TFs and of their binding site when available [11, 12].

Here, we show that ARACNe inferred transcriptional regulation maps can be effectively interrogated for the unbiased inference of TFs that may induce or suppress specific gene signatures associated with poor prognosis in breast cancer. Furthermore, we show that Master Regulator is a better and more robust classifier than the large gene signatures proposed in [4] and [13].

2. Results

To identify and compare transcriptional regulators that are determinants of the gene expression signatures associated with rapid progression to metastasis in breast cancer, we used two datasets of 295 [3] and 286 [4] breast cancer samples each (hereafter denoted as the NKI and Wang datasets). The authors had previously used supervised analysis methods on these datasets to determine a 70-gene and a 76-gene prognostic signatures, respectively, (S_{NKI} and S_{Wang}). While these signatures were able to accurately classify the original datasets in cross-fold validation tests, using a Support Vector Machine (SVM) classifier [14], classification performance of the $\mathrm{S}_{\mathrm{NKI}}$ signature on the Wang data or of the $\mathrm{S}_{\mathrm{Wang}}$ signature on the NKI data was shown to be quite poor.

We performed the following analysis independently on the NKI and on the Wang dataset (For simplicity, we illustrate it only for the NKI data): (a) First, the NKI dataset was processed by ARACNe to infer a genome-wide transcriptional interaction network N_{NKI}. (b) Then, the network was interrogated using the Master Regulator Analysis (MRA), see Methods, to identify a repertoire of candidate TFs, i.e. master regulators MR_i, whose regulons R_i, (transcriptional target set) was highly overlapping with the signature S_{NKI}. (c) Finally, the $\mathrm{MR_i}$ genes were tested as classifiers

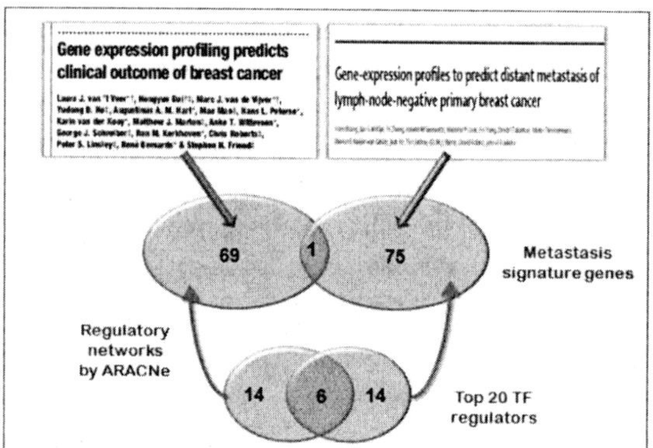

Figure 1. Graphical representation of the overlap of metastasis signature genes and the top 20 TF regulators between the two datasets.

Table 1 Top 20 Master Regulators inferred by MRA independently from NKI and Wang datasets.

GID_{NKI}	SYM_{NKI}	FET p-val	GID_{WANG}	SYM_{WANG}	FET p-val
9232	PTTG1	4.5×10^{-16}	7534	YWHAZ	4.9×10^{-5}
2305	FOXM1	6.0×10^{-11}	5933	RBL1	1.1×10^{-4}
7832	BTG2	4.1×10^{-6}	9232	PTTG1	5.3×10^{-4}
7534	YWHAZ	1.1×10^{-5}	7027	TFDP1	7.4×10^{-4}
7704	ZBTB16	1.2×10^{-5}	10736	SIX2	2.7×10^{-3}
4782	NFIC	1.4×10^{-5}	3148	HMGB2	4.1×10^{-3}
1054	CEBPG	1.6×10^{-5}	4208	MEF2C	5.0×10^{-3}
9735	KNTC1	1.9×10^{-5}	4605	MYBL2	5.7×10^{-3}
2737	GLI3	4.5×10^{-5}	4488	MSX2	6.2×10^{-3}
8522	GAS7	4.5×10^{-5}	1869	E2F1	6.7×10^{-3}
7027	TFDP1	5.9×10^{-5}	6095	RORA	1.3×10^{-2}
147912	SIX5	6.9×10^{-5}	2305	FOXM1	1.3×10^{-2}
1869	E2F1	1.2×10^{-4}	3607	FOXK2	1.4×10^{-2}
3608	ILF2	2.4×10^{-4}	6936	C2orf3	1.5×10^{-2}
4904	YBX1	2.4×10^{-4}	4603	MYBL1	1.5×10^{-2}
3223	HOXC6	5.2×10^{-4}	23054	NCOA6	1.7×10^{-2}
1746	DLX2	5.8×10^{-4}	7528	YY1	2.2×10^{-2}
51123	ZNF706	7.0×10^{-4}	677	ZFP36L1	2.2×10^{-2}
3148	HMGB2	8.6×10^{-4}	171017	ZNF384	2.3×10^{-2}
3066	HDAC2	9.3×10^{-4}	10520	ZNF211	2.4×10^{-2}

outcome against the S_{NKI} signature, both in the NKI and Wang set using five-fold cross-validation based on an SVM classifier.

Surprisingly, from an unbiased list of 852 TFs (determined by Gene Ontology annotation) shared by the two datasets, the analysis produced a 30% overlap among the top 20 inferred master regulator TFs ($p = 2\times10^{-6}$, by FET), which is especially significant when compared to the 1 gene overlap between the two signatures ($p = 0.25$, FET) (Figure 1). Note that both ARACNe and the MRA were performed independently on each of the two datasets. The top 20 TFs are listed in Table 1.

To study the stability of MR genes as a function of the training set, we repeated this procedure using different GEP subsets from the NKI dataset to infer prognostic signatures. We generated 100 sample sets by randomly selecting 34 out of 78 NKI samples with poor outcome and 44 out of 118 samples with good outcome in each training set, preserving the ratio of poor to good prognosis samples in the original study. We then computed the Pearson correlation between the prognostic groups

Table 2. Average overlap between S_{NKI} signatures and *top 20 Master Regulators* inferred from 100 GEP sample sets.

Predictor	Overlap
Signature genes	8.6%
Top 20 regulators	47.8%

and the expression level for all informative genes across the 78 samples [13]. The top 70 genes with the highest absolute correlation coefficient were selected as the corresponding S_{NKI}, and the top 20 Master Regulator TFs were identified using the MRA analysis. Finally, we compared the overlap of signature genes and regulators derived from each set to the ones from the original set, and reported the average overlap percentage in Table 2. Results clearly show that while the signature gene selection is highly affected by the training sample choice, with only 8.6% average overlap between signatures, the inferred master regulator TFs were much more consistent, with a 47.8% overlap despite the highly discordant signatures that were used to infer them. Although the causal role of the inferred regulators in the phenotype remains to be validated, this observation suggests that master regulators are less sensitive to training set selection and experimental variability.

In order to study whether the signature regulators possess the same or better prognostic capabilities than the signature genes, we used an SVM classifier to differentiate patients that developed distant metastasis within 5 years from patients that were metastasis-free for at least 5 years. We used both the inferred Master Regulators and the NKI/Wang signatures as the features for SVM-based classification. Gaussian Radial Basis Function kernel was employed in the SVM classifier along with the default parameters specified in MATLAB Bioinformatics Toolbox. Five-fold cross-validation was used to evaluate the performance in predicting the patients' prognostic group. Master regulators were tested independently, using lists of increasing size from the single best MR to a list of the top 70 MRs.

Two methods were used to assess the enrichment of TFs' regulon genes in genes that are differentially expressed between the two prognostic groups. The first one used the Fisher Exact Test (FET) to determine the statistical significance of the overlap between TFs' regulon genes and the prognostic signatures. The second used the Kolmogorov-Smirnov test, as implemented by the Gene Set Enrichment Analysis (GSEA) method, to assess enrichment of the TFs' regulon genes in

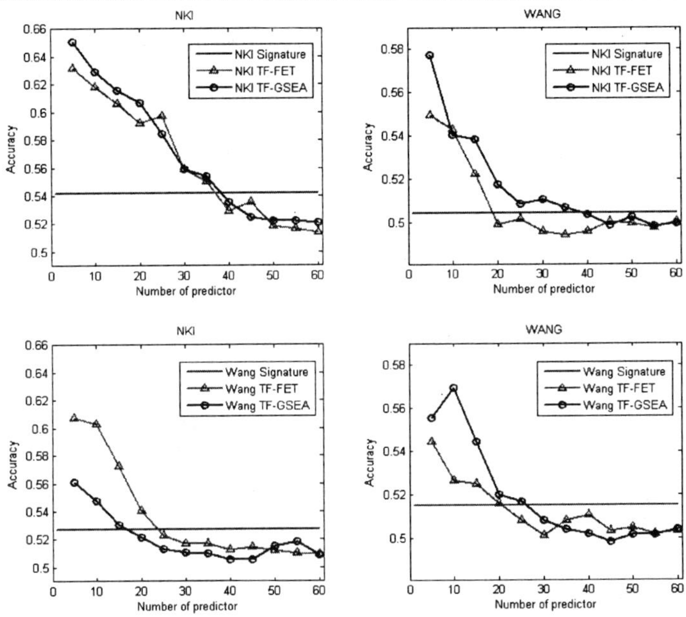

Figure 2: Accuracy of metastasis classification using signature genes and top TFs derived from (a) NKI dataset, and (b) Wang dataset. Predictions were made on NKI (left) and on Wang dataset (right) in both cases.

differentially expressed genes without having to select a specific threshold (see Methods). Both methods produced similar performance.

Results of prognosis classification for the SVM classifier using either (a) the S_{NKI} or S_{Wang} signature genes, (b) Master Regulator TFs ranked by FET p-value, and (c) Master Regulator TFs ranked by GSEA are shown in Figure 2. Since protein activity is not necessarily proportional to the mRNA concentration of the corresponding genes, due to post-transcriptional/translational modifications, we used the regulon of a TF as an indicator of the TF's activity. In particular, gene expressions were first converted into z-score. In each sample, all genes were sorted by the z-score and then used as the reference list in GSEA. The TF's activity level was approximated by the Normalized Enrichment Score (NES) computed for its regulon.

Classification performance was assessed by five-fold cross-validation. The analysis shows that even a handful of Master Regulators can predict progression to metastatic tumors with higher accuracy than the ≥70-gene signatures. Indeed, performance of the Master Regulator classifier invariably decreased as more genes were used. Classification accuracy using the gene-signatures drops when the S_{NKI} is used to classify Wang samples or vice-versa (i.e. in cross-dataset studies), while Master Regulator prediction remained highly predictive. The decrease in classification power as a function of the number of Master Regulators used in the classifier also suggests that the method produces informative ranking of the TFs', with more significant Master Regulators producing more accurate classification. In summary, the most significant master regulators were much better conserved across training sets (48% vs. 9%), consistently outperformed differentially expressed signatures in five-fold cross-validation studies, and were better able to classify samples in independent studies.

3. Methods

3.1. *ARACNe Network Inference*

ARACNe uses an information theoretic approach to dissect physical transcriptional interactions between TFs and their targets [12]. Briefly, the algorithm first uses a large GEP dataset to distinguish candidate interactions between a TF and other genes in the GEP by computing pairwise mutual information (MI). It employs a computationally efficient Gaussian kernel estimator to estimate the MI as:

$$I^*[X;Y] = \frac{1}{M}\sum_i \log\frac{f(x_i, y_i)}{f(x_i)f(y_i)} \tag{1}$$

where the 1-dimensional and 2-dimensional kernel estimators with kernel K and the position-dependent kernel width, h, are defined by Equations (2) and (3) respectively.

$$f(x) = \frac{1}{M}\sum_j K\left\{\frac{(x-x_j)^2}{h^2}\right\} \tag{2}$$

$$f(x,y) = \frac{1}{M}\sum_j K\left\{\frac{(x-x_j)^2 + (y-y_j)^2}{h^2}\right\} \tag{3}$$

ARACNe first eliminates interactions that are below a minimum MI threshold, defined based on statistical significance threshold. Then, the Data Processing Inequality (DPI) theorem from information theory is used to eliminate the vast majority of interactions that are likely mediated by another TF.

Bootstrap sampling, a nonparametric technique for statistical inference, is employed in the networks reconstruction process in order to accommodate the noise in microarray data and the error in MI estimation. The sampling procedure generates bootstrap datasets that are obtained by randomly selecting samples with replacement from the original dataset. One hundred bootstrap datasets are used to create the *bootstrap networks*. Due to the sampling procedure with replacement, the generated bootstrap datasets will consist of some replicated samples, and thus increase MI values for all edges computed from these datasets. To avoid this artifact, an infinitesimal amount of uniformly distributed noise is added on the bootstrap samples so that all the repeats with identical values would produce a randomized order. A consensus network is then constructed by retaining edges supported across a significant number of the bootstrap networks.

3.2. *Gene Sets Enrichment Analysis*

GSEA uses the Kolmogorov-Smirnov statistical test to assess whether a predefined gene set (in this case the Master Regulator set) is statistically enriched in genes that are the two extremes of a list ranked by differential expression between two biological states [15]. The algorithm is very useful to detect differential expression of a set of genes as a whole, even though the fold-change may be small for each individual gene.

Since gene regulons include both TF-activated and TF-repressed genes, GSEA was extended to assess enrichment of two complementary gene sets against N ranked genes. For instance, suppose that is expected to be a Master Regulator capable of activating a signature. Then one would expect the TF-activated gene set to be enriched in genes that are upregulated, while the TF-repressed gene set should be enriched in genes that are downregulated. Standard GSEA would dilute the evidence supporting TFs that can function both as activators and repressors.

The extended GSEA, for two complementary gene sets, proceeds as follows: (a) Compute signal-to-noise ratio (S2N), the difference of means scaled by the standard deviation $(\mu_A-\mu_B)/(\sigma_A-\sigma_B)$, between phenotype A and B, for each of the N genes in microarray. Order the N genes by S2N from the most positive to the most negative values, denoted by R; (b) identify hits independently for the positive gene set $S+$ in R, and the negative gene set $S-$ in $\bar{R}$, in which $\bar{R}$ is the inversed ranking of R with the inverted S2N values; (c) Combine R and $\bar{R}$ and reorder the S2N values by keeping the hits for both *S+* and *S-*, denoted as R_C (see Figure 3); (d) Compute a running score by walking down the combined ranking R_C. The score will increase by $|r_i|^p/\Sigma_{i\epsilon S}|r|^p$ if the ith gene is a hit, or otherwise decrease by *1/(2N-S)*, where *S* is the combined total number of genes in *S+* and *S-*. Finally, (e) Enrichment Score (ES) is determined as sum of the maximum and the minimum deviation from zero along the running score.

We randomly permuted the phenotype labels and repeated steps (a – d) for 10,000 times to compute the ES null distribution. Statistical significance of the ES can be computed by comparing the observed ES to the null distribution. A MATLAB function implementing the extended GSEA described above is available from http://www.dbmi.columbia.edu/~wkl7001/gsea2.m.

3.3. *Master Regulator Analysis (MRA)*

Given two phenotypes A and B and a transcriptional interaction network model, the Master Regulator analysis attempts to identify TFs that may induce the transition from A to B.

For each TF in the transcriptional interaction model, we test whether activation or inhibition of the TF may affect genes to produce a change

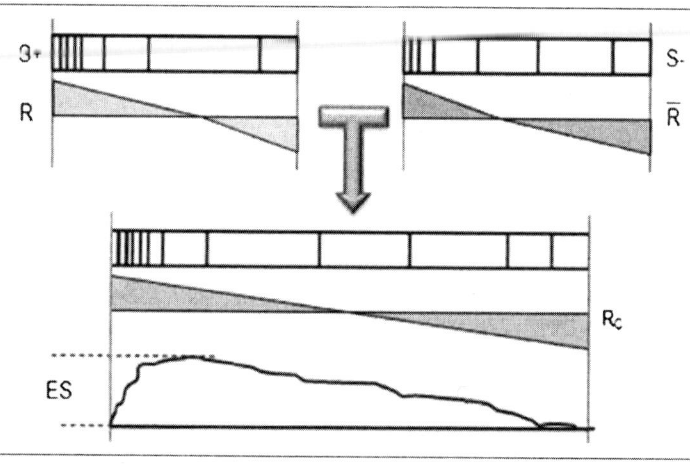

Figure 3. Extended GSEA assesses two complementary gene sets at once. Hits are identified independently for the positive gene set *S+* in *R*, and the negative gene set *S-* in $\bar{R}$, in which $\bar{R}$ is the inversed ranking of *R* with the inverted S2N values. The rankings *R* and $\bar{R}$ are then combined and reordered according to the S2N values, denoted as R_C. All the identified hits, for both *S+* and *S-*, are remained throughout the reordering process. ES is defined as sum of the maximum and the minimum deviation from zero along the running score.

as similar as possible to that observed between B and A. Specifically, we first assume that TF activation may lead to the transition. In this case, we test whether TF-activated targets (the R+ regulon of the TF) are enriched in the genes that are overexpressed in B and whether the TF-repressed targets (the R- regulon of the TF) are enriched in genes that are underexpressed genes in B. This is accomplished with one of two tests:

FET Analysis: In this analysis, genes that are differentially expressed in B vs. A are first filtered by a statistical significance threshold (e.g., 5% FDR by t-test or u-test) and then divided into two groups, S+ and S-, containing respectively genes that are overexpressed or underexpressed in B vs. A. The statistical significance p^+ of the intersection between R+ and S+ as well as the statistical significance p^- of the intersection between R- and S- is computed, given the total number of considered genes, using the FET test. Since the two tests are statistically independent, the final p-value is assessed as $p = p^+ \times p^-$.

Gene Set Enrichment Analysis: In this test, rather than selecting gene signatures S+ and S-, the GEP genes are ranked in a list $L_{A \to B}$, from the most underexpressed to the most overexpressed in B vs. A. Then, the extended GSEA analysis is used to simultaneously assess whether R+ and R- are respectively enriched in genes that are overexpressed and underexpressed in $L_{A \to B}$. The procedure is then repeated to test whether TF silencing may lead to the transition. In this case, the role of R+ and R- is inverted but the procedure is not affected. Finally, Master Regulators are sorted by statistical significance (p-value).

4. Discussion

In this manuscript, we have presented an approach for identifying robust prognostic markers using transcriptional regulatory networks in the breast cancer context. Rather than establishing signatures of genes that are differentially expressed in poor prognosis vs. good prognosis samples, the method attempts to identify the upstream transcriptional regulators of the signature that are consistent with the network topology.

We have shown that top Master Regulator genes, independently inferred from different datasets or from different subsets of the same dataset, are far more stable and robust than top genes correlated to disease outcome. For instance, overlap of the top 70 correlated genes after subsampling was only 8.6% while overlap of the top 20 Master Regulators inferred from the same datasets was 46.8% on average.

Finally, we have shown that in independent testing using two large datasets for which prognostic signatures were previously published, the top inferred Master Regulators consistently and substantially outperformed the signatures in five-fold cross-validation tests using an SVM-based classifier. This was especially visible when the signature or Master Regulators inferred from one dataset were used to classify samples from the other.

Since they are inferred from a network model that explicitly encodes the regulation logic (i.e. edges are directed), these findings also generate testable hypotheses and rational models for understanding the oncogenic processes leading to the phenotypic difference between poor and good prognosis samples.

References

1. Alizadeh, A. A., Eisen, M. B., Davis, R. E., Ma, C., Lossos, I. S., Rosenwald, A., Boldrick, J. C., Sabet, H., Tran, T., Yu, X., Powell, J. I., Yang, L., Marti, G. E., Moore, T., Hudson, J., Jr., Lu, L., Lewis, D. B., Tibshirani, R., Sherlock, G., Chan, W. C., Greiner, T. C., Weisenburger, D. D., Armitage, J. O., Warnke, R., Levy, R., Wilson, W., Grever, M. R., Byrd, J. C., Botstein, D., Brown, P. O. & Staudt, L. M. (2000) *Nature* **403,** 503-11.
2. Golub, T. R., Slonim, D. K., Tamayo, P., Huard, C., Gaasenbeek, M., Mesirov, J. P., Coller, H., Loh, M. L., Downing, J. R., Caligiuri, M. A., Bloomfield, C. D. & Lander, E. S. (1999) *Science* **286,** 531-7.
3. van de Vijver, M. J., He, Y. D., van't Veer, L. J., Dai, H., Hart, A. A., Voskuil, D. W., Schreiber, G. J., Peterse, J. L., Roberts, C., Marton, M. J., Parrish, M., Atsma, D., Witteveen, A., Glas, A., Delahaye, L., van der Velde, T., Bartelink, H., Rodenhuis, S., Rutgers, E. T., Friend, S. H. & Bernards, R. (2002) *N Engl J Med* **347,** 1999-2009.
4. Wang, Y., Klijn, J. G., Zhang, Y., Sieuwerts, A. M., Look, M. P., Yang, F., Talantov, D., Timmermans, M., Meijer-van Gelder, M. E., Yu, J., Jatkoe, T., Berns, E. M., Atkins, D. & Foekens, J. A. (2005) *Lancet* **365,** 671-9.
5. Ein-Dor, L., Kela, I., Getz, G., Givol, D. & Domany, E. (2005) *Bioinformatics* **21,** 171-8.
6. Michiels, S., Koscielny, S. & Hill, C. (2005) *Lancet* **365,** 488-92.
7. Tomlins, S. A., Rhodes, D. R., Perner, S., Dhanasekaran, S. M., Mehra, R., Sun, X. W., Varambally, S., Cao, X., Tchinda, J., Kuefer, R., Lee, C., Montie, J. E., Shah, R. B., Pienta, K. J., Rubin, M. A. & Chinnaiyan, A. M. (2005) *Science* **310,** 644-8.

8. Rhodes, D. R. & Chinnaiyan, A. M. (2005) *Nat Genet* **37 Suppl,** S31-7.
9. Basso, K., Margolin, A. A., Stolovitzky, G., Klein, U., Dalla-Favera, R. & Califano, A. (2005) *Nat Genet* **37,** 382-90.
10. Palomero, T., Lim, W. K., Odom, D. T., Sulis, M. L., Real, P. J., Margolin, A., Barnes, K. C., O'Neil, J., Neuberg, D., Weng, A. P., Aster, J. C., Sigaux, F., Soulier, J., Look, A. T., Young, R. A., Califano, A. & Ferrando, A. A. (2006) *Proc Natl Acad Sci U S A* **103,** 18261-6.
11. Mani, K. M., Lefebvre, C., Wang, K., Lim, W. K., Basso, K., Dalla-Favera, R. & Califano, A. (2008) *Mol Syst Biol* **4,** 169.
12. Margolin, A. A., Wang, K., Lim, W. K., Kustagi, M., Nemenman, I. & Califano, A. (2006) *Nat. Protocols* **1,** 662-671.
13. van 't Veer, L. J., Dai, H., van de Vijver, M. J., He, Y. D., Hart, A. A., Mao, M., Peterse, H. L., van der Kooy, K., Marton, M. J., Witteveen, A. T., Schreiber, G. J., Kerkhoven, R. M., Roberts, C., Linsley, P. S., Bernards, R. & Friend, S. H. (2002) *Nature* **415,** 530-6.
14. Vapnik, V. (1995) *The Nature of Statistical Learning Theory* (Springer.
15. Subramanian, A., Tamayo, P., Mootha, V. K., Mukherjee, S., Ebert, B. L., Gillette, M. A., Paulovich, A., Pomeroy, S. L., Golub, T. R., Lander, E. S. & Mesirov, J. P. (2005) *Proc Natl Acad Sci U S A* **102,** 15545-50.

PREDICTION OF INTERACTIONS BETWEEN HIV-1 AND HUMAN PROTEINS BY INFORMATION INTEGRATION

OZNUR TASTAN[1], YANJUN QI[1], JAIME G. CARBONELL[1] AND JUDITH KLEIN-SEETHARAMAN[1, 2†]

[1]School of Computer Science, Carnegie Mellon University,15213 and [2]Department of Structural Biology, School of Medicine, University of Pittsburgh, 15260, Pittsburgh, PA USA.

Human immunodeficiency virus-1 (HIV-1) in acquired immune deficiency syndrome (AIDS) relies on human host cell proteins in virtually every aspect of its life cycle. Knowledge of the set of interacting human and viral proteins would greatly contribute to our understanding of the mechanisms of infection and subsequently to the design of new therapeutic approaches. This work is the first attempt to predict the global set of interactions between HIV-1 and human host cellular proteins. We propose a supervised learning framework, where multiple information data sources are utilized, including co-occurrence of functional motifs and their interaction domains and protein classes, gene ontology annotations, posttranslational modifications, tissue distributions and gene expression profiles, topological properties of the human protein in the interaction network and the similarity of HIV-1 proteins to human proteins' known binding partners. We trained and tested a Random Forest (RF) classifier with this extensive feature set. The model's predictions achieved an average Mean Average Precision (MAP) score of 23%. Among the predicted interactions was for example the pair, HIV-1 protein tat and human vitamin D receptor. This interaction had recently been independently validated experimentally. The rank-ordered lists of predicted interacting pairs are a rich source for generating biological hypotheses. Amongst the novel predictions, transcription regulator activity, immune system process and macromolecular complex were the top most significant molecular function, process and cellular compartments, respectively. Supplementary material is available at URL www.cs.cmu.edu/~oznur/hiv/hivPPI.html

1. Introduction

1.1. *Motivation*

Human immunodeficiency virus-1 (HIV-1) is the etiologic agent of acquired immune deficiency syndrome (AIDS) and continues to be a major health threat [1, 2]. The number of AIDS-related deaths was ~2.1 million in 2007 alone [3];

† Corresponding author

an estimated ~33.2 million people worldwide are infected [3]. HIV-1 contains a single stranded RNA genome, which codes for only 15 proteins; thus, it relies on human cellular functions [4]. The virus exploits the host cell's machinery such that it can successfully produce its progeny while at the same time avoid the immune system. Protein-protein interactions (PPIs) between HIV-1 and its host are vital at every step of the virus life cycle [1]. A recent functional genomic screen using siRNA technology revealed several hundred proteins critical for HIV-1 function [5]. An extensive survey of individual interactions described in the literature [6] has retrieved ~2500 pairs, of which ~1000 are likely direct physical interactions. Here, we propose to use data integration methods on these reported interactions in conjunction with a variety of different biological information sources to predict new PPIs between human and HIV-1.

1.2. *Related Work*

In order to identify PPIs in general, many experimental methods are available [7]. Computational methods have been useful in assisting the experimental efforts by either prioritizing PPIs to be tested by subsequent experiments or by validating (or refuting) high-throughput screens [8]. Such computational methods include those based on over-representation of domain or motif pairs observed in interacting protein partners [9-13], or the conservation of gene neighborhood and gene order [14], gene fusion events [15, 16], or the co-evolution of interacting protein pair sequences [17, 18]. Others designed kernels specific for PPIs that make use of the sequence signatures in interacting pairs [19] and protein structural information [20-22]. The large amount and heterogeneity of these multiple indirect and direct information sources suggests to integrate them in a supervised learning framework [23-26].

The computational methods above were applied to predicting PPIs within a single organism ("intra-species prediction"). In contrast, computational work on predicting PPIs between organisms ("inter-species prediction"), especially between host and pathogens has been limited. Dyer *et al.* [27] studied human-*Plasmodium falciparum* interactions and estimated the probability of interaction based solely on sequence signature information [11]. Davis *et al.* [28] studied ten host-pathogen PPIs (not including HIV-1) using structural evidence with a comparative modeling approach: the host, pathogen protein pairs that share similarity to protein complexes with known structures are used to build 3d structural models of putative complexes and pairs with high quality models are filtered by functional and genomic experimental information. The applicability of these methods in the specific case of HIV-1, human prediction task is limited: most of the HIV-1 proteins do not contain domains, statistics on which would be required in [27] and not all HIV-1 proteins have high similarity to proteins with known structure which is necessary for [28].

1.3. *Approach*

Here, we propose a supervised learning framework to predict PPIs between HIV-1 and human proteins by integrating multiple biological information sources. This framework has been found useful in integrating heterogeneous biological information to predict intra-species interactions [23-26]. However, the inter-species prediction task is comparatively more challenging: high-throughput interaction datasets from yeast two hybrid and mass spectrometry based screens are not yet available publicly and many other useful sources such as co-expression pattern of genes [24] are not directly applicable. Nevertheless, there is another potentially exploitable source of information: the human protein interactome (set of intra-human interactions) itself. The proteins that the pathogen will target should in principle depend on interaction relationships between human proteins because the virus makes use of the existing communication pathways within the cell. Capitalizing on this source of information, we developed a number of features incorporating existing knowledge of human intra-PPIs and integrated these features together with other available information sources to predict HIV-1, human interactions.

2. Data and Methods

Details can be found in the supplementary online material Methods section.

2.1. *Classification Framework*

The task is formulated as a binary classification problem, where each protein pair belongs to either the 'interaction' or 'non-interaction' class. A protein pair is described by a feature vector, where each feature is derived from one or more biological information sources. We solve the classification problem utilizing the Random Forest classifier (RF) [29]. We chose the RF based on its robustness in scenarios where the features are noisy and redundant, as is the case here. The RF method has been highly successful in predicting intra-species PPIs [24, 26]. Initial tests comparing other classifiers for the HIV-1, human inter-species PPI prediction task also suggested the RF to be suited best (data not shown).

2.2. *Dataset*

2.2.1 ***Interacting protein pairs:*** Interactions between HIV-1 and cellular proteins reported in the scientific literature were retrieved from the NIAID database [6]. The dataset included 2512 interactions involving 1406 human proteins. Note that in addition to the 15 HIV-1 proteins, the database includes interactions for the precursors of the envelope (env gp160) and gag (gag pr55) and the gag product, p1. We excluded gag p1 due to the limited information available for this protein. Each interaction in the database is represented by one or more descriptive key phrases. Some entries are more likely associated with direct physical PPIs than others (e.g. "interacts with" as compared to "causes accumulation of"). Based on the keywords we grouped the interactions into two exclusive groups: Group 1 represents most likely direct physical interactions. In Group 2, interactions may also be indirect (Supplementary Table S1). Group 1 interactions included 1063 protein pairs involving 721 human proteins. Group 2 included 1447 pairs involving 914 human proteins. The Group 1 interactions constituted the "interaction class" and were used in model building and testing, whereas Group 2 was used to mine the final predictions.

2.2.2 ***Non-interacting protein pairs:*** Since it cannot be proven that two proteins do not interact, there is no "gold standard" negative set available. For training and testing purposes in the PPI prediction task it is therefore common to choose protein pairs uniformly at random from the set of protein pairs which are not known to interact, and treat them as negative interactions [30]. This is rationalized by the fact that the probability that two randomly chosen proteins interact is small and most methods are able to handle contamination from the small number of potential false negative (FN) matches. In the testing phase, the negative to positive ratio was set as 100:1, a value chosen based on the average number of interactions involving HIV-1 proteins. In training, the negative to positive example ratio is optimized for each cross validation step on the training data.

2.3. *Features*

We devised a total of 35 features (Supplementary Table S3). Some features are specific to HIV-1, human protein pairs, while others are related only to human or HIV-1 proteins, and some are derived from the human interactome (Fig. 1).

2.3.1 ***GO similarity features:*** The Gene Ontology (GO) [31] provides a defined vocabulary of protein attributes for molecular function, cellular component and biological process; proteins are annotated with one or more descriptive GO terms. For each of the three ontologies we developed two features: 'pairwise GO similarity' measures the similarity between the HIV-1 and human proteins in a pair, while 'neighbor GO similarity' refers to the similarity between the HIV-1 proteins and the human protein's human interactors (Fig. 1).

2.3.2 ***Graph properties of the human interactome:*** Three graph property features were derived from topological properties of the human intra-PPI network: degree, clustering coefficient and betweenness centrality. The degree of a node in a network is the number of its neighbors; whereas clustering coefficient [32] is the ratio of the edges present among its neighbors to the all possible edges that could be present between them. Betweenness centrality for a node is calculated as the fraction of shortest paths between node pairs that pass through the node of interest [33]. A node with a high betweenness centrality is located in a 'bottleneck' and has control over the information flow between other nodes.

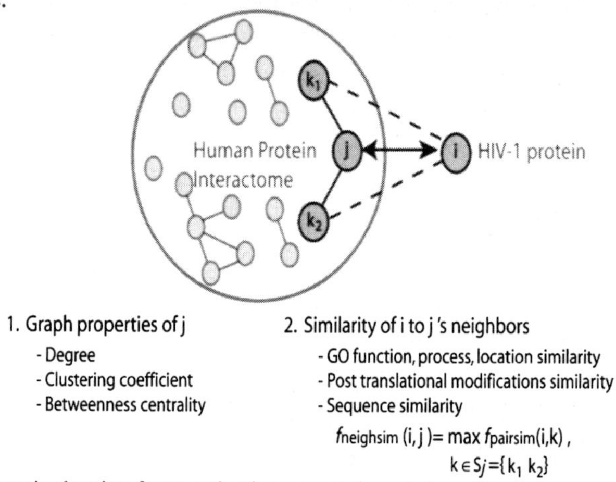

Figure 1. Schematic showing features that incorporate knowledge of the human protein interactome. These features can be conceptually grouped into two categories: 1) graph properties of human protein j in human protein interaction network, which include degree, clustering coefficient and betweenness centrality of node j 2) the similarity of the HIV-1 protein, i, to human protein j's interaction partners denoted by f_{neigh} (i,j) in the figure. The maximal similarity to the neighbors is used. Five features are derived; GO function, process and location similarity in addition to post translational modification and sequence similarity.

2.3.3 ***ELM-ligand feature:*** Functional sequence motifs which mediate binding were downloaded from the Eukaryotic Linear Motif (ELM) database [34]. For example, the sequence pattern PXXDY (ELM id: LIG_SH3_5) is recognized by SH3 domains. The feature evaluates whether an ELM motif is found in a given HIV-1 sequence and its ligand domain is present in the human protein. The HIV-1 sequences mutate rapidly and some motifs are very short. To avoid false positive matches, we only consider a motif match if it is conserved in multiple HIV-1 sequences; the feature value is weighted with the specificity of the motif.
2.3.4 ***Gene expression features:*** Four features reflect differential expression patterns of human genes across HIV-1 infected vs. uninfected samples.
2.3.5 ***Tissue feature:*** This feature encodes whether the tissues that the human proteins are expressed in are susceptible to HIV-1 infection or not.

2.3.6 ***Sequence similarity features:*** For each pair, two sequence similarity features are employed, pairwise sequence similarity and similarity of the HIV-1 protein sequence to human protein's human binding partners.

2.3.7 ***Posttranslational modification similarity to neighbor:*** Some PPIs require binding partners to be in a certain posttranslationally modified state and some HIV-1 proteins mimic the posttranslational modification of the human protein's interaction partner [35]. This feature captures if the HIV-1 protein shares any modification with at least one of the interaction partners of the human protein.

2.3.8 ***HIV-1 protein type features (ptf):*** We included a set of features (one for each HIV-1 protein, with values 1 for pairs that include a particular HIV-1 protein and 0, otherwise) to include the information of how likely an HIV-1 protein is to be in an interaction with a human protein.

2.4. *Feature Importance*

In constructing trees in the RF, at each node the attribute causing the highest decrease in the Gini index is chosen as split. Let p denote the fraction of interacting pairs assigned to node i and $1-p$ the fraction of the non-interacting pairs, the Gini index is computed as $G_i = 2p(1-p)$ [29]. Gini feature importance is derived from the Gini index and is the sum of all decreases in the forest due to a given feature, normalized by the number of trees in the forest.

2.5. *Performance Evaluation*

Classifier performance was evaluated with 3-fold cross validation in 10 repeat runs to obtain average values. When evaluating the performance of a classifier on an imbalanced test set such as is the case here, computing accuracy is not useful because a high true-negative (TN) rate can easily be obtained by chance. Therefore, we evaluated the quality of our predictive model using two figures of merits which ignore the success on the TN rate: the receiver operating curve (ROC) and precision vs. recall curve [36]. We employ the Mean Average Precision (MAP) score to summarize the precision vs. recall curve and the area under the ROC (AUC) to summarize the ROC curve as a scalar score which ranges between 0 and 1 [36]. Since the low FP region of the ROC curve is of particular interest in the PPI prediction task, the partial AUC scores R50, R100, R200 and R300 were determined, measuring the area under the ROC curve until reaching 50, 100, 200 and 300 FP predictions, respectively.

2.6. *GO Enrichment*

GO enrichment of the human proteins involved in the predicted interactions was identified using Ontologizer 2.0 [37] using the child-term parent intersection method [38] and using Bonferroni correction for multiple hypothesis testing.

3. Results and Discussion

3.1. *Classifier Performance*

We trained an RF classifier with a rich feature set derived from several biological information sources. The performance of the model was evaluated through 10 repeated 3-fold cross validation experiments. The average precision *vs.* recall curve of these experiments is given in Fig. 2 (solid line). Table 1 lists the average MAP, AUC and partial AUC scores of the model. The model achieves an average MAP score of 0.23 (±0.02) indicating that, on average, of all pairs predicted as interacting, 23% are TP. For PPI predictions, this is a very good performance: because of the highly skewed class distribution the probability of predicting a TP is not 0.5 as when positive and negative pairs are equally distributed, but is ~0.01 (for a ratio positive: negative pairs of 1:100).

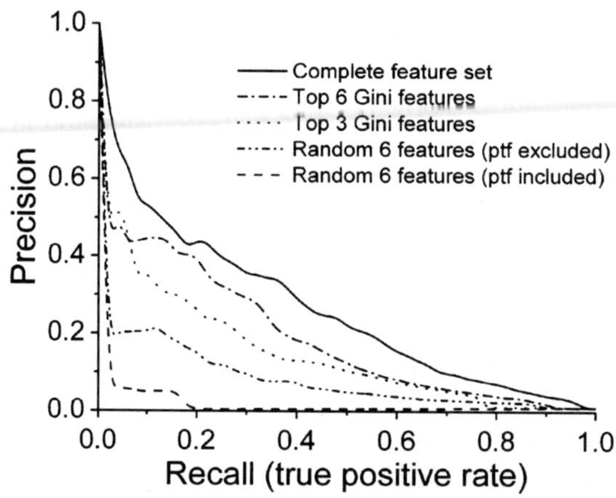

Figure 2. The average precision *vs.* recall curve of the Random Forest model trained on the complete feature set, in comparison to models trained with a subset of features. The top 3 Gini features are degree, betweenness centrality, and GO neighbor process similarity features. The top 6 Gini features are the top 3 Gini features plus clustering coefficient, GO neighbor function, and location features. These are compared to two baseline classifiers, where 6 features are randomly selected from the set of features that does not include the top 6 Gini features, with and without protein type features (ptf).

Table 1. Averages (Avg) and standard deviations (Std) of MAP, AUC and partial AUC scores.

	MAP	AUC	R50	R100	R200	R300
Avg	0.2300	0.9150	0.0670	0.1073	0.1682	0.2156
Std	0.0217	0.0120	0.0135	0.0169	0.0204	0.0230

3.2. *Feature Importance*

Biologically, it is of interest to identify the features that contribute the most to the classification of protein pairs. This not only helps reveal relationships between different data sources, but can also suggest which data should be generated by experiments to find novel interactions in this and other host-pathogen systems. We assessed feature importance based on the Gini importance of the RF classifier (see Methods). Strikingly, the graph property and the GO neighbor similarity features are ranked at the top (Fig. 3).

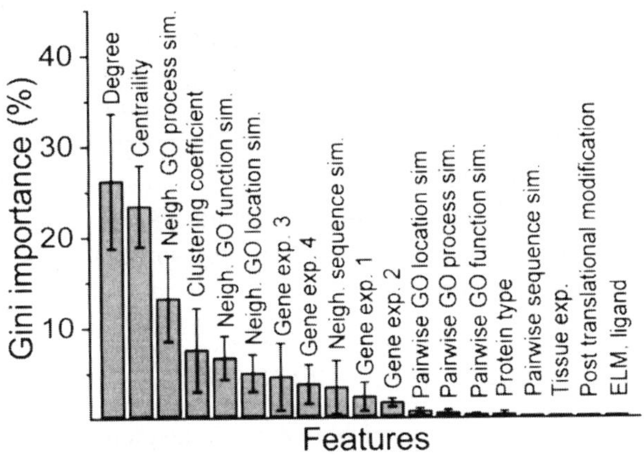

Figure 3. RF Gini importance measures for each feature. Protein type features are grouped together.

To assess the extent to which these features are predictive, we built models using the same train/test data splits as before with only the top 3 and top 6 Gini features (Fig. 2). The top 3 Gini features are degree, betweenness centrality and neighbor GO process similarity (Fig. 3) and the top 6 Gini features in addition include clustering coefficient, neighbor GO function, and cellular location similarities. These models are compared to two baseline models. In the first, the RF classifiers were trained with 6 features selected randomly from the set of features excluding the top 6 Gini features. These random feature sets include the 17 protein type features (ptf), one for each HIV-1 protein. Since these vectors alone do not contain much information, this model forms a weak baseline. A second stronger baseline was built, where the 6 features are randomly selected from the set of features excluding ptf and the top 6 Gini features. Fig. 2 compares the performance of the above 5 models. The top 6 Gini model performs quite strongly compared to both baselines. However, this model is not as good as the model built using the complete feature set The top 3 Gini model performs significantly worse than the top 6 Gini model, but significantly better than the two baseline models suggesting that the additional top 3 features contain independent and complementary information. Statistical significance of these differences (Fig. 2) was confirmed based on paired t-test comparison of the 30 experiments' MAP scores (at a significance level of 0.05).

The above analysis reveals that the graph features and neighbor similarity features are very informative confirming our intuition in incorporating human interactome knowledge into the model. Graph properties have also been found previously useful in the intra-PPI network prediction task [39-42]. Furthermore, it has been proposed earlier that pathogens exploit network properties of the human interactome: it was shown that the Epstein–Barr virus targets high degree human proteins [43] and it was found recently that pathogens tend to interact with host proteins with high degrees and betweenness centrality [44].

The significant performance difference between the top 6 Gini model and the complete model (Fig. 2) indicates that the low ranked features also contribute to the final performance. For example, the removal of protein type features levels off the precision *vs.* recall curve with respect to the complete feature set (Supplementary Fig. S2). The reason why some of these features' Gini importance scores are very low could be due to their low coverage (Supplementary Table S3).

3.3. *Mining Predicted Pairs with siRNA and in Virion Data*

A final model was trained with all available Group 1 interactions, according to the standard methodology that typically the larger the training data, the better the model. We then ranked all HIV-1, human pairs according to their RF score. The score measures the difference between positive and negative votes from the decision trees in the trained RF model and reflects the likelihood of an interaction. The derived ranked order list is available in the supplementary online material. The set of predicted interactions depends on the chosen RF score threshold; lowering the threshold will increase the TP rate at the expense of a higher FP rate. Table 2 (top) presents the number of predicted interactions for different cutoff values. At the lowest threshold we considered (0), 2100 novel interactions are predicted, of which 1 in 5 interactions is expected to be true based on precision measured on the hold out set.

The unknown predicted pairs (Group 2 and novel) were examined in light of the 281 human genes that have been reported in the siRNA screen to have an effect on HIV-1 infection upon silencing [5] and 314 human proteins highjacked by HIV-1 in its virion [45]. Table 2 (bottom) gives the size of the overlap of our Group 2 and novel predictions with these two datasets. Although the comparison cannot provide means to verify the predictions; the overlapping pairs would be of interest to HIV-1 virologists: the siRNA data provide experimental evidence pointing at their functional relevance and the in virion overlapping set could help differentiate between mere by-stander human in virion proteins from those with functional roles for the virus.

Table 2. Number of predicted pairs at different choices of RF score cutoff. Average recall and precision was calculated on the held-out test sets in cross-validation experiments. The second part of the table presents the overlap (the number of the predicted pairs including the reported human gene) between the new predictions (Group 2 and Novel) and siRNA [5] and in Virion [45] datasets (for details, see text). 'Interactor' refers to the predicted interactions, where the human protein is at least one of the siRNA reported human protein's interaction partner.

Cutoff	Predictions	Group 1	Group 2	Novel	Recall	Precision
≥ 0.00	3372	1040	232	2100	0.51	0.20
≥ 0.50	1942	1034	141	767	0.37	0.29
≥ 1.00	1440	1023	68	349	0.26	0.36
≥ 1.50	1085	894	34	157	0.18	0.41
≥ 2.00	622	538	15	69	0.13	0.47
≥ 2.50	279	243	8	28	0.09	0.47
	Group2			**Novel**		
	in Virion	**siRNA**	**Interactor**	**in Virion**	**siRNA**	**Interactor**
≥ 0.00	34	4	120	246	46	1064
≥ 0.50	24	3	83	101	13	441
≥ 1.00	10	1	43	48	5	212
≥ 1.50	5	1	24	17	2	99
≥ 2.00	3	1	13	8	1	49
≥ 2.50	2	0	7	4	0	25

3.4. *Functions of Predicted Interacting Pairs*

The rank-ordered lists of predicted interacting pairs are a rich source for generating biological hypotheses that can be experimentally validated. For example, we predict that the HIV-1 protein tat and the vitamin D receptor (VDR) interact with a high RF score of 1.96. Tat is a regulatory protein of HIV-1, its main role is to transactivate HIV-1 transcription from the viral long-terminal repeat (LTR) promoter [47]. VDRs are members of nuclear receptors, which act as ligand-inducible transcription factors in response to hormones [48]. Variations at the VDR locus are associated with susceptibility and progression of AIDS and other immune diseases [49, 50]. The only interaction reported in the NIAID for VDR is a Group 2 interaction with env gp120. Recently, it was reported that tat acts with VDR in a synergistic manner as a stimulator for HIV-1 LTR activity, providing an independent validation of the functional significance of our prediction [51].

A global analysis of the predicted interactions by assessing the enrichment of GO functional terms in predicted Group 2 or novel interactions revealed 31 molecular processes, 19 biological functions and 14 cellular components (Supplementary Tables S5-S7) at significance level 0.01. For example, transcription regulator-, ligand-dependent nuclear receptor-, MHC class I receptor-, and protein kinase C activities are highly enriched molecular functions, while immune system process and response to stimulus are highly represented processes and macromolecular complex, membrane-enclosed lumen and plasma membrane were the top most significant cellular compartments.

4. Conclusions

Computational methods can be very effective in assisting experimental efforts in identifying interacting protein pairs within a single organism. This paper extends these methods to predicting interactions between hosts and pathogens, here human and HIV-1. Features derived from multiple genomic and functional data sources and exploiting our knowledge of the human protein interactome were integrated in a supervised learning framework. We inspected our predictions with independent biological datasets to identify the most promising interacting pairs. By providing these new testable hypotheses we hope that our predictions will accelerate experimental efforts to define a reliable network of HIV-1, human protein interactions.

Acknowledgments

This work is supported in part by National Institutes of Health grants P50-GM082251 and LM07994-01 and National Science Foundation grants EIA0225656/0225636, CAREER CC044917, and the PA Department of Health.

References

1. A. Trkola, *Curr Opin Microbiol* **7**, 555-9 (2004).
2. D.D. Ho and P.D. Bieniasz, *Cell* **133**, 561-5 (2008).
3. UNAIDS. *2007 AIDS epidemic update.* (2007).
4. A.D. Frankel and J.A. Young, *Annu Rev Biochem* **67**, 1-25 (1998).
5. A.L. Brass *et al*, *Science* **319,** 921-6 (2008).
6. *NIAID HIV-1, Human Protein Interaction Database.* [http://www.ncbi.nlm.nih.gov/RefSeq/HIVInteractions/].

7. B.A. Shoemaker and A.R. Panchenko, *PLoS Comput Biol* **3**, e42 (2007).
8. B.A. Shoemaker and A.R. Panchenko, *PLoS Comput Biol* **3**, e43 (2007).
9. Y. Liu, N. Liu and H. Zhao, *Bioinformatics* **21**, 3279-85 (2005).
10. W.K. Kim, J. Park and J.K. Suh, *Genome Inform* **13**, 42-50 (2002).
11. E. Sprinzak and H. Margalit, *J Mol Biol* **311**, 681-92 (2001).
12. S.M. Gomez, W.S. Noble and A. Rzhetsky, *Bioinformatics* **19**, 1875-81 (2003).
13. T.M. Nye *et al., Bioinformatics* **21**, 993-1001 (2005).
14. T. Dandekar *et al., Trends Biochem Sci* **23**, 324-8 (1998).
15. E.M. Marcotte *et al., Science* **285**, 751-3 (1999).
16. A.J. Enright *et al., Nature* **402**, 86-90 (1999).
17. F. Pazos and A. Valencia, *Protein Eng* **14**, 609-14 (2001).
18. C.S. Goh *et al., J Mol Biol* **299**, 283-93 (2000).
19. S. Martin, D. Roe and J.L. Faulon, *Bioinformatics* **21**, 218-26 (2005).
20. P. Aloy and R.B. Russell, *Bioinformatics* **19**, 161-2 (2003).
21. R. Singh, J. Xu and B. Berger, *Pac Symp Biocomput*, 403-14 (2006).
22. L. Lu, H. Lu and J. Skolnick, *Proteins* **49**, 350-64 (2002).
23. R. Jansen *et al., Science* **302**, 449-53 (2003).
24. Y. Qi, Z. Bar-Joseph and J. Klein-Seetharaman, *Proteins* **63**, 490-500 (2006).
25. X.W. Chen and M. Liu, *Bioinformatics* **21**, 4394-400 (2005).
26. N. Lin *et al., BMC Bioinformatics* **5**, 154 (2004).
27. M.D. Dyer, T.M. Murali and B.W. Sobral, *Bioinformatics* **23**, i159-66 (2007).
28. F.P. Davis *et al., Protein Sci* **16**, 2585-96 (2007).
29. L. Breiman, *Random Forests*, in *Machine Learning*. 2001. p. 5-32.
30. A. Ben-Hur and W.S. Noble, *BMC Bioinformatics* **7 Suppl 1**, S2 (2006).
31. M. Ashburner *et al., Nat Genet* **25**, 25-9 (2000).
32. D.J. Watts and S.H. Strogatz, *Nature* **393**, 440-2 (1998).
33. L.C. Freeman, *Sociometry* **40**, 35-41 (1977).
34. P. Puntervoll *et al., Nucleic Acids Res* **31**, 3625-30 (2003).
35. M. Matsubara *et al., Protein Sci* **14**, 494-503 (2005).
36. R.R. Baeza-Yates and B.d.A. Neto, *Modern information retrieval.* 1999, New York Harlow, England: ACM Press; Addison-Wesley.
37. S. Bauer *et al., Bioinformatics* **24**, 1650-1 (2008).
38. S. Grossmann *et al., Bioinformatics* **23**, 3024-31 (2007).
39. H.B. Fraser *et al., Science* **296**, 750-2 (2002).
40. H. Jeong *et al., Nature* **411**, 41-2 (2001).
41. H. Liang and W.H. Li, *Trends Genet* **23**, 375-8 (2007).
42. H. Yu *et al., PLoS Comput Biol* **3**, e59 (2007).
43. M.A. Calderwood *et al., Proc Natl Acad Sci U S A* **104**, 7606-11 (2007).
44. M.D. Dyer, T.M. Murali and B.W. Sobral, *PLoS Pathog* **4**, e32 (2008).
45. D.E. Ott, *Rev Med Virol* **18**, 159-75 (2008).
46. S. Mathivanan *et al., Nat Biotechnol* **26**, 164-7 (2008).
47. H. Kato *et al., Genes Dev* **6**, 655-66 (1992).
48. D.J. Mangelsdorf *et al., Cell* **83**, 835-9 (1995).
49. C.H. Wu *et al., Nucleic Acids Res* **34**, D187-91 (2006).
50. G. Nieto *et al., J Steroid Biochem Mol Biol* **89-90**, 199-207 (2004).
51. J. Nevado *et al., J Mol Endocrinol* **38**, 587-601 (2007).

FASTCHI: AN EFFICIENT ALGORITHM FOR ANALYZING GENE-GENE INTERACTIONS

XIANG ZHANG [1], FEI ZOU [2] AND WEI WANG [1]

[1] *Department of Computer Science,* [2] *Department of Biostatistics*
University of North Carolina at Chapel Hill

Recent advances in high-throughput genotyping have inspired increasing research interests in genome-wide association study for diseases. To understand underlying biological mechanisms of many diseases, we need to consider simultaneously the genetic effects across multiple loci. The large number of SNPs often makes multilocus association study very computationally challenging because it needs to explicitly enumerate all possible SNP combinations at the genome-wide scale. Moreover, with the large number of SNPs correlated, permutation procedure is often needed for properly controlling family-wise error rates. This makes the problem even more computationally demanding, since the test procedure needs to be repeated for each permuted data. In this paper, we present FastChi, an exhaustive yet efficient algorithm for genome-wide two-locus chi-square test. FastChi utilizes an upper bound of the two-locus chi-square test, which can be expressed as the sum of two terms – both are efficient to compute: the first term is based on the single-locus chi-square test for the given phenotype; and the second term only depends on the genotypes and is independent of the phenotype. This upper bound enables the algorithm to only perform the two-locus chi-square test on a small number of candidate SNP pairs without the risk of missing any significant ones. Since the second part of the upper bound only needs to be precomputed once and stored for subsequence uses, the advantage is more prominent in large permutation tests. Extensive experimental results demonstrate that our method is an order of magnitude faster than the brute force alternative.

1. Introduction

Disease association study analyzes genetic variation across a population consisting of diseased and healthy individuals. The most abundant source of genetic variation in mammalian genome is represented by single nucleotide polymorphisms (SNPs), which account for heritable inter-individual differences in complex phenotypes. The allele differences at these single base sites are usually represented as binary variables (e.g. inbred mice) or ternary variables (e.g. human subjects). Recent advancement of the technologies that enable genotyping a vast number of genetic polymorphism has made genome wide association study possible. Initial reports on genome-wide searching for disease associated genes are appearing in the literature [6,10,15].

Most existing analytical methods consider each genetic marker individually [20]. In many cases, however, the diseases are complex traits, that is, they are likely due to the interactions among multiple genes [2,17]. In order to understand their underlying biological mechanisms, we need to consider simultaneously the joint effects of genotypes across multiple loci. Various machine learning models have been adopted to study the interactions among genes, such as neural networks [3,18] and classification and regression trees (CART)[12,22]. Under the assumption that the number of SNPs is small, exhaustive algorithms that explicitly enumerate all possible SNP combinations have been developed [7,13]. Since these methods explicitly enumerate all possible SNP combinations, they are not suitable for genome-wide association studies.

The number of SNPs in public datasets ranges from thousands to hundreds of thousands[1,21]. The computational burden of searching for interactions among the large number of SNPs often makes the complete genome wide association study intractable. SNP tagging [5,16] have been widely used to reduce the number of SNPs to be analyzed. The goal of SNP tagging is to select a subset of SNPs that can be used as proxies for all SNPs in the genome. The tagged SNPs are then used in the association study. These methods are not complete because some important SNPs may not be tagged.

The computational challenge of genome-wide association study is also caused by another problem known as multiple testing problem. It can be described as the potential increase in Type I error when statistical tests are performed multiple times. Let α be the significant level for each independent test. If n independent comparisons are performed, the family-wise error α' is given by $\alpha' = 1 - (1-\alpha)^n$. For example, if $\alpha = 0.05$ and we test twenty null hypotheses, then we have probability $\alpha' = 1 - 0.95^{20} = 0.64$ to get at least one spurious result. Permutation testing has been the gold standard for assessing significance levels in association studies using multiple markers. However, it is time consuming since the test procedure needs to be repeated for every permutation. To make this process feasible, other correction methods have been proposed [8,9]. Some recent work [23] addresses the problem of two-locus quantitative phenotype association mapping when large permutation tests are needed. However, this method focuses on the case where the phenotypes are continuous variables, hence is not readily applicable to case-control study of diseases.

Theoretically well studied, the chi-square test has been widely used in association studies [11]. In this paper, we examine the ***computational aspect*** of the chi-square test. We present an efficient algorithm, FastChi, and show that the standard chi-square test can be applied in the genome-wide scale for two-locus association study even when large permutation tests are performed. Different from the algorithms applying heuristics, tagging SNPs, or adopting other correction methods, FastChi is an *exhaustive* algorithm. It guarantees to find the optimal solution. Yet, FastChi does not need to explicitly compute the chi-square value for every SNP pair. It utilizes an upper bound of the two-locus chi-square test value, which is the sum of two terms: one based on the single-locus chi-square test, and the other based on the pair-wise SNP genotypes. Using this bound, a large portion of the SNP-pairs are pruned without performing the tests. Due to space limitation, in this paper, we mainly focus on the case where the SNPs are binary variables which are encoded using $\{0, 1\}$. We also have similar results for the case where the SNPs are ternary variables, which will be discussed in Section 5.

2. Problem Definition

Let $\{X_1, X_2, \cdots, X_N\}$ be the set of all biallelic SNPs , and Y be the binary phenotype of interest (e.g., disease or non-disease). For any SNP X_i $(1 \leq i \leq N)$, we represent its chi-square test value with Y as $\chi^2(X_i, Y)$. For any SNP-pair X_i and X_j, the chi-square test value is denoted as $\chi^2(X_i X_j, Y)$. We formalize the problem as follows. Given the set of N SNPs and a phenotype Y for a set of M individuals, let $Y' = \{Y_1, Y_2, \cdots, Y_K\}$ be the set of K permutations of Y. There are two possible cases:

(1) For a single pass association study, i.e., no permutation correction needed: find all SNP-pairs $(X_i X_j)$ such that $\chi^2(X_i X_j, Y) \geq \theta$.

(2) If there are multiple phenotype permutations: for each $Y_k \in Y'$, find all SNP-pairs $(X_i X_j)$ such that $\chi^2(X_i X_j, Y_k) \geq \theta, (1 \leq k \leq K)$.

Our problem formalization can also be applied in other problem settings. For example, it is easy to modify this problem definition to find the top-k SNP-pairs that have the largest chi-square test values among all SNP-pairs. In this scenario, θ would be a dynamic value, i.e., the k-th largest chi-square test value identified by the algorithm so far.

3. The FastChi Algorithm

We first present the upper bound of the two-locus chi-square test value in Section 3.1. Then we show how our algorithm FastChi utilizes the upper bound to achieve efficient two-locus chi-square testing. In Section 3.2, we

Table 1. **Notations used in the upper bound**

Symbols	Formulas
T_1	$\frac{M^2}{(O_A+O_B)(O_A+O_C)(O_C+O_D)}$
S_1	$\max\{O_A^2, O_C^2\}$
$\mathscr{R}_1$	$\min\{\frac{O_{X_j=1}}{O_{X_j=0}}\|X_i=0, \frac{O_{X_j=0}}{O_{X_j=1}}\|X_i=0\}$
T_2	$\frac{M^2}{(O_A+O_B)(O_B+O_D)(O_C+O_D)}$
S_2	$\max\{O_B^2, O_D^2\}$
$\mathscr{R}_2$	$\min\{\frac{O_{X_j=1}}{O_{X_j=0}}\|X_i=1, \frac{O_{X_j=0}}{O_{X_j=1}}\|X_i=1\}$

describe the method for a single phenotype Y. In Section 3.3, we discuss how FastChi performs under permutation procedure.

3.1. *The Upper Bound*

Let A, B, C, D represent the following events respectively: $Y = 0 \wedge X_i = 0$; $Y = 0 \wedge X_i = 1$; $Y = 1 \wedge X_i = 0$; $Y = 1 \wedge X_i = 1$. Let E_{event} and O_{event} denote the expected value and observed value of an event. T_1, T_2, S_1, S_2, $\mathscr{R}_1$, and $\mathscr{R}_2$ represent the formulas shown in Table 1. We have the upper bound of $\chi^2(X_iX_j, Y)$ stated in Theorem 3.1. The derivation of the upper bound is omitted due to space limitation.

Theorem 3.1. *(Upper bound of $\chi^2(X_iX_j, Y)$)*

$$\chi^2(X_iX_j, Y) \le \chi^2(X_i, Y) + T_1S_1\mathscr{R}_1 + T_2S_2\mathscr{R}_2.$$

3.2. *A Single Phenotype*

It is obvious that, if the upper bound of $\chi^2(X_iX_j, Y)$ is less than θ, there is no need to calculate the exact value of $\chi^2(X_iX_j, Y)$, which is guaranteed to be smaller than θ. We now discuss this idea in further detail.

For every X_i $(1 \le i \le N)$, let $AP(X_i) = \{(X_iX_j)|i+1 \le j \le N\}$ be the SNP-pairs with X_i being the SNP of lower index value. For all SNP-pairs in $AP(X_i)$, the phenotype Y and SNP X_i do not vary, thus O_A, O_B, O_C and O_D are constants for all SNP-pairs in $AP(X_i)$. The number of individuals, M, is also a constant. Thus, in the upper bound, T_1S_1 and T_2S_2 are constants. Moreover, $\chi^2(X_i, Y)$ is a constant for a given X_i, and θ is given too. Therefore, $\mathscr{R}_1$ and $\mathscr{R}_2$ are the only variables that depend on X_j and may vary for different SNP-pairs $(X_iX_j) \in AP(X_i)$. Thus for a given X_i, we can treat equation $\chi^2(X_i, Y) + T_1S_1\mathscr{R}_1 + T_2S_2\mathscr{R}_2 = \theta$ as a *straight line* in the 2-D space of $\mathscr{R}_1$ and $\mathscr{R}_2$.

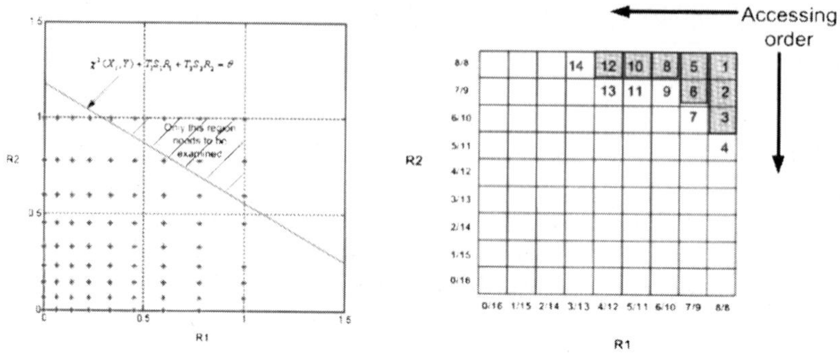

(a) Pruning SNP-pairs in $AP(X_i)$ (b) Candidate retrieval from $Array(X_i)$

Figure 1. **Applying the upper bound**

From now on, we use $\mathscr{R}_1(X_iX_j)$ and $\mathscr{R}_2(X_iX_j)$ to represent the specific values of $\mathscr{R}_1$ and $\mathscr{R}_2$ for the SNP-pair (X_iX_j). The following proposition specifies the values that $\mathscr{R}_1(X_iX_j)$ and $\mathscr{R}_2(X_iX_j)$ can take.

Proposition 3.1. *If there are m 0's and $(M-m)$ 1's in X_i, then for any $(X_iX_j) \in AP(X_i)$, the possible values that $\mathscr{R}_1(X_iX_j)$ can take are: $\{\frac{0}{m}, \frac{1}{m-1}, \frac{2}{m-2}, \cdots, \frac{\lfloor m/2 \rfloor}{\lceil m/2 \rceil}\}$. The possible values that $\mathscr{R}_2(X_iX_j)$ can take are: $\{\frac{0}{M-m}, \frac{1}{M-m-1}, \frac{2}{M-m-2}, \cdots, \frac{\lfloor (M-m)/2 \rfloor}{\lceil (M-m)/2 \rceil}\}$.*

Therefore, for all $(X_iX_j) \in AP(X_i)$, in the 2-D space of $\mathscr{R}_1$ and $\mathscr{R}_2$, $(\mathscr{R}_1(X_iX_j), \mathscr{R}_2(X_iX_j))$ falls in the region $[0,1] \times [0,1]$. The line $\chi^2(X_i, Y) + T_1S_1\mathscr{R}_1 + T_2S_2\mathscr{R}_2 = \theta$ divides this region into two parts: one above the line and one below it. Among the SNP-pairs in $AP(X_i)$, we only need to perform the test for those ones whose $(\mathscr{R}_1(X_iX_j), \mathscr{R}_2(X_iX_j))$ values are above the line, i.e., whose upper bounds are greater than the threshold θ. We refer to such SNP-pairs as *candidate* SNP-pairs.

Example 3.1. Suppose that there are 32 individuals, half alleles of X_i are 0's, and half are 1's. Thus, for the SNP-pairs in $AP(X_i)$, the possible values of $\mathscr{R}_1(X_iX_j)$ (and $\mathscr{R}_2(X_iX_j)$) are $\{\frac{0}{16}, \frac{1}{15}, \frac{2}{14}, \frac{3}{13}, \frac{4}{12}, \frac{5}{11}, \frac{6}{10}, \frac{7}{9}, \frac{8}{8}\}$. Figure 1(a) shows the 2-D space of $\mathscr{R}_1$ and $\mathscr{R}_2$. The blue stars represent the values that $(\mathscr{R}_1(X_iX_j), \mathscr{R}_2(X_iX_j))$ can take. The line $\chi^2(X_i, Y) + T_1S_1\mathscr{R}_1 + T_2S_2\mathscr{R}_2 = \theta$ is also plotted in the figure. The candidate SNP-pairs are those whose $(\mathscr{R}_1(X_iX_j), \mathscr{R}_2(X_iX_j))$ values are in the shaded region. The ones whose $(\mathscr{R}_1(X_iX_j), \mathscr{R}_2(X_iX_j))$ values fall below the line can be pruned without any further test.

To efficiently retrieve the candidates, SNP-pairs (X_iX_j) in $AP(X_i)$ are grouped by their $(\mathscr{R}_1(X_iX_j), \mathscr{R}_2(X_iX_j))$ values and indexed in a 2D array, referred to as $Array(X_i)$.

Example 3.2. Following Example 3.1, Figure 1(b) shows the 9×9 array, $Array(X_i)$, whose entries represent the possible values of $(\mathscr{R}_1(X_iX_j), \mathscr{R}_2(X_iX_j))$ for SNP-pairs $(X_iX_j) \in AP(X_i)$. The $\mathscr{R}_1(X_iX_j)$ $(\mathscr{R}_2(X_iX_j))$ value of each column (row) is noted beneath (left to) each column (row). Each entry of the array is a pointer to the SNP-pairs having the corresponding $(\mathscr{R}_1(X_iX_j), \mathscr{R}_2(X_iX_j))$ values.

In order to find the candidates SNP-pairs whose upper bounds are greater than θ, we start from the right most column of the array, i.e., the entries having the largest $\mathscr{R}_1(X_iX_j)$ value. We scan this column from the top (entries with larger $\mathscr{R}_2(X_iX_j)$ values) towards the bottom (entries with smaller $\mathscr{R}_2(X_iX_j)$ values). If an entry satisfies the inequality $\chi^2(X_i, Y) + T_1S_1\mathscr{R}_1 + T_2S_2\mathscr{R}_2 \geq \theta$, then the SNP-pairs indexed by it are the candidates subject to the chi-square tests. Once we reach an entry violating the inequality, we stop searching the current column, since the remaining entries in the column will not satisfy the inequality. We then move to the top entry of the column left to it and repeat the same scanning process. This whole process terminates when (1) we finish examining all columns or (2) we reach a column whose top entry does not satisfy the inequality.

Example 3.3. Continuing with Examples 3.1 and 3.2, the entries numbered from 1 to 14 in Figure 1(b) are the ones visited by the scanning process. The numbers show the order in which the entries are visited. Only the SNP-pairs indexed by shaded entries need to be evaluated by chi-square tests. The SNP-pairs indexed by the blank entries, including the entries on the boundary can be safely pruned.

3.3. *Permuting the Phenotype*

Let $Y' = \{Y_1, Y_2, \cdots, Y_K\}$ be the K permutations of the phenotype Y. The upper bound in Theorem 3.1 can be easily incorporated in the algorithm to handle the permutations: For any $(X_iX_j) \in AP(X_i)$, its $(\mathscr{R}_1(X_iX_j), \mathscr{R}_2(X_iX_j))$ value does not change over different permutations. That is, for every SNP X_i, the indexing structure $Array(X_i)$ is *independent* of permutations in Y'. Thus, for each X_i, once we get $Array(X_i)$, it can be reused in all permutations.

The FastChi algorithm is described in Algorithm 1. For each X_i, FastChi first indexes $(X_iX_j) \in AP(X_i)$ using $Array(X_i)$. Then it finds

Algorithm 1: FastChi

```
Input: SNPs X' = {X_1, X_2, ··· , X_N}, phenotype permutations
       Y' = {Y_1, Y_2, ··· , Y_K}, and input parameter θ
Output: for every Y_k ∈ Y', find the set of SNP-pairs
        Result(Y_k) = {(X_i X_j) | χ²(X_i X_j, Y_k) ≥ θ, 1 ≤ i < j ≤ N}

for every X_i ∈ X', do
    index (X_i X_j) ∈ AP(X_i) by Array(X_i);
    for every Y_k ∈ Y', do
        access Array(X_i) to find the candidate SNP-pairs and store them in
        Cand(X_i, Y_k);
        for every (X_i X_j) ∈ Cand(X_i, Y_k) do
            if χ²(X_i X_j, Y_k) ≥ θ then
                Result(Y_k) ← (X_i X_j);
            end
        end
    end
end
Return Result(Y_k) for all Y_k ∈ Y'.
```

the set of candidate SNP-pairs $Cand(X_i, Y_k)$ by accessing $Array(X_i)$ for every phenotype permutation Y_k. The candidates in $Cand(X_i, Y_k)$ are then evaluated for their chi-square test values. The candidates whose chi-square test values are greater than or equal to θ are reported by the algorithm.

Time complexity: The complexity to build the indexing structure for all SNPs is $O(N^2M)$. The worst case for accessing all $Array(X_i)$ for all permutations is $O(KNM^2)$. Let $C = \sum_{i,k} |Cand(X_i, Y_k)|$ be the total number of candidates. The time complexity of FastChi is $O(N^2M+KNM^2+CM)$. Note that the time complexity of the brute force approach is $O(KN^2M)$. The number of SNPs N is the dominant factor here. **Space complexity**: The dataset size is $O((N + K)M)$. The size of the $Array(X_i)$ is $O(M^2 + N)$. For each X_i, once the evaluation process is over for all permutations, $Array(X_i)$ can be cleared from the memory. Therefore, the space complexity of FastChi is $O((N + K)M) + O(M^2 + N)$. Since M is usually much smaller than N, this space complexity is linear to the dataset size.

4. Experimental Results

We present extensive experimental results on evaluating the performance of FastChi. FastChi is implemented in C++. The experiments are performed on a 2.4 GHz PC with 1G memory running WindowsXP system.

The SNP dataset used in the experiments is extracted from a set of combined SNPs from the 140k Broad/MIT mouse dataset [21] and 10k GNF [1] mouse dataset. This merged dataset has 156,525 SNPs for 71 mouse strains.

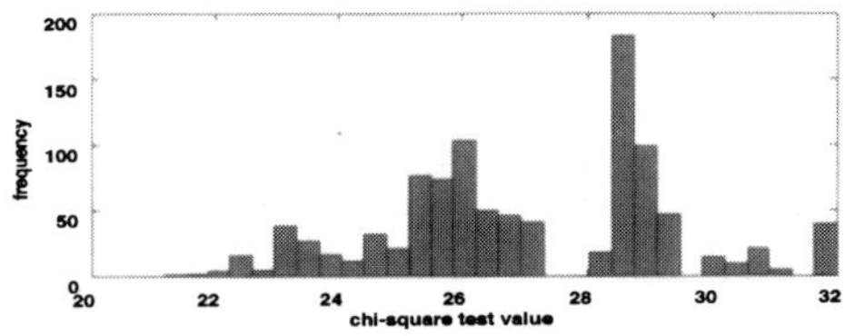

Figure 2. **Distribution of the maximum chi-square test values**

The missing values in the dataset are imputed using NPUTE [14]. The default setting of the experiments are as follows: the phenotypes are random permutations of binary variable with half 0's and half 1's, #individuals = 32, #SNPs=8k, #permutations=20. There are 60,970 unique SNPs for these 32 mice strains. To find the appropriate threshold value, we permute the phenotypes 1000 times. Figure 2 shows the distribution of the maximum chi-square test values of the 1000 permutations. Using a critical significance level of 1%, we set the default threshold value of θ to be 32.

Note that these experimental settings are chosen to demonstrate the performance gain and enhanced scalability offered by FastChi over the brute force approach. In real utility, one may use larger SNP panels and/or more permutation tests. The performance of FastChi is expected to follow the same trends presented in the remainder of this section.

FastChi v.s. the brute force approach: As far as we know, FastChi is the first algorithm addressing the problem of how to scale up the complete two-locus Chi-square test involving large permutation test. For comparison, we show the runtime of FastChi versus the runtime of the brute force approach. The implementation of the brute force approach includes the computation of two-locus chi-square test for every SNP pairs. Figures 3(a) to 3(d) show the running time comparison under various parameter settings. The numbers below the runtime line of FastChi indicate the ratio of the runtime of the brute force approach and the runtime of FastChi. Figure 3(a) shows that the runtime of FastChi dramatically decreases as θ increases. FastChi offers 3.9 fold speedup when $\theta = 26$ and 16.3 fold speedup when $\theta = 34$. Figure 3(b) shows that FastChi is an order of magnitude faster than the brute force approach. Figure 3(c) shows that the runtime of FastChi increases as the number of individuals increases. This is because more SNPs-pairs are expected to have larger chi-square values when the number of individuals increases. Their upper bounds will also increase accordingly. In practice, it is reasonable to set higher threshold

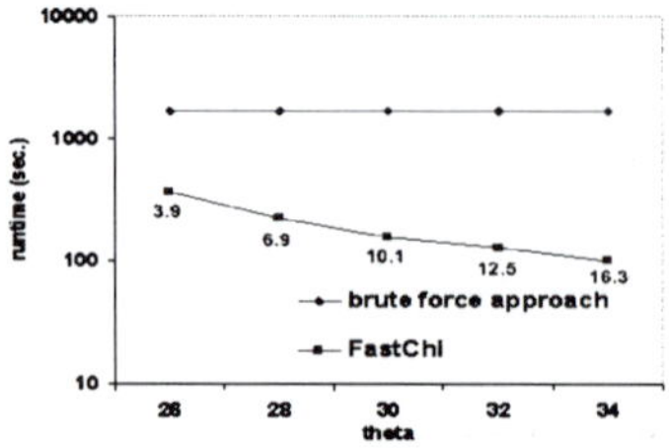

(a) Varying θ

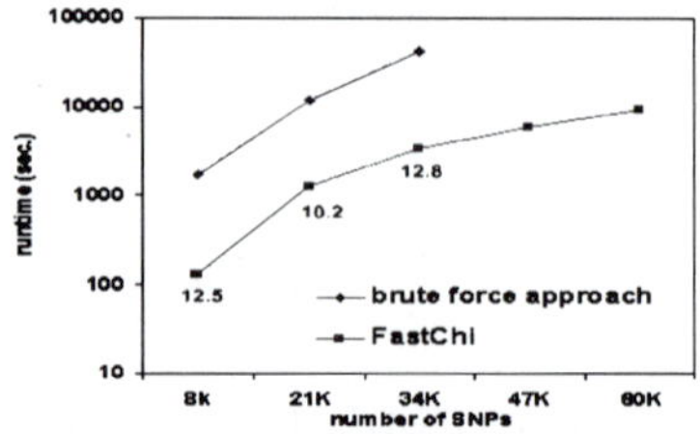

(b) Varying #SNPs

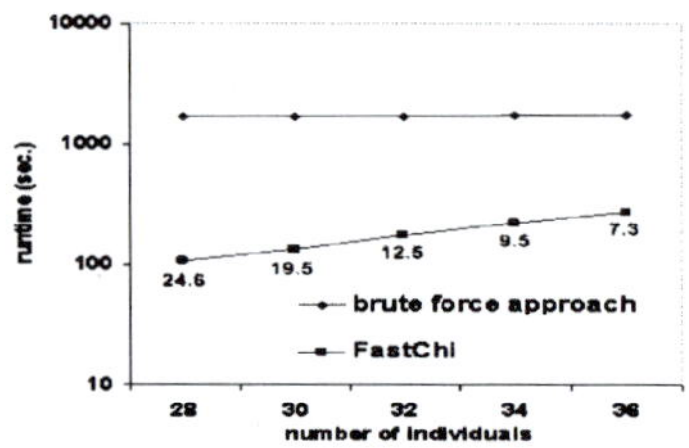

(c) Varying #individuals

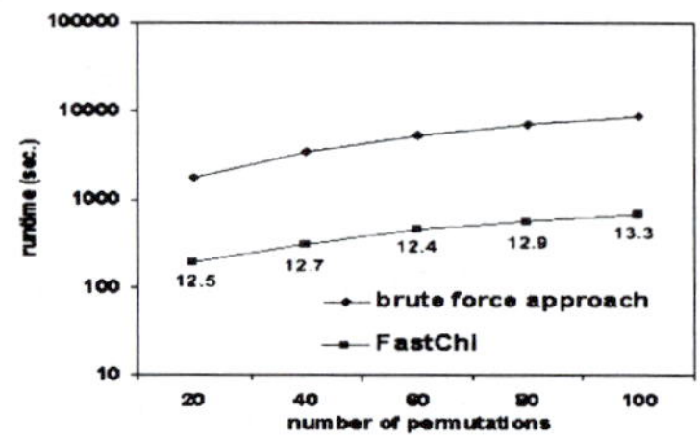

(d) Varying #permutations

Figure 3. **Comparisons between FastChi and the brute force approach**

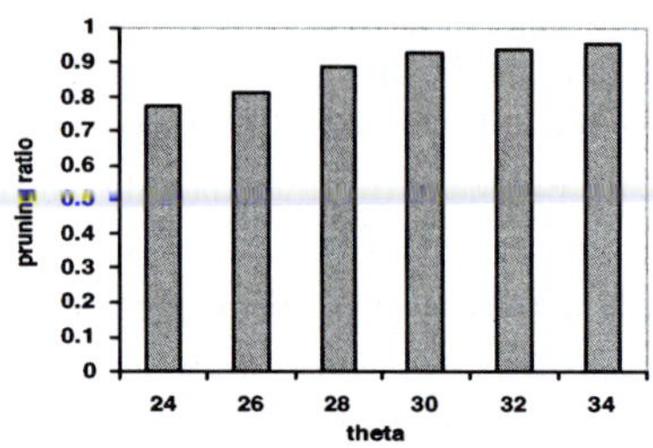

(a) Varying threshold values

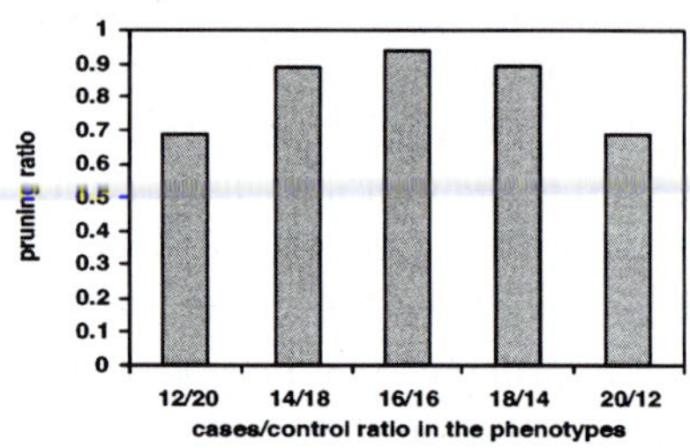

(b) Varing case/control ratios

Figure 4. **Pruning effect of the upper bound**

values for the datasets containing more individuals. Figure 3(d) shows that FastChi is consistently an order of magnitude faster than the brute force approach in permutation tests.

Pruning effect of the upper bound: Figure 4(a) shows the fraction of SNP-pairs pruned under different thresholds. The pruning ratio is averaged over 20 random phenotype permutations. The datasets contain half

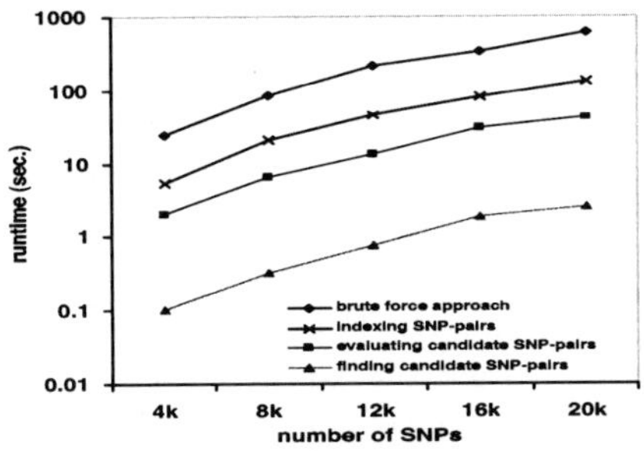

Figure 5. **Computational cost of each component of FastChi**

cases (diseased individuals) and half controls (healthy individuals). A large portion of the SNP-pairs are pruned even when the threshold is low. Figure 4(b) show the pruning ratio of the SNP-pairs when the case/control ratio varies, while the total number of individuals is fixed. Clearly, the pruning effect reaches the maximum power when there are 16 cases and 16 controls, which demonstrates that FastChi is more suitable for balanced study.

Computational cost of each component of FastChi: FastChi has three major components: building the indexing structure $Array(X_i)$ for every SNP X_i, accessing $Array(X_i)$ to find the candidate SNP-pairs whose upper bounds are greater or equal to the threshold, and performing chi-square tests on these candidates. Figure 5 shows the runtime of these three components when the number of SNPs increases. We also plot the runtime of the brute force approach for reference, which is the top line. Note that the runtimes in this figure are for a single permutation. As we can see, the most time consuming component of FastChi is building the index structures. Yet, its runtime is about 1/5 of the time required to perform the two-locus chi-square tests on all SNP pairs in one permutation. Note that when the number of permutations is large, the cost on building the index structures is negligible since they only need to be built once and can be reused in all permutations. Thus the performance gain of FastChi is more prominent for large permutation tests.

5. Discussion

In this paper, we present the FastChi algorithm for genome-wide two-locus chi-square test. FastChi is an exhaustive method which guarantees to find the optimal solution. It utilizes an upper bound of the two-locus chi-square test value to prune a majority of the SNP-pairs. The upper bound developed in this paper can be easily incorporated in the algorithm for SNP-pair pruning and candidates retrieval. By eliminating redundant computation of the invariant units in each permutation, FastChi is even more effective than the brute force method in large permutation tests.

So far, we have described the method for given θ. The main goal of permutation test is to find threshold θ for a given family-wise error α'. FastChi can be easily modified for this task: For each permutation Y_k, we use a parameter θ_k (initially 0) to track the largest chi-square value identified so far by the algorithm. The non-decreasing θ_k is then used as the threshold to prune the search space when the remaining SNP-pairs are examined.

FastChi can also benefit the two-stage genome-wide association study. The idea of the two-stage approache [4,19] is to first select a subset of important SNPs according to some criteria. Then in the second step, an exhaustive search is performed to find the interactions among the selected SNPs. FastChi can dramatically speed up the interaction analysis procedure in the second step. A much larger number of SNPs can now be selected in the first step for the subsequent interaction analysis.

In this paper, we mainly focus on the biallelic SNPs. For the heterozygous case (where SNPs are encoded using $\{0,1,2\}$), we can also derive a similar upper bound. Let A, B, E, C, D, F represent the following events respectively: $Y = 0 \wedge X_i = 0$; $Y = 0 \wedge X_i = 1$; $Y = 0 \wedge X_i = 2$; $Y = 1 \wedge X_i = 0$; $Y = 1 \wedge X_i = 1$; $Y = 1 \wedge X_i = 2$. The upper bound for $\chi^2(X_iX_j, Y)$ is: $\chi^2(X_iX_j, Y) \leq \chi^2(X_i, Y) + T_1S_1\mathscr{R}_0 + T_2S_2\mathscr{R}_1 + T_3S_3\mathscr{R}_2$. In the upper bound, $\mathscr{R}_a = \{[\min\{\frac{O_{X_j=1}}{O_{X_j=0}}, \frac{O_{X_j=0}}{O_{X_j=1}}\} + \min\{\frac{O_{X_j=2}}{O_{X_j=0}}, \frac{O_{X_j=0}}{O_{X_j=2}}\} + \min\{\frac{O_{X_j=2}}{O_{X_j=1}}, \frac{O_{X_j=1}}{O_{X_j=2}}\}] | X_i = a\}$, where $a \in \{0,1,2\}$, $T_1 = L/(O_A + O_C)$, $T_2 = L/(O_B + O_D)$, $T_3 = L/(O_E + O_F)$, (where $L = M^2/[(O_A + O_B + O_E)(O_C + O_D + O_F)]$), $S_1 = \max\{O_A^2, O_C^2\}$ $S_2 = \max\{O_B^2, O_D^2\}$ $S_3 = \max\{O_E^2, O_F^2\}$.

In our future work, we will investigate association study involving more than two SNPs following the same principle discussed in this paper.

References

1. *http://www.gnf.org/.*
2. C. S. Carlson, M. A. Eberle, L. Kruglyak, and D. A. Nickerson. Mapping complex disease loci in whole-genome association studies. *Nature*, 429, 2004.
3. D. Curtis, B. V. North, and P. C. Sham. Use of an artificial neural network to detect association between a disease and multiple marker genotypes. *Ann. Hum. Genet.*, 65:95–107, 2001.
4. D. M. Evans, J. Marchini, A. P. Morris, and L. R. Cardon. Two-stage two-locus models in genome-wide association. *PLoS Genet.*, 2: e157, 2006.

5. E. Halperin, G. Kimmel, and R. Shamir. Tag snp selection in genotype data for maximizing snp prediction accuracy. In *Proceedings of the International Conference on Intelligent Systems for Molecular Biology*, 2005.
6. A. Herbert and et. al. A common genetic variant is associated with adult and childhood obesity. *Science*, 312:279–284, 2006.
7. M. R. Nelson, S. L. Kardia, R. E. Ferrell, and C. F. Sing. A combinatorial partitioning method to identify multilocus genotypic partitions that predict quantitative trait variation. *Genome Research*, 11:458–470, 2001.
8. K. K. Nicodemus, W. Liu, G. A. Chase, Y.-Y. Tsai, and M. D. Fallin. Comparison of type i error for multiple test corrections in large single-nucleotide polymorphism studies using principal components versus haplotype blocking algorithms. *BMC Genet.*, 6(Suppl 1):S78, 2005.
9. D. R. Nyholt. Simple correction for multiple testing for single-nucleotide polymorphisms in linkage disequilibrium with each other. *Am. J. Hum. Genet.*, 74(4):765–769, 2003.
10. K. Ozaki and et. al. Functional snps in the lymphotoxin-alpha gene that are associated with susceptibility to myocardial infarction. *Nat. Genet.*, 32, 2002.
11. M. Pagano and K. Gauvreau. *Principles of Biostatistics*. Pacific Grove, CA: Duxbury Press, 2000.
12. M. A. Province, W. D. Shannon, and D. C. Rao. Classification methods for confronting heterogeneity. *Adv. Genet.*, 42:273–286, 2001.
13. M. D. Ritchie, L. W. Hahn, N. Roodi, L. R. Bailey, W. D. Dupont, F. F. Parl, and J. H. Moore. Multifactor-dimensionality reduction reveals high-order interactions among estrogen-metabolism genes in sporadic breast cancer. *American Journal of Human Genetics*, 69:138–147, 2001.
14. A. Roberts, L. McMillan, W. Wang, J. Parker, I. Rusyn, and D. Threadgill. Inferring missing genotypes in large snp panels using fast nearest-neighbor searches over sliding windows. In *ISMB*, 2007.
15. A. Roses. The genome era begins. *Nat. Genet.*, 33(Supp2):217, 2003.
16. P. Sebastiani, R. Lazarus, S. T. Weiss, L. M. Kunkel, I. S. Kohane, and M. F. Ramoni. Minimal haplotype tagging. *PNAS*, 100(17), 2003.
17. D. Segr, A. DeLuna, G. M. Church, and R. Kishony. Modular epistasis in yeast metabolism. *Nat. Genet.*, 37:77–83, 2005.
18. A. Sherriff and J. Ott. Applications of neural networks for gene finding. *Adv. Genet.*, 42:287–297, 2001.
19. J. Storey, J. Akey, and L. Kruglyak. Multiple locus linkage analysis of genomewide expression in yeast. *PLoS Biology*, 8: e267, 2005.
20. D. C. Thomas. *Statistical methods in genetic epidemiology*. Oxford Univeristy Press, Oxford, 2004.
21. C. M. Wade and M. J. Daly. Genetic variation in laboratory mice. *Nat. Genet.*, 37:1175–1180, 2005.
22. H. Zhang and G. Bonney. Use of classification trees for association studies. *Genet. Epidemiol.*, 19:323–332, 2000.
23. X. Zhang, F. Zou, and W. Wang. Fastanova: an efficient algorithm for genome-wide association study. In *KDD*, 2008.

OPEN SCIENCE: TOOLS, APPROACHES, AND IMPLICATIONS *

CAMERON NEYLON

STFC Rutherford Appleton Laboratory, Harwell Science and Innovation Campus
Didcot, OX11 0QX, United Kingdom

SHIRLEY WU

Program in Biomedical Informatics, Stanford University
Stanford, CA 94305 USA

Open Science is gathering pace both as a grass roots effort amongst scientists to enable them to share the outputs of their research more effectively, and as a policy initiative for research funders to gain a greater return on their investment. In this workshop, we will discuss the current state of the art in collaborative research tools, the social challenges facing those adopting and advocating more openness, and the development of standards, policies and best practices for Open Science.

1. Introduction

Openness is arguably the great strength of the scientific method. Through open examination and critical analysis, models can be refined, improved, or rejected. Conflicting data can be compared and the underlying experiments and methodology investigated to identify which, if any, is more reliable. While individuals may not always adhere to the highest standards, community mechanisms of review over time have proved effective in developing coherent and practically useful models of the physical world around us. As Lee Smolin has put it, "*we argue in good faith from shared evidence to shared conclusions*".[1]

The Internet and the World Wide Web provide the technical ability to share a much wider range of the evidence, argument and conclusions driving modern research. Data, methodology, and interpretation can also be made available online at lower costs and with lower barriers to access than has traditionally been the case. Wikis and blogs enable geographically and temporally widespread collaborations, the traditional journal club can now span continents, and the smallest details of what is happening in a laboratory can be shared.

---

* This workshop is supported by Burroughs Wellcome Fund.

The potential of online tools to revolutionize scientific communication and their ability to open up the details of the scientific enterprise so that a wider range of people can participate is clear. In practice, however, the reality has fallen far behind the potential. Although this is in part due to a need for tools that are specifically designed with scientific workflows in mind and to the inertia of infrastructure providers with pre-Internet business models, it is predominantly due to cultural and social barriers within the scientific community.

There will always be places where complete openness is not appropriate – for example, where personal patient records may be identifiable or where research is likely to lead to patentable results. These, however are special instances for which exceptional cases can be made, not the general case across the whole of the global research effort. Along with funders, government, and special interest groups there is a growing community of scientists interested in adopting more open practices in their research, and this community is developing as a strong voice in discussions of science policy, funding, and publication.

2. The case for open science

The case for taxpayer access to the taxpayer funded peer reviewed literature was made personally and directly in Jonathan Eisen's first editorial for PLoS Biology. [2] As a scientist in a small institution, he was unable to access the general medical literature for the information he desperately sought for an urgent family medical situation. More generally, as a US taxpayer he was unable to access the outputs of US government funded research or indeed of research funded by the governments of other countries.

A similar case can also be made for research data. Andrew Vickers, in a New York Times essay[3], dissected the reasons that scientists gave for not making cancer treatment data available – data that could enhance patient survival times and quality of life. In other fields, the case for data sharing may seem less clear. There is little obvious damage done to the general public by not making the details of research available; however, in the non-clinical sciences, aggregation and re-analysis can also lead to new insights, more effective analysis, and even new types of analysis. Sharing enables more effective and more efficient science. And this really is the crux of the matter: making the data, process, and conclusions available is nothing more than doing good science.

3. Tools for Open Science

The rapid expansion and development of tools that are loosely categorised under the banner of 'Web2.0' is what makes the sharing of research material practical. Many of the generic tools available have been adopted and used by a wide range of researchers but often these tools do not fit into the existing workflows of researchers.

Tools, whether they be social networking sites, electronic laboratory notebooks, or controlled vocabularies, must be built to help scientists do what they are already doing, not what the tool designer feels they should be doing. In the current environment, combined with the requirements and desire to provide access to laboratory data, the most obvious target is tools that make it easier to capture the research record so that it can be incorporated into and linked from papers. Once it is electronic, one can choose to make the record public at any stage.

4. Social issues and measures of success

Scientists are inherently rather conservative in their adoption of new approaches and tools. New methodologies often struggle to be accepted until the evidence of their superiority is overwhelming. The wider community is waiting for evidence of benefits before adopting either open access publication or open data policies. This actually provides the opportunity for individuals to take a first mover advantage.

Quantitatively measuring success in the application of open approaches relative to traditional approaches is a challenge, as demonstrated by the continuing controversy over the citation advantage of open access articles. However, Open Science has a clear public relations advantage as the examples are out in the open for anyone to see.

There are both benefits and risks associated with open practice in research. Often the discussion with researchers is focused on the disadvantages and risks. These concerns should not be dismissed, but taken seriously and considered. Radical change never comes without casualties, and while some concerns may be misplaced there are many real risks. What is important is providing information to enable people to balance the risks and benefits of any approach they take.

5. Standards for Open Science

Two approaches to standards for Open Science are currently being discussed. The first of these is 'the fully supported paper'. In essence this is the idea that the claims made in a peer reviewed paper in the conventional literature should be fully supported by a publicly accessible record of all the material that contributes to those claims. The technical challenges of delivering such a record are substantial; however, it is difficult to argue that this shouldn't be available. While the target is challenging, it is simply a proposal to do good science, properly communicated.

While the fully supported paper would be a massive social and technical step forward it in many ways is no more open than the current system. It does not deal with the problem of unpublished or unsuccessful studies that may never find a home in a traditional peer reviewed paper. What then are the standards that need to be met before an organisation can claim they are doing Open Science? Science Commons have published four 'Principles for Open Science' [4] which focus on the accessibility of published literature, research tools and data, and the development of cyberinfrastructure to make this possible. These principles provide a set of criteria that could form the basis of standards. The details are important, and will take time to work out. In the short term it is, therefore, arguably more effective to identify and celebrate examples of best practice and observe how it works in the real world. This will raise the profile of Open Science without making it the exclusive preserve of those who can immediately change their practice.

6. Summary

The development of deposition and data sharing mandates by a range of research funders show that real progress is being made in increasing access to both the finished products of research and the materials that support them. There is, however, a risk of over-enthusiasm driving expectations that cannot be delivered and of alienating the mainstream community that we wish to draw in. The fears and concerns of researchers in widening access to their work need to be addressed sensitively and seriously, pointing out the benefits but also acknowledging the risks involved in adopting these practices.

Now is the right time to find examples of best practice; to celebrate these and to see what can be learnt from them. Now is the right time to be clearly articulating specific aspirations, and to provide targets that we can work towards. Now is the right time to organize as a community. The fully supported paper and the Science Commons principles are useful starting points, but the community will benefit from a concerted effort to develop and actualize additional goals, standards, and resources. Open Science is gathering momentum, and that is a good thing. But equally it is a good time to take stock, identify the best course forward, and effect change as widely as possible.

References

1. L Smolin, *Perimeter Institute Recorded Seminar Archive.* **PIRSA#08090035,** http://pirsa.org/08090035/ (2008).

2. J.A. Eisen, *PLoS Biology.* **6(2)**, e48 (2008)

3. A. Vickers, *New York Times.* **January 22** (2008).

4. Science Commons, http://sciencecommons.org/resources/ readingroom/principles-for-open-science/ (2008)

POST-TRANSCRIPTIONAL GENE REGULATION: RNA-PROTEIN INTERACTIONS, RNA PROCESSING, MRNA STABILITY AND LOCALIZATION

BENJAMIN BLENCOWE[1,2], STEVEN BRENNER[3], TIMOTHY HUGHES[1,2], and QUAID MORRIS[1,2,4]

[1]*Department of Molecular Genetics, University of Toronto, 1 King's College Rd., Toronto, ON, M5S 1A8, Canada*

[2]*Banting and Best Department of Medical Research, University of Toronto, 160 College Street, Toronto, ON, M5G 1L6, Canada*

[2]*Departments of Molecular and Cell Biology and Plant and Microbial Biology, University of California, Berkeley, California 94720, USA.*

[4]*Department of Computer Science, University of Toronto, 10 King's College Road, Toronto, ON, M5S 3G4, Canada*

Background:

Regulation of gene expression at the post-transcriptional level, including the control of splicing, mRNA transcript stability, localization, and translation, is widespread in eukaryotes. An explosion of recent findings underscores both the predominance and complexity of PTR, including the discovery of ubiquitous but previously unknown regulatory mechanisms, such as microRNAs[1] (miRNAs); the observation that RNA recognition motif (RRM) and Kelch homology (KH) RNA-binding domains (RBDs) are among the most numerous protein domains in metazoan genomes, including the human genome[2, 3]; and the finding that, in several cases, RNA-binding proteins (RBPs) bind and presumably co-regulate sets of functionally-related transcripts[4]. An overall hypothesis has emerged that mRNA-containing ribonucleoprotein complexes (mRNPs) are the functional equivalent of "post-transcriptional operons"[5]. There are, however, relatively few examples of such "operons"; many putative PTR *cis-* and *trans-*regulatory factors were discovered by computational analysis of genome sequences, and have not been subjected to individual directed experimentation. Nonetheless, the sheer number of miRNAs (>500 in humans) and apparent sequence-specific RBPs (i.e. those carrying RRM, KH, and other RBDs) suggests that PTR mechanisms are as abundant as transcriptional regulatory mechanisms.

The study of PTR regulation presents a number of experimental challenges. Two of the major modes of PTR (control of subcellular localization and translation) cannot be assayed by analysis of total RNA extracted from cells; rather, they must be analyzed either *in situ* or via extraction of intact macromolecular structures (e.g. organelles, P-bodies, polysomes). Another major mode, transcript degradation rate, must be separated from transcript synthesis rate in order to make accurate measurements.

Coupled with these experimental challenges are computational challenges both in analysis of experimental data and in the modeling of *cis*-regulatory elements involved in PTR. For example, though alternative splicing can be assayed by analyzing total RNA using microarrays or next-generation sequencing, because alternatively spliced isoforms share most of their RNA sequence, estimating relative abundance levels from these data requires careful computational modeling [6-8]. Unlike DNA-binding proteins, many RBPs have both sequence and RNA secondary structure binding preferences – even among RRM and KH domains, which primarily recognize unpaired bases, roughly half of the proteins that have been characterized have a preference or requirement that the primary sequence is embedded in a particular type of secondary structure[3, 9, 10]. As such, motif models of RBP binding preference should include both primary sequence and secondary structure features. However, most of the recent work in describing secondary-structure motifs has been focused on use of Covariance Models (CMs)[11, 12] to model families of non-coding RNA (ncRNA)[13, 14] using stochastic context-free grammars[15]. However, fitting CMs to collections of sequences bound by an RBP can be difficult because the procedures rely heavily on sequence alignment and compensatory mutations among paired bases[16] to detect features of RNA secondary structure. As such, it is not clear that these representations are appropriate for protein binding sites, which tend to be much smaller than ncRNAs and may be subject to different constraints. So, despite an increase variety of experiment methods available for identifying sequences bound by RNA-binding protein, either *in vivo* (e.g. RIP-chip[17] and CLIP[18, 19]) or *in vitro* (e.g., SELEX [20]), substantial effort is still needed to develop motif models that capture RBP binding preferences and can be easily learned from sequence data.

There also remain open questions in understanding how trans-acting regulators of PTR identify their targets. For example, though microRNA targeting has been extensively studied in the last few years, until recently, the most accurate predictors of miRNA target sites were conserved matches to the ~7bp seed region near the 5' end of the miRNA[21, 22]. Accessibility of target sites is now known to be an even more accurate predictor[23] though the target predictions of miRNAs are still far from complete and new features of target mRNA sequence continue to be discovered [24].

Summary

The goal of our workshop is to introduce some recent work in the area of post-transcriptional regulation to a wider computational community, discuss some of the unique computational problems faced in this area, and to present some preliminary solutions to these problems. In particular, we will focus on emerging computational and large-scale experimental strategies (e.g. microarray and deep sequencing) for investigating aspects of gene regulation at the post-transcriptional level, with an emphasis on the identification and characterization of the cis- and trans-acting RNA and protein components involved. We will also be exploring new developments in computational methods to detect and characterize cis-regulatory signals encoded in mRNAs.

Bibliography

1. Bartel DP: **MicroRNAs: genomics, biogenesis, mechanism, and function**. *Cell* 2004, **116**(2):281-297.
2. Anantharaman V, Koonin EV, Aravind L: **Comparative genomics and evolution of proteins involved in RNA metabolism**. *Nucleic Acids Res* 2002, **30**(7):1427-1464.
3. Clery A, Blatter M, Allain FH: **RNA recognition motifs: boring? Not quite**. *Curr Opin Struct Biol* 2008, **18**(3):290-298.
4. Gerber AP, Herschlag D, Brown PO: **Extensive association of functionally and cytotopically related mRNAs with Puf family RNA-binding proteins in yeast**. *PLoS Biol* 2004, **2**(3):E79.
5. Keene JD: **RNA regulons: coordination of post-transcriptional events**. *Nat Rev Genet* 2007, **8**(7):533-543.
6. Anton MA, Gorostiaga D, Guruceaga E, Segura V, Carmona-Saez P, Pascual-Montano A, Pio R, Montuenga LM, Rubio A: **SPACE: an algorithm to predict and quantify alternatively spliced isoforms using microarrays**. *Genome Biol* 2008, **9**(2):R46.
7. Shai O, Morris QD, Blencowe BJ, Frey BJ: **Inferring global levels of alternative splicing isoforms using a generative model of microarray data**. *Bioinformatics* 2006, **22**(5):606-613.
8. Xing Y, Stoilov P, Kapur K, Han A, Jiang H, Shen S, Black DL, Wong WH: **MADS: a new and improved method for analysis of differential alternative splicing by exon-tiling microarrays**. *Rna* 2008, **14**(8):1470-1479.
9. Auweter SD, Oberstrass FC, Allain FH: **Sequence-specific binding of single-stranded RNA: is there a code for recognition?** *Nucleic Acids Res* 2006, **34**(17):4943-4959.
10. Valverde R, Edwards L, Regan L: **Structure and function of KH domains**. *Febs J* 2008, **275**(11):2712-2726.
11. Eddy SR, Durbin R: **RNA sequence analysis using covariance models**. *Nucleic Acids Res* 1994, **22**(11):2079-2088.
12. Sakakibara Y, Brown M, Hughey R, Mian IS, Sjolander K, Underwood RC, Haussler D: **Stochastic context-free grammars for tRNA modeling**. *Nucleic Acids Res* 1994, **22**(23):5112-5120.

13. Griffiths-Jones S, Bateman A, Marshall M, Khanna A, Eddy SR: **Rfam: an RNA family database**. *Nucleic Acids Res* 2003, **31**(1):439-441.
14. Griffiths-Jones S, Moxon S, Marshall M, Khanna A, Eddy SR, Bateman A: **Rfam: annotating non-coding RNAs in complete genomes**. *Nucleic Acids Res* 2005, **33**(Database issue):D121-124.
15. Yao Z, Weinberg Z, Ruzzo WL: **CMfinder--a covariance model based RNA motif finding algorithm**. *Bioinformatics* 2006, **22**(4):445-452.
16. Pedersen JS, Bejerano G, Siepel A, Rosenbloom K, Lindblad-Toh K, Lander ES, Kent J, Miller W, Haussler D: **Identification and classification of conserved RNA secondary structures in the human genome**. *PLoS Comput Biol* 2006, **2**(4):e33.
17. Keene JD, Komisarow JM, Friedersdorf MB: **RIP-Chip: the isolation and identification of mRNAs, microRNAs and protein components of ribonucleoprotein complexes from cell extracts**. *Nat Protoc* 2006, **1**(1):302-307.
18. Ule J, Jensen KB, Ruggiu M, Mele A, Ule A, Darnell RB: **CLIP identifies Nova-regulated RNA networks in the brain**. *Science* 2003, **302**(5648):1212-1215.
19. Ule J, Jensen K, Mele A, Darnell RB: **CLIP: a method for identifying protein-RNA interaction sites in living cells**. *Methods (San Diego, Calif* 2005, **37**(4):376-386.
20. Schneider D, Tuerk C, Gold L: **Selection of high affinity RNA ligands to the bacteriophage R17 coat protein**. *J Mol Biol* 1992, **228**(3):862-869.
21. Krek A, Grun D, Poy MN, Wolf R, Rosenberg L, Epstein EJ, MacMenamin P, da Piedade I, Gunsalus KC, Stoffel M *et al*: **Combinatorial microRNA target predictions**. *Nat Genet* 2005, **37**(5):495-500.
22. Lewis BP, Burge CB, Bartel DP: **Conserved seed pairing, often flanked by adenosines, indicates that thousands of human genes are microRNA targets**. *Cell* 2005, **120**(1):15-20.
23. Kertesz M, Iovino N, Unnerstall U, Gaul U, Segal E: **The role of site accessibility in microRNA target recognition**. *Nat Genet* 2007, **39**(10):1278-1284.
24. Grimson A, Farh KK, Johnston WK, Garrett-Engele P, Lim LP, Bartel DP: **MicroRNA targeting specificity in mammals: determinants beyond seed pairing**. *Mol Cell* 2007, **27**(1):91-105.